Flower Seeds

Biology and Technology

Edited by

Miller B. McDonald

Department of Horticulture and Crop Science
Ohio State University
USA

and

Francis Y. Kwong

PanAmerican Seed Company
West Chicago
Illinois
USA

CABI Publishing

CABI Publishing is a division of CAB International

CABI Publishing
CAB International
Wallingford
Oxfordshire OX10 8DE
UK

Tel: +44 (0)1491 832111
Fax: +44 (0)1491 833508
E-mail: cabi@cabi.org
Website: www.cabi-publishing.org

CABI Publishing
875 Massachusetts Avenue
7th Floor
Cambridge, MA 02139
USA

Tel: +1 617 395 4056
Fax: +1 617 354 6875
E-mail: cabi-nao@cabi.org

A catalogue record for this book is available from the British Library, London, UK.

Library of Congress Cataloging-in-Publication Data
Flower seeds : biology and technology / edited by Miller B. McDonald and Francis Y. Kwong.
        p. cm.
    Includes bibliographical references and index.
    ISBN 0-85199-906-9 (alk. paper)
    1. Seeds.   2. Flowers--Seeds.   I. McDonald, M. B.   II. Kwong, Francis Y.   III. Title.

    SB117.F59 2005
    635.9'1521--dc22

                                                                2004007865

ISBN 0 85199 906 9

Typeset by AMA DataSet Ltd, UK.
Printed and bound in the UK by Biddles Ltd, King's Lynn.

Contents

_____

Contributors

Anderson, N.O., *Department of Horticultural Science, University of Minnesota, 1970 Folwell Avenue, St Paul, MN 55108, USA.*

Baskin, C.C., *Department of Biology, University of Kentucky, Lexington, KY 40506-0225, USA and Department of Agronomy, University of Kentucky, Lexington, KY 40546-0321, USA.*

Baskin, J.M., *Department of Biology, University of Kentucky, Lexington, KY 40506-0225, USA.*

Bruggink, G.T., *Syngenta Seeds B.V., Westeinde 62, 1601 BK Enkhuizen, The Netherlands.*

Erwin, J., *Department of Horticultural Science, University of Minnesota, 1970 Folwell Avenue, St Paul, MN 55108, USA.*

Geneve, R.L., *Department of Horticulture, N318 Agriculture Science North, University of Kentucky, Lexington, KY 40546-0091, USA.*

Hamrick, D., *FloraCulture International, 2960 Claremont Road, Raleigh, NC 27608, USA.*

Jalink, H., *Plant Research International, Bornsesteeg 65, 6708 PD Wageningen, The Netherlands.*

Koivula, N., *All-America Selections® and the National Garden Bureau, 1311 Butterfield Road, Suite 310, Downers Grove, IL 60515, USA.*

Kwong, F.Y., *PanAmerican Seed Company, 622 Town Road, West Chicago, IL 60185-2698, USA.*

Lionakis Meyer, D.J., *Seed Laboratory, Plant Pest Diagnostics Center, California Department of Food and Agriculture, 3294 Meadowview Road, Sacramento, CA 95832-1448, USA.*

Mari, J., *PanAmerican Seed Company, 622 Town Road, West Chicago, IL 60185-2698, USA.*

McDonald, M.B, *Seed Biology Program, Department of Horticulture and Crop Science, Ohio State University, 2021 Coffey Road, Columbus, OH 43210-1086, USA.*

Miller, A., *USDA/ARS National Center for Genetic Resources Preservation, 1111 South Mason Street., Fort Collins, CO 80521-4500, USA.*

Milstein, G.P., *Applewood Seed Company, 5380 Vivian Street, Arvada, CO 80002, USA.*

Sellman, R.L., *PanAmerican Seed Company, 622 Town Road, West Chicago, IL 60185-2698, USA.*

Stephenson, M., *California Department of Food and Agriculture, 3294 Meadowview Road, Sacramento, CA 95832, USA.*

Tay, D., *Ornamental Plant Germplasm Center, 670 Vernon Tharp Street, Columbus, OH 43210-1086, USA.*

van der Schoor, R., *Plant Research International, Bornsesteeg 65, 6708 PD Wageningen, The Netherlands.*

Acknowledgements

The editors and publishers wish to thank the Ball Horticultural/PanAmerican Seed Company for a contribution towards the cost of printing colour plates in this book.

1 Introduction to Flower Seeds and the Flower Seed Industry

Miller B. McDonald[1] and Francis Y. Kwong[2]
*[1]Seed Biology Program, Department of Horticulture and Crop Science,
Ohio State University, 2021 Coffey Road, Columbus, OH 43210-1086, USA;
[2]PanAmerican Seed Company, 622 Town Road, West Chicago, IL 60185-2698, USA*

Floriculture is a small, but significant, component of the global economy. A recent European study (Hannover University, 2003) estimated the worldwide production value of flowers to be around €60 billion. The North American production totalled about 10% of this amount. The retail floriculture industry has been steadily increasing in size and importance. For example, in the USA, retail sales of flowers have increased from US$4.7 billion in 1980 to US$19 billion in 2002 (Fig. 1.1). In comparison with overall agricultural income, the floral industry represented 7.3% in 1994 and increased to 9.8% in 2002 (Fig. 1.2).

Great diversity exists in the use of flowers both from a geographical and from a historic perspective. The incorporation of bedding plants into our gardens can be traced back to the early 19th century when colourful annuals and tender ornamentals were imported to Europe from the Americas, South Africa and Asia. From the onset, domestication of exotic plant species has been a mainstay in the

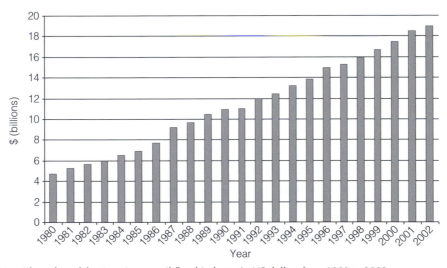

Fig. 1.1. The value of the American retail floral industry in US dollars from 1980 to 2002.

Fig. 1.2.   The percentage contribution of the floral industry to American agriculture from 1994 to 2002.

development of garden designs (Hobhouse, 1992).

Seeds are perfect natural packages that facilitate the migration of plant species across land and sea. Early developments of the flower and seed industries generally involved importing exotic plant materials and improving them through selection in accordance with local climates and prevalent tastes. Starting in the 18th century, mass migration of peoples throughout the world contributed to the broader distribution of novel ornamental varieties. New trends in desirable flower types and how flowers are used continue to develop, and the flower seed trade evolves to accommodate these trends. The vast diversity of flower crops is an important, unique feature of the flower seed industry and this presents greater technical challenges to it than to other agricultural industries.

Special Features and Challenges of the Flower Seed Industry

The flower seed trade developed from small-scale sales of selected varieties by speciality nurseries two centuries ago to a highly technical, vibrant segment of the seed industry today. There are a number of unique characters that set the flower seed trade apart from its crop and vegetable seed counterparts:

1.   *Product range.* There are hundreds of species commonly used as flowers, compared with only a few major food staples and tens of vegetables. Existing flower seeds come from a broad botanical taxonomic base, and new crops are being introduced every year. Most of the popular flowers are annuals, but there are also many biennials and perennials. They are commonly used as bedding plants, pot plants and cut flowers. Some flowers are edible and are often used in regional gourmet cuisines or as garnishes in salads. This very broad product range demands a very broad knowledge base for those involved in the industry.

2.   *Continual quest for new products.* New flowers are selected mainly for their ornamental value rather than crop yield, which is the primary selection factor in food crops. Being a qualitative trait, ornamental value is largely subjective and depends greatly on cultural changes in society. Novelty captures the attention of the consumers and there is a continual need for a different look, a newer colour. Flower breeders have to keep abreast of significant cultural changes in society and the successful ones are trendsetters, much like designers in the fashion industries. Identifying exotic plant materials from foreign countries and adapting them to local requirements through breeding techniques is just as important today as it was more than 200 years ago.

3. *Serving highly sophisticated customers.* The growth of the greenhouse industry is an important factor contributing to the growth of the floricultural industry. Indoor production under a controlled environment allows many of the warm-climate plants to be produced out of season in colder climates. Since greenhouse production is capital intensive, successful operators are generally both skilful growers and businessmen. It is a highly industrialized segment of the agricultural sector today and greenhouse growers have very exact product requirements for their seed suppliers. Many of these requirements can be summarized in one phrase: programmable production. The growers need to know the technical details of the crop culture and information on how to force a flower crop for special holidays when the demands for fresh flowers are high. The seed companies must ensure that seeds of the desirable varieties are available at the required time. They also need to develop cultural guidelines for optimal crop performance and provide information on how to induce flowering at the desired period. This information is particularly key to the successful introduction of new crops.

4. *High seed germination requirement.* For the flower seed industry, the single most important development in greenhouse production in the last 20 years is the emergence of 'plug' growers. These are growers who specialize in the production of young plants that are then sold to 'finishers' who bring the young plants to flower for the market. Standardized sized trays containing different numbers of individual cells ('plugs'), each planted with one seed, are used for young plant production. The plug growers sell these trays to finishers with a guaranteed minimum plant count. The finishers then transplant the young plants into larger containers by hand or, more commonly, by computer-driven automatic transplanting machines. To optimize production efficiency, growers demand that each seed must germinate and produce a seedling, otherwise valuable greenhouse space and nutrients are wasted. Transplanting labour is a major cost in production.

Moreover, these seeds must germinate uniformly because bedding plant operations function by 'staging' trays containing the plugs. Everything is moved on schedule and a non-uniform tray is both visually undesirable and less valuable from a production perspective. Seed companies have to produce flower seeds of the highest quality in order to satisfy these growers. Seed viability testing procedures that were developed decades ago are no longer sensitive enough to differentiate seed lots of varying quality. Vigour tests that correlate well to practical conditions have been developed for this purpose. But the diversity of crops has made it difficult to forge agreement that one universal vigour test functions ideally for all ornamental crops.

5. *Advances in seed technology.* Because growers have high expectations for flower seed performance, they are willing to pay increased seed costs that are relatively high compared with their agronomic and vegetable counterparts. Fortunately, these increased prices also allow the industry to practise seed enhancement technologies not commonly encountered in other crops. For example, many flower seeds are routinely primed for improved performance or pelleted to either permit greater ease of handling or increase precision planting using mechanical seeders. Each of these techniques must be researched for optimum performance according to crop, which requires specific equipment, expertise and knowledge.

Why this Book?

The ornamental industry has changed substantially in the last 20 years. The science of flower seed technology has also advanced rapidly to keep up with new market demands. There is much information scattered in various scientific and trade publications, and more is being kept in the proprietary domain of seed companies. Anyone searching for information on flower seeds finds it in short supply and seldom collated into one comprehensive reference. The objective of this book is to update the reader on the current state of knowledge about flower seeds in one common resource. Because the topic is broad and complex, no one individual could have authored such a treatise. As a result, we have taken the approach of identifying experts knowledgeable in specific areas of flower seeds. As editors, our role has been twofold: (i) to ensure that each author provided comprehensive technological detail about each subject; and (ii) to ensure a consistency in editing style and presentation for the reader.

The book is arranged in three sections. The first section (Chapters 2–4) provides an introduction to the flower seed industry and identifies the users of flower seeds. These chapters provide a historical treatment of the industry and emphasize the challenges and opportunities for the ornamental and bedding plant and wildflower industries. The second section (Chapters 5–10) considers the science of flower seed production. How do seed companies establish successful breeding programmes? What factors control flower induction so that not only the highest quality seeds are produced, but yields are maximized? What is the structure and anatomy of flower seeds? What physiological events govern seed germination, dormancy imposition and release, and seed deterioration? These important topics are considered in detail. The third section (Chapters 11–17) discusses the technology of flower seeds. This includes chapters on cleaning, enhancements, germination, viability and vigour testing of flower seeds. Here, for the first time, are specific guidelines and information to accomplish these tasks. A chapter is also provided describing the activities of the Ornamental Plant Germplasm Center – the only governmental organization dedicated to conserving germplasm for the future genetic improvement of flower crops.

We are pleased with the new, comprehensive, and detailed discussions concerning flower seeds presented in this text. As editors, however, we acknowledge a major deficiency in the lack of coverage of flower seed pathogens. This is an important topic that must be incorporated in future books on this subject. Seeds can be major carriers of disease organisms. As a result, it is important to provide information on common pathogens, their epidemiology, disease traits and

approaches to testing for pathogen presence. Increasingly, as flower seeds are produced in one country and used in another, certification standards for seedborne infections will be necessary. These should be established on the basis of sound scientific principles and knowledge.

Another challenge for this book was the inherent diversity of ornamental crops. Should we consider all ornamental crops or emphasize specific ones? Fortunately, a select committee composed of industry representatives, growers and academics was formed to identify the 20 priority genera for the Ornamental Plant Germplasm Center. These were selected on the basis of their potential economic value and impact for the ornamental industry in the future. When authors were uncertain which ornamental crops to emphasize, we encouraged them to select from this list. By so doing, the chapters permit the development of common concepts and themes rather than a compilation of disparate bits of unrelated information.

Throughout, we have kept our audience in mind. This book is intended as a reference resource written at the advanced level. Authors were encouraged to provide comprehensive literature citations in each chapter so the reader could have the opportunity to glean additional information when desired. Anyone interested in flowers will find this book useful. However, it is those directly involved with the utilization of flower seeds, their production and quality control who will find this reference most beneficial. This includes students, growers and researchers in both the public and private sectors as well as seed technologists.

Much of the information presented in this text is derived from original research. Where feasible, authors were encouraged to develop high quality figures depicting important points. This comes at increased publication cost. Royalties generated from the purchase of this book will be used to offset these costs and reduce the purchase price to the reader. Any remaining royalties will be donated to the Research Committee of the International Society of Seed Technologists to support continued research in flower seeds. By so doing, further advancements in our understanding of the biology and technology of flower seeds will be possible. We thank each author for his or her outstanding contributions in time and knowledge to this initiative. We hope the reader enjoys these presentations and learns more about flower seeds.

References

Hannover University. 2003. *International Statistics: Flowers and Plants*. Institut für Gartenbau-okonomie der Universität Hannover, Vol. 51.

Hobhouse, P. 1992. *Plants in Garden History*. Pavilion Books, London.

2 History of the Flower Seed Industry

Nona Koivula

*Executive Director, All-America Selections® and the National Garden Bureau,
1311 Butterfield Road, Suite 310, Downers Grove, IL 60515, USA*

American Seed Industry

The origins of the United States' flower seed industry can be traced back to 1784 when the David Landreth Company opened a retail store in Philadelphia, Pennsylvania. It was the first store offering seed, bulbs and related garden supplies to consumers.

In addition to retail stores, there are three branches of the flower seed industry in the USA. While related, they serve different customers. The oldest two branches serve home gardeners as customers. The first branch is the mail-order seed catalogue group, which sells flower seeds to gardeners. The second branch is the packet seed industry, which buys flower seed, puts it in a packet and sells the packets through retail stores. The third branch is the most recently founded: the branch consisting of seed companies that sell to the professional bedding plant or pot plant growers. The three branches are not mutually exclusive. For example, W. Atlee Burpee distributes seed through mail-order catalogues, in packets at retail stores, and another division serves the professional bedding plant grower.

Origins of the American Seed Industry

The focus of this chapter is to provide information about the professional growing industry and this branch will be emphasized in this brief history. The origins of the packet seed industry are unique and are included as a reference.

One of the oldest branches was born of a distinct religious group, the Shaking Quakers, aka Shakers. The Shakers were an industrious group of religious separatists from Manchester, England, whose spiritual leader was Ann Lee, a factory worker and infirmary cook (Beale and Boswell, 1991). Shakers settled in New Lebanon, New York, in 1784. This village was selling herb seed in 1789. The network of 17 independent villages across the north-east, Ohio and Kentucky gave the Shakers a distinct advantage in the seed business. They learned various seed growing, harvesting and marketing techniques from each other. The American seed packet industry began with Shakers placing seeds in small hand-made paper envelopes. The labelled envelopes were assembled into wooden boxes and placed on horse-drawn wagons. The men travelled extensively from town to town dropping off the Shaker box of seed at general stores. There was an agreement that the store did not have to pay for the seed until packets were sold. At the end of the growing season, the Shaker would return to the store. The store owner would then pay for the seed packets sold and return the unsold seed. Amazing as it may sound, the packet seed industry remains the same today – a cash flow nightmare for any seasonal business.

The Shakers had the packet seed industry all to themselves until about 1888. The market shifted from plain hand-made envelopes to packets printed with wood engravings of flowers or vegetables. This new marketing style led to other seed companies gaining market share in the home garden packet seed business.

Brief History of Major Seed Companies

American seed companies

George W. Park

Seven existing seed companies trace their origins to between 1868 and 1905. In 1868 George W. Park published his first home garden mail-order catalogue, which contained only eight pages and two illustrations. He grew his business in Libonia, Pennsylvania, until he married a South Carolinian and moved his business to Greenwood, South Carolina. The Geo. W. Park Seed Company is well known for offering unusual flower species to home gardeners and bedding plant growers. The trial ground covers 9 acres where varieties are tested before being added to their product line. Geo. W. Park Seed Company has grown into a strong merchandizing company. It is still owned and operated by the founding family. Karen Park Jennings leads the company into the 21st century.

W. Atlee Burpee

In 1876, W. Atlee Burpee began his mail-order business in Philadelphia, Pennsylvania. While offering flower and vegetable seed, a large part of the business was farm livestock. The 1888 Burpee catalogue was entitled *'Burpee's Farm Annual'* (Fig. 2.1). The Rural Free Delivery system, recently formed by the federal government, helped Mr Burpee sell his wares, since it ensured mail delivery to almost every family. Burpee was a consummate seedsman, looking for new varieties around the world. Unlike flowers, several of the Burpee vegetable varieties introduced in the 1880s are still available on the market,

such as the sweetcorn 'Golden Bantam'. This variety single-handedly changed the favourite edible corn kernel colour from white to yellow. Still available are Burpee's 'Bush' Lima bean (1890), 'Rocky Ford' melon (1881) and the now famous 'Iceberg' (1894) lettuce. In 1888, Burpee offered the petunia 'Blackthroated Superbissima' as a new variety (Fig. 2.2). It looks like a relative of dianthus.

Vaughan's Seed Co.

In the basement of a building on LaSalle Street in downtown Chicago, J.C. Vaughan established a store selling nursery stock and vegetable seed. In 1876, vegetable farmers raised crops on the outskirts of Chicago and sold fresh vegetables to the city markets. Vaughan's Seed Company supplied the vegetable seed to local farmers. At the Columbia Exposition in Chicago, Vaughan's gained international recognition by winning 23 awards for its horticultural displays (Whitmore, 1990). Vaughan's diversified into mail-order catalogue sales and packet seed sales. After World War II, the company expanded into the greenhouse grower trade. The Vaughan family sold the business to Sandoz. It evolved into Syngenta and S&G Seeds, a strong company in the bedding plant business today.

Joseph Harris Company

Joseph Harris travelled from England to the Genesee Valley near Rochester, New York. He found the area agreeable and sent for his family. He rented a farm and in 1863 bought property, which he named Moreton Farm after his birthplace in England (Harris Seeds, 1979). Mr Harris operated a highly diverse farm and, as a gifted writer, edited or published numerous magazines and textbooks. In 1879, the writer/farmer issued a 44-page seed catalogue offering hundreds of flower and vegetable varieties as well as purebred sheep and pigs. The seed business grew and ownership was transferred to Selah Harris, the founder's son. Joseph Harris, grandson of the founder, became president in 1949, the third generation to own the business. Under his direction the seed business expanded into

Fig. 2.1. 'Burpee's 1888 Farm Annual' – W. Atlee Burpee catalogue cover.

flower and vegetable breeding. A network of salesmen sold Harris Seeds to bedding plant growers. A gifted plant breeder, Fred Statt, became one of the major petunia breeders in the 1960s. As a flower and vegetable seed breeder and producer, Harris Seeds was an important seed supplier in the 1970s and 1980s. While ownership changed, the Harris Seed Company remains in Rochester, New York, selling seed to bedding plant growers, market gardeners and consumers.

Bodger Seeds, Ltd

In Santa Paula, California, John Bodger was working for a seed company and later began his own, Bodger Seeds, Ltd, in 1890. Bodger Seeds contributed significantly to the flower seed industry as a flower seed breeding and producing company. Working with the University of Wisconsin, Bodger Seeds introduced the first wilt-resistant asters, which were an important cut flower crop. In 1919, Bodger Seeds introduced the first dahlia-flowered zinnia. The founder's son, John C. Bodger, spotted an unusual, sweet-smelling double nasturtium in a bouquet of flowers on his table. His wife informed him that the flowers were a gift from her friend. Mr Bodger investigated the source of the double nasturtium and secured some seed from his wife's friend, who was a gardener. The double nasturtium was bred and entered into the new testing programme, All-America Selections®. The variety was clearly superior to other nasturtiums as a double bloom. It was introduced in 1933 as a Gold Medal AAS winner, 'Gleam Mixture'. Even during the depths of the Depression, this new variety was in demand and sold for up to 5 cents per seed (Fig. 2.3).

Fig. 2.2. Petunia 'Blackthroated Superbissima' – sample of petunias available in 1888.

Bodger Seeds, Ltd remains firmly rooted in southern California. The company is still owned and managed by the Bodger family, continuing the tradition of breeding and producing improved flower seeds for the bedding plant market and home gardeners.

Henry F. Michell

Henry F. Michell began a seed company in the 1890s originally in Philadelphia, Pennsylvania. His company conducted business in German or English and targeted home gardeners (Nau, 1999). He assisted their garden planning and actually planted estate gardens in the area. The company exists today in King of Prussia, Pennsylvania, focusing the business on sales to bedding plant and nursery companies. It remains within the Michell family with Rick Michell as president.

Geo. J. Ball, Inc.

Born in Milford, Ohio, George Jacob Ball began working in Galloway's Greenhouse, Cincinnati, Ohio, when he was 14 years old.

He moved to the Chicago, Illinois, area, continuing to work in greenhouses. He realized there were improvements that could be made to florists' crops such as sweet peas grown as cut flowers. In 1905, he began his seed business but with a different group of customers. Most other seed companies opened their businesses with home gardeners as the immediate customers but not Mr Ball. He focused entirely on growers. At that time, greenhouse growers produced cut flowers for the numerous florist shops in the developing urban area. Mr. Ball selected flowers and produced seed specifically for growers' needs, focusing on crops such as asters and calendula. The small company grew, selling improved cut flower seed. In 1927, he purchased land and established the Geo. J. Ball business in West Chicago, Illinois. The founder's son, George K. Ball, joined the company and became president in 1949 after the founder's death. In the 1960s, the company diversified, purchasing PanAmerican Seed Co. and Petoseed Co. Two companies were formed to serve the bedding plant growers better. They were

Fig. 2.3. Nasturtium 'Gleam Mixture' – introduced in 1935; the first all double flowered nasturtiums.

PanAmerican Plant Company and Ball Pacific. Today, Ball Horticultural Company breeds, produces, distributes and markets seed and other horticultural products worldwide. Anna Caroline Ball is the third-generation leader of the Ball Corporation.

International flower seed companies

Some diplomats, like Henry Kissinger, achieve international fame and recognition. Seed companies are similar. Some have world-class breeding programmes or marketing accomplishments that are noticed in a global community. There are five seed companies that have achieved this global status in the bedding plant seed industry.

S&G Flowers

A significant flower seed breeding company traces its origins back to Nanne Jansz Groot who began selling cabbage seed in the Netherlands in 1818. Joined by his grand-sons, Nanne Sluis, Nanne Groot and Simon Groot, a cooperative was formed to sell vegetable seeds. The company was named Sluis and Groot (S&G) and was founded in 1867. Their export business expanded inter-nationally with seed sales to the USA and Russia. In 1872, S&G moved their offices and warehouses to Enkhuizen, where S&G headquarters remain today. At the turn of the 19th century, S&G initiated a flower seed department, which contributed significant new varieties to the bedding plant industry in North America.

After World War II, there was a housing boom and the economy strengthened. S&G identified a growing consumer demand for flowers in bloom. Their research was focused on uniformity, hybridization and seed quality. Breeding flowers in Enkhuizen, the new varieties were tested across North America and were well received by seed companies and bedding plant growers. S&G Flowers introduced geraniums propagated from seed, and commands a large proportion of that market today. In addition, new varieties of pansy, impatiens, petunia, begonia and verbena poured into the North American bedding plant market during the 1970s, 1980s and 1990s. S&G Flowers grew into the global horticulture marketplace with improved flower cultivars, seed production and quality.

Sakata Seed Corporation

Sakata was founded by Takeo Sakata in 1913 and, like the Sluis and Groot grandsons, he began his business venture by exporting seeds. The original office headquarters were in Yokohama, Japan, and remain there today. Sakata Seed Corporation began breeding flowers in the late 1920s and 1930s. An ambitious breeding goal was formulated to breed a genuine fully double petunia propagated from seed. In the seed business, breeding is half the picture. Once a new variety is bred, the seed must be produced to complete the picture and the transaction – the seed sale. In 1933, Sakata Seed Corporation entered into a newly founded testing organization, All-America Selections®. A fully double flowered petunia was entered and Sakata won an AAS award in 1934 for their breeding achievement. They also received international recognition for this breeding accomplishment. The variety was introduced as petunia 'All Double Victorious Mix' (see Fig. 2.4).

Sakata Seed Corporation focused on breeding improved cultivars for the bedding plant industry. The varieties introduced by Sakata Seed Corporation gained a reputation in the USA for high quality, meeting the bedding plant growers' needs. Sakata Seed continued breeding petunias, but added new classes such as dianthus, primula polyanthus and celosia to their product line. The owners of Sakata Seed Corporation, the Sakata family, understood the value of an award given to a new variety, particularly in North America

Fig. 2.4. Petunia 'All Double Victorious Mix' (courtesy of Sakata Seed Corporation).

where the bedding plant industry developed rapidly. Sakata Seed Corporation entered celosia into AAS trials and received seven AAS awards. Sakata Seed bred all three carnation AAS award winners. In 1966 a large flowered pansy was introduced as 'Majestic Giants Mix', an AAS winner, bred by Sakata Seed Corporation. Bedding plant growers have grown this class of pansy for the last 40 years. This is an example of world-class breeding that withstands the test of time. Recently, Sakata won three Gold Medal AAS awards for their breeding breakthrough in zinnias, 'Profusion Cherry, Orange and White'. As a result of innovative breeding and international recognition, Sakata Seed Corporation is one of the top four flower breeding companies in the world.

Takii Seed

The history of Takii Seed began in 1835 when Mr Jisaburo Ohmoriya grew vegetables and distributed seed to farmers. The seed business expanded and a mail order catalogue was published in 1905. Breeding plants became a priority in 1935 when an experimental station was established in Kyoto, Japan. Takii was an innovator and led the way in breeding hybrid brassica. Hybrid cabbage and cauliflower varieties received international recognition with AAS awards. Takii expanded by establishing breeding stations and adding scientists to their staff.

North American recognition for their hybrid flower breeding began in 1978 with two AAS awards. Dianthus 'Snow Fire' and zinnia 'Red Sun' achieved award-winning status. Takii expanded their flower breeding programmes into celosia, pansies, snapdragons, stock, ornamental cabbage, ornamental kale, osteospermum and nierembergia. Bedding plant growers accepted these flower classes as the industry produced more flowering plants.

Colegrave Seeds

Some people refer to the UK as a country of gardeners. David Colegrave began his English flower seed business in 1962. It grew from a humble garage operation in the village of West Adderbury, England, to a global family-owned horticultural business. Mr Colegrave introduced hybrid flower seeds to commercial bedding plant growers in the UK. He realized the value of plant uniformity and consistency. He thought the growers would pay more for hybrid flower seed, and they did. Colegrave Seeds grew rapidly and expanded distribution operations into France and Australia. Expansion continued with Colegrave Seeds purchasing Waller Flowerseeds in Guadalupe, California, and establishing a breeding/research station in the Netherlands with friends Glenn Goldsmith and Johan Meewisse. When David Colegrave died on 3 May 1992, the flower seed industry suffered a great loss. Colegrave Seeds continued expansion with joint ventures and acquisitions in Germany, France, Italy, China and Portugal. In 2001, Ball Horticultural Company, a company Mr Colegrave had used as a pattern almost 40 years earlier, purchased Colegrave Seeds.

E. Benary Seed Growers

Since 1843 E. Benary Seed Growers has bred and produced flower seed. More remarkable – the company is still owned by the Benary family. Established in Enfurt, Germany, E. Benary Seed Growers became an international player early in 1845. There was an emphasis on flowers, not vegetables. The first Benary-bred cultivar was a perennial flower, the *Lychnis* hybrid 'Haageana'.

Customer service has always been a primary business goal. They note that Gregor Mendel, the scientist, was a Benary customer in 1873. Friedrich Benary established a subsidiary in Hann Münden, Germany, which is the corporate headquarters today.

Benary has long been associated with annual wax begonias, tuber begonias and perennials. In 1972, the introduction of 'Nonstop' tuberous rooted begonias placed Benary in a strong position in the North American bedding plant industry. These were the first tuberous rooted begonias that were true-to-type from seed. The cocktail series of wax begonias, 'Gin', 'Vodka', 'Whiskey', 'Brandy', and 'Rum', were market leaders for decades. Benary expanded their breeding programme into pansies, rudbeckia, sunflowers, verbena, zinnias, and other cut flowers. They sell more than 2000 varieties to seed companies in 100 countries. They have achieved international recognition and accepted 19 AAS awards since 1939.

Associations serving the seed trade

American Seed Trade Association

As seed companies grew, they needed a voice to address common problems and find solutions. The American Seed Trade Association (ASTA) was founded in 1883. Its motto, 'First – the Seed', was a reminder to all in agriculture or horticulture that the origins traced back to seed. The Association remains strong today, headquartered in the Washington, DC area, since ASTA is involved with regulatory and legislative issues.

National Garden Bureau

In the wake of World War I, James H. Burdett convinced several home garden seed companies, such as Vaughan's, Bodger Seeds and Northrup King Seed Co., that a central office for garden communications would benefit the industry. Mr Burdett was a newspaper journalist and knew the importance of press releases to the media. In 1920, the National Garden Bureau was founded to disseminate accurate information on gardening to the garden media. The seed companies did not have their own public relations departments and allowed the Bureau to represent the seed industry to the garden media. The Bureau exists today with approximately 55 seed company members supporting the media campaign.

All-America Selections®

A southern gentleman, W. Ray Hastings grew up in the family business, Hastings Seed Co. He had an idea to conduct an independent testing programme that would encourage breeders to select innovative traits and receive recognition for this breeding with a national award. He proposed his idea to the Southern Seedsman's Association in Atlanta, Georgia in 1932. With approval and seed money from this Association, the All-America Selections® testing organization was born. The non-profit organization exists today and developed in a parallel course with the bedding plant seed industry.

Three Major Influences – Breeding, Plugs and Universities

There were three major influences on the budding bedding plant seed industry. They were innovative breeding, greenhouse plug production systems and university research. During the last 100 years, the bedding plant industry germinated, grew slowly, and then built rapidly during the 1970s, 1980s and 1990s to approach maturity in the new millennium.

Breeding

The first major influence that helped grow an industry was innovative breeding. 'Mendel's laws were published in 1866 but their application to plant breeding was not generally recognized until 1900', commented Lyman N. White in his book *Heirlooms and Genetics* (White, 1981). As with many other innovations, the universities were the first to apply the new science and the first hybrid was a vegetable not a flower. The first hybrid was a sweetcorn cryptically named 'Redgreen', which was released by the Connecticut Agricultural Experiment Station in 1924. The first commercially successful hybrid sweetcorn was 'Golden Cross Bantam' introduced in 1932 by Purdue University. Vegetables led farmers to understand hybridization and the resulting benefits of the controlled cross.

Petunias

Early innovative flower breeding leads to Japan where people had been selecting chrysanthemums for thousands of years. Breeding flowers was an ancient Japanese tradition and art. Sakata Seed Corporation in

Yokohama, Japan, bred the first consistently double flowered petunia prior to 1931. Lyman White (White, 1981) noted, 'Sakata had made a major breakthrough, leading the way to our modern hybrid grandiflora and multiflora petunias'. The variety was petunia hybrid 'All Double'. Sakata won the prestigious All-America Selections® award in 1934 for 'All Double Victorious Mix' (1997) (Fig. 2.4). The importance of these early achievements was that the excitement generated newspaper publicity and spurred sales of seed to home gardeners. Not many home gardeners had the ability to grow petunias from seed, since the bedding plant industry was only just beginning. These new flowers created a consumer demand for plants.

If one had to pick a time period for the origins of the bedding plant industry, it would have to be the 1950s. If one crop had to be chosen to signify the development, petunia was that crop. Petunias led the way to new colours and flower forms and coined the term 'pack performance' for the neophyte bedding plant industry. In 1950, Bodger Seeds was recognized for their innovative breeding of an orange/red petunia, 'Fire Chief', winner of an AAS award.

Charlie Weddle was breeding petunias and became an important American breeder. Mr Weddle and W.D. Holly formed Pan-American Seed Company (PAS) in 1946. The first truly red petunia, 'Comanche', was introduced in 1953 as an AAS winner (Fig. 2.5). It was bred by PanAmerican Seed Co., located in Paonia, Colorado (Ball Catalog, 1953). Lyman White (White, 1981) provides insight into the early years of PanAmerican Seed. He remarked, 'They [PAS] went on from there to produce a broad line of hybrid doubles and singles, both grandiflora and multiflora, that revolutionized the bedding plant industry in the early 1950s'.

Fred Statt, breeder for the Joseph Harris Seed Co., bred many petunia varieties with excellent qualities. The 'Pearl' series were extremely free-flowering plants, exhibiting exceptional recovery from severe weather. It was noted that the Harris petunias were selected under low light conditions in Rochester, New York, which may have influenced the plant response when grown under mid-western spring conditions. The Harris petunias were not necessarily early to bloom in packs but, overall, the garden performance was superior.

Breeding company cooperation

The flower seed industry was small in the 1960s. If a person worked in the industry for 2 or 3 years, that was sufficient time to meet everyone. There were probably no more than 150 people in the business. Most companies were family owned and operated. Owners of companies, even if competitors, knew each other. It was a time when people became friends for life and it didn't matter if friends worked for competing companies. Gentlemen's agreements could be relied upon as if in writing. William John Park, of Park Seed Co., and John Gale, owner of Stokes Seeds, sat down to share stories and refreshments during American Seed Trade Association conventions in spite of their companies' rivalry.

Charlie Weddle and Fred Statt were friends and maintained their relationship over many years. Both were using florist quality forcing snapdragon hybrids as breeding material to improve garden performance of snapdragons. As colleagues, they probably shared information about inbred lines. In 1958, PanAmerican Seed entered six snapdragons in the AAS trials. They all won AAS awards for garden performance and Charlie Weddle decided to share the glory of the award with Fred Statt. PanAmerican Seed and Harris Seed Company sold the new AAS winners. The 'Rocket' snapdragons were a symbol of the spirit of cooperation within the seed industry based on the strength of the Weddle and Statt friendship (NGB, 1994) (Fig. 2.6).

New breeding companies

LINDA VISTA.   In the same year that Smithers Laboratory introduced 'OASIS® floral foam', a young Texan, Claude Hope, founded his production facility 'Linda Vista' in Cartago, Costa Rica (Martinez, 1997). It was 1953 and international travel was stressful at best. Mr Hope was intimately aware that F_1 hybrid production of any flower seed was extremely

COMANCHE PETUNIA

CHIEF OF THE MULTIFLORAS

ALL-AMERICA Ⓐ SELECTIONS

Newly improved — Better than ever
Like all the better petunias
Bred and Grown
by

Pan-American Seeds, Inc.

SPECIALTY FLOWER SEEDS _Scientifically Bred & Grown_

Fig. 2.5. Petunia 'Comanche' sales and promotional flyer, 1953 – first red petunia.

labour-intensive. The USA had an educated workforce but it was becoming increasingly expensive. His production farm in Costa Rica had a local, less expensive, labour pool that was searching for work. Linda Vista was the first offshore seed production facility. Most other seed companies followed Mr Hope's lead, establishing offshore production sites in the 1960s and 1970s.

GOLDSMITH SEEDS.   In the 1960s, another plant breeder, Glenn Goldsmith, was breeding for PanAmerican Seed. When PanAmerican Seed was purchased by the Geo. Ball Seed Co. in 1962, Mr Goldsmith saw this as a perfect opportunity to begin his own breeding seed company, Goldsmith Seeds. He started his

company at Hirasaki Ranch in the remote cowboy town of Gilroy, California. He found other seed companies willing to invest in his company. He began breeding petunias and selecting plants for traits specifically designed for pack performance; traits such as earliness to bloom, compact branching habit, and overall attractive plants in packs. His breeding objectives were tied to the needs of the bedding plant industry.

Goldsmith Seeds grew rapidly as the bedding plant industry developed. In 1968, Mr Goldsmith decided to grow his new varieties in packs and pots until they flowered. He invited his customers to look at the Goldsmith products in bloom in the Gilroy greenhouses. He called this event 'Pack Trials', which

Fig. 2.6. Snapdragon 'Rocket Mix', 1960 – example of cooperation among friends in the flower seed industry.

continue every spring in a slightly more elaborate venue. Goldsmith Seeds is owned and managed by the founder's three sons: Joel, Richard and Jim Goldsmith.

In the 1960s there were basically four major American flower plant breeding companies for the emerging bedding plant industry. They were Bodger Seeds, Goldsmith Seeds, Harris Seeds and PanAmerican Seed Co.

As previously mentioned, the most popular and important crop for bedding plant growers beginning in the 1950s was the petunia. In the 1960s, every colour of the rainbow was introduced, yellow being one of the latest. After breeders expanded the colour range, they moved on to white star shapes and other bicolours. Petunias reigned supreme during the 1960s and 1970s but within 10 years a relatively unknown class would knock petunias out of the number one spot as the most popular bedding plant.

Impatiens

Bob Rieman of PanAmerican Seed Company was breeding *Impatiens walleriana*. In the 1960s, Mr Rieman bred hybrid impatiens 'Salmon Jewel', 'Red Jewel' and 'Pink Sprite'

(NGB, 1995). He gave the inbred parent lines to Claude Hope since most of the Pan-American Seed product line was produced at Linda Vista. Claude Hope was no stranger to impatiens. He noticed wild impatiens growing in Costa Rican fencerows in 1944. He and Mr Rieman had both been breeding impatiens plants. The magic of their breeding programmes combined at Linda Vista. Mr Hope's primary breeding goals were to select symmetrical plants that produced flowers like a canopy completely covering the mounded plant. Mr Hope sent his hybrid test lines to two universities for trial before introduction. Both universities tested the lines and informed Mr Hope that they were worthy of introduction. In 1968, the 'Elfin' series of eight colours was introduced by Pan-American Seed Co. (Ball, 1968b).

The 'Elfin' series was improved and expanded to 15 colours and bicolours. Popularity increased since the plants were fairly easy for bedding plant growers to produce in packs for spring flowering. The plants were even easier for home gardeners to grow. 'Elfin' impatiens flourished in almost every shade garden. Vic Ball remarked in 1977 that impatiens were vying for the number one spot held by petunias (Ball, 1977). By the 1980s, impatiens had pushed petunias out of the way and took the 'top' position as the most popular bedding plant grown.

Claude Hope has become a legend and is known as the 'father of the modern impatiens' but that does not recognize his numerous innovations in seed breeding and production. Mr Hope had an insatiable scientific curiosity, which resulted in continuous innovation in breeding and production. For example, he created an enclosed plastic growing chamber for the female impatiens parent lines. The enclosure was necessary to capture the impatiens seed as it was thrown in all directions by the 'spring loaded' seedpod. The propulsion of the seedpod is the reason for the genus name, *Impatiens*, and the common name, busy Lizzie.

Perennials

Another trend became apparent in the 1980s: an increasing consumption of perennials. While the most popular perennials, day lilies and hostas, were vegetatively propagated, some could be grown from seed. In 1989, an AAS Gold Medal winner was introduced, coreopsis 'Early Sunrise' (Fig. 2.7). It was a breeding breakthrough because it was the first perennial coreopsis that consistently bloomed the first year from seed. This variety was bred by W. Atlee Burpee Company at its Santa Paula, California, research station. The significance was that a perennial could be improved to flower consistently and early the first year from seed. This was a new breeding objective that transformed a perennial into performing as if it were an annual. This objective has been used by many seed companies to improve and introduce perennials as annuals.

During the 1990s, impatiens continued to hold the number one spot in the popularity charts. There were thousands of cultivars on the market and designer colour blends. Even a light blue impatiens was introduced.

Petunias – a new wave

In 1992, a purple spreading petunia was entered into the AAS trial. The habit was unique, a virtual ground cover plant that

Fig. 2.7. Coreopsis 'Early Sunrise' – 1989 AAS Gold Medal winner; example of breeding objective to change a perennial into an annual.

rooted at nodes. The plant flowered continuously regardless of severe weather conditions and it was tough; some AAS Judges even walked on it. This new petunia earned sufficiently high points to be considered for an award. It was a holdover, waiting for introduction. The breeder came as a complete surprise to the bedding plant industry; it was Kirin Brewery Co. Ltd, Plant Laboratory, Tochigi-Ken, Japan. The largest brewery in Japan bred an innovative new petunia that rocked the North American bedding plant seed industry. One can never predict where competition will enter the market.

After three years of negotiations, Pan-American Seed Company (PAS) came to an agreement with Kirin Brewery to market the new AAS winner as Petunia 'Purple Wave' in North America (Fig. 2.8). PAS realized that the plant, introduced in 1995, was not easy for bedding plant growers to produce. After research, they published cultural information to grow 'Purple Wave' in pots. Most petunias were still produced in packs or 10-cm pots but 'Purple Wave' needed larger pots such as

Petunia F$_1$ Purple Wave
All-America Selections
Winner 1995

All-America Selections
1311 Butterfield Rd., Ste. 310
Downers Grove, IL 60515

Fig. 2.8. Petunia 'Purple Wave' – 1995 AAS winner; the first petunia ground cover, bred by Kirin Brewery Co. Ltd.

12.5-cm. PanAmerican Seed decided to invest in a pull-through marketing campaign to consumers. Consumers were already informed of this new product as an AAS winner because of AAS consumer publicity. 'Purple Wave' became the first annual to have its own website: www.wave-rave.com. PAS commissioned a spokesperson to be a guest on TV news programmes to promote 'Purple Wave', again reaching more consumers with the Wave story. PAS invested in a purple pot designed to hold 'Purple Wave' plants and expanded the point-of-purchase promotional materials. Growers soon felt the tug of consumer demand and larger numbers of 'Purple Wave' pots were produced. PAS continued the campaign by informing the bedding plant growers of their marketing to consumers. PanAmerican Seed developed the first exceptionally successful consumer and grower trade marketing campaign for a new petunia. They established a new standard for marketing annuals. Other companies have imitated this programme but PanAmerican Seed deserves recognition as the first to accomplish a well-executed consumer and trade marketing programme.

Breeding for disease tolerance

Another major breeding objective that has been only moderately successful in floriculture is disease tolerance or resistance. This seems to have been a major obstacle for many breeding programmes. One company attained a giant leap forward with zinnia disease tolerance. Sakata Seed Corporation introduced two new varieties as AAS Gold Medal winners in 1999 (Figs 2.9 and 2.10). They were 'Profusion Cherry' and 'Profusion Orange', the first zinnias with disease tolerance to powdery mildew and bacterial leaf spot. AAS judges recognized these and other improvements, including continuous flower production and ease of growing. The new varieties were embraced by bedding plant growers who in the past had had such difficulty growing zinnia plants due to disease that many had dropped them from their product line.

The breeding developments of petunias, impatiens, pansies and other annuals encouraged bedding plant growers to introduce new varieties. These new classes or species held the consumers' interest as they learned

 1999 AAS Winner
Zinnia 'Profusion Orange'

Fig. 2.9. Zinnia 'Profusion Orange'.

 1999 AAS Winner
Zinnia 'Profusion Cherry'

Fig. 2.10. Zinnia 'Profusion Cherry' – 1999 AAS Gold Medal winners; varieties bred with disease tolerance to powdery mildew and bacterial leaf spot.

to grow these new plants and decorate their outdoor living spaces with easy-to-grow annuals. The bedding plant market enjoyed continued growth from the 1950s through to the 1990s.

Greenhouse plug production

Equally as important as innovative breeding is the development of automated machinery and advanced products to grow bedding plants more efficiently. A starting point to understand the 'dark ages' of growing is to consider the growing media. It may be difficult to imagine, but growers actually used 'dirt' from land and had to steam-sterilize it to kill pathogens before using it to grow plants. There were numerous difficulties in growing plants in soil, inconsistency being one of the most obvious. An innovation that greatly improved the quality of plants was the introduction of a commercial, soilless medium named Jiffy Mix. The mix was a collaborative effort by Cornell University, W.R. Grace and Jiffy Products of America. The first coast-to-coast distribution of Jiffy Mix, a uniformly balanced medium, occurred in 1964.

Development of plugs

SPEEDLING. At about the same time, expandable polystyrene was invented. A young vegetable grower, George K. Todd, started experimenting with moulded containers made from polystyrene to grow tomato transplants. He tried different shapes and found the square 'V'-shaped cell produced transplants that resulted in an earlier harvest and higher yield. In 1969, he founded Speedling, Inc. in Sun City, Florida and called his transplants 'speedy seedlings' or 'speedlings' (Onofrey, 2001). The first crops he grew were tomatoes, peppers, celery and sugarcane. Many innovations were first used in vegetables and then applied to flowers later. Mr Todd created growing systems to produce vegetables but these systems were easily applied to growing annual flower transplants as well. One of the first important Speedling flower crops was geraniums grown from seed.

BLACKMORE CO. In Belleville, Michigan, another company was experimenting with small cells in which to grow plants. Blackmore Co., another family-owned business, created vacuum-formed rigid plastic 'waffle'-like trays in which to sow seed. Fred Blackmore knew one of the most difficult tasks was placing one or two seeds into the small sized waffle. He engineered a seed-sowing machine that matched the waffle trays. In 1970, Mr Blackmore introduced the singulated seedling

tray and seed-sowing machine to mechanize the process. The idea of growing a singulated seedling that could easily be picked up by hand and transplanted into the pack was innovative. Only a year later, Mr Blackmore introduced another equally innovative machine, the automatic transplanter, which punched the small seedlings through the waffle tray into the flat below. While this mechanical transplanter caught the eye of many bedding plant growers, it was slow to be utilized. Vic Ball noted in 1977 that 95% of all bedding plants were still transplanted by hand (Ball, 1977).

Seed technology

Both Fred Blackmore and George Todd looked closely at individual seeds as sown one at a time in a single cell. The seed was fairly inconsistent in size, germination time and seedling vigour. Some seed never germinated, which caused skips in the cell, a labour-intensive problem to solve. These two engineering innovators approached the seed companies, at different times, with the same message, 'Clean up your seed!'

Flower seed companies looked to the vegetable seed industry for help in cleaning the seed. The vegetable seed breeding and producing companies had already developed the techniques to size seed and test for seedling vigour. The technology existed to coat or pellet seed for uniform size and to improve visibility. Many seed companies hired seed technologists from vegetable seed companies to begin to solve the flower seed problems. No-one realized the seed-sowing machine would result in the birth of flower seed technology and seed quality control.

Most flower seed companies jumped into seed quality and treatments based on the demand from growers using the mechanized sowing systems. Seed technologists learned how to expertly clean the seed by removing dead seed, which is not easy when the seed, such as petunia, is the size of dust. There are about 100,000 petunia seeds per gram (Fig. 2.11). An innovative idea based on improving seed quality was an assurance of germination. PanAmerican Seed was one of the first seed companies to guarantee 90% usable seedlings

Fig. 2.11. Petunia seed – 10,000 seeds per gram (courtesy of National Garden Bureau).

from their seed. Seed technology grew rapidly to offer growers pregerminated seed, film-coated seed and pelleted seed of a uniform size and seedling vigour (S&G Flowers, 2002). The seed quality assurances inform growers how many usable seedlings to expect from a variety. This helps growers predict the number of flats or pots that can be grown from an amount of seed.

Emerging plug industry

John Holden, flower seed manager for Ball Seed Company, predicted that growers could not place seed into individual cells, expect excellent germination and grow plants. He was proven incorrect. The waffle cells or speedling flats created a new industry to be coined: the 'plug' industry. These growers became specialized to sow seed, germinate in chambers and grow a seedling or young plant. The plants were called plugs and Speedling was one of the first to offer plugs for sale to growers. Plug growers now use the most automated growing systems and computer technology to produce young plants. The list of automated equipment is long. There are seeders, flat fillers, transplanters, taggers, potting machines, conveyor systems, watering systems and robotics to grow plants.

In 2001, one grower produced over 380 million plugs. This advanced technology resulted in the ability to grow billions of bedding plants, which was exactly what the mass merchandizers such as Wal-Mart, K-Mart and Home Depot wanted.

Universities

The third critical influence on the development of the bedding plant flower seed industry was scientific research and education. Many universities conducted flower research in the 1950s and 1960s, but three provided the most significant research and were the most influential in the industry: Pennsylvania State University, Michigan State University and Ohio State University; all land grant colleges. There were many unknowns in the early days of growing annuals. Fertilization, soil media, water quality, light levels and temperatures had to be determined in the correct combination to grow a quality annual crop. Each crop had its own pathogens, pests and 'things that went bump in the night'. There were thousands of ways to kill a crop. The horticultural scientists had to find a way to keep the plants alive.

Michigan State University

The Department of Horticulture at Michigan State University (MSU) stepped into this educational vacuum by conducting the first National Bedding Plant Conference in East Lansing, Michigan, 16–18 September 1968 (Ball, 1968a). Vic Ball attended the conference and reported that the educational sessions were excellent but only 170 people attended (Ball, 1968c). William H. Carlson, professor of horticulture at MSU, was a driving force for research and education. The second national conference was held in 1969 and the same year a new trade association was founded: Bedding Plants, Inc. (BPI). This association grew rapidly. Within 10 years there were 2900 members who were bedding plant or pot plant growers, educators or allied trade people (BPI, 1979). By 1980, there were 3600 members (BPI, 1980).

Development of BPI

The annual educational programme evolved into the BPI International Conference. The BPI objectives became multifaceted. There were six primary objectives. The first was critical to the association and emphasized the educational component. It was 'the sponsorship of an annual meeting to provide a forum for public discussion, lectures, seminars, research reporting, and whenever practical, a tradeshowing of the products and services allied to this industry.' Another objective was 'the collection and preservation of valuable and useful information relating to the business of growing and selling bedding and container plants'. The association also promoted 'the economic studies and the fostering of research into the production, marketing, sales, and distribution of these floricultural crops'. Lastly, the formulation and passage of useful legislation and leadership were also objectives.

The BPI list of officers, directors, and past presidents reads like a 'who's who' in the bedding plant industry. George K. Todd was president from 1976 to 1978. In the 1970s and 1980s, everyone in the bedding plant industry attended the international conference. For example, in 1980, the conference and first trade show held in Atlanta, Georgia, were organized by a local arrangements committee consisting of 38 people. Volunteers were everywhere to help build and grow the organization. There was no other meeting that offered so much information and networking for bedding plant growers. Every new trend was identified and discussed at the conference.

Because of successful conferences, the association was financially strong. In 1985, the BPI office was located in the historic Turner House on Hamilton Road in Okemos, Michigan (BPI, 1985). The staff of eight included Will Carlson as executive vice president and Terry Humfeld as executive director. The association provided leadership for the rapidly growing number of growers. One of the BPI educational programmes utilized computers. Computers were not in every household in 1985, but BPI created a computerized information system named Spartan Ornamental Network. The purpose was to offer a growers' guide containing up-to-date information on the cultural requirements for 40 floricultural crops. The network also offered technical advice on chemical treatments, rate of application and schedule of use for diseases that affected floricultural crops. It was an innovative approach to disseminating information to all growers, not exclusive to members.

In the late 1990s the wind started to diminish from the BPI sails. For many reasons, the association began to lose members and the international conference attracted fewer people. Many regional educational programmes were available, such as the Southeast Conference. Numerous growers could easily attend a conference within driving distance, which further reduced attendance at BPI conferences. The organization was dissolved in 2001 but led to the birth of America in Bloom.

OFA

Another organization born from a university horticulture department was the Ohio Florists' Association (OFA). Originally started in 1928–1929, graduates from the Ohio State University horticulture department met annually to remain current. Later, faculty and students organized OFA and planned a formal educational activity known as the Ohio State Short Course. Membership consisted primarily of growers in Ohio and the surrounding states. The Short Course was regional in nature at first and expanded into cultural and technological issues common to all greenhouse growers. The Ohio State Short Course grew in the 1970s and 1980s.

At the turn of the millennium, OFA was the premier grower trade association. There were 3504 members of the association. After 11 September 2001, business travel was greatly reduced. This did not affect the 2002 OFA Short Course, which attracted 10,499 attendees, breaking another attendance record. Every July, everyone who is involved in the bedding plant or pot plant growing business travels to Columbus, Ohio, for the educational programme and trade show held there annually.

Economics

History

Like a heavily loaded wheelbarrow being pushed downhill, the bedding plant industry gathered momentum from 1959 to 2000. This industry relied on a Professor of Agricultural Education, Alvi O. Voigt, Pennsylvania State University, to gather data and report statistics. In 1976, Mr Voigt reported on the 'Status of the Industry'. 'Between 1949 and 1959, the wholesale value of bedding plants increased 94% from US$16.9 million to US$32.8 million. Between 1959 and 1970, another 88% value increase was registered from US$32.8 to US$61.6 million' (Mastalerz, 1976). He used the Census of Agriculture special reports as the basis for the data. From 1970 to 1974, Mr. Voigt relied on BPI member surveys to report that 'production intentions each year seemingly have moved higher by about 10% per year'.

These reports indicate unbridled sales increasing annually. There was continuously increasing consumer demand for bedding plants from 1949 to 2000. There are many reasons for this continued growth: lifestyles, economic growth, and media personalities

such as Martha Stewart. Often overlooked are the progressive improvements of bedding plants by breeders. Annuals are easier to grow. Breeders have created hybrid or improved plants that exhibit more tolerance to heat, drought and disease. Plants produce more flowers but need less care so that even a neophyte gardener can derive great pleasure and satisfaction from the gardening experience.

Recent statistics

The industry now uses the US Department of Agriculture's National Agricultural Statistics Service (USDA/NASS) Floriculture Crops Summary for statistical information. It is published annually. The 2001 summary showed that the wholesale value of floriculture sales was US$4.739 billion for the 36 leading states (Miller, 2002). When unfinished plants are removed the value drops to US$4.129 billion, which is an increase of 2.9% over the corresponding value for 2000. Sales of bedding/garden plants, while measured in billions of dollars, have slowed. While bedding plants and garden plants remain the largest segment of floriculture sales at 52.7%, the overall market demand for live plants experiences only modest growth.

There are many reasons for modest growth. They include the sluggish economy, fear of war, terrorist acts and abnormal weather patterns. Seed sales are declining for three primary reasons. Seed technology has greatly improved seed quality so growers consume less seed. They can produce more plants with less seed. There is less waste. The increasing size of containers consumes less seed. Packs/flats production is declining. Growers are producing hanging baskets, combination planters and 12.5- to 20-cm pots overflowing with summer annuals. There is a market trend towards summer plants that are vegetatively propagated, not grown from seed. These factors have reduced seed consumption in the last 5 years.

Under current market conditions, all companies need to be marketing orientated to distinguish themselves and their products from others. The emphasis on point-of-purchase materials is an example of the shift from sales to marketing.

Conclusion

Small but statistically measurable industry growth will occur. Consumers want large plants in full bloom so that the trend towards larger plants in larger containers will continue.

Two related influences will be major factors in the next few decades; they are plant pathogens and government agencies. As plant material is shipped globally, pathogens travel for free. There will be outbreaks of pathogens identified as biohazardous, not the result of terrorism but due to the lack of sanitary conditions. The threat of pathogens can damage the entire industry. The safety systems to protect plant crops will require significant resources in the future.

Governmental agencies such as USDA, Animal Plant and Health Inspection Service (APHIS) and the Department of Homeland Security, as well as equivalent agencies in other countries, will have to coordinate their efforts to identify pathogens and quarantine areas, and inspect and certify compliance on a timely basis. The industry relies on agency oversight and intervention to protect crops from venturesome, hazardous pathogens. The industry has been relatively free from government intervention. I suggest this will not be the case in the future. Working closely with governmental agencies will be necessary for most companies in the business of selling or growing plants.

Consumers will always want to buy plants regardless of size or shape, flowers or vegetables. The bedding/garden industry will continue with slower growth.

References and further reading

American Seed Trade Association. 2002. *About ASTA*. 8 November, www.amseed.org/about.asp

Ball Catalog. 1953.

Ball, V. (ed.). 1968a. Michigan Bedding Plant Meeting. *GrowerTalks*. August, p. 18.

Ball, V. (ed.). 1968b. Important News. *GrowerTalks*. September, back cover.

Ball, V. (ed.). 1968c. National Bedding Plant Conference. *GrowerTalks*. October, p. 1.

Ball, V. 1977. *Ball Bedding Book, a Guide for Growing Bedding Plants*. Geo. J. Ball, Inc., West Chicago, Illinois.

Beale, G. and M.R. Boswell. 1991. *The Earth Shall Blossom, Shaker Herbs and Gardening*. The Countryman Press, Inc., Woodstock, Vermont.

[BPI] Bedding Plants, Inc. 1979. Twelfth International Bedding Plant Conference, Chicago 1979. BPI, Okemos, Michigan.

[BPI] Bedding Plants, Inc. 1980. Thirteenth International Bedding Plant Conference and First Trade Show, Atlanta 1980. BPI, Okemos, Michigan.

[BPI] Bedding Plants, Inc. 1985. *Membership Directory and Buyers Guide 1985*. BPI, Okemos, Michigan.

Harris Seeds. 1979. *A History of the Joseph Harris Company, Growing for a Century, 1879–1979*. Joseph Harris Co., Inc., Rochester, New York.

Martinez, R.A. 1997. *The Master of Seeds, Life and Work of Claude Hope*, 2nd edn. Burpee Books, Cartago, Costa Rica.

Mastalerz, J.W. 1976. *Bedding Plants – a Manual*. Pennsylvania Flower Growers, University Park, Pennsylvania.

Miller, M. 2002. Floriculture's Changing Face. *GrowerTalks*. July, pp. 86–120.

[NGB] National Garden Bureau. 1994. *Year of the Snapdragon* fact sheet. NGB, Downers Grove, Illinois.

[NGB] National Garden Bureau. 1995. *Year of the Impatiens* fact sheet. NGB, Downers Grove, Illinois.

[NGB] National Garden Bureau. 1997. *Year of the Petunia* fact sheet. NGB, Downers Grove, Illinois.

Nau, J. 1999. *American Seed Houses in the 18th and 19th Century*. West Chicago, Illinois.

Onofrey, D. 2001. Plug Pioneer's Perspective. *Greenhouse Grower*. Mid-September Bonus Edition, pp. 16–21.

S&G Flowers. 2002. *Plug Into a World of Support*. GMPro. November, p. 32.

White, L.N. 1981. *Heirlooms and Genetics, 100 Years of Vegetables and Flowers*. Cambridge Seed Packet, Cambridge, New York.

Whitmore, L. 1990. Vaughan's Corporate Culture Evolves. *Seed World*. September, pp. 23–24.

3 Ornamental Bedding Plant Industry and Plug Production

Debbie Hamrick

FloraCulture International, 2960 Claremont Rd, Raleigh, NC 27608, USA

The world of flowers from seed evolved rapidly. What started decades ago with packet and bulk seed sales of mail-order seeds, mainly to home gardeners, exploded into the multinational, multibillion-dollar bedding plant industry of today. Growing flowers from seed is an important component of commercial floriculture production. Commercial sales of flower seeds drive the flower seed industry. Within flower seed sales, bedding plant crops are the most important seed-raised crops as a total worldwide.

Why would a commercial flower grower choose to produce a crop from seed rather than a vegetatively propagated clone? There are a number of reasons.

First, for many crops, seed is the only or best alternative. For example, marigolds (*Tagetes*) are generally not produced by cuttings. They are seed grown. Sometimes growers even direct-sow marigolds into the final growing container. In other instances, seed is superior to a vegetatively produced crop. For example, pot gerbera may be produced from tissue culture or from plug seedlings. However, in nearly every instance, the grower will choose the seed crop because, while plants may be a bit more uniform in time to flower from tissue culture, the expense is high and the resulting uniformity does not offset the economics under most conditions. Cut flower gerberas are almost exclusively propagated from tissue culture since the plants can be cropped for 1 to 2 years, making the input cost less significant to the economics of the crop.

Second, seed-raised crops get their start 'clean'. Plants that are grown from seed initiate growth free of latent viruses, mycoplasmas or bacterial disease. Not necessarily so for a vegetatively propagated crop, which must be maintained in a rigorous clean stock programme that involves indexing of elite stock and maintenance of stock plants in a clean environment.

The third reason a grower would choose to grow a crop from seed rather than cuttings is ease of handling. Seed is portable, takes up little space, can be easily transported from one location to another and has a shelf life. Every day, flower seed travels across and between continents. Growers routinely access flower seed genetics from around the world simply by ordering from their local distributor sales representative. All of this means that a small bedding plant or pot plant grower in Portland, Oregon, can produce varieties that were bred in the UK, Germany, Japan, the Netherlands and the USA. The hybrid seed produced from this breeding may have been grown and harvested in Chile, Costa Rica, Guatemala, India, Indonesia, Kenya, France, the Netherlands or the USA.

Fourth is the basic premise that flower growers will chose to grow a crop from seed whenever they can because it's cheaper to do so. In general, a crop grown from seed will always be less expensive to produce than a crop grown from cuttings.

Other reasons growers choose to grow seed crops include inherent plant characteristics such as plant habit. Seed-raised crops branch more freely than the same crop produced from cuttings. An excellent example of this is verbena, where vegetative crops need pinching or treatment with a branching agent such as Florel to cause them to branch. There are even some seedsmen who argue that with proper crop culture, a seed-grown crop allowed to obtain its genetic potential will outperform a vegetative crop of the same species.

Floriculture Production from Seed

World wholesale floriculture production is estimated to be about US$22 billion (USDA, 2003; VBN, 2002). The three leading producing countries, Japan, the USA and the Netherlands, produce about 55% of the world's production value (Table 3.1). Figure 3.1 provides a look at the type of floriculture production value in the three leading producers. The USA's industry is primarily composed of bedding plant production (53%), while in Japan and the Netherlands cut flower and cut foliage production is more prevalent. Interestingly, over the past decade, the American floriculture pie has

become more and more dependent on bedding plants, rising to 53% of all floricultural production in 2002 – up from 35% of all floricultural production in 1991 (Fig. 3.2). The exact percentage of total floricultural crops originating as a seed, a cutting or other propagation forms is unknown.

Bedding plants

Bedding plants are the most important commercial floricultural crop produced from seed. In the world's three most important floricultural production countries, Japan, the Netherlands and the USA, bedding plant production is valued at approximately US$3.5 billion wholesale. The US bedding plant production market is by far the largest, valued at about US$2.5 billion wholesale for all growers with US$10,000 or more in wholesale production. Figure 3.3 shows the rapid rise of bedding plant production in the USA from 1988 to 2002.

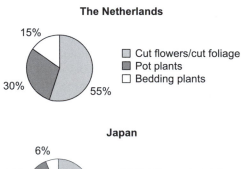

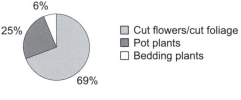

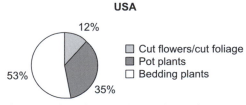

Table 3.1. Comparison of the top three floricultural production countries.

	ha*	Production (US$)	Value/ha (US$)
USA[a]	24,092	4,880,000,000	202,556
Japan[b]	24,525	4,309,547,000	175,720
The Netherlands[c]	8,898	4,069,800,000**	457,383

[a]USDA Floriculture Crops 2002 Summary, April 2003.
[b]MAFF, Japan. www.maff.go.jp, accessed February 2003. 'Planted area and shipment of flowers in 2000', and 'Summary of the wholesale market for flowers 2000'.
[c]Flower Council of Holland. Facts & Figures 2002.
*1 ha (hectare) = 2.4 acres.
**Euros 3,420,000,000 converted at a rate of US$1.19.

Fig. 3.1. Floriculture production for top three countries.

The most often produced bedding plant is pansy (*Viola × wittrockiana*, Fig. 3.4). It unseated *Impatiens walleriana* (Fig. 3.5) in most units produced recently. While pansies have been the most important bedding plant in the UK market for years, it wasn't until the 1990s that Americans discovered how well pansies overwintered. In the USA, pansy production is valued at US$140 million.

Impatiens are the next most important bedding plant worldwide. The USA consumes more impatiens than any other country, with a wholesale crop valued at about US$157 million, making it the most important US seed-grown bedding plant (Table 3.2).

Other important bedding plants grown from seed include petunias, marigolds, begonias, salvia, dianthus and catharanthus. While seed crops have dominated bedding plant production in the past, the trend is that seed crop production of bedding plants is stagnant, while clonal varieties are increasing in importance.

Floricultural crop production pie 2001

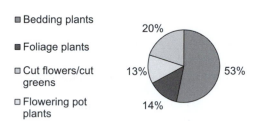

- Bedding plants
- Foliage plants
- Cut flowers/cut greens
- Flowering pot plants

20% 13% 53% 14%

Floricultural crop production pie 1991

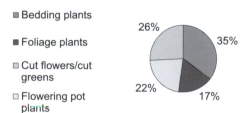

- Bedding plants
- Foliage plants
- Cut flowers/cut greens
- Flowering pot plants

26% 35% 22% 17%

Fig. 3.2. Floricultural crop production. Source: United States Department of Agriculture Floriculture Crops Summary.

Cut flowers

At the turn of the last century, the most important cut flower crops were produced from seed. Crops like calendula, sweet peas (*Lathyrus odoratus*), godetia (*Godetia whitney/Clarkia amoena*) and snapdragons (*Antirrhinum majus*) were the rage in Victorian flower arrangements. Over time, clonally propagated crops edged them out and carnations (*Dianthus caryophyllus*) and roses became the leaders.

As commercial floriculture continued to evolve, seed-grown cut flower crops have

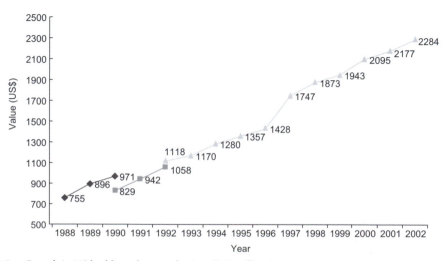

Fig. 3.3. Growth in US bedding plant production ($US millions).

Fig. 3.4.   Pansies (*Viola* × *wittrockiana*) have become the most valuable bedding plant from seed world-wide thanks in part to the fact that they enjoy dual seasonality: growers produce pansies for autumn and spring sales. Seed enhancements such as priming have also raised the value of worldwide pansy seed sales.

Fig. 3.5.   *Impatiens wallerina*, the most popular bedding plant grown from seed worldwide. The Stardust flower pattern is shown here.

reemerged as important crops for domestic and local producers and as an important way for flower producers in developing nations to expand their export flower product assortment.

Today, many cut flower growers world-wide routinely produce seed-grown crops such as asters (*Callistephus chinensis*), snap-dragons (*A. majus*), delphinium (*Delphinium* × *belladonna/Delphinium* × *elatum*), lisianthus (*Eustoma grandiflorum*) and sunflowers

(*Helianthus annuus*). In addition to these, there are a number of smaller crops produced indoors under minimal cover or outdoors as 'summer flowers'.

Potted plants/foliage plants

Several important flowering potted plants are grown from seed. The most important of

Table 3.2. Breakdown of bedding plant crops produced in the USA.

	US$1,000s (Total)
Begonias	94,550
Geraniums (seed)	48,970
Impatiens	156,765
Marigolds	57,631
Pansies	139,664
Petunias[a]	141,444
Other bedding[a]	729,953

[a]Also includes vegetatively propagated plants.

these in value is cyclamen. This autumn and winter flowering pot plant is important to holiday sales in both the North American and European markets and is an important plant for the winter gift season in Japan. Crop time is very long (24 weeks or longer), making plants among the highest in value at wholesale.

Other flowering potted plants grown from seed enjoy seasonal popularity such as gloxinia (*Sinningia speciosa*), cineraria (*Pericallis cruenta*), calceolaria, and primula (*Primula* sp.).

Foliage plants are primarily grown from clones propagated via cuttings or tissue culture. A few crops, however, are routinely grown from seed. For example, many palm species, such as the areca and kentia palms, are grown for their stately fronds. Asparagus fern (*Asparagus sprengerii*) and dracaena spike (*Cordyline indivisa*) are popular seed-grown foliage plants for use with annuals in combination planters. Other crops, such as hypoestes or coleus, are seed crops grown for their colourful foliage. While both may also be propagated by cuttings, propagating the crop from seed is less costly. Many cacti and succulents are propagated from seed by specialist growers.

Perennials

While perennial production is growing quickly in the USA, the Netherlands and Japan, only a few perennial species are routinely produced from seed, as clones tend

to be more stable and produce true-to-type plants. Other factors limit the number of perennials grown from seed too, such as difficulty in germination. Perennials as a class of floricultural plant encompass hundreds of genera. Germination techniques range from cold stratification to scarification. It has only been recently, as perennial production has expanded, that breeding companies have begun to focus on improving perennial germination and marketing cultivars and seed treatments to improve germination percentages for growers. As a result, more perennials are being produced alongside bedding plants for spring and summer sales. Researchers have also gained better understanding of the flowering mechanisms of a wide range of perennials, which has helped increase perennial production. The phenomenon of producing perennial crops and selling them in flower as you would an annual has helped make perennials the fastest growing segment of floricultural production in the USA.

The other main limiting factor for perennial production from seed is the need for commercial producers to offer plants that flower in their first year. Many perennials do not flower in the first year from seed. Other perennial cultivars may flower in the first year from seed but might have lower garden performance than their cloned relatives. Other seed perennials, while marketed as perennials, are better treated as annuals.

The three leading perennials, hosta, hemerocallis and iris, are all vegetatively propagated for commercial production, although breeders frequently grow seedlings as they strive to develop improved cultivars. Once the breeder finds a plant with desirable traits, the plant is propagated vegetatively and slowly makes its way into commercial trade as plants are multiplied each year.

Some perennials, such as coreopsis, delphinium and rudbeckia, are routinely grown from both seed and cuttings. In this case, the seed varieties tend to be restricted for production in smaller container sizes such as bedding plant packs or 10-cm pots. The vegetatively propagated varieties are produced as liners or bareroot plants that might be used for 4-litre or larger containerized production or even planted in the field or greenhouse for cut flower production.

The outlook for seed-grown perennials is bright as more and more traditional flower seed hybridizers focus on developing cultivars that flower in their first year from seed and germinate in quantities that allow them to be grown economically.

Food uses of flower seed

Several flowers grown from seed are edible! Increasingly, speciality grocery stores and upscale restaurants offer seed-grown flowers such as nasturtium, marigolds or pansies/violas as plate or salad garnishes. These edible flowers are produced pesticide-free.

The largest non-ornamental use of flowers grown from flower seed is for processing into an additive for chicken feed. The lutein in African marigolds (*Tagetes erecta*) makes the yolks of eggs bright and gives chicken skin its golden glow. Marigold farmers produce the blossoms from select open-pollinated varieties. The blossoms are then sold for processing where the lutein is chemically extracted.

The fastest growing edible segment of marigold lutein use is in the dietary supplement industry. Research has revealed that lutein slowed age-related macular degeneration in humans. As a result, dietary supplement manufacturers have begun to incorporate lutein extracted from African marigolds into their supplements.

Plugs/Plug Industry

The most important single driver that has had an impact on flower seed use in the past century has been the development of the 'plug' as a propagation unit. Plugs were first used for vegetable transplants (cauliflower). It was a new way to grow field transplants pioneered by George Todd at Speedling Inc., Sun City, Florida, in the late 1960s (Miller and Smith, 1980).

While plugs got their start in vegetables, it was the application of this concept to flowers that caused the boom in plug technology and

later propelled the development of bedding plant production. This flower seed plug industry technology has spread to other areas of agriculture such as forestry, back to vegetables and even to tobacco seedling production.

In just 10 years after George Todd's first cauliflower plugs made history, the Florida plug industry grew to US$6.25 million in sales across 19 companies who produced 346 million plug units in 1978/79. Seven of these firms also began growing ornamental plugs.

The first commercial floricultural greenhouse in the USA to adapt plugs to production of flowering crops was not in Florida, but in the north-east. Kube-Pak, Allentown, New Jersey, began researching a 'Precision Seed-Feeding machine' to sow seeds (the world's first drum seeder) back in 1969 with inventor and engineer J.V. Bertrand. The model Kube-Pak developed was capable of sowing 1200 flats per hour into 50×34 cm ($19\frac{1}{2} \times 13\frac{1}{4}$-inch) trays with 600 cells 1×1 cm ($\frac{1}{2} \times \frac{1}{2}$-inch) tall (Swanekamp, personal interview and correspondence). That original machine was in use until the late 1990s, when it was retired in favour of newer models that were more flexible in the tray sizes they accommodated. Kube-Pak is credited with taking the idea of the plug and developing it into a system for growing bedding plant transplants.

As bedding plants increasingly were sold as F_1 hybrids, seed became so expensive that growers could no longer afford to broadcast-sow their crops. As a result, growers rapidly shifted their bedding plant production out of open fields from which plants were 'dug and wrapped', and into cell packs, which allowed one seedling or a seedling clump the space it needed to grow and develop.

About the time that plugs were making their way into larger production operations, seed companies continued steaming ahead on the F_1 track, further increasing the price of seed, which further fuelled the shift to plugs.

Today, the revolution continues. With some hybrid varieties selling for $0.05 per seed, growers continue to increase the size of the finished product. High-end seed crops are often produced in 10-cm (4-in) or larger containers for the best economic returns.

What is a 'plug'?

A plug is a seedling that has been produced in a small container called a 'cell' in a plug tray (Styer and Koranski, 1997). One plug tray can contain hundreds of these cells, or 'plugs'. Seed is sown in plug trays that may contain from 72 to 800 cells per standard 55 × 28 cm (21½ × 11-in) or common element (CE) 25 × 51 cm (10 × 20-in) tray. Different plug (cell) sizes are selected for varying reasons. For example, bedding plants that will be grown in flats are typically produced in smaller cell sizes such as 512s, 384s or 288s. Bedding plant plugs that can be used to plant hanging baskets might be 288s, 162s, 105s or even 72s. Perennials tend to be grown in larger cells. Some growers choose their plug tray because they will be using it for automatic transplanting, or in the case of the X-tray produced in Europe by Syngenta, because of superior root development and faster plant establishment in the final growing container.

Whenever two growers speak to one another about their production systems, cell size of the plug seedlings they use is always included in the conversation because it has important implications for finishing time and how the crop will be produced. For the plug producer, the smaller the cell size, the faster the plug will finish out as a plug. For example, a 512-impatiens crop can take from 34 to 43 days. A 288 of that same crop would take about 14 days longer.

Bedding plant growers who use plugs to finish the crop also talk widely about the plug sizes they prefer. A general rule of thumb is that for every decrease in plug cell size, if all other factors are the same and growing temperatures and light are ideal, crop time will increase by 1 week. So, a 512 plug will take 1 week longer to finish than a 384, which will take 1 week longer to finish than a 288, and so on. Most bedding plant growers would begin their first crop of bedding plant flats from 512 plugs about 6–8 weeks before the sale date. This crop is the most expensive crop to produce as transplanting occurs in the late winter, the crop requires heat, and additional light must be supplied since ambient outdoor light levels are low. A second crop can be planted from larger plugs such as 288s and turned in just 2–3 weeks. A third turn, or 10, 15, 20 cm (4, 6 or 8-in) pot annual production for sales in the summer, can be grown from 162s or 105s and also turned in 2–3 weeks.

The plug size affects not only finishing times for the final grower, but also the ease with which the crop is finished. Smaller celled bedding plants have more delicate roots as plants are not as mature. Starting off a bedding plant crop with small-celled plugs such as 512s requires more grower attention to detail than starting the same crop from 288s or 128s. The decision on which size plug to use is based primarily on economics. In general, many bedding plant growers will plant their first bedding plant turn using the smallest celled plugs they can because they are less expensive to purchase. As the season progresses, temperatures warm up and growers strive to maximize the turns from their space by increasing their plug sizes to 384s and/or 288s, which will finish faster. The best growers use various sized plugs for different crops through the year. In general, using larger sized plugs means shorter crop times and higher turns, which results in greater profit per area of production. Figure 3.6 shows that

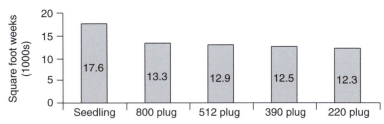

Fig. 3.6. Begonia crop space requirement per 1000 flats (36-cell pack). From: Gary Szmurlo, Ball Seed Co., West Chicago, Illinois.

1000 flats of begonias in 36-cell packs grown from seedlings takes 17,600 sq. ft weeks. The same number of flats produced from 390 plugs takes just 12,500 sq. ft weeks – 30% less time.

Plugs revolutionized bedding plant production, gaining rapid and widespread acceptance for the following reasons:

- **Speedier transplanting**. Old-style bedding plants were produced by sowing seed into open seed trays, allowing the seedlings to attain a size that could be handled, and then painstakingly and gently pulling them apart and planting them into the packs and trays filled with medium. Plugs simplified and speeded this process, perhaps by much as 50% depending on the crop. Instead of working to pull apart seedlings, workers could simply pop out a plug, dibble a small hole using their finger or a dowel and plant it.
- **Less transplant shock**. Plugs allow growers to achieve a higher quality, more uniform finished crop. Disease can also be less of a factor because roots are not damaged during transplant.
- **Better overall space use**. Plugs can be held at a high plant density for weeks during the plug production cycle, which means better space use economics. Although plug trays take up significantly more space than open seed flats, seedlings from open seed flats take much longer to finish. Figure 3.7 shows that growing out a begonia crop in 36-cell packs takes about 15 weeks in total. However, using plugs, most of that time is pre-transplant growing as a plug. From bareroot seedlings, a begonia crop takes about 10 weeks finishing. However, from 390s, that same crop spends only 6 weeks finishing.

The plug 'industry'

Plug production has morphed into a mini industry within the overall auspices of the floriculture/ornamental horticulture industry. Several dozen plug producers worldwide form the manufacturing base for the majority of all flower plugs produced. These growers sell their plugs to other growers who in turn finish the crop. Table 3.3 presents a list of the top ten plug growers in the USA.

When the idea of plugs first spread through the bedding plant industry, growers everywhere rushed to give plug production a try. However, growing plants in what are in essence 'miniature containers' is very difficult. Due to the small size of the medium and root mass, any problems with water, pH or nutrition are magnified.

Over time, most growers abandoned growing plugs in favour of buying plugs from specialist growers. Even those growers who continued their own plug production purchase plugs of difficult-to-grow or slow-growing plug crops such as begonias, pansies (*V. wittrockiana*) or vinca (*Catharanthus roseus*).

While the plug production industry began with bedding plants, it has spread to other crops within floriculture/ornamental horticulture. Cut flower plugs have enabled cut flower producers to have a regular, high quality year-round supply of crops such as asters, snapdragons, lisianthus and delphinium without the difficulty of growing their own seedlings. Many pot plant growers found that they also could produce potted plug crops from seed such as cyclamen, gloxinia, calceolaria, hypoestes and gerbera. More recently, plug production has been widely adapted to perennials. Several perennial plug growers offer multiple plug sizes and even cooled

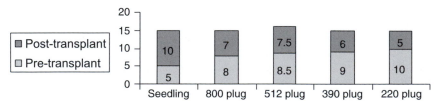

Fig. 3.7. Begonia crop time (weeks) in 36-cell packs. From: Gary Szmurlo, Ball Seed Co., West Chicago, Illinois.

Table 3.3. Top ten US plug growers.

Grower	Production units (millions)	
	2002	2001
1. Tagawa Greenhouses, Colorado and California	500	350
2. Green Circle Growers, Ohio	350+	380
3. The Plug Connection, California	275	250
4. Speedling Inc., California, Florida and Georgia	250–300	250
5. C. Raker & Sons Inc., Michigan	165	165
6. Plainview Growers, New Jersey	155	155
7. Floral Plant Growers, Wisconsin and Maryland	150	210[a]
7. Wagner Greenhouses, Minnesota	150[a]	150+[a]
9. Knox Nursery, Florida	130–150[a]	130–150[a]
10. Kube-Pak, New Jersey	130	130
10. Van de Wetering, New York	130	150+

[a]Estimates.
Reprinted with permission from *Greenhouse Grower*, 'Making Choices: Top 10 Plug Growers', mid-September 2002, Volume 20, No. 11, page 14. Meister Publishing Co., Willoughby, Ohio.

(vernalized) plugs that are ready-to-pot and force for spring sales.

The switch to bought-in plugs from open seed flats sown by each grower has not been entirely positive. Probably the biggest negative is that plug producers limit the assortment of varieties and species on the market simply because they do not grow them. There are two major negative ramifications regarding this. First, a new variety from a breeder must have distribution by being offered as a plug before it will be widely adapted by a large number of growers. While most plug producers offer several series of each main crop on their plug list, it is impossible to offer all cultivars. Growers who wish to request specific varieties not on the supplier's list can place a special order, provided they meet minimum order requirements. Second, many miscellaneous crops have such a small market share that seed companies cannot economically justify investing in higher seed quality. As a result, germination rates can be as low as 30% or 40%, making them very expensive as potential plug crops.

Plug grower suppliers address each of these limitations with 'custom sowing'. Most plug suppliers will grow just about any seed-grown crop as long as the grower meets a minimum order (usually ten or 12 trays, although some require less). Unfortunately, the minimum order is too large for smaller and medium sized growers.

Technology development

The development of flower plug production had ripple effects throughout the entire floricultural industry, spurring booms in technology. Because producing plugs is so technical, it also spurred development of more sophisticated environmental controls and equipment designed to aid growers in manipulating the environment. Typically, the plug range is heavy on equipment and sophisticated under-bench heating, boom irrigation or sophisticated mist, overhead curtains and high intensity discharge lamps. For most greenhouses producing plugs in addition to finished crops, the plug range has the highest per square foot capital outlay, with prices ranging from US$12 to US$25 or more per square foot.

Over time, as plugs became more and more popular, equipment suppliers in the industry developed special equipment for sowing, handling and transplanting plugs (see Fig. 3.8). Until plugs became commonplace, mechanization for bedding plant growers was virtually non-existent. Today, specialized seeders, automatic tray and flat fillers and

Fig. 3.8. Plugs enabled development of automatic transplanters to plant bedding plants and pot plants, thus eliminating much of the handwork involved in bedding plant production.

watering conveyors are standard head house production equipment in plug greenhouses. When plugs are ready to transplant, machines such as dibblers press holes in the medium, so all that remains is for the worker to place the plug into the dibbled hole. Transplanting lines, which consist of a conveyor belt with planting stations set up on either side, allow workers to sit as they work, and significantly speed the transplanting process. Watering conveyors irrigate newly planted flats before they are taken to the greenhouse for finishing. Today's largest bedding plant operations now use automatic transplanters that remove plugs from the plug tray and place them into the finishing flat insert or pot. The process is not without flaws since workers still must 'patch' skips in automatically transplanted pots and trays as they emerge from the transplanter. However, using a modern transplanter, growers can transplant as many as 1200 or more flats of 36-cell packs (606) per hour. Smaller, manually operated transplanters have been introduced which enable growers to transplant 300 flats per hour.

The most advanced technology introduced for plug production to date is 'patching' plug trays during plug production to increase crop uniformity and thus final per square foot yield. Many of the world's largest flower plug specialists use special plug transplanters

during the plug production process. The machines first scan plug trays with video cameras that send a digitized signal to a computer. The computer instructs the transplanter to literally 'blow out' skips or under-performing seedlings and place a good plant in its place (see Figs 3.9 and 3.10). Trays are returned to the greenhouse to tone and finish. After automatic grading during the plug crops' final days, bench space yield for the plug growers is 100%. The rationale for the extra expense is strictly economic. Typically, plugs are sold by count. For example, a 288-plug tray that has 288 holes for plugs may be sold as 250 or more. For plug growers supplying the plugs, by patching to 100%, the plug grower adds 38 plugs to the tray, increasing the sale by increasing the value of the box (four plug trays) by 15%. If the tray is sold as a 270, yield is an additional 6%. The economics are also better for the grower buying in the plugs: freight charges are less expensive per unit since there are now more plugs per box. Patching allows the plug producer to maximize both the greenhouse production bench and the shipping carton.

Development of plug technology did not stop with equipment. Plugs forced seed suppliers to focus on seed technology as well. Germination rates of 90% on blotting paper in a laboratory did not correlate with real

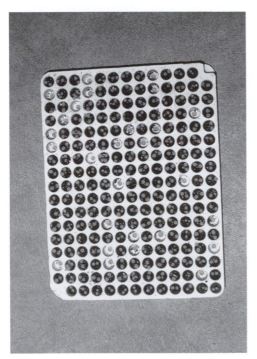

Fig. 3.9. Grading seedlings during plug production enables plug growers to make more efficient use of their production space and to ship higher value boxes or racks. This photo shows a plug tray in the Netherlands after it passed through one of the earliest commercial plug grading machines in 1989 at Zaadunie De Lier (Syngenta). Underperforming seedlings were literally blown out.

world experience on the plug bench. Because seed was singulated into a single growing cell, growers became painfully aware of every cell that was empty. Not only that, but some seedlings were underperforming and lagged behind others. Seedling vigour and germination on soil became the new paradigm for seed suppliers. 'Usable plants' became more important to the plug grower than overall germination. Leading suppliers supplanted germination on blotting paper in a laboratory chamber and replaced it with soil germination under controlled growing conditions and sophisticated seedling rating systems, such as the Ball Vigor Index (BVI) and others.

Advances in seed technology also facilitated production increases for some crops. Pansies, for example, are difficult to germinate in the summer due to heat. As a result, only growers with specialized cooling chambers or in cool regions of the country were able to grow high quality pansy plugs. As pre-germinated seed was introduced, more and more plug growers were able to produce pansies, thus increasing the supply of high quality plugs to finishing growers.

Other improvements also increased germination at the grower level. Many plug producers were building their own germination chambers in which near ideal environmental

Fig. 3.10. At this plug grading line at Hamer Bloemzaden in the Netherlands workers manually patch the empty cells before trays are returned to the greenhouse.

conditions were duplicated to germinate diffi-cult crops. At the same time, seed handling knowledge and techniques improved at the breeder, distribution and grower levels. Today, refrigerated seed storage is standard at any plug production facility.

The cultural knowledge base that was established because of plug production extended into other crops as well. Principles on how plugs respond to water, media and nutrition apply equally as well to crops such as poinsettias or chrysanthemums. The most interesting application for the seed industry of the cultural technology developed through plug production was that it made entry into rooted cuttings for plug producers almost a seamless transition. Each year, floricultural crops grown from seed decline in total market share in favour of vegetatively propagated crops.

Conclusion

While seed-grown floricultural crops con-tinue to be an important part of the total floricultural supply, they are experiencing

a declining total market share in well-developed bedding plant consumption markets such as the USA. The floriculture industry owes much of the advancements in production technology and cultural techniques made over the past two decades to the invention of the plug seedling.

References

Miller, M.N. and Smith, C.N. 1980. *The Containerized Vegetable Transplant Industry*. Institute of Food and Agricultural Sciences, University of Florida, Gainesville, Economic Impact Report 129.
Styer, R.C. and Koranski, D.S. 1997. *Plug and Transplant Production: a Grower's Guide*. Ball Publishing, Batavia, Illinois.
Szmurlo, G. 1994. *Plug Economics and Scheduling*, a seminar during GrowerExpo 1994, Chicago, Illinois.
[USDA] United States Department of Agriculture. April 2003. *Floriculture Crops 2002 Summary*. National Agricultural Statistics Service, Washington, DC.
[VBN] Vereniging van Bloemenveilingen in Nederland. 2002 *Statistiekboek 2002*, February 2003. VBN, Leiden, The Netherlands.

4 The Uses and Potential of Wildflower Seed in Landscaping

Gene P. Milstein

Applewood Seed Company, 5380 Vivian Street, Arvada, CO 80002, USA

History of the Industry

In the 1960s, many forces were at work that influenced the development of the wildflower seed industry. The environmental movement was just beginning. An awareness of air and water pollution in the big cities motivated people to look around and consider the condition of the world their children would inherit. They also began to see an accelerating loss of plant and animal species that would leave a sterile environment for future generations. As a result, there was an increased use of wild flowers with the intention of improving and enhancing our environment.

Prior to that time, the sale of wildflower seed and plants occurred through mail order catalogues of seed collectors such as, in the USA, Clyde Robin Seed Company, and mail order nurserymen such as Mid-West Native Plants. Retail native plant nurseries were few and wildflower seed packets in garden centres rare. Occasionally, Alaskan seed collectors would travel to the lower 48 states and offer their seed to garden centres and nurseries. In 1967, Applewood Seed Company began collecting wildflower seeds and offered them in retail packets to the gift and garden markets around the USA. In 1973, Bodger Seed established a new division, Environmental Seed

Producers (ESP). Environmental Seed Producers, Applewood and other California seed companies such as S&S Seeds began offering bulk wildflower seed to the landscape market. In the 1970s and 1980s, more companies entered the market and wild flowers, along with grass, shrub and tree seeds, were included in the increasingly common re-vegetation and reclamation projects. Various niches developed such as prairie restoration and highway beautification. In 1982, Lady Bird Johnson opened the National Wildflower Research Center in Austin, Texas.[1] She was instrumental in assuring the inclusion of a provision into a Federal Highway Act that required a percentage of all federal highway funds to be spent on re-vegetation with wild flowers. This caused a surge in highway beautification and a commensurate increase in wildflower seed demand.

Retail sales of wildflower seed in packets and other products expanded rapidly with many large companies entering the market, including Burpee, NK Lawn and Garden, Ferry-Morse, Pennington and others. The retail market grew from independent retailers to include mass merchants and discount chains. Nearly everyone selling packet seed began offering wildflower seed as well.

During the 1980s, substantial quantities of wildflower seed were sold in retail markets

---

[1] In 1989, the *National Wildflower Research Center's Wildflower Handbook* was published.

in various containers besides packets, including bags, pouches, boxes and cans. Meadow-in-a-Can was probably the best known of these products. This marketing effort peaked with the introduction of the 'mulch bag concept', which used hundreds of thousands of kilograms of wildflower seed per year. This product included a small amount of seed mixed with a large quantity of recycled paper mulch as a spreading agent. It was very popular with the home gardener for planting areas of 5–10 square metres (approximately 50–100 square feet).

Wild flowers are now used in a very wide range of situations including habitat restoration, parks, residential and commercial developments, golf courses and amusement parks, to name a few (see Plate 4.1). The concept has spread to most developed countries in varying degrees. Europe and Japan have been significant users for 10–20 years.

Wild flowers represent one of the most colourful and attractive aspects of nature. They symbolize a part of the natural environment that we do not often see in our daily lives, particularly in the suburbs and inner city. They are a form of beauty that home gardeners can easily replicate in their backyards. For some, wild flowers represent a sense of simplicity, freedom, a casual life-style, a feeling of wildness in a complex, demanding, fast-paced society. Above all, they provide colour and beauty in a world of glass, brick and concrete. There is nothing more refreshing on the front page of a magazine or a website than a field or a bouquet of wild flowers.

Current Issues in the Wildflower Seed Industry

There are many issues confronting the wildflower seed industry. For example, what is the definition of a wild flower? No-one agrees and the question has been discussed for hours. *Webster's Dictionary* defines a wild flower as the flower of a wild or uncultivated

plant.[2] This seems simplistic in today's complex world.

One definition by a botanist is, 'Any showy or attractive flowering plant that occurs naturally in a given geographical area or region; it could be either native or introduced.[3] Applewood Seed Company defines a wild flower as 'any flower that has not been hybridized or chemically altered'. Still other definitions exist depending on the point of view. For example, 'It's not an American wild flower unless it's indigenous or endemic here.'

The use of natives vs. non-natives is also the source of controversy, especially in discussions with botanists and ecologists. Historically, there have been three categories of wild flowers: native, non-native and naturalized. The term 'native' is as hard to define as 'wild flower'. Webster defines native as 'living or growing naturally in a given region; indigenous or endemic'.[2] But since the true origin of most species is generally unknown, the term native is often applied to species present prior to 'the arrival of white man or Europeans'. Non-native according to Webster means 'exotic, foreign or introduced'. Dr Hartman[3] defines native as 'any plant found in a given geographical area or region which was there prior to any recent modification of the environment caused by humans (that is, prior to AD 1600). A naturalized species refers to one that comes from another state, country or region and has been able to establish itself and reproduce outside of cultivation without the aid of man. Recently, environmental groups have preferred not to recognize this category and refer to naturalized species as introduced' or exotic species. This would mean that in places where only 'wild flowers' are allowed, such as highways, commonly planted species like *Achillea millefolium* (white yarrow), *Daucus carota* (Queen Anne's lace) and *Linum perenne lewisii* (wild blue flax, APAR) could no longer be planted.

In recent years, a number of wild flowers in the USA have been labelled invasive species and this issue has become so important that President Clinton established the National

2   Webster's 3rd New International Dictionary.
3   Dr R.L. Hartman, Director of the Rocky Mountain Herbarium, University of Wyoming.

Invasive Species Council in Washington, DC in 1999. Which species should go on an invasive or a state 'Noxious Weed List' is a question that has consumed many hours of discussion with government officials. This has caused much confusion in the public arena because many people have treated invasive as synonymous with noxious. However, 'noxious' is a legal term since each state has a Noxious Weed Law that lists 'noxious' species that are controlled or prohibited by law. Invasive weeds and lists of invasive species are at present only the subjective opinions of the individuals or organizations that prepare them. They have no legal standing and cannot be enforced by any governmental agency. State agencies may decide that a species is so invasive that it should be put on the Noxious Weed List. But, until that occurs, no 'invasive' species can be regulated or prohibited.

Industry has been impacted because many consumers, companies and public agencies believe they cannot plant species that are on an invasive list, even though they have been planted regularly in landscaping for generations. Therefore, the sale of many commonly used species has been prohibited or reduced. Some lists include crops such as wheat, rye, oats, Kentucky blue grass and California poppy, crown vetch, dame's rocket, vinca and blue flax.

The latest issue involves both the state and federal governments who are entering the business of producing wildflower seed and competing with private industry. Some states, for example, Iowa, Minnesota, Florida and Missouri, and federal agencies such as the Forest Service and the Bureau of Land Management have been using government employees or prison inmates to produce species or ecotypes they say they cannot procure from industry. The wildflower seed industry believes that the demand for local ecotypes is inappropriate because there have been no scientific studies to prove that local ecotypes are necessary with wild flowers. The industry is small and even regional ecotypes are uneconomical to produce. From experience, it is clear that the cost of collection or of producing small quantities of locally grown seed lots will be so high that no-one will be able to afford them. If the government produces them, it will be an extravagant misuse of taxpayers' money.

The American Seed Trade Association and the American Association of Nurserymen have taken an active role in almost all of these issues, especially because they affect major segments of both industries.

Current Challenges Facing the Wildflower Seed Industry

Key challenges facing the wildflower industry involve keeping up with the diverse markets that use wildflower seed, including the many niche markets. Providing adequate personnel to address these small markets is difficult and the cost of travel to the majority of the customers becomes prohibitive. Maintaining profitability in a maturing, highly competitive market is also a significant challenge. Thirty years of increasing competition has made the wildflower business like any other highly competitive business. Smaller margins mean that companies must become smarter and more efficient. As more companies compete for a limited amount of business, maintaining high quality seed in the industry also becomes a challenge.

In addition, producing popular regional items at competitive prices is an ongoing challenge. Highway departments are a source of substantial contracts. Especially when there are large budgets, the contracts go to the lowest bidder. Other governmental business comes from the Park Service, Forest Service, the Bureau of Land Management, and city, county and state landscaping projects.

Balancing the demand for low-priced seed with the request for locally grown ecotypes can be a significant challenge. Locally grown seed or local ecotypes are generally very expensive and low-cost seed can be produced in other countries with low labour costs. Trying to satisfy both of these demands can be difficult.

Uses of Wildflower Seed

Many different individuals and organizations use wildflower seed, including home gardeners, landscapers, landscape architects, golf courses, amusement and theme parks, city, county, state and federal parks, highway departments, grass, flower and packet seed companies, universities, conservation groups, zoos, botanic gardens and restoration, re-vegetation and reclamation projects such as mined lands, prairie and wetlands restoration (see Plate 4.1).

There are 300–500 individual species of wild flowers in the US landscape trade, and still more are listed in some mail order catalogues. Altogether the listed species make up no more than 3–5% of those native to North America. There are many thousands more around the world. The listed species represent only a small fraction of those in the wild. In this respect, the industry has a large untapped resource. However, less than 100 species account for the majority of the seed sold each year. These species are primarily the low-cost, easy-to-grow species with a wide range of versatility.

A typical wildflower seed catalogue provides the information a landscaper needs to make good decisions concerning seed selection. This information includes: scientific and common plant names; plant type: annual, biennial or perennial; cultural requirements such as sun exposure and moisture needs, e.g. full sunlight or shade; moist conditions, which require over 75 cm (30 inches) rainfall per year, or dry conditions requiring 25–75 cm (10–30 inches) rainfall per year; bloom colour; blooming period; seeds per pound (× 2.2 for seeds/kg).

Even though the number of species is small compared with the potential, many different mixtures are developed from 100 to 200 individual species. From specialized mixtures for golf courses to highway departments to knee-high to bird and butterfly mixtures for home gardeners, there are blends for almost every conceivable purpose or theme. Custom mixtures are an important service of the industry. Most seed companies have a large selection in their computer, since it is not uncommon for customers to request custom mixes reserved for their use only.

Table 4.1 illustrates some common regional mixtures in the USA and elsewhere, frequently offered special use mixtures, and various theme mixtures in the trade.

Some of these mixtures are 'kitchen sink' mixes with every inexpensive species included, while others are refined blends that have required years to perfect a desirable balance of species. Some companies conduct extensive field trials every year to develop new mixtures for new purposes. These can take several years to perfect. Each application has special needs, and if the volume of seed required is large enough, a special mixture will be designed. For example, a golf course might use a different blend near a tee or green from what it would use in the rough or outlying areas. Available irrigation is also a consideration in determining which seeds to include in the mixture.

The seed quality of individual species or mixtures is important. Table 4.2 lists the minimum state germination standards for many wildflower species. Minimum standards were primarily established for the packet seed trade and they are quite low. Most seed sold in the landscape trade is usually 10–20% higher.

The main uses of wild flowers are for beautification and erosion control. Wildflower seed is planted by home gardeners as an alternative to bedding plants because they are less formal, more economical and require less water and maintenance. Annuals are effective for a 1-year display, while perennials are planted as long-term flower gardens. The seed may be planted as individual species for blocks or swaths of colour or alternatively as mixtures in different themes: tall, low growing, shade, sun, moist or dry conditions, mountains, coastal areas, regional, colour blends and many other parameters. Plate 4.1 illustrates different mixtures and various types of projects where wildflower seed is used.

Landscape contractors and architects use wildflowers in residential and commercial projects to bring the native look back into the urban environment. Golf courses use them around tees and greens for a touch of colour,

Table 4.1. Regional, special use and theme mixtures common in the wildflower industry.

Common regional mixtures in the USA:
Pacific Coastal Mix
Gulf Coast/Caribbean Mix
High Plains Mix
Midwest Mix
Northeast Mix
Northwest Mix
Rocky Mountain Mix
Southeast Mix
Southwest Mix
Western Mix
Other regional mixtures:
Central Japan Mix
European Flower Mix
Israeli Flowers
Mediterranean Flowers
Northern China Mix
North American Flower Mix
South African Flowers
South India Mix
South Korea Mix
Selected special use mixtures:
Annuals for Sun
Bird and Butterfly Mix
Deer Resistant Mix
Golf Course Mix
Hummingbird Mix
Knee-Hi Mix
Low Growing Mix
Native Perennial Mix
Salt Tolerant
Shade Mix
Super-short Mix
Xeriscape Mix
Wildlife Mix
Theme mixtures:
Beneficial Insect Mix
Cool Colors Mix
Cottage Garden Mix
Cutflower Mix
Fragrant Mix
Jewel Tones Mix
Moonlight Garden
Orange Flowers
Pastel Colors Mix
Red, White & Blue Mix
Red & Gold Mix
Warm Colors Mix
White Flowers Mix
Yellow Flowers

and they use them in out-of-play areas to recreate the natural environment. Wild flowers are especially helpful in reducing water use. Amusement and theme parks often use wild flowers to provide inexpensive, colourful displays to enhance the look of their entertainment facilities.

City, county, state and federal parks frequently use native flowers to enhance or replicate the native species, often requesting local ecotypes. Maintaining the natural ecology is a common goal of many public and non-profit organizations. Prairie restoration is a focus of many of these entities, particularly in mid-western states. But a number of private individuals and corporations also wish to recreate the natural prairie or native landscape. Highway departments request wild flowers for erosion control and beautification. Some use natives to create an authentic local ecosystem, others focus on beautification to make the highways exciting and interesting for the traveller. Many newspaper articles have been written about these beautiful displays. Conservation groups such as the Nature Conservancy and various habitat groups have botanists on their staff to recommend the use of particular native species and local ecotypes to enhance the appearance of their land.

Universities, botanic gardens, zoos and other institutions have created wildflower display gardens to show the public how wild flowers can be used in home and commercial landscapes. Retail garden products often incorporate wildflower seeds to satisfy the home consumer's desire for a native environment in their backyard.

Wild flowers have their own cultural needs and maintenance requirements. Many people still assume that scattering wildflower seed will result in a beautiful display with little or no effort. In most cases, this does not occur. Wildflower installations, like all horticultural plantings, require care and attention, but less than many other types of plantings. Attention must be paid to weed control, water availability, and balance of species, especially when mixtures are used. The care and maintenance of wild flowers is different from

Table 4.2.   Minimum germination standards for wild flowers.[a]

Species	% Germination
Achillea spp.	50
Dimorphoteca aurantica, Dimorphoteca sinuata	55
Lobularia maritima	60
Anagallis arvensis	60
Arabis alpina	60
Aubrieta deltoidea	45
Rudbeckia hirta	60
Brachycome iberidifolia	60
Calendula officinalis	65
Eschscholzia californica	60
Campanula rotundifolia	50
Iberis umbellata	65
Iberis sempervirens	55
Centaurea cyanus	60
Cerastium biebersteinii, Cerastium tomentosum	65
Chrysanthemum carinatum, Chrysanthemum coronarium	40
Chrysanthemum leucanthemum	60
Chrysanthemum maximum, Chrysanthemum superbum	65
Clarkia unguiculata	65
Aquilegia spp.	50
Echinacea purpurea	60
Coreopsis lanceolata	40
Coreopsis tinctoria (Caliopsis bicolor)	65
Cosmos bipinnatus, Cosmos sulphureus	65
Delphinium ajacis (Consolida ajacis)	60
Dianthus deltoides	60
Dianthus barbatus	70
Bellis perennis	55
Linum grandiflorum, Linum lewisii, Linum perenne	60
Gaillardia aristata, Gaillardia pulchella	45
Gila spp.	65
Heliopsis helianthoides	55
Hesperis matronalis	65
Linaria spp.	65
Lupinus spp.	65
Myosotis spp.	50
Nemophila menziesii	70
Nemophila maculata	60
Penstemon spp.	60
Phacelia campanularia	65
Phlox drummondii	55
Papaver rhoeas, Papaver nudicaule	60
Saponaria ocymoides, Vaccaria hispanica	60
Helianthus spp.	70
Machaeranthera tanacetifolia	60
Thunbergia alata	60
Vinca roseus (Catharanthus rosea)	60
Zinnia linearis	50

[a]These standards are found in many state seed laws and are adopted from the RUSSL Amendments prepared by the AASCO (American Association of Seed Control Officials).

traditional bedding plants because they are usually planted in large areas where intensive care is unavailable or impractical. In general, wildflowers thrive and produce a colourful display if they receive over 75 cm (30 inches) of rainfall during the growing season. If this is not available, 1.25 cm (½ inch) per week of irrigation will suffice. Wild flowers are usually planted in areas requiring a colourful display but a lower degree of expense and maintenance. They meet this expectation as long as the appropriate cultural practices are followed.

Wildflower Seed Usage Globally

Wild flowers are used widely in developed countries that have the resources, especially the time and money, for beautification and maintenance of the natural environment. Less developed societies tend to focus on food, housing and survival. Highly populated regions in developed countries are more conscious of wild flowers because they are aware of the loss of natural habitat. Therefore they attempt to incorporate wild flowers into their surroundings. In addition, people living in cooler, more temperate climates are more interested in growing and cultivating flowers of all kinds. In tropical and sub-tropical areas where many plants grow easily, there is less interest in wild flowers. The USA and Canada, in general, show more interest in wild flowers than Europe partly because there is still more undeveloped land on which to recreate the native ecosystem. Wild flowers are used in varying degrees in other developed countries such as Japan, South Korea and China, mostly for parks and highway plantings. Plate 4.2 shows a canal restoration project using wild flowers in Tokyo, Japan.

The use of wildflower seed in Asia has been much slower to develop than in North America and only a few countries are active at this time. The countries planting wild flowers from seed include Japan, China, South Korea, Australia and New Zealand. Australia and New Zealand have shown much interest in the concept because of their close cultural connection with the USA and Europe and because of their interest in preserving the natural environment.

In general, people in Australia and New Zealand have observed the usage of wild flowers in western countries for the last 30 years and have used wild flowers in similar ways. Australia, in particular, has been especially sensitive to the introduction of alien species because introduced species can get established and spread so easily in its mild climates. Thus, Australia has tended to use its native species more for re-vegetation and reclamation projects. Since New Zealand has a great diversity of climates, a much broader range of species of wild flowers has been used in this country and in similar markets as the USA. Australia's plant species, as with its animal species, are so unusual and unique in the world that the seeds of many of them have been sold to tourists and gardeners around the world.

In Japan, China and South Korea the concept of gardening and landscaping with wild flowers is relatively new. Japan began using wildflower seed within the last 20 years because they provided a 'new look' for parks, commercial developments, golf courses and along the highways. On roadsides, wild flowers provided a source of colour as well as erosion control. They have been used extensively in residential and amusement parks because they provide a less formal, casual look compared with traditional bedding plants. Recently there has been an increasing interest in using Japanese natives as protection for the native environment.

The market for wildflower seed in China and South Korea has started to grow in the last 5 years mainly because it provides an inexpensive source of colour in the landscape along roadsides, on golf courses and in parks. Particularly in China, horticulture is being relearned through the university system, since the art was almost wiped out during the Cultural Revolution. South Korea's market is primarily on golf courses at the present time, because wild flowers provide an inexpensive method for bringing colour around greens, tees, the clubhouse and along fairways especially in out of-play areas. Many golf courses are being built in these countries for the emerging middle class and tourists from Japan.

The markets for wildflower seed in Europe are similar to the markets in the USA, except smaller. Wild flowers have been used for several decades, but there is less undeveloped land in Europe; therefore reclamation and re-vegetation are less important markets.

Wild flowers are defined differently in most western European countries. Flower seeds sown along the highways, in parks, on golf courses and in residential and commercial developments are referred to as 'meadow flowers'. These same flowers are offered in retail products such as packets, cans, boxes and bags for the home gardener. The uses and markets for 'meadow flowers' are similar to the markets for wild flowers in the USA and Canada, but smaller because people own less land around their homes. The species used here are more native to Europe, whereas those in the USA and Canada tend to be more North American natives. However, there is considerable overlap because the flower seed breeders and the bedding plant trade have introduced many of the same species to both markets over the last 150 years.

In Europe, the term 'wild flowers' is reserved for species collected in the wild, often at high elevations in the mountains. They are very expensive seeds and only limited quantities are available. Thus, these species are generally offered only in small seed packets primarily for the tourist trade. Seeds grown on seed farms would not qualify for the wildflower labelling in these countries.

The Future of the Wildflower Industry

The wildflower seed industry has been in existence for about 30 years, enjoying major growth during the first 20 years. It is now a maturing industry with many competitors at various levels filling different niches. It remains a small industry compared with the bedding plant or turf industries, but a stable one with many different aspects and interests. There are several markets to satisfy. Some people simply want a beautiful display and are not concerned about species native to their area. They want low-cost, easy-to-grow seed. Others have very specific expectations and will only use species native in their region. Some even require that the seed be produced locally within 90–180 km (50–100 miles) of where they are to be planted. These expectations keep wildflower seed companies on their toes (some say off-balance). Because traditional seed companies find it difficult to respond to these specific demands, non-traditional, local and regional seed companies have emerged with the support of local and state governments.

How these issues are resolved has an impact on the direction of the traditional wildflower seed industry. How many of these niches the seed companies can supply will determine how much the industry grows in the future. Therefore, the future depends a great deal on how much each of the member companies get involved with resolving these issues. It is important that wildflower seed companies express their views as these issues are ultimately defined and resolved.

Differences Between Wildflowers and Bedding Plants and Other Ornamental Plants

The wildflower seed industry is a smaller industry than the bedding plant/F_1 hybrid seed industry. The seed is, in general, less expensive and open-pollinated annual, biennial and perennial species. The wildflower seed industry seeks to identify attractive, easy-to-grow original native species wherever they can be found. The goal is to produce the seed at the lowest cost with the highest quality possible. It requires 5–10 years to go from initial seed stock to a saleable crop. This is similar to plant breeding and developing an F_1 hybrid. Finding the correct climate to produce high quality wildflower seed can be a major challenge.

Just because a crop is native in a certain climate is no guarantee that it will produce good seed or a high yield in that location.

For example, New England aster seldom produces a high yield or good quality seed in New England. Surprisingly, it does produce abundant, high quality seed in the semi-desert climate of Colorado. Although butterfly weed (*Asclepias tuberosa*) is native in many states,[4] so far only Michigan has yielded good production, even though it has been tested in many places.

The wildflower seed industry is made up of many small markets, whereas the bedding plant industry is composed of fewer but larger markets. Because the markets are different, the channels of distribution are also different. Most wildflower seed is sold for direct seeding on site. Therefore, very little of it is used by bedding plant growers. Highway department, re-vegetation and reclamation officials contract directly with seed companies or their distributors, which are different companies from those selling bedding plant seed. Thus, each industry has separate channels of distribution.

Proper care of wildflower seed mixtures differs significantly from other flower and ornamental seed crops. Wild flowers require less water and maintenance. But, even more important, the care of wildflower seed mixtures is different because there is competition among species in the mixture. Maintaining a balance of species is especially important after the first year of growth. Several techniques exist for doing this, such as mowing, selective removal of seed heads and over-seeding. Weed control is also an issue with wild flowers because they are often used in larger areas where hand weeding is impractical. It is essential to eliminate both the existing vegetation and weed seed in the soil before planting because it almost impossible to differentiate weed seedlings from wildflower seedlings. Providing adequate irrigation is also different with wild flowers because they are often planted where traditional sprinkler systems are not used. Since direct seeding with wild flowers is most common, the planting techniques are different, often using cyclone seeders, grass drills and hydro-seeding equipment.

Collection and Production of Wildflower Seed

Prior to 1970, most wildflower seed was obtained from collectors or produced by European seed growers who had harvested original stock seed during their travels and produced it on seed farms for sale to the European garden market. Farm production of wild flowers began in the USA in the early 1970s, mostly in California (see Plate 4.3). Collectors harvested either by hand or with machines such as combines.

Hand collection involves dumping or breaking the seed stalks with seed pods or heads into paper bags or plastic containers. The seed is usually dried on a plastic film or a concrete floor. The pods or heads may be crushed if they do not shatter on their own. Cleaning the seed requires similar seed cleaning equipment as used for grass, vegetable and herb seeds. The main purpose of such cleaning is to remove other crop seed and inert material such as chaff, stalks, seed pods, hulls and other plant parts.

Seed production is usually a three-step process common to most grass seed:

1. *Seed production* may take 1, 2 or 3 years or more until the plants flower, depending on whether the crop is an annual, biennial or perennial. The success of production is dependent on the soil and climate, including rainfall, sunlight, day length, temperatures and length of growing season. Each species has different requirements and determining the appropriate location for production is largely a trial and error process. There will be some consistency within a species, but not necessarily within a genus.
2. *Seed harvesting* may be done by hand or with mechanical equipment. Large crops are usually harvested with combines except in countries where labour is less expensive.

---

[4] See Hortus Third, L.H. Bailey Hortorium, Cornell University.

Small crops of 0.25 ha (½ acre) or less are generally harvested by hand.

3. *Seed cleaning* involves a range of equipment as noted above to remove dust, soil, rock, weed and other crop seeds, stalks, leaves, hulls, chaff and other foreign matter. Some of the equipment used for seed cleaning includes air screen machinery, gravity tables, spirals, indent cylinders, brushing and scalping equipment.

Seed companies typically work with farmers in a range of climates to produce wild-flower seed as insurance against crop failure. Seed maturity and high quality are usually best in arid climates where the grower can control the water the plant receives through irrigation and avoid rains during the seed setting and harvest time.

However, there is a demand for locally grown and regional ecotypes. This causes the risks and costs to increase substantially. Often the yield and seed quality will be low when regional production is a requirement. One of main stimulants to the wildflower industry in the 1970s and 1980s was lower priced, farm-produced seed. But the new environmental and ecological interests have reversed that trend. Local ecotype production is often accomplished by smaller and less experienced growers. Also, seed testing and seed quality are often ignored or omitted on small lots. Without testing, the seed may receive insufficient cleaning. Therefore, the seed quality is typically poor or unknown.

Wildflower Seed Testing

Germination and purity testing are the primary tests conducted on wildflower seed. Sometimes, tetrazolium (TZ) tests (see Chapter 15, this volume) are performed on hard-to-germinate species or when a quick indication of viability is desired. All of these tests are performed by registered or certified seed analysts in private or state laboratories according to the official *Rules for Seed Testing* as established by the Association of Official Seed Analysts (AOSA) or the International Seed Testing Association (ISTA). See also Chapters 14 and 15, this volume.

Germination tests involve counting 100–400 seeds, placing them on a sterile growing medium (such as pH-balanced blotting paper), placing them in a germinator for a specified period, keeping them at optimum temperatures and moisture level and then counting and evaluating the seedlings as they germinate. There are specific rules for testing many wildflower seed species to help standardize this process throughout all laboratories in the world. According to law, these tests must be repeated every 6 months.

Purity tests determine the percentage of pure seed compared with crop and weed seed, soil, chaff and other foreign material. These tests are performed by skilled purity analysts and greatly affect the value and selling price of the wildflower seed. Noxious weeds are a particularly important contaminant, and prohibited noxious weeds can make any given lot of seed unmarketable.

Tetrazolium (TZ) tests are chemical tests done by cutting the seed in half to expose the embryo and endosperm. The seed is then soaked in a tetrazolium solution for several hours. Pink-stained tissue indicates viability. TZ tests are not considered legal tests in most cases and are generally used as indicators of potential germination for those seed lots that are highly dormant.

Quality Control

Wildflower seed, like all flower seed, is subject to minimum germination standards and labelling laws in many states. Many species have their own minimum germination percentage standard. If no such standard exists, then a minimum of 50% is usually applied. Companies may have their own purity standards, but the flower seed laws require labelling for the presence of any noxious weeds. Many public and private landscaping architects and contractors require the listing of the Pure Live Seed (PLS) for each seed lot. PLS is determined by multiplying the decimal version of percentage germination by the percentage purity. For example, if the purity is 98% and the germination is 78%, multiply 0.98×0.78 which = 0.7644 or 76.44% PLS.

This tells a customer what percentage of the seed will be viable seed.

Poorly adjusted cleaning and harvesting equipment or improper climate and storage conditions can harm the seed and reduce germination. High humidity is the principal factor causing loss of viability, but mechanical damage can also affect seed quality and vigour. Proper storage includes temperature and humidity control and special seed rooms are often constructed for this purpose. Insect control can also be an issue, although insects are less problematic with wildflower seed compared with other crops such as grains and beans.

Rare or exotic wildflower species are often not tested and the seed quality is unknown. In this case, buyers must understand they are dealing with an untested product. Sometimes a TZ or cut test is conducted. The methodology is to cut the seed in half to determine whether the seed is filled with a white appearing embryo and endosperm.

Seed testing laboratories can be either state or private commercial laboratories. Some specialize in wildflower, native tree and shrub seeds. An increasing number of wildflower species are getting specific rules and testing techniques so that tests around the world are standardized. National and international associations are encouraging AOSA and the Society of Commercial Seed Technologists (SCST) to develop proper testing methods for commonly encountered wild flowers. Most state laboratories are situated on land grant universities. Most well-established wildflower seed companies are familiar with both private and state laboratories and use them regularly.

As with any new aspect of the wildflower seed industry, production and collection of seed are the first priorities. Marketers, customers and governmental agencies were often tolerant of low seed quality in the early years. But as the wildflower seed industry grew and matured, normal testing and quality standards have become expected. The wildflower seed industry is an example of a mature industry and the emerging wetland seed industry is an example of a small industry that is receiving tolerance in terms of quality and testing. Usually, it is 5–10 years after

producers and collectors introduce new wildflower species to the landscape market before appropriate testing techniques and standards are established by seed testing laboratories and regulatory agencies. Since the wildflower seed industry is 30–40 years old, the quality expectations are similar to those for grass, vegetable and field seed. Seed of locally grown ecotype wildflowers is an exception, and much of that seed is still untested or does not meet typical wildflower standards. Although there are no specific wildflower regulations, general flower seed laws that exist in many states are applied to wild flowers. When wild flowers are mixed with grasses, they are subject to grass seed labelling requirements. Minimum germination standards for each species are common, and 50% is the minimum for those without a specific standard. The minimum germination standards for the most commonly used wildflower species in the landscape trade are provided in Table 4.2.

Crop and noxious weed seed in wild flowers are becoming an important issue, particularly prohibited noxious weed seed. With over 10 million ha (25 million acres) of land in the western USA overrun with noxious weeds, it is critical not to allow more of them to spread through re-vegetation and reclamation projects.

Introduction of New Wildflower Species

The new introduction of wildflower species involves selecting, testing and producing new species from the 50,000 or so native wildflower species. Each state, region, and country has its own indigenous wildflowers. Selections are made on the basis of popularity, ease of culture, seed yield and cost of production, and versatility in use and climate range. The question of whether to introduce a species native to a given region or a well-adapted one native to some other region, state or country is a common question for seed companies. As Stephen Jay Gould said, 'Exotics are often far more adapted than natives'. The answer is usually dependent on how many appropriate native species are available. For example, in the mid-west,

Rocky Mountain region and the north-eastern part of the USA, there are few native annual wildflower species. Therefore, if annuals are desired in a mixture for first year colour, it is essential to select species from California, Oregon, Washington or countries in Europe or Africa. In general, it is difficult to prepare native mixtures that will be suitable for a whole country. For this reason, most such mixtures have between 20 and 25 species to assure performance in all areas. Selection of appropriate species often takes years of trials in different climates.

Preliminary field trials are often conducted before deciding where and when to do a large production of 1–2 ha (2–5 acres). Both germination and purity tests are essential before making a transition from test to production. At least 2 years of trials may be necessary to confirm the climate that will reliably yield high quality seed.

The equipment for cleaning and harvesting can be major investments, depending on the crop and size of production. Most crops in the USA are mechanically planted and harvested, especially fields over 0.25 ha (½ acre) in size.

Other factors involved in selecting a new species are colour, height, blooming period, aggressiveness and general aesthetic appearance and beauty. Cost of production, establishment time and durability of the crop and seed viability are additional issues considered when making a selection. In summary, new species introduction is a long, complicated process with many unknowns and surprises.

Future Potential of the Wildflower Seed Industry

There are many uses for wildflower seed from golf courses, home garden and commercial landscaping to mine land reclamation and highway beautification. Each niche has its own requirements and potential. New uses are showing up all the time from cemeteries to wetlands to xeriscape gardens. In that sense, the future of the wildflower seed industry is bright and unlimited. Competition has also become intense as the wildflower market matures. Seeking out new markets becomes more challenging. The US, Canadian, European and Japanese markets seem stable, and are subject to the fluctuations of their economies. But other countries in Asia, Africa and South America are just beginning to discover the charm and benefits of wild flowers.

Many different people and organizations plant wildflower seed including home gardeners, commercial landscapers, city, county, state and federal governments, conservation groups, zoos and botanic gardens. The list of applications and types of people interested in their use is endless and continues to grow and change.

One thing can be said for certain about the wildflower seed industry: it is never boring!

References and Further Reading

Art, H.W. 1986. *A Garden of Wildflowers*, Storey Communications Inc., Pownal, Vermont.

Art, H.W. 1990/1. *Wildflower Gardener's Guides*
Vol 1. *Pacific Northwest, Rocky Mountain & Western Canada* (1990).
Vol 2. *Midwest, Great Plains, Canadian Prairies* (1991).
Vol 3. *California, Desert Southwest, & Northern Mexico* (1990).
Vol 4. *Northeast, Mid-Atlantic, Great Lakes & Eastern Canada* (1990).
Storey Communications Inc., Pownal, Vermont.

Brooklyn Botanic Garden Record. 1990. *Gardening with Wildflowers and Native Plants*. Revised edn. of Vol. 45, No. 1. Brooklyn Botanical Gardens, Brooklyn, New York.

Bruce, H. 1982. *How to Grow Wildflowers & Wild Shrubs & Trees in Your Own Garden*. Van Nostrand Reinhold Company, New York.

Cox, J. 1991. *Landscaping with Nature*. Rodale Press, Emmaus, Pennsylvania.

Diekelman, J. and Schuster, R. 1982. *Natural Landscaping: Designing with Native Plant Communities*. McGraw-Hill Book Company, New York.

Martin, L.C. 1990. *The Wildflower Meadow Book*. The Globe Pequot Press, Chester, Connecticut.

Phillips, H.R. 1985. *Growing and Propagating Wildflowers*. University of North Carolina, Chapel Hill, North Carolina.

Rock, H.W. 1981. *Prairie Propagation Handbook.* 6th edn. Wehr Nature Center, Hales Corners, Wisconsin.

Smith, J.R. and Smith, B.S. 1980. *The Prairie Garden: Seventy Native Plants You Can Grow in Town or Country.* University of Wisconsin Press, Madison, Wisconsin.

Sperka, M. 1984. *Growing Wildflowers, A Gardener's Guide.* Charles Scribner's Sons, New York.

Steffek, E.F. 1983. *The New Wild Flowers and How to Grow Them.* Timber Press, Portland, Oregon.

Taylor, K.S. and S.F. Hamblin, 1976. *Handbook of Wildflower Cultivation.* Collier Books, Macmillan, New York.

Wilson, J. 1992. *Landscaping with Wildflowers.* Houghton Mifflin Co., Boston, Massachusetts.

Wilson, W.H.W. 1984. *Landscaping with Wildflowers and Native Plants.* Ortho Books, Chevron Chemical Co., San Ramon, California.

5 Breeding Flower Seed Crops

Neil O. Anderson

Department of Horticultural Science, University of Minnesota,
1970 Folwell Avenue, St Paul, MN 55108, USA

Background

Flowers have many beneficial components for the consumer that are created, enhanced, or improved by flower breeding programmes using classical or molecular techniques. Some new crops may initially be direct selections from the wild with inherent traits of interest. Once such a new crop has been successfully commercialized and shows market potential, however, flower breeders begin the arduous task of crop domestication and directed plant breeding. Many attributes of floriculture crops enable continued popularity with the gardening public, such as flower power, convenience, therapeutic value, fragrance, colour, enhanced longevity (postharvest life), expression of emotions, artistic value, or multiple uses in a variety of environments (floral designs, holiday decorations, flowering potted plants, colourful foliage, colour containers, hanging baskets, annual bedding plants, perennials; Plate 5.1). Flower power is the most common attribute that attracts consumers, providing continuous flowering throughout the growing season in the case of annuals, colour integration of interiorscapes or exterior landscape plantings with architectural surroundings, fragrance, exotic appeal, therapeutic value and the promotion of human health. Convenience refers to the relative ease of propagation, production, flowering and care or maintenance in the landscape. This range in floral value and usage ensures a continued need for flower breeding and crop enhancement.

Product groups

An array of floral products provides market diversity for the industry. These include flowering potted plants, foliage plants, cut flowers, cut foliage, bedding and garden plants (annuals, biennials, perennials), and herbaceous perennials. The wholesale value of US floricultural crops was US$4.74 billion in 2001 for all growers with ≥ US$10,000 in sales and 931 million ft^2 in production (USDA, 2002). While this category is now the third largest within the US Department of Agriculture, surpassed only by maize and poultry, it contains multiple commodity groups with hundreds of crops. A flower seed breeder has a wide variety of commodities and crops on which to focus. The largest floriculture commodity is bedding and garden plants (US$2.18 billion, 45.9%), followed by potted flowering plants (US$832 million, 17.6%), potted foliage (US$585 million, 12.3%), herbaceous perennials (US$488 million, 10.3%), cut flowers (US$424 million, 8.9%), propagative material (US$151 million, 3.2%) and cut foliage or greens (US$111 million, 2.3%). It is noteworthy that the top five cut flowers, all cut and potted foliage crops, the top eight potted flowering plants, and the top two herbaceous perennials are vegetatively propagated (Table 5.1). The rarity of seed-propagated floral crops in the top five ranking of these commodities indicates that there is great potential for a shift in propagation mode. Such a shift has been made in the bedding plants, where seed-propagated geraniums (*Pelargonium × hortorum*; Craig and Laughner, 1985) are ranked in the top ten along with their vegetative counterparts (Table 5.2). In contrast with all other commodities, most of the top nine bedding plant crops are seed-propagated (Table 5.2).

Potted plants (flowers, foliage) provide a diversity of colour and fragrance in the home for special events or holidays. The major holiday crops are potted plants, i.e. poinsettia (*Euphorbia pulcherrima*) for Christmas; Easter lily (*Lilium longiflorum*), hydrangea (*Hydrangea macrophylla*), florist's azalea (*Rhododendron* hybrids), bulbous crops (*Tulipa gesneriana, Crocus vernus, Hyacinthus orientalis, Narcissus pseudonarcissus*) for Easter and Mother's Day; chrysanthemums (*Dendranthema × grandiflora*) and ornamental peppers (*Capsicum annuum*) for Thanksgiving. Other potted crops are not

Table 5.1. Ranked value and propagation method(s) for cut, potted flower/foliage crops and herbaceous perennials (seed, vegetative) according to the 2001 USDA sales statistics (USDA, 2002).

Rank	Crop	Propagation method	Sales value (w) in US$ millions
Cut flowers			
1	Roses	Vegetative	67.656
2	Lilies	Vegetative	57.452
3	Tulips	Vegetative	26.265
4	Gladiolus	Vegetative	24.183
5	Iris	Vegetative	20.117
6	Gerbera	Primarily vegetative	20.045
7	Snapdragons	Seed	17.249
8	Chrysanthemums, pompons	Vegetative	16.578
9	Delphinium, larkspur	Seed	10.807
10	Lisianthus	Seed	10.101
11	Orchids	Seed, vegetative	8.883
12	Alstroemerias	Vegetative	5.206
13	Carnations, standard	Vegetative	4.571
–	Other (miscellaneous)	Vegetative, seed	135.143
Cut foliage			
1	Leatherleaf fern	Vegetative	54.283
–	Other (miscellaneous)	Vegetative	56.783
Potted flowering plants			
1	Poinsettias	Vegetative	256.211
2	Orchids	Seed, vegetative	99.514
3	Chrysanthemums	Vegetative	77.262
4	Florist's azaleas	Vegetative	63.029
5	Spring bulbs	Vegetative	45.959
6	Easter lilies	Vegetative	38.521
7	Mini-roses	Vegetative	26.412
8	African violets	Vegetative	18.564
–	Other (miscellaneous)	Seed, vegetative	206.427
Potted foliage plants			
1	Indoor, patio use	Vegetative	495.093
2	Hanging baskets	Vegetative	90.231
Herbaceous perennials			
1	Garden chrysanthemums	Vegetative	103.240
2	Hostas	Vegetative	39.021
–	Other (miscellaneous)	Vegetative, seed	345.483

4.1 A.

4.1 B.

Plate 4.1. Examples of various wildflower projects. **A.** Commercial parking lot, Utah, USA. Median, early season and late season perennials. **B.** Mountain residence, Evergreen, Colorado, USA. Knee-Hi Mix, first year.

4.1 C.

4.1 D.

Plate 4.1. (*Continued*) Examples of various wildflower projects. **C.** Golf course, Port Ludlow, Washington, USA. Custom mix, first year. **D.** Amusement park, Denver, Colorado, USA. Annuals for Sun mix.

Plate 4.2. A canal restoration project using wild flowers in Tokyo, Japan (courtesy of Kyowa Seed Co.). Upper left: 15 October 1990, bare soil. Upper right: 15 October 1990, top layer of peat moss with seed now added. Mid-left: 30 October 1990, seedlings becoming established. Mid-right: 17 January 1991, white flowers starting to bloom. Lower left: 19 April 1991, early spring bloom. Lower right: 11 May 1991, late spring bloom.

4.3 A.

4.3 B.

Plate 4.3. Wildflower seed production. **A.** California poppy/cosmos seed production field. **B.** New England aster seed production field.

5.1

A.

B.

C.

D.

Plate 5.1. Examples of seed- or vegetative-propagated floricultural crops. **A.** Bedding plants – primarily annuals. **B.** Herbaceous perennials. **C.** Flowering potted plants. **D.** Cut flowers for floral designing.

5.2 A.

Plate 5.2. Flower colours within flower seed crops. **A.** Colour variation within various flower seed crops.

Plate 5.2. (*Continued*) Flower patterns within flower seed crops. **B.** Star pattern. **C.** Picotee pattern. **D.** Blotch pattern in *Viola x wittrockiana*. **E.** Blush pattern in New Guinea Impatiens 'Kokomo Pink Blush'. **F.** Frost pattern in New Guinea Impatiens 'Infinity Orange Frost'. **G.** Eye pattern in New Guinea Impatiens 'Kokomo White with Eye'. **H.** Swirl pattern. **I.** Whiskers patterns in *V. x wittrockiana*.

5.2 B.

C.

D.

E.

F.

G.

H.

I.

5.3

5.4 A.

5.4 B.

5.5

Plate 5.3. Rapid generation cycling setup for chrysanthemum laboratory seed ripening to facilitate reduced seed development/ripening time and maximize seed set.

Plate 5.4. Upright plant habit (**A**) versus the revolutionary prostrate or wave habit (**B**) in petunias.

Plate 5.5. Example heat/drought tolerant *Pelargonium domesticum* flowering in the wilds of South Africa. Photo courtesy of Mark Galatowitsch and Dr Susan Galatowitsch.

Table 5.2. Ranked value and propagation method for annual bedding plant crops (flowers, vegetables) according to the 2001 USDA sales statistics (USDA, 2002).

Crop	Propagation method(s)	Flats Sales value (w) (US$ millions)	Rank	Potted Sales value (w) (US$ millions)	Rank	Hanging baskets Sales value (w) (US$ millions)	Rank	Totals per crop Sales value (w) (US$ millions)	Rank
Impatiens	Seed	117.476	1	26.598	5	19.036	3	163.110	1
Geraniums	Veg	12.240	7	111.870	1	28.749	2	152.859	2
Petunias	Seed, veg	96.408	3	22.124	6	18.524	5	137.056	3
Pansy/viola	Seed	103.179	2	20.855	8	2.725	7	126.759	4
Vegetables (all)	Seed	72.021	4	28.697	4	–	–	100.718	5
Begonias	Seed, veg	60.374	5	21.362	7	18.847	4	100.583	6
N.Guinea impatiens	Veg, seed	5.723	9	38.530	2	29.540	1	73.793	7
Marigolds	Seed	51.152	6	5.740	9	0.650	8	57.542	8
Geraniums	Seed	8.135	8	37.986	3	3.748	6	49.869	9
Other (misc.)	Seed, veg	340.489	–	290.217	–	96.089	–	726.795	–

associated with any holiday and are available throughout the year, for instance potted orchids (a variety of genera including *Phalaenopsis, Cattleya, Dendrobium, Oncidium, Miltonia*) and kalanchoe (*Kalanchoe blossfeldiana*) (Dole and Wilkins, 1999). The major potted crops (poinsettia, orchids, chrysanthemums) are 100% vegetatively propagated whereas many minor crops, e.g. cyclamen and ornamental peppers, are seed-propagated (Table 5.1; Dole and Wilkins, 1999). Flower breeders may yet have the potential to create seed-propagated cultivars of these crops, although the small market provides little incentive for monetary return. Minor potted crops, such as dwarf alstroemeria, have been transformed into seed-propagated products (*Alstroemeria* 'Jazze' series) as a result of focused flower breeding efforts (PanAmerican Seed Co., 1999). Orchids are a major potted crop, yet amateur breeders have developed all of the cultivars and no major North American distributor carries these products. Clearly, there are opportunities for flower breeding to transform this situation.

Cut flowers are used throughout the year for personal celebrations (birthdays, anniversaries, weddings, proms), to promote well-being (get-well bouquets), provide sympathy (funerals, memorial services) and for holidays (Hunter, 2000). Roses (*Rosa* × *hybrida*) are commonly associated with Valentine's Day whereas orchids (*Cattleya* hybrids) are often used for Mother's Day. Here again, the major cut flower crops are vegetatively propagated, e.g. roses, lilies (*Lilium* hybrids), tulips, and gladiolus (*Gladiolus* × *hybridus*) (Table 5.1). There is increased interest in seed-propagated cut flowers, given the continued popularity of F_1 hybrids such as snapdragons (*Antirrhinum majus*), lisianthus (*Eustoma grandiflora*), stock (*Matthiola incana*), sweet pea (*Lathyrus odoratus*), China asters (*Callistephus chinensis*), delphinium (*Delphinium elatum*), larkspur (*Delphinium consolida*), statice (*Limonium sinuatum*), strawflower (*Helichrysum* spp.), and sunflowers (*Helianthus annuus*). The cut flower market has remained fairly static in past years (USDA, 2002), although new crops are continuously being developed with industry support by the Cut Flower Growers Association and national trials (Dole, 2002).

Bedding plants have traditionally been annual or biennial crops (Carlson and Johnson, 1985). Most bedding plants have been seed-propagated, although a small percentage are vegetative, e.g. geranium (*Pelargonium* × *hortorum, Pelargonium domesticum, Pelargonium peltatum*) and New Guinea impatiens (*Impatiens hawkeri*) (Table 5.2; Dole and Wilkins, 1999). Seed-propagated annual bedding plants include impatiens (*Impatiens walleriana, Impatiens balsamina*), pansies/violets (*Viola* × *wittrockiana, Viola tricolor hortensis*), petunias (*Petunia* × *hybrida*), vegetables (a wide array of crops and genera), begonias (*Begonia semperflorens*), marigolds (*Tagetes erecta, Tagetes patula*), seed geraniums, New Guinea impatiens (*I. hawkeri*), vinca (*Catharanthus roseus*), alyssum (*Lobularia maritima*), floss flower (*Ageratum blossfeldiana*), cosmos (*Cosmos bipinnatus, Cosmos sulphureus*), nasturtium (*Tropaeolum majus*), salvia (*Salvia elegans, Salvia farinacea*), zinnia (*Zinnia elegans, Zinnia angustifolia*), and verbena (*Verbena* × *hybrida*). Biennials propagated by seed include honesty plant (*Lunaria annua*), Dame's rocket (*Hesperis matronalis*), and Sweet William (*Dianthus barbatus*). A significant innovation from the geranium breeding programme at Pennsylvania State University was the creation of F_1 hybrid, seed-propagated cultivars, the first of which was 'Nittany Lion' (Craig and Laughner, 1985). Seed-propagated geraniums have achieved ranking in the top ten bedding plants (currently ranked as No. 9), although they never surpassed the vegetative types (current ranking is No. 2) (Table 5.2). Other bedding plant crops have had a similar transition during domestication, such as seed-propagated New Guinea impatiens, *I. hawkeri* 'Java' series, canna (*Canna* × *hybrida* 'Tropical'), and flower maple (*Abutilon* × *hybridum* 'Bella') (PanAmerican Seed Co., 1999; Ball Seed Co., 2002). Currently, an increasing percentage of bedding plant crops are vegetative items due to the rapid pace of new crop development or retailoring of old crops (Wilkins, 1985).

Herbaceous perennials, long a mainstay of vegetatively propagated favourites such as day lily (diploid *Hemerocallis* hybrids, tetraploid *Hemerocallis fulva*), hosta (*Hosta* spp. and hybrids), and garden chrysanthemum

(*D. × grandiflora*), have experienced a resurgence in popularity (Callaway and Callaway, 2000; Fausey *et al.*, 2000). Historically, all seed-propagated perennials took one or more years from seed to flower. Accompanying this interest has been flower breeding focused on the creation of perennials that could be grown as bedding plants with plug production. The resultant 'annualized' perennials created a new class that flowered during the first year from seed, yet retained their perenniality (Finical *et al.*, 2000). Examples of annualized perennials include *Campanula longistyla* 'Isabella Blue' (Goldsmith Seed Co., 2002), *Campanula carpatica* 'Blue Chips' (Finical *et al.*, 2000), *Delphinium grandiflorum* 'Blue Mirror' (Frane *et al.*, 2000), *Hibiscus moscheutos* 'Disco Belle' (Wang *et al.*, 2000), *Aquilegia × hybrida* 'Songbird' (PanAmerican Seed Co., 1999), and *Verbascum × hybridum* 'Southern Charm' (PanAmerican Seed Co., 1999). Herbaceous perennial breeding programmes often neglect selection for winter hardiness in northern climates. This lack of reliability in winter survival has prompted a resurgence of winter hardiness as a breeding objective (Anderson and Gesick, 2004).

Public vs. private sector flower breeding

The two major categories of professional flower breeders are employed in public and private sector programmes. Amateur flower breeders may also be found for virtually every important floricultural crop.

Amateur breeders have made significant contributions to cultivar development, product enhancement and the creation of interspecific hybrids. Commercial hosta, day lily, lily, orchid, iris and gladiolus cultivars were primarily bred by amateur breeders; in some instances, professional flower breeders may have also enhanced the initial products. Most products developed by amateur breeders are vegetative, rather than seed-propagated, due to the lengthy research and development costs and time involved in inbred parent development.

Public and private sector breeding programmes overlap in many areas, but differ in several significant ways (Table 5.3). Public sector flower breeders are found within universities/colleges (academic professors), public/private botanical gardens and the USDA. Academic breeding programmes have as their mission the training/education of undergraduate and graduate flower breeding/genetics students. The crops being bred at an institution of higher learning serve as the tools for educational development of the students by conducting basic and applied research (Table 5.3). Research publications from these programmes have served to establish important breeding and genetic information for floricultural crops, since private sector

Table 5.3. Flower breeding objectives in the public and private sector programmes.

Objective	Public sector	Private sector
Basic research	Extensive; primary area	Limited
Applied research	Limited	Extensive; primary area
Commercial cultivars	Limited; by-products of research	Extensive; primary area
Improvements within a series[a]	Rarely	Primarily
New series creation[a]	Rarely	Primarily
Distinguishable products[a]	Frequently	Limited to frequently
Innovative products[a]	Primarily	Limited
Long-term, high-risk research	Primarily	Rarely
Primary influencing factors	Genetic/breeding research	Profitability
Breeding objectives	Knowledge base expansion, educational opportunities, competitive grants	Competition (market forces)
Publications	Essential to project survival and notoriety, create published knowledge base for industry breeding	None (proprietary), except for production information

[a]For further explanation on the relative time and value placed on these products, refer to Fig. 5.1.

breeding information is proprietary, strictly confidential, and rarely, if ever, published. Commercial products may be a tangible outcome of public sector research (Table 5.4), although the primary research focus is on long-term, high-risk research objectives that private sector breeders are not able to pursue. As a result of the complexity of the breeding objectives and student education, completion of breeding objectives usually takes considerably longer in public sector programmes than in the private industry. Most outcomes from this long-term research would be classified as innovative or distinguishable products (Fig. 5.1). Public sector breeders may also be hired as contract breeders to work on products for the private industry where the primary focus might be the creation of a new series with representative cultivars in all of the flower colour classes. Examples of current public sector academic breeding programmes include The Pennsylvania State University (Richard Craig, *Pelargonium*, *Exacum*), University of Minnesota (Neil Anderson, *Dendranthema*, *Gaura*, *Lilium*), University of Wyoming (Karen Panter, *Castilleja*), Cornell University

(Mark Bridgen, *Alstroemeria*, Chilean geophytes), University of Florida (Zhanao Deng, *Caladium*; Brent Harbaugh, *Eustoma*, *Lycopersicon*; Rick Henny, *Anthurium*, *Aglaonema*), University of Nebraska (Dale Lindgren, *Penstemon*), University of Wisconsin (Dennis Stimart, *Zinnia*, *Antirrhinum*), University of Hawaii (D. Oka, *Leucospermum*), and the University of Massachusetts (Thomas Boyle, *Schlumbergera*). Botanic garden flower breeding programmes include the Chicago Botanic Garden (James Ault, *Echinacea*, *Lilium*) and Longwood Gardens (James Harbage, *Pericallis*); USDA flower breeding programmes encompass *Petunia* (Robert Griesbach), *Clethra*, *Hydrangea* (Sandra Reed) and *Lagerstroemia* (Margaret Pooler).

In contrast with public sector breeding, the private sector flower breeding programmes conduct limited basic research, focusing primarily on applied research and the development of products (Table 5.3). Particularly with the top-ranked products in each commodity group (Tables 5.1 and 5.2), i.e. the 'cash cows', considerable effort is devoted to maintaining market share with the

Table 5.4. Public sector flower breeding programme products that have revolutionized the floriculture industry.

Crop	Features	Date	Source
Castilleja linariifolia	Hemiparasitism, germination	2000	Univ. of Wyoming
Dendranthema × *grandiflora*	Cushion habit	1950s	Univ. of Minnesota
	F$_1$ hybrid seed-propagated	1980s	Univ. of Minnesota
	Day neutrality	1980s	Univ. of Minnesota
	Winter hardiness	2000	Univ. of Minnesota
	Shrub habit	1990s	Univ. of Minnesota
Echinacea	Interspecific hybrids with new flower colours	2003	Chicago Botanic Garden
Eustoma grandiflora	First dwarf lisianthus for bedding/pot culture	1990s	Univ. of Florida, USDA
Exacum affine	Breeding	1980s	Penn. State Univ.
Gaura lindheimeri	Winter hardiness, selection against invasiveness	2000	Univ. of Minnesota
Impatiens hawkeri	Collection, screening of germplasm, cytogenetics	1970s	USDA, Longwood Gardens
Oxalis spp.	Sterile selections, large flowers	1990s	USDA
Pelargonium × *hortorum* 'Nittany Lion'	Seed-propagated	1960s	Penn. State Univ.
Petunia hybrida	Transformation, insect resistance	1990s	USDA, Univ. of Florida
Schlumbergera spp.	New hybrids, earlier flowering	1990s	Univ. of Massachusetts
Zinnia angustifolia	Breeding, germplasm release, mildew tolerant	1980s	Univ. of Wisconsin

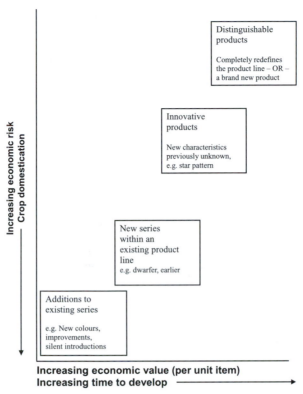

Fig. 5.1. Relationship between risks and rewards for new floricultural introductions.

continuous development of new cultivars (improvements within a series) and the creation of new series (Fig. 5.1). Creation of distinguishable and innovative products is important, but private sector flower breeders typically devote less effort to such long-term objectives, as they may take years to accomplish. Examples of long-term breeding successes from private sector flower breeders include a change in propagation mode, non-pungent fruit, or plant habit modifications (Table 5.5). Thus, the bulk of the flower breeder's effort must be spent on objectives that may be accomplished within a short period in order to maximize profitability and competitive ability.

Basic breeding objectives

Regardless of the sector (public or private) in which a flower breeder may operate, the most important axiom for a successful breeding programme is 'Know thy crop'! Prior to commencing breeding and genetic research on any crop, it is essential to spend the time thoroughly searching the literature (scientific literature and trade journals) to lay a solid foundation of factual information. This allows for the creation of practical, yet visionary, breeding objectives. Assemble this information in a template that can be used for all crops being bred (Table 5.6). By assessing the published taxonomic descriptions, geographical distribution and the native habitat, a flower breeder may deduce the ideal crop phenotype that may be obtained (the crop ideotype) and the important factors of plant growth to ensure production of a vigorous and healthy product (Table 5.6). Commercial cultivars or open-pollinated seed may also be available. If so, obtain this germplasm for screening with wild populations and add any published production information to the crop template. If no published information exists,

Table 5.5. Examples of recent private sector flower breeding programme product innovations.

Crop	Features
Abutilon × *hybrida* 'Bella' series	Seed-propagated, dwarf, open flowers
Alstroemeria hybrida 'Jazze'	First seed-propagated, dwarf alstroemeria
Begonia × *hybrida* 'Dragon Wing'	First seed-propagated angel wing
Campanula longistylla 'Isabella Blue'	Seed-propagated bell flower
Canna × *hybrida* 'Tropical'	Dwarf, seed-propagated
Capsicum annuum 'Medusa'	First non-pungent ornamental pepper
Dendranthema × *grandiflora* 'Fleurette'	First dwarf, daisy pot mums
Dianthus caryophyllus 'Monarch'	Dwarf, scented carnations
Dimorphotheca 'Spring Flash'	Vibrant colours, seed-propagated
Euphorbia pulcherrima 'Winter Rose'	New rose flower type
Euphorbia pulcherrima 'Plum Pudding'	New purple flower colour
Impatiens hawkeri 'Java'	Seed-propagated New Guinea
Lilium × *formolongo* 'Raizan No. 1'	First seed-propagated Easter lily
Melampodium 'Million Gold'	First dwarf, seed-propagated
Osteospermum hybrida 'Passion Mix'	First seed-propagated, dwarf cultivar
Petunia hybrida 'Purple Wave'	First prostrate petunia released into the market

initial screening of the germplasm will aid in the identification of areas where genetic improvement is required.

The next most important accomplishment is the acquisition of germplasm and characterization of the genetic variation found within the crop species. First and foremost, access a wide sampling of the germplasm to ensure that the full genetic potential is accessible for use. Consult the USDA Germplasm Resource Information Network (GRIN) database to determine whether any wild populations or cultivars are being maintained (http://www.usda/gov); if so, they are accessible to any flower breeder. Also, seek advice from the USDA Ornamental Plant Germplasm Center dedicated to floricultural crops, located at the Ohio State University (Columbus, Ohio), to determine whether there may be an opportunity to collect plants in the wild (http://hcs.osu.edu/opgc/).

It is also mandatory that every flower breeder conduct a collection trip(s) to the geographical region(s) where the crop species is native, i.e. the centres of origin and diversity for the genus. This will allow for the collection of a wide germplasm base and provide expansive opportunities for the breeding objectives. Historically, flower breeders have not conducted their own collecting expeditions and, as a result, the majority of floriculture crops have a narrow genetic base and the full

potential of many crops has never been realized. For instance, the Easter lily market has a single vegetative cultivar, 'Nellie White', for all of North America. While the Easter Lily Research Foundation conducts a breeding programme to develop new cultivars, none has matched the performance of 'Nellie White' in comparison trials (Lee Riddle, personal communication, 2002). Additionally, had initial collections been made of the wild *Pelargonium* species in South Africa, a wider array of geranium phenotypes and environmental tolerance would be available. A wide germplasm base may provide the essential genes necessary to overcome production problems with existing cultivars. Anderson *et al.* (2003) found a wide range in genetic variability within and between *Gaura* species collected in their native habitats. Observation of new plant phenotypes or previously undocumented traits may readily expand the breeding objectives for a particular crop.

The crops assigned to, or chosen by, the flower breeder, and their corresponding breeding objectives will be affected both by market demand (short-term objectives) and visionary product enhancement and retooling. A flower breeder must determine which areas of research to focus on to create new, improved cultivars within a short- and/or long-term time-frame (Fig. 5.1). Market value (Tables 5.1 and 5.2) may also influence the

Table 5.6. Background information required for the breeding and commercialization of new flower crops.

Taxonomy
  Scientific name (genus, species, subspecies):
  Synonym(s):
  Common name(s):
  Family:
Geographical distribution
  Continent(s):
  Country(-ies):
  State(s)/province(s)/region(s):
  Latitudinal range(s):
  Altitude:
  General climactic conditions:
  Tendency to naturalize or become invasive:
Native habitat
  Habitat (climactic factors):
  Plant community:
Taxonomic description
  Overall plant habit/description:
  Root system type:
  Presence/type of underground storage organs:
  Leaves:
  Flower:
  Season of bloom:
  Use(s) by indigenous people:
  Other uses:
  Additional notes:
Name and description of cultivars already
  commercialized (if any)
Propagation method(s)
  Vegetative vs. seed:
  If veg., plant tissue source(s):
  If veg., proposed propagation method(s) and
    temperatures:
  If seed, no. of seeds/flower:
  If seed, seed dormancy?
  If seed, germination temperatures/duration:
Product specifications
  Crop ideotype (the ideal phenotype that a
    marketable cultivar will possess):
Market niche – identification and justification
  Target sales date(s):
  Potential holiday(s) for this product:
  Programmability, i.e. could this be forced
    year-round?
  Crops with which this will compete in the
    market:
  What kind of 'story' can be told about this
    product?
  Will this ever be a major crop (why or why not)?
  What will be the initial crop limitations/
    problems?
  Is this product already identifiable to the
    growers and consumers?

Table 5.6. *Continued.*

How soon could this product be available?
Anticipated cultural requirements
  Winter hardiness (USDA zones):
  Heat/drought tolerance:
  Temperature (day/night):
  Light quantity, quality, duration; photoperiod
    response:
  Nutrition:
  Soil:
  Plant growth regulators:
  Container size (through entire production
    cycle):
  Disease resistance/susceptibility:
  Fungicides, insecticides:
  Other:
Production schedule (from seed or cuttings)
  Estimated no. of weeks from planting to flower
    bud initiation, flower development, and
    shipping:
  Estimated time, type, and quantity of special
    treatment applications:
  Target sales date: Mother's Day sales
    (mid-May):
Needs assessment for genetic improvement
  Based on the production schedule you have
    assembled, assess the need for crop
    improvement using standard breeding
    methodology or genetic transformation:
Relevant literature

amount of effort devoted to a specific product by a private sector breeder. Critical product changes desired by the commercial grower may not be appreciated by the retail consumer and vice versa, as will be discussed later in greater detail. Many different inputs should be sought from a variety of sources to prioritize the breeding objectives.

The relationship between risks and rewards in breeding floral crops depends on a variety of factors (Fig. 5.1). Crop domestication begins with the introduction of a new crop into the market as a distinguishable product; as domestication progresses, the ease of product creation increases. Crop domestication is inversely related to increasing economic risk, i.e. as a crop becomes more domesticated, it becomes easier to create new cultivars (additions to existing series) or phenotypes (flower pattern changes, dwarfism, increased branching, earlier flowering) within an existing product line to create a new series. This is due to competitors and, thus, more flower breeders are working to capitalize on the crop's genetic resources and each breeding programme can speed product development due to genetic knowledge (linked traits, dominance patterns, inbred parent development, etc.). The 'cash cow' crops remain in the lower left-hand corner of the graph with minor improvements until new germplasm is accessed, interspecific hybridizations made, or new genes inserted with molecular techniques. Dramatic alterations to crop phenotypes result in a recapitulation of innovative or distinguishable product lines (Fig. 5.1). An excellent example of these alterations is the invention of 'wave', or prostrate growth habit, petunias. Prior to the creation and introduction of 'Purple Wave' (http://www.wave-rave.com), the petunia market was essentially flat for several decades (*cf*. USDA sales data pre-1995); the only innovations had been additions to existing series or new series. 'Wave' petunias sold for as much as US$0.10/seed (w) whereas standard petunias sold for fractions of a cent. Thus, per unit economic value is greater with distinguishable products than additions to existing series. 'Cash cow' crops have value derived from the sheer volume of units sold, despite the significantly lower per unit price.

In order for new products or old crop revisions to be acceptable in the market, they must match production specifications. These specifications include total production time, as well as product size, postharvest duration, and amenability to fit into existing crop production recipes. Crop production (maximum finishing time) for bedding plants is 16 weeks, whereas potted plants require 16–18 weeks and cut flowers a total of 18–20 weeks (W. Healey, personal communication, 1997). While there are a few exceptions to these time-frames, e.g. standard *Cyclamen persicum* exceed the 18 week period, it would be inadvisable to breed any new crop that falls outside of this range. Crop height and width dimensions must adhere to the aesthetic ratio for the container size, i.e. a total height that is ≤ 2.6 × container diameter. Potted crops that are too tall, for instance, will neither fit on to standard carts for transport prior to shipping nor into standard packing containers.

An additional important rule for all flower breeders is to be ruthless in selection. As Allard (1960) so aptly wrote (paraphrased to remain politically correct):

> The beginning plant breeder must develop an attitude of ruthlessness in selection. S/he must dispel the feeling that among the plants s/he discards may be the one that will lead to the variety s/he has in mind. Unless s/he keeps this natural inclination under control, s/he will shortly find they are overwhelmed with materials, and his/her effectiveness as a plant breeder may be impaired or lost.

Open-pollinated crops

Prior to the advent of F_1 hybrid seed-propagated flowers (particularly bedding plants), most seed-propagated bedding plants, potted plants and cut flowers were open-pollinated cultivars (Craig and Laughner, 1985). This meant that, in most cases, hand pollination was not required by a producer company, allowing pollen vectors (wind or insects) to accomplish pollination. Seeds were also cheaper to produce than with hand pollinating. If products did not have inbred lines developed, then open-pollinated products were mixes rather than straight colours. Many flower seed crops are still open-pollinated (Table 5.7).

F_1 hybrid crops

The establishment of PanAmerican Seed Company in Colorado during the 1940s and the focus on creating the first F_1 hybrid, seed-propagated bedding plant crops (impatiens and petunia) paved the way for the advent of crop uniformity, plug technology, and greenhouse mechanization (Craig and Laughner, 1985). Currently, most seed-propagated crops have F_1 hybrid products, which may be supplemented with F_2 hybrids, open-pollinated, or vegetative cultivars (Table 5.7). To develop F_1 hybrids, a flower breeder spends considerable effort developing inbreds with good general or specific combining ability to serve as parents. Such an effort requires an

Table 5.7. Sample commercial floricultural crop classification by propagation method, after release to the market (Craig and Laughner, 1985; Anderson, 2001).

| Crop | Seed | | | Vegetative |
	F_1	F_2	O.P.[a]	
Abutilon	x			x
Ageratum	x			
Alstroemeria	x		x	x
Alyssum	x[b]		x	
Begonia	x		x	x
Celosia			x	
Cleome	x		x	
Cosmos			x	
Chrysanthemum	x	x		x
Cuphea			x	
Cyclamen	x	x	x	
Daffodil, narcissus				x
Dahlia		x		x
Eustoma, lisianthus	x			
Freesia				x
Fuchsia			x	x
Geranium	x	x	x	x
Gerbera				x
Hibiscus	x		x	x
Impatiens, bedding and New Guinea	x	x	x	x
			x	
Liatris				x
Lily	x			x
Marigold	x		x	
Ornamental pepper	x		x	
Pansy, viola	x	x	x	
Pentas	x		x	
Petunia	x	x	x	x
Poinsettia				x
Portulaca	x	x	x	
Primula	x			
Rhododendron, florist's azalea				x
Rose				x
Salvia			x	
Snapdragon	x			x
Torenia	x			x
Tulip				x
Verbena			x	x
Vinca	x		x	
Zinnia	x		x	

[a]O.P. denotes open-pollinated.
[b]'F_1 hybrids' are actually inbreds.

investment of time and labour over several years with extensive trials (segregating – inbred, hybrid) over years and locations to ensure product stability and equal or superior performance in comparison with market standards. Both open-pollinated and F_1 hybrid cultivars will also undergo extensive germination (75–95%+) and yield potential tests to ensure that each product is within range of market standards. For extensive reviews of breeding methodology consult Allard (1960), Craig and Laughner (1985), Callaway and Callaway (2000). Many new crops possess characteristic traits that must be eliminated through selection or completely circumvented, as described below.

Characteristic Traits – Behind Flower Power and Convenience

Beyond the major phenotypic traits that characterize flower seed crops, a variety of additional traits are also expressed. Most horticultural crops are highly heterozygous, outcrossing species, and frequently polyploid, all of which facilitate easier vegetative propagation. Seed propagation is more difficult, as uniformity may be impossible to achieve if high levels of seed set, germination, and/or yield potential can be achieved. A multitude of reproductive barriers exist in floricultural crops, necessitating flower breeders to characterize, manipulate, or select against them in the process of domestication.

Reproductive barriers

Self incompatibility

At least 71 plant families (nearly 50% of the major crops and ornamental species in the world) and more than 250 genera have species with self incompatibility (SI) which prevents fertilization (selfs or crosses) when S alleles are matched, as well as crosses between related individuals (Ascher, 1976; de Nettancourt, 1977; Anderson and Ascher, 2000). Self incompatibility was first discovered in a flower crop, *Verbascum phoeniceum*, when self pollinations of a fertile plant did not set seed (Kölreuter, 1761–1766). Originally termed 'self sterility', it is now referred to as SI since the phenomenon operates in fertile plants (Stout, 1916). Correns (1912–1913) first applied Mendelian genetics to SI research of *Cardamine pratensis*, meadow cress. Interestingly, most of the early application of genetics, following the rediscovery of Mendel, were on flower crops (East and Mangelsdorf, 1925).

SI systems show a characteristic lack of pollen germination or stigmatic penetration (sporophytic SI) or the slowed growth of pollen tubes in the style relative to compatible crosses (gametophytic SI) (Ascher, 1976). Thus, SI is a prefertilization reproductive barrier controlled by a single locus, *S*, with multiple alleles that encodes the specificities (East and Mangelsdorf, 1925). There are both homomorphic and heteromorphic SI systems. With homomorphic SI, all plants in a population have the same floral morphology, i.e. no discernible difference in anther and style lengths between genotypes. Heteromorphic SI systems occur in plants with two (distyly) or three (tristyly) flower forms or morphs. In distyly, two different anther and style lengths, short (thrum) and long (pin), are found. Tristylous plants possess three style lengths, shorts, mids, and long; flowers of each genotype have the same style (one) and anther morphs (two). Heterostyly is not always linked with SI in all species. Darwin (1895) conducted extensive research on SI and heterostyly in flower crops such as purple loosestrife (*Lythrum salicaria*), primrose (*Primula veris*, *Primula vulgaris*), flax (*Linum grandiflorum*) and *Oxalis* spp. Extensive reviews of both types of SI system have been reported elsewhere (Ascher, 1976; de Nettancourt, 1977; Liedl and Anderson, 1994).

In diploids, SI is controlled by a single gene, *S*; polyploids may have additional genes. For instance, hexaploid chrysanthemums possess three genes controlling SI (Drewlow *et al.*, 1973; Zagorski *et al.*, 1983; Boase *et al.*, 1997), all of which interact such that outcross seed set between unrelated genotypes ranges from 36% to 71% (Ronald, 1974).

Since it operates prior to fertilization, SI has often been regarded as a resource-inexpensive reproductive barrier

that regulates offspring quality by preventing mating of related individuals and eliminating inbreeding depression (Darwin, 1876; Lande and Schemske, 1985; Winsor *et al.*, 1987). If SI is a tight system, then one would expect that SI species would be highly outbred and self-compatible species would be inbred (Ruiz and Arroyo, 1978). This is rarely the case, however, since most reproductive barriers are phenotypically plastic in their expression (Kress, 1983; Stephenson and Bertin, 1983). Too tight an SI system would mean that a population could become extinct if the founding population size is small (at or near one individual). Pseudo-self compatibility (PSC) is expressed when the SI system is plastic, where limited to full complements of seed are produced despite the presence of SI. Such plasticity of expression has suggested that a tight SI system is only one portion of the continuum that ends with complete self compatibility (SC) (Liedl and Anderson, 1986a, 1993).

Many flower seed crops possess SI systems; an additionally equally high percentage of vegetative flower crops are SI. Flower crops with gametophytic SI systems include petunia (*P. × hybrida*; Flaschenriem and Ascher, 1979), lily (all *Lilium* species with the exception of *Lilium formosanum*; Ascher and Peloquin, 1968), gaura (*Gaura lindheimeri*), evening primrose (*Oenothera* spp.) and heliconia (*Heliconia*) (Kress, 1983; Peters, 2002). Sporophytic SI systems operate in floricultural crops such as ageratum (*Ageratum houstonianum*; Reimann-Philipp and Fuchs, 1971), chrysanthemum (*D. × grandiflora*; Drewlow *et al.*, 1973; Zagorski *et al.*, 1983), zinnia (*Z. elegans*, *Z. angustifolia*; Boyle and Wicke, 1996), marigold (*T. patula*, *T. erecta*), gerbera (*Gerbera jamesonii*), Dame's rocket (*H. matronalis*), dahlia (*Dahlia × hybrida*), sweet potato vine (*Ipomoea batatas*; Martin, 1972), and alyssum (*L. maritima*). Floral crops with heteromorphic SI systems include narcissus/daffodils (*N. pseudonarcissus*), purple loosestrife (*L. salicaria*), primroses (*Primula* spp.), and oxalis (*Oxalis* spp.) (Darwin, 1895).

Self incompatibility is disadvantageous in flower breeding programmes when *S* alleles are limited and inbred lines are a breeding objective (Anderson and Ascher, 2000). Thus, a flower breeder must select against SI expression (while maintaining its existence) and for PSC (Liedl and Anderson, 1993). In contrast, at the end of the inbred line development, the flower breeder must select for SI inbreds to prevent inbred contamination in hybrid flower seed. Strong SI inbreds facilitate less expensive flower seed production since costly hand emasculation or pollination is avoided. In this scenario, two SI, but cross-compatible, fertile inbreds with high combining ability are used as parents.

PSC is a phenomenon related to SI. Mather (1943) noted that limited to full complements of self seed could occur in SI plants under certain conditions. Darwin (1876) first reported PSC in California poppy (*Eschscholtzia californica*) and flowering maples (*Abutilon darwinii*). These SI species set self seed in England, but were SI when pollinations were performed in Brazil. Since these initial reports of PSC, it is has been reported in virtually all SI species if one screens sufficient genotypes and populations (Ascher, 1976). PSC and self-compatible species may appear identical on the macroscopic level. However, PSC genotypes still possess an operative SI system. The primary components of PSC are genetic and environmental (Denward, 1963). Environmental modifications that trigger the expression of PSC in a self-incompatible genotype include high temperature treatments (Ronald and Ascher, 1975), application of carbon dioxide (Palloix *et al.*, 1985), mutilation of the stigmatic surface (Roggen, 1982) and end of season stresses (Lawson and Williams, 1976).

Percentage PSC of individual genotypes can be calculated using the formula (Ascher, 1976):

$$\%\text{PSC} = [(\text{mean self seed set}) / (\text{mean outcross seed set})] \times 100$$

This formula ensures that fertility decreases due to inbreeding depression (from selfing) are accounted for by dividing the self seed set by outcross seed production (from compatible crosses). Pseudo-self compatibility may be selected for and, frequently, stable PSC expression is not found until the percentage PSC exceeds a critical threshold.

The genetics of PSC are complex and vary depending on the SI system, the taxa, and qualitative/quantitative gene action (Liedl and Anderson, 1994). Pollen- and stylar-part mutants (Lewis, 1949; Lewis and Crowe, 1954) and the S_f gene (Wergin, 1936) are examples of qualitative PSC. S_f genes have been identified in two Solanaceae crops, *Petunia* (Mather, 1943) and *Nicotiana* (East, 1932). Dana and Ascher (1986a,b) found that the S locus and a major PSC gene controlling stylar inactivation in *P. × hybrida* were 20 map units apart. *Nemesia strumosa* has a threshold effect where homozygous alleles must occur at a critical number of loci before PSC can be expressed (Henny and Ascher, 1976; Robacker and Ascher, 1978). In chrysanthemum, a similar threshold effect was found (Anderson and Ascher, 1996).

Two specific PSC phenotypes are expressed in *Nemesia* (Robacker and Ascher, 1982) and *Petunia* (Flaschenriem and Ascher, 1979; Dana and Ascher, 1985): pollen-mediated PSC (PMPSC) and discriminating styles (DS). In some instances, when PMPSC is in operation, seed set following incompatible pollinations are related to the pollen parent's PSC level (Flaschenriem and Ascher, 1979). Genotypes with DS discriminate among incompatible pollen tubes such that some grow as though they were compatible. Both PMPSC and DS interact, although they are not controlled by the same genes. In *Petunia*, a threshold of 27–31% PSC is necessary for the expression of DS (Liedl and Anderson, 1994). Quantitative PSC has been reported, determined by several additive genes being expressed in concert, e.g. *Petunia* (Mather, 1943; Takahashi, 1973). Regardless of the type of PSC in operation, most SI flower crops can be inbred until the desired levels of homozygosity have been reached. Then the flower breeder must select within the inbred line for genotypes with strong SI expression to be used as parents in the creation of F_1 hybrids.

Anderson *et al.* (1989) noted that plants show a range of compatibility from complete SI to PSC to complete SC. This range can be influenced by factors other than just SI, e.g. inbreeding depression, incongruity, and other post-fertilization phenomena. A statistic was developed, known as the coefficient of crossability; calculated as 1 − [(potential − actual)/potential], where 'potential' is the greatest amount of seed that can be produced and 'actual' is the amount of seed produced in a cross. They distinguished male (MCC) and female (FCC) coefficient of crossability by which column or row the potential value occurs in a diallel. Once these two values have been determined, a 1:1 χ^2 test is performed to determine if they are significantly different. Then MCC is plotted against FCC and linear regression is performed on the results. Anderson *et al.* (1989) demonstrated that a pure SI system (without outside influences) would show two classes of data points, one at 0,0 for complete SI, and one at 1,1 for complete compatibility, and a regression line with the equation $Y = 0.0 + 1.0 X$. As more and more modifiers such as inbreeding, interspecific crossing, incongruity, etc., are introduced, the regression line moves away from the ideal. They created a chart showing the gradation from SI to SC with outlying areas showing incongruity and extreme points showing inbreeding depression. Values lying within the expected range of the SI system show the system is stable. It also shows the system does not seem to be modified by outside factors, or these factors are balanced between male and female parents. This system is useful for determining how stable a system is and if other factors are affecting the system, and to what extent the system is affected.

Self compatibility

After Darwin (1876) established the existence of deleterious effects from selfing (inbreeding) plants, a philosophy prevailed that inbreeding was inherently disadvantageous (Anderson *et al.*, 1992a). This view was later modified, following the discovery of plant species where selfing in natural populations confers adaptation and perpetuation of the species, even though there may be reductions in mean fitness (Schemske and Lande, 1985). Clearly, selfing – due to the expression of self compatibility – may have adaptive advantages. Likewise, self compatibility is also highly desirable in domesticated, seed-propagated flower crops. Most horticultural species outcross in the wild, due to the existence of reproductive barriers. The

process of domestication, however, frequently produces flower crops that evolve from outcrossing to becoming self-compatible. For example, commercial petunia (*Petunia × hybrida*) is essentially self-compatible while all of its wild progenitor species are self-incompatible (Flaschenriem and Ascher, 1979; Dana and Ascher, 1986b).

Inbreeding depression

Hybrids are the first filial (F_1) generation of progeny derived from crossing two inbred parents that differ in ≥ 1 gene(s) (Allard, 1960). Hybrids typically display hybrid vigour (heterosis) with superior performance over either the better parent or midparent values (Anderson *et al.*, 1993). However, many flower crops express reproductive barriers that prevent self-pollination and the creation of homozygous inbred lines. An historic alternative has been to vegetatively propagate heterozygous genotypes (Reimann-Philipp, 1969).

Traditionally, F_1 hybrid flower breeding has been applied primarily to annual bedding plant species, although an increasing number of potted and cut crops are seed-propagated. For instance, the dominant flowering potted crops, poinsettia (*E. pulcherrima*), chrysanthemum (*D. × grandiflora*), Easter lily (*L. longiflorum*), florist's azalea (*Rhododendron*), hydrangea (*Hydrangea*), and bulbous crops (*T. gesneriana, C. vernus, Iris, Narcissus, Hyacinthus orientalis*) are vegetatively-propagated. Other potted crops are seed-propagated, notably *C. persicum, Pericallis × hybrida* (= *Senecio cruentus*), *Calceolaria herbeohybrida, G. jamesonii, Streptocarpus × hybridus,* and *Begonia tuberhybrida* (Reimann-Philipp, 1983; Schmidt and Erickson, 1981; Wellensiek, 1959). Cut flower crops are also predominantly vegetatively propagated, such as lily of the Incas (*Alstroemeria* hybrids), gladiolus (*Gladiolus × hybridus*), gay feather (*Liatris*), chrysanthemum (*D. × grandiflora*), rose (*Rosa × hybrida*), all bulbous crops (*T. gesneriana, Crocus, Iris, Narcissus, Hyacinthus*), and lilies (Oriental and Asiatic *Lilium* hybrids). In the past few decades, there has been renewed interest in the development of seed-propagated hybrids that once were considered vegetative crops (Meynet, 1978; Reimann-Philipp, 1983; Satory, 1986; Anderson, 1987). Typically, self incompatibility is the first barrier that must be overcome in the development of seed-propagated cultivars, which effectively prevents the use of selfing to create homozygous inbred parents (Niwa, 1931; Mulford, 1937; Drewlow et al., 1973). Selection ensues for pseudo-self-compatible inbreds to continually create homozygous inbred lines. Once SI has been overcome, the next common reproductive barrier is inbreeding depression.

Inbreeding is defined as the mating of individuals more closely related than random matings within an infinitely large population (Allard et al., 1968). Mathematically, inbreeding is expressed as F, Fisher's coefficient of inbreeding, where $F = 0$ for outcrossers and $F = 1$ for homozygosity at the selected loci (Kempthorne, 1957; Boucher, 1988). The purpose of inbreeding is to increase homozygosity for the genes under selection to create inbred parents for use in hybrid seed production. The effects of inbreeding are to reduce the genetic variation within inbred families and a corollary increase in variation between families (Johannsen, 1926). Inbreeding depression decreases the fitness and vigour resulting from inbreeding (selfing, full-sib mating, half-sib mating, etc.) taxa that are normally outcrossers, due to the expression of deleterious recessive mutations in homozygotes (Campbell, 1988). Inbreeding depression is one of the major factors maintaining outcrossing (Schemske and Lande, 1985).

Inbreeding depression has been reported in many flower crops with a wide range of phenotypes. It has been studied extensively in cultivated allohexaploid chrysanthemums, D. × grandiflora, where inbreeding depression significantly decreased percentage germination, yield potential, the proportion of individuals reaching reproductive maturity (percentage survivorship), seed set and pollen stainability (Niwa, 1931; Mulford, 1937; Anderson et al., 1992a,b; Anderson and Ascher, 2000). Percentage germination and survivorship were correlated, suggesting no independence of lethality in stages of the life cycle (Anderson et al., 1992a). Inbred families at various levels of inbreeding, from $F = 0.5$ to $F = 0.995$, expressed lethal and sublethal phenotypes. Similar findings have been reported in diploid and tetraploid flower crops. In diploid, self-incompatible Borago officinalis, germination and embryo abortion were negatively affected by inbreeding depression (Crowe, 1970). Traits such as fecundity, germination, establishment, fruit number and seeds/fruit were reduced due to inbreeding depression in Gilia achilleifolia, a diploid self-incompatible annual (Schoen, 1983). Similarly, diploid and self-incompatible Phlox drummondii also had reduced fecundity, germination and survivorship (Levin and Bulinska-Radomska, 1988). Tetraploid C. persicum plants were inbred for five selfed generations; flower number decreased in a linear fashion (Wellensiek, 1959). Diploid H. annuus were selfed for 18 generations, with F values ranging from $F = 0.5$ to $F = 0.999996185$ (Schuster, 1970). Seed yield, plant height and head diameter reductions occurred in the early selfed generations for 'Hesa'. Diploid Lilium × formolongi (L. formosanum × L. longiflorum) F_2 inbreds ($F = 0.5$) had significant levels of inbreeding depression for most morphological traits, although the levels varied between inbred families (Anderson and Dunn, 2003).

Selection for fertility in inbred lines is always challenging, yet imperative (Martin, 1972; Anderson and Ascher, 2000). Inbred parents for hybrid seed production must have high fertility for the product to be economically competitive. Fertility poses problems during inbred line development since inbreeding depression decreases fertility and is not correlated with PSC levels (Anderson and Ascher, 2000). Inbred families will often differ significantly in their fertility loss due to inbreeding depression. As a result, a flower breeder will often select two or more inbred lines with adequate fertility (Anderson and Ascher, 2000). Despite significant inbreeding depression in flower crops, flower breeders have selected for fertility and yield (seed set, germination, yield potential) and must, therefore, select for inbred parents to successfully create F_1 hybrid seed cultivars.

Recombinant inbreeding

Inbreeding depression can be a formidable barrier, being severe in most flower crop species and occurring in all possible life cycle stages (Anderson *et al.*, 1993). However, the creation of recombinant inbreds has been one useful method to circumvent inbreeding depression. Recombinant inbreeding involves the formation of favourable recombinations (hybrid × hybrid, hybrid × inbred, or unrelated inbred × inbred, inbred backcrosses, double-first-cousin matings) followed by several successive generations of inbreeding (selfing) (Sullivan and Bliss, 1983a,b; Templeton and Read, 1984; Werner *et al.*, 1989). The process is repeated every time inbreeding depression begins to surface in the inbred lines being developed (Bailey, 1971). Thus, crossing and inbreeding are juxtaposed every four to five generations, on average.

This method of inbreeding fixes the chance recombination of favourable genes in decreasing amounts, as homozygosity increases in advanced inbred generations. Such recombinant inbreds display marked 'heterosis' despite the fact that they are not hybrids *per se*, although a pedigree may trace to a common F_1 hybrid ancestor. Unusual 'heterosis' may be due to partly dominant alleles being dispersed throughout the genome (Jinks, 1983). In 61 inbred chrysanthemum families where inbreeding depression was common, recombinant inbreeding resulted in greater gains and heterosis (Anderson *et al.*, 1993). Recombinant inbreds out-performed nonrecombinants for at least two generations of selfing. By the end of four selfed generations, recombinant inbreds dominated the chrysanthemum inbred families. Other crops have also used recombinant inbreeding to circumvent inbreeding depression, e.g. *Brassica* (Werner *et al.*, 1989) and *Phaseolus* (Sullivan and Bliss, 1983a,b).

It would be advantageous for flower breeders to use recombinant inbreeding to achieve greater gains than would otherwise be possible (Campbell, 1988; Anderson *et al.*, 1993). In fact, recombinant inbreeding may be a primary factor responsible for the significant and continued increase in hybrid corn performance since the 1950s (Anderson *et al.*,

1993). Hybrid corn yield had no increase in performance (yield) over open-pollinated varieties (Duvick, 1984). Since competing corn breeding companies could not access their competitors' inbreds, they had to use their hybrids. Inbreeding the competitors' hybrids, followed by crossing these inbreds with the companies' hybrids or unrelated inbreds, resulted in recombinant inbreds. Flower breeding programmes frequently implement a similar strategy to maintain their competitive edge, as well as circumventing inbreeding depression.

Incongruity

Incongruity is a pre- and/or post-zygotic reproductive barrier occurring when previously isolated taxa are hybridized (Hogenboom, 1973). This results in the failure of intimate partner relationships because of a lack of genetic information in one parent about the critical factors of the other (Haghighi and Ascher, 1988). Incongruity may occur in both intra- and interspecific hybridizations.

Intraspecific crosses in *Phaseolus vulgaris* from different areas of the world have led to F_1 hybrid weakness (Gepts and Bliss, 1985). They analysed the various crosses performed by other researchers where 'crippled' plants or other abnormalities occurred and found that the genotypes in these crosses had different phaseolin patterns (major seed storage protein). Large-seeded beans typically had 'S' phaseolin, while small seeded beans had 'T' or 'C' phaseolin. Hybrid weakness exhibited by these crosses may be due to geographical isolation between the areas of origin for the different types of beans. The centre of origin for these species was identified as Middle America or the Andes.

While incongruity acts as a barrier to gene exchange, it evolves passively and is not selected. Phenotypic incongruity symptoms may match other reproductive barriers such as SI (lack of seed set), inbreeding depression and genetic load (substandard growth or vigour) etc. This reproductive barrier has also been extensively reviewed elsewhere (Liedl and Anderson, 1994). Many different methods have been proposed to overcome incongruity mechanisms in wide crosses (de Nettancourt, 1977) and two new ones will be discussed below as they have the greatest promise for creating fertile hybrids.

CONGRUITY BACKCROSSING.   Haghighi and Ascher (1988) proposed congruity backcrossing as a method of overcoming incongruity. Congruity backcrossing is defined as recurrent backcrossing to each parent in alternate generations (Allard, 1960). There are two reasons to backcross an interspecific hybrid with each parental species in alternating generations: (i) this may be necessary when the hybrid has low fertility and the chance of obtaining self- or full-sib seed is low; (ii) to increase gene flow between both species via interspecific hybridization (known as introgression). Haghighi and Ascher (1988) compared recurrent backcrossing with congruity backcrossing in interspecific *Phaseolus* (beans). Hybrids derived from recurrent backcrossing were fertile but eventually lost all traits from the non-recurrent parent, while those from congruity backcrossing had increasing fertility with new traits (differing from the parents). This has been extensively documented in beans (Anderson *et al.*, 1996) and is being successfully attempted in some flowers (Anderson, unpublished data). Congruity backcrossing may be a useful tool in promoting gene transfer for many additional floricultural crops.

RECURRENT F$_1$ BRIDGING.   In this method, incongruity is overcome by using the primary F$_1$, interspecific hybrid as a bridge to transfer the nucleus of one species into the cytoplasm of another (Mutschler and Cobb, 1985). This has been successfully implemented with tomato, *Lycopersicon esculentum*. Crosses of this crop with wild *Lycopersicon pennellii* are unidirectional, with seed set or cultural embryos occurring only when the crop is the female parent. Recurrent F$_1$ bridging may be useful in cases where reciprocal differences occur in interspecific hybridizations.

Polyploidy

Most plants are diploid ($2n = 2x$); however, polyploid crops and species may also occur with three or more sets of chromosomes ($2n > 2x$). Several flower and foliage crops are polyploid, e.g. *Dendranthema*, *Dahlia* and homosporous ferns (Sparnaaij, 1979). Polyploidy is considered a 'special arithmetic relationship between the chromosome numbers of related organisms which possess different numbers' and can be defined as 'the presence of three or more chromosome sets in an organism' (Grant, 1981; Haufler, 1987). Polyploidy denotes genome multiplication. Depending on meiotic pairing mechanisms (homologous or homoeologous), polyploidy may complicate the rate of progress in flower breeding programmes (selfing or outcrossing), but also create opportunities for phenotypic change and enhancement (MacKey, 1970).

Increased DNA content, due to polyploidy or larger genome sizes, and species distribution are correlated such that a higher occurrence of polyploidy is found at higher latitudes (Price, 1976, 1988). No research has investigated whether this increase in DNA content is also correlated with winter or cold hardiness. Herbaceous perennial breeders may find it useful to increase ploidy levels to extend the range of perenniality.

Autopolyploidy

Autopolyploids arise from duplication of chromosome sets within a single species (Poehlman, 1979). While naturally occurring autopolyploids are not as common as allopolyploids, chromosome doubling is a useful tool for flower breeding. Many wide, interspecific crosses – particularly with parents differing in ploidy levels ($2x \times 4x$ crosses) or from differing generic sections, subtribes, or tribes – result in hybrid sterility (Van Tuyl *et al.*, 2000). The use of chromosome doubling techniques to create autotetraploids may restore fertility. Haploid production for 'instant inbreds' requires spontaneous or induced chromosome doubling before they can be used as diploids in a crossing programme, although these are technically not autopolyploids. If haploid derivation occurs with tetraploid or higher parents, chromosome doubling of such haploids produces autotetraploids.

Synthetic chromosome duplication is accomplished with the use of chemical agents, i.e. colchicine, amiprophos-methyl (APM), pronamide, oryzalin and trifluralin (van Tuyl *et al.*, 1992). Wan *et al.* (1991) had the greatest success with APM and pronamide, which did not inhibit plant regeneration capacity. Van Tuyl *et al.* (1992) compared the effects of oryzalin with colchicine for chromosome doubling in *Nerine*. Colchicine was lethal, whereas oryzalin doubled the chromosomes with few of the side effects normally seen with colchicine. Zlesak *et al.* (2002) examined the effects of colchicine (0.5%) and two levels of trifluralin (0.0086% and 0.086%) on rose seedlings. The higher rate of trifluralin had the highest rate of mortality, but also the highest frequency of polyploidization. Colchicine had low rates of mortality, but did not affect the treated meristem. Autopolyploids often showed roughened leaf surfaces, enlarged stomata, and larger flower, floral characteristics, leaves and plant size when compared with diploid plants. Polyploidy must be confirmed on these plants through cytogenetic analyses (chromosome counts and/or flow cytometry).

Allopolyploidy

Allopolyploidy occurs when two or more species combine their entire genomes to form a species with all of the genomes (Poehlman, 1979). A wide array of floricultural crops are interspecific hybrids. These may be vegetative- or seed-propagated crops. Allopolyploidy is an important mechanism in flower crop domestication to create new phenotypes.

Harlan and DeWet (1975) proposed three mechanisms to create allopolyploids. The first mechanism, Class I polyploids, is through $2n$ gametes in the F_1 generation, and is the most common. Sugarcane (*Saccharum*) is an example of this class since it produces many unreduced gametes (primarily eggs) when used in interspecific crosses. Other genera known to have this type of unreduced gametes include *Brassica, Nicotiana, Oenothera, Petunia, Primula* and *Solanum*. Class II polyploids occur in the F_2 and are caused by meiotic disturbances in the F_1. In this case, the

F_1 that is produced has the normal number of gametes. However, when the plant is selfed or backcrossed to a parent, normal meiosis does not occur and the resulting progeny are often polyploid. This phenomenon results from interspecific crosses to produce a 'normal' F_1, but hybrid breakdown (incongruity) occurs and the selfed progeny from the F_1 cannot undergo normal meiosis. Many of the genera found in Class I are also in this class, including *Gossypium, Oenothera* and *Solanum*. The third mechanism, Class III, occurs via somatic doubling and is very rare.

Flower colour

Flower colour is one of the most important phenotypes that flower breeders focus a significant portion of their time on developing. Once a superior new crop has been identified, the next major phenotypic directive is additional flower colours – usually as many as possible – to create a series (Fig. 5.1). It is desirable that cultivars within a series be phenotypically identical, differing only for flower colour. This is accomplished by selecting a female parent with high general combining ability, which is then crossed with a variety of male parents to create each flower colour.

The genetics of flower colour inheritance are often proprietary and few genetic maps for flower colour exist. Older flower crops, e.g. *Petunia, Cyclamen, Impatiens, Dendranthema*, have a wide array of flower colours available within any given series. For these crops, classical flower breeding has frequently reached a limit in the creation of novel flower colours (Mol *et al.*, 1998). In the future, additional basic and applied research will need to focus on adding molecular techniques, coupled with biochemistry and genetics, to enhance flower colour possibilities.

A variety of factors are involved in floral pigmentation, including pigment types, co-pigments, vacuolar pH, regulatory genes and viruses. Pigment types include water-soluble compounds – derived from the anthocyanin biosynthetic pathway, e.g. anthocyanins, flavones, flavonols, and fat-soluble compounds – from the carotenoid biosynthetic pathway, e.g. carotenoids and lycopenes (Asen, 1976).

Co-pigmentation of compounds forming a chemical complex with anthocyanins results in the visible flower colours (Asen *et al.*, 1972; Brouillard, 1988; Kondo *et al.*, 1992). Vacuolar pH enables a wide range of flower colours with identical anthocyanin compounds (Asen *et al.*, 1971; Griesbach, 1996). Several pH genes have been identified, primarily in *Petunia* with pleiotropic effects and co-dominant inheritance (de Vlaming *et al.*, 1983; Griesbach, 1998). Regulatory genes have been identified in *Petunia*, although their genetics are not well understood (de Vetten *et al.*, 1997; Spelt *et al.*, 2000). Viruses have played an important historic role in flower colour and market appeal, particularly with the genus *Tulipa*. The 'colour breaks' caused by the tulip breaking virus may impact the anthocyanin pathway (Dekker *et al.*, 1993). Future work on flower colour will require a substantial investment in basic research before commercial applications can be realized.

Once sufficient new flower colours have been commercialized, flower patterns are usually added to create an expansive mix of series offerings. Common flower patterns include star, picotee, eyes, blush, stripes and mottling (see Plate 5.2).

Fragrance

Floral fragrance is an important breeding objective for some, but not all, flower crops. Consumers place a high value on desirable fragrances in floricultural crops (Vainstein *et al.*, 2001). Traditionally, scented flowering potted plant and cut flowers are the two major categories where fragrance is desirable. Potted plants such as Oriental and Asiatic lilies (*Lilium* hybrids), Easter lily (*L. longiflorum*), hyacinth (*H. orientalis*), miniature roses (*Rosa* × *hybrida*), paperwhite narcissus (*Narcissus*), and cyclamen (*C. persicum*) are scented (Vainstein *et al.*, 2001; Heil, 2002; Ishizaka *et al.*, 2002; Pouw, 2002). Scented cut flowers include lilies (*Lilium* hybrids), lilac (*Syringa vulgaris*), hyacinth (*H. orientalis*), some roses (*R.* × *hybrida*), freesia (*Freesia* × *hybrida*), tuberose (*Polyanthes tuberosa*), carnation (*Dianthus caryophyllus*), and stephanotis (*Stephanotis*). Some fragrant cut flowers

were developed, but never gained commercial acceptance. For instance, Luther Burbank (Whitson *et al.*, 1914) bred fragrant calla lilies, *Zantedeschia aethiopica*, although the current cut callas are non-fragrant. Fragrant herbaceous perennials are the next most common commodity with scented cultivars, e.g. *R.* × *hybrida*, *D. barbatus*, *Paeonia lactiflora*, *Iris germanica*, *Iris pallida*, and *Lilium* spp. and hybrids (Firmin *et al.*, 1998). Historically, annual bedding plants have been the least common commodity with scented products. There are a few exceptions with 100% fragrant cultivars, such as Dame's rocket (*H. matronalis*) and alyssum (*Lobularia maritima*).

Recently, additional crops have value-added fragrance in new cultivars. Several fragrant *Begonia* cultivars have been bred, e.g. '77WL' has a pendulous habit with a distinctive floral fragrance described as 'sweet and lingering but subtle and not overpowering' (Snow, 2002). Fragrance production in this begonia is temperature-dependent, with the highest fragrance produced at 18.3–26.7°C (65–80°F). Fragrant, wild species of *Nemesia* have greatly expanded product offerings in this crop as well.

Since the production of volatile compounds composing fragrance are expensive resource sinks for flowering plants, elimination of fragrance would allow more resources to enhance floral longevity and postharvest life (Vainstein *et al.*, 2001; Ishizaka *et al.*, 2002). Fragrant cut roses with cytotoxic monoterpene alcohols or β-damascene have a short vase life while cultivars without these compounds have longer postharvest life (Miller, 1954; Oka *et al.*, 1999; Pouw, 2002). The creation of new, non-fragrant rose hybrids may be due to unintentional selection for flower quality and postharvest life and such selected cultivars may coincidentally have no fragrance. An absolute correlation between fragrance and longevity does not always exist. For instance, postharvest shipping and handling environments have significant effects on reducing or eliminating scent in cut roses (Helsper *et al.*, 1998; Laws, 1999). Cut and potted carnations may have both fragrant flowers and post-production longevity, e.g. *D. caryophyllus* 'Sunflor Triton'

(Heil, 2002). New potted *Amaryllis* cultivars have been created with 'intense fragrance', excellent postharvest life, high temperature tolerance, as well as resistance to red scorch fungus (Meerow, 2002).

Edible, fragrant flowers are a new innovation with wide appeal to consumers and the culinary industry. Kelley *et al.* (2001) conducted two surveys assessing consumer and professional chefs' perceptions of edible-flower species, *Viola tricolor* 'Helen Mount', *B. officinalis*, and *T. majus* 'Jewel Mix', for taste fragrance, and visual appeal. A majority of consumers (76%) rated all flowers as acceptable; a slightly lower rating of 66% was found with professional chefs (Kelley *et al.*, 2001). All flowers were rated favourably for fragrance.

Numerous difficulties exist in breeding for increased fragrance due to the large number of volatiles that constitute a single fragrance, changes in which may significantly alter floral appeal (Firmin *et al.*, 1998; Kim *et al.*, 2000; Vainstein *et al.*, 2001; Ishizaka *et al.*, 2002). For instance, in *Anthurium armeniense*, *Anthurium fragrantissimum*, *Anthurium lindenianum*, *Anthurium ochranthum* and *Anthurium roseospadix*, 28 compounds were identified that contributed to fragrance (Kuehnle, 1996; Kuanprasert *et al.*, 1998). Identification of volatiles is expensive and complicated, involving the use of gas chromatography (GC) and mass spectroscopy (MS). Flower breeders typically avoid GC/MS analysis and select fragrant cultivars that are appealing, based on consumer preference surveys. The genetics of fragrance are complicated, due to the number of volatiles involved as well as regulatory genes invoked in the chemical pathways. Several laboratories have tried, with varying success, to use genetic engineering to up-regulate volatile production or down-regulate production of secondary compounds (Pellegrineschi *et al.*, 1994).

Disease/insect resistance

In recent years, the importance of breeding for insect and/or disease resistance has become a funding priority for national floriculture agencies, e.g. The Fred Gloeckner

Foundation, the American Floral Endowment. The reasons for this new interest are varied, from environmental concerns (groundwater contamination, consumer awareness), and a decreasing palette of effective chemical controls (i.e. removal of methyl bromide for soil fumigation) to the relatively small floriculture crop significance (making floral crop-specific chemical registration costly). Strategies for disease and insect resistance breeding in flower crops has been extensively reviewed elsewhere (Sparnaaij, 1991).

There are two categories of customers interested in disease- and insect-resistant floral crops, i.e. the commercial growers and retail consumers, which determine the relative worth of product development in these areas. Commercial greenhouse and field growers place a high priority on superior crop quality since it is correlated with pricing and market value (consumer acceptance). A retail customer may or may not also have a similar interest in resistance. For instance, if a flower breeder develops white-fly (*Trialeuroides vaporariorum*) resistant cut snapdragons (*A. majus*) the greatest impact will be on the commercial grower, who will have fewer input costs into cut snap dragon production. However, the value of white-fly resistance in snapdragons arranged in a floral design is unnoticed by the retail consumer. When there is no apparent difference(s) in resistant versus non-resistant cultivars, the value-added benefit is only for the grower. Retail consumers would not pay a higher price for disease or insect resistance when there is no tangible value. Due to this dilemma, few flower breeding programmes emphasize resistance as a major breeding objective. Example research on disease or insect resistance that benefits primarily growers includes nematode resistance in Easter lily (*L. longiflorum*; Westerdahl *et al.*, 1998) and thrips resistance for impatiens (*I. walleriana*; Cloyd *et al.*, 2001).

Cost is an additional factor that has limited flower breeding for disease- and insect-resistant crops. If a breeding programme has an objective of the development of resistance, considerable investment in breeding facilities (greenhouses fitted with insect-proof screening on all vents, isolation chambers or houses to keep the diseases/insect secluded from adjacent breeding work) and personnel (each segregating or hybrid progeny must be carefully inoculated and screened to ensure accurate data collection) is required. Failure to invest in such facilities can result in significantly less selection opportunity and progress in developing resistant strains. For instance, if a breeding programme relied on field infestations of powdery mildew to screen for resistance, no progress would be made in years when the environmental conditions did not favour its development and dissemination. Additionally, during years of heavy infestation, there is no guarantee that an uninfected genotype has true resistance rather than the lack of infection being due to chance isolation or other factors. Once resistant cultivars have been created, extensive testing in multiple environments is also a necessity. As for other important floricultural traits, a disease- or insect-resistant cultivar must display stability across years and environments (Allard, 1960).

Disease- or insect-resistant flower crops that provide tangible and visible benefits have been developed, for which home-gardeners pay higher prices. Powdery mildew tolerant or resistant cultivars of bee balm (*Monarda didyma* 'Petite Delight', 'Pink Delight') and zinnia (*Z. angustifolia*) did not have white mildew on leaf surfaces or lose their leaves during the summer months (Boyle and Wicke, 1996). Black-spot resistant roses (*Rosa × hybrida*) are being developed which would share a similar enthusiastic response by home gardeners (Quarles, 1998). The release of a scented geranium (*Pelargonium*) that may discourage mosquitoes provides an additional benefit to humans (Raver, 1991). In the case of flowering potted plants and hanging baskets a greater acceptance exists for disease and insect tolerance by the retail consumer. For this reason, *Protea* potted plant breeders have disease resistance as an objective (Brits, 1995). Potted *Amaryllis* plants have been developed that are resistant to red scorch fungus, a common disease that blemishes the foliage (Meerow, 2002). Consumers would also benefit from white-fly resistant fuchsia (*Fuchsia × hybrida*) or borer-resistant *Iris* species. There is considerable opportunity for development of disease and insect resistance in flower seed crops.

Tools to Facilitate Genetic Recombination

Obtaining progeny with the desired phenotype(s) frequently challenges flower seed breeders. Genetic recombination of favourable alleles may be prevented in intra- and interspecific cross pollinations due to a variety of pre- and post-fertilization reproductive barriers. In such instances, flower breeders must employ a variety of techniques to circumvent the disrupted gene transfer, including breeding methodologies, methods of rapid generation cycling and *in vitro* culture (embryo rescue), as well as regeneration and transformation systems.

Breeding methodology

Interspecific hybridization

Interspecific hybridizations, crosses between taxa within the same genus (denoted with an '×' between the genus and the specific epithet), are frequently performed to transfer desirable traits from wild species into the domesticated taxon. Many floricultural crops are complex hybrids, the result of many generations of interspecific hybridization, backcrossing, selection and integration of additional taxa, e.g. *P. × hortorum*, *D. × grandiflora*, *Lilium × hybridum*, *Petunia × hybrida*. These tend to be primarily vegetatively propagated, often due to polyploidy, inbreeding depression, reduced fertility and heterozygosity. Complex intraspecific hybrids are common in seed-propagated crops such as *I. walleriana*.

In many floral crops, interspecific hybrids are a critical source of genetic variation (Van Tuyl, 1997). Following pollination, pollen tube growth is commonly inhibited at differing regions within the style (Ascher and Peloquin, 1968). In several crops, notably *Lilium* and *Alstroemeria*, a cut style technique may circumvent this stylar barrier to obtaining interspecific hybrids (De Jeu, 2000). The cut style technique involves removal of the stigma and the upper two-thirds of the style, whereupon stigmatic exudate is loaded on to the cut stylar stump and the desired pollen applied. To avoid pollen dehydration during germination and early pollen tube growth, a gelatin cap or parafilm may be used to wrap around the cut stump. Style grafting has also been used in lily where a cross-compatible stigma/style is grafted on to a cross-incompatible stylar base and ovary. After the application of pollen which is compatible with the grafted style, pollen tube growth continues through the graft union and into the cross-incompatible style to ensure fertilization (Van Tuyl et al., 1991).

Laboratory seed ripening and rapid generation cycling

During the product development phase of domestication, it is beneficial to implement laboratory seed ripening and/or rapid generation cycling techniques, particularly when working with crops that are perennial, have depressed fertility levels, or are particularly prone to seed reduction due to environmental stress or wide crosses.

Herbaceous and woody perennial outcrossing taxa have traits that impede flower breeding, the derivation of hybrids from wide crosses and parental development (Anderson et al., 1990). Perennials have reduced seed set, annual threshold seed production levels, long-term source–sink interactions that transcend annual species, self incompatibility, severe inbreeding depression or genetic load and lengthy seed development (embryogenesis) and juvenility periods (Sorenson, 1969; Franklin, 1972; Wiens, 1984; Schemske and Lande, 1985; Klekowski, 1988). Since hardy perennials survive beyond one growing season, plant survival is more critical than seed production. The converse is true of annuals. If the perennial species has a lengthy embryogenesis and seed development period, it may be possible to reduce either or both of these with the use of laboratory seed ripening and embryo rescue.

Laboratory seed ripening has been used on a variety of flower crops from perennial chrysanthemums (Scott, 1957; Cumming, 1964; Anderson et al., 1990) to annuals such as petunias (Flaschenriem, 1974). The primary advantage of this technique is removal of the inflorescence from the maternal plant, thereby eliminating the source–sink interactions on the plant (Anderson et al., 1990). This results in the developing seed being the only sink for carbohydrate reserves, thus maximizing seed set potential (Anderson and Ascher, 2000). Flower stems are removed from the parent, placed in a nutritive solution and pollinated under laboratory (29°C, constant) conditions (Plate 5.3). Nutritive solutions range from 200 ppm 8-hydroxyquinoline citrate + 1% sucrose (Flaschenriem, 1974; Wilkins, 1979; Anderson et al., 1990) to the use of commercial floral preservatives (all of which contain sucrose, a pH adjustor and bactericides). Light is not necessary for crossing under these conditions, as all supplemental carbon is supplied by the nutritive solution in which the stems are placed, rather than photosynthesis. With chrysanthemums, the use of laboratory seed ripening shortened the seed development cycle by 50% (from 52+ days to 25 days), hastening the onset of heart-stage embryos (by 9 days) such that embryo rescue could be implemented earlier than with comparable in situ ripened seed (Anderson et al., 1990).

Rapid generation cycling techniques encompass the use of laboratory seed ripening with embryo rescue (at the heart stage or later). The use of rapid generation cycling successfully reduced the generation time in intraspecific chrysanthemum such that a complete generation time (from seed to seed) encompassed as little as 100 days, depending on the parental genotypes and short day response period (Anderson et al., 1990). Such a flower breeding programme could generate three generations/year, compared with only one or less using in situ techniques (Anderson et al., 1990). Variations in rapid generation cycling may be necessary to obtain the best system for the flower crop being bred. Other useful benefits from implementing both techniques for rapid generation cycling include faster inbred line development, achieving homozygosity more quickly, and eliminating years of selection necessary to obtain germplasm that is genetically fast cycling, as has been done with rapid cycling Brassica (Williams and Hill, 1986) or annualized perennials.

Tissue culture

Given the preponderance of heterozygosity, polyploidy, and reproductive barriers (particularly SI) in floricultural crops, it is frequently necessary to implement tissue culture or *in vitro* techniques to facilitate gene transfer (De Jeu, 2000; Hvoslef-Eide and Vik, 2000). While several crossing methods exist to overcome incongruity, *in vitro* techniques are frequently employed to rescue embryos from wide crosses that otherwise abort due to nonfunctional endosperm or embryos, endosperm balance number (EBN), or other hybrid breakdown phenomena (Hogenboom, 1973). Many thorough reviews have been written on the subject of techniques for overcoming pre- and post-fertilization barriers (Liedl and Anderson, 1993; Van Tuyl and De Jeu, 1997; De Jeu, 2000).

In vitro techniques may be used early in crop domestication to facilitate gene transfer between wild × wild or crop × wild crosses or later when new traits are identified in taxa closely or distantly related to the crop, i.e. incorporation of disease/pest resistance, fragrance, floral traits or other phenotypic characters. Pre-fertilization techniques are usually accomplished *in situ*, e.g. cut styles and stylar grafts, as discussed earlier. Post-fertilization barriers are more easily overcome or manipulated with *in vitro* systems.

Defective endosperm growth and embryo abortion have been documented in a variety of floricultural crops, including *Alstroemeria* (De Jeu and Garriga Calderé, 1997), *Cyclamen* (Ishizaka and Uematsu, 1992) and *Lupinus* (Busmann-Loock *et al.*, 1992). Both of these phenomena occur after the first abnormal zygotic division. Many different embryo rescue techniques have been successfully developed and employed to circumvent defective endosperm and abortive embryos in a variety of crops (Davies, 1960; Sanders and Ziebur, 1963; Stimart and Ascher, 1974; Watanabe, 1977; Haghighi and Ascher, 1988). The embryo rescue techniques can be divided into two broad categories of embryo development: (i) globular and (ii) heart or later stages. Each category requires a different type of tissue culture medium to ensure successful embryo rescue.

Generally, globular embryos may be rescued on a liquid Norstog medium (Norstog, 1973). Norstog medium poses difficulties to novice tissue culturists, as some medium constituents are not autoclaved. Improper handling of the ingredients can result in rapid contamination.

Embryo rescue of heart or later embryonic stages is more easily accomplished than globular stages and, consequently, is the most commonly employed *in vitro* technique. The variations in embryo rescue techniques are crop-dependent, based on the rate of embryo development (how soon the heart stage occurs), fruit/inflorescence type, specific medium requirements (lack of plant growth regulators, carbon sources and levels) and germination requirements (light vs. dark). All such techniques use solid media, most commonly Murashige-Skoog (MS) (Murashige and Skoog, 1962). *Alstroemeria* embryo rescue techniques are implemented 2 days postpollination using a hormone-free MS medium with 9% saccharose (Buitendijk *et al.*, 1995; De Jeu and Jacobsen, 1995). Wide, interspecific hybrids with faster germination and growth (in comparison with *in situ*-generated hybrids) have been recovered using this embryo rescue technique. In intraspecific chrysanthemum, embryo rescue significantly reduced the generation time (compared with *in situ* seed development), increased percentage germination and the percentage of progeny reaching anthesis (Watanabe 1977; Anderson *et al.* 1990). Embryo rescue techniques are routinely employed in flower breeding programmes of *Alstroemeria* (Burchi *et al.* 2000), *Cyclamen* (Ishizaka and Uematsu, 1992), *Delphinium* (Honda and Tsutsui, 1997), *Dendranthema* (Anderson *et al.*, 1990; Anderson and Ascher, 2000), *Freesia* (Resier and Ziessler, 1989), *Lilium* (Van Tuyl *et al.*, 1992) and *Tulipa* (Van Creij, 1997).

High sucrose concentrations should not be used in the MS media because it may impede the transition from *in vitro* culture into greenhouse culture, e.g. hybrids successfully embryo rescued have died and were unable to photosynthesize or root (Haghighi, 1986; Haghighi and Ascher, 1988). In all of these techniques it is important to either completely remove the seed coat or scarify it in some manner (nick with a scalpel). Failure to do so may result in embryos that do not perceive the rescue, fail to germinate, and continue with embryogenesis as though they were still *in situ*. Whether rescued embryos are germinated in light or darkness can be determined by the colour of the embryo after the seed coat has been removed. If the embryo surface is white (lacking pigmentation), it most probably will germinate the best and fastest in darkness, whereas embryos with chlorophyll need light.

Other embryo culture techniques have been used when routine embryo rescue procedures were not successful. *Lilium* ovaries may be pollinated, cultured *in vitro*, and later embryo rescued (Van Tuyl *et al.*, 1991). If ovaries are held *in situ* until a later embryonic stage (torpedo), they may be removed from the plant and sliced (leaving the placental tissues intact). Ovary slicing is commonly used in monocot floral crops such as *Lilium*, *Nerine* and *Tulipa* (Van Tuyl *et al.*, 1993).

In vitro fertilization techniques have been attempted in a few flower crops, such as *Alstroemeria* (De Jeu, 2000) and *Lilium* (Janson *et al.*, 1993), although most have been unsuccessful. The difficulties are poor pollen germination or erratic pollen tube growth near the micropylar entrance. Since *in vitro* gamete fusion methodology has been developed in agronomic *Nicotiana tabacum* (Tian and Russell, 1997), flower breeders may find this approach applicable to floral crops in the Solanaceae, i.e. flowering tobacco (*Nicotiana landsdorfii*, *Nicotiana alata*), ornamental peppers (*Capsicum annuum*), Jerusalem cherry (*Solanum pseudocapsicum*) and petunia (*P. × hybrida*).

Protoplast or somatic fusion has been attempted with limited success in crops where *in vitro* embryo rescue, fertilization or gamete fusion failed. A somatic hybrid between *Dianthus chinensis* and *D. barbatus* was produced (Nakano and Mii, 1993). The resulting hybrid(s) matched the expected interspecific phenotypic and cytogenetic profile. Other laboratories have had less successful protoplast fusion between *Lavatera thuringiaca* and *Hibiscus rosa-sinensis*, although this may be attributable to intergeneric incongruity (Vazquez-Thello *et al.*, 1996).

Regeneration and transformation systems

Classical flower breeding has many limitations, based on the available genes and physiological pathways existing in the crop species or genera. For instance, no classical flower breeder has succeeded in developing either a true blue-flowered rose or chrysanthemum, a pure white marigold, or an orange petunia (Mol *et al.*, 1989, 1995). However, the ability to introduce individual genes from unrelated taxa via molecular technology has increased the probability of creating such novel phenotypes (Bryant, 1982).

The advent of modern biotechnology has dramatically shifted plant breeding research in many seed-propagated crops. Different biochemical and molecular markers are used to improve plant breeding efficiency (Tanksley and Orton, 1983), which have been extensively reviewed in relation to ornamental breeding (Arus, 2000). *In vitro* regeneration and transformation systems must be developed prior to the use of genetic transformation. With seed-propagated flower crops, most developmental work is performed *in situ* with the exception of embryo rescue and other aforementioned methods to rescue abortive embryos. *In vitro* production of flower seed crops is only used in instances where virus-free inbreds or stock plants replace the usual stock seed sources for hybrid seed production, e.g. double petunia inbred parents. In contrast, most stock of vegetatively propagated crops is maintained *in vitro*. Other *in vitro* uses include cutting production (particularly foliage crops) and thermotherapy (heat treatments) to reduce or eliminate virus titre, which is often coupled with meristem-tip culture. Virus-free plants obtained via these methods are easily reinfected, if they are not resistant, necessitating the constant maintenance of virus-free *in vitro* stock. There is considerable opportunity in this area for flower breeders to develop virus-resistant crops, particularly with the use of molecular techniques.

Tissue culture regeneration systems are largely undeveloped for flower seed crops and the difficulties posed in their development often thwarts the potential use of plant biotechnology techniques to achieve crop transformation. A few reviews have been published that define the techniques available for genetic transformation of floricultural crops (Mol *et al.*, 1995). Examples of well-developed systems in flower seed crops are provided for illustrative purposes. The primary difference between classical flower breeding and molecular techniques is that with the latter, one or a few target genes are incorporated and the remaining cells and DNA are left essentially intact (De Jong, 2000). Classical flower breeding, in contrast, exploits an infinitely greater array of genetic potential. Molecular techniques are similar to the use of mutation breeding in vegetative crops, where a superior clone with market value is irradiated to produce a new trait, e.g. flower colour changes (Broertjes and Verboom, 1974). These two techniques differ in that molecular techniques rely on a higher level of precision instead of chance.

Flower crops with the greatest amount of genetic transformation include *Petunia × hybrida* (Oud *et al.*, 1995), *Antirrhinum majus* (Sommer and Saedler, 1986; Cui *et al.*, 2000) and *Dendrathema × grandiflora* (Courtney-Gutterson, 1994; Fukai *et al.*, 1995, Yepes *et al.*, 1995; Mitiouchkina and Dolgov, 2000). Other crops that have been successfully transformed include *Alstroemeria* (Lin *et al.*, 2000), *Anthurium andreanum* (Kuehnle *et al.*, 1993), *Dendrobium* (Kuehnle and Sugii, 1992), *Eustoma grandiflora* (Deroles *et al.*, 1995), *Gerbera jamesonii* (Elomaa *et al.*, 1993), *Osteospermum* (Allavena *et al.*, 2000) and *Rosa × hybrida* (Souq *et al.*, 1996). Transgenic applications for flowering crops have included flower colour changes (chrysanthemums, gerberas, lisianthus, petunias, roses), enhanced postharvest life (carnation), disease resistance (*Anthurium, Osteospermum, Petunia*), plant architecture and enhanced branching using *Agrobacterium rhizogenes* (*Dendranthema, Eustoma*). The continued use and applicability of transformation techniques will be dependent on regeneration/transformation system development in the crop of interest, as well as consumer acceptance of genetically modified floral crops (Mol *et al.*, 1995).

Ideotype Breeding

Each flower breeding programme has specific objectives to be achieved in crop domestication or continued development. The short- and/or long-term objectives provide for the creation of new phenotypes by the flower breeder, depending on the status of the crop, its continued market potential and the competition. Thus, a flower seed breeder may be creating new colours to add to a series, creating an entirely new series, or developing innovative or distinguishable products (Fig. 5.1). Since the flower market is highly competitive and the number of flower breeding programmes small, market-driven competition often prevents private sector breeding programmes from envisioning long-term crop potential and the possible directions in which a crop could be expanded or phenotypically transformed.

Visionary flower breeding, however, must focus beyond the immediate product potential of a new or old crop to proactively create continually expanding phenotypes (Table 5.7). An excellent illustration of this is the 'wave' or prostrate habit petunia (PanAmerican Seed Co., 1999). During the 20 years prior to the conception and release of wave petunias, the petunia market was essentially not changing in the USA bedding plant industry. Petunia breeders had been focusing on improvements within a series or the continual creation of new series of petunias to fit the various product classes (Grandifloras, Millifloras, Doubles, Floribundas, etc.) with little monetary return. Little attention had been directed by petunia breeders to creating innovative or distinguishable products (Fig. 5.1). Indeed, the wave petunia habit was created by a non-traditional flower breeding operation, the Japanese Kirin Brewing Company, by interspecific hybridization with wild species. This habit had actually been created with the identical interspecific cross by a public sector breeding programme at the University of Minnesota in the 1970s, but commercial petunia breeders dismissed the phenotype as 'too different, having no market value' (P.D. Ascher, 1997, personal communications). Indeed, if petunia breeders had spent sufficient time analysing taxonomic data, along with occasional collection trips to observe wild germplasm, they would have seen wild petunia species with the 'wave' phenotype (Plate 5.4).

The best method to encourage flower breeders to think beyond the current crop phenotype is to reiterate the importance and absolute necessity that each and every flower breeder must go and collect germplasm in its wild, native habitat. The prudent flower breeder should spend time observing natural populations and find variant phenotypes at the outer limits of the species range(s). Had this been done years ago during the domestication of the various product classes of geranium (*P. × hortorum, P. domesticum,* etc.) a much wider range of crop phenotypes would be market classics today. For instance, heat and drought tolerant *P. domesticum* (Martha Washington geraniums) exist in wild South African populations (Plate 5.5) and could still be used to revolutionize the current production and use of this crop.

After the breeder has gone to the native habitats and collected the range of variation naturally selected for and present in the wild species and ancestors, the data should be used to create a crop ideotype. Plant ideotypes are models for predictable plant growth in a prescribed environment (Donald, 1968). In a successful plant breeding programme, an ideotype may be used to select plants with a suite of traits modelled in the prescribed environments (the commercial production requirements). Ideotypes are commonly used in most agronomic seed-propagated crops, yet they are nearly unknown in flower breeding programmes. An ideotype created for a specific crop can provide tangible targets to direct breeding programme objectives (Table 5.7). It could be a long-range plan to integrate multiple traits from wild germplasm into the crop or create new plant habits or flowering phenotypes previously unknown. At least three ideotypes have been created by public sector flower breeders for cultivated greenhouse and garden chrysanthemums (*D. × grandiflora*). Such ideotypes have been useful in transforming cut spray types (Langton and Cockshull, 1979), changing the method of propagation in F_1 hybrid seed-propagated garden types (Ascher, 1986) and incorporating

novel traits of day neutrality, heat-delay insensitivity in garden chrysanthemums (Anderson and Ascher, 2001, 2004). Ideotypes are working tools, which may be modified in response to market, technological, and knowledge changes. They are the means of keeping your breeding programme leagues ahead of your competitors' and creating revolutionary change.

Summary

The future of flower seed breeding is bright and filled with opportunities or challenges. The multitude of product groups and flower crops on the market, as well as the continued domestication of 'new' ornamental species across the plant kingdom, ensures job security for flower breeders in both the public and private sectors. There is tremendous opportunity for transforming vegetative crops into flower seed products, particularly with the top ten flowering potted plants and cut flowers. Domestication of flower seed crops is a slow process with potential for failure but the promise of successful genetic transformation is great.

It is imperative that current and future flower breeders learn from the mistakes of previous research and avoid duplication of effort. A productive and successful flower breeding programme must be based on scientific literature, correct taxonomy and thorough sampling of genetic variation. A species that may be 'new' to a flower breeder may have been cultivated in past decades or centuries. Buyer interest, gardening demands, flower colours, and preferred crops are cyclic in their appeal. For example, recent 'new crops' such as *Alonsoa, Arctotis, Brachychome, Collinsia, Datura, Diascia, Dimorphotheca, Mimulus, Nemophila, Penstemon, Perilla, Salvia farinacea, Salvia patens, Scabiosa* and *Torenia,* were cultivated as annuals or herbaceous perennials in the 1920s (Bailey, 1927; Hottes, 1937a,b). This fact has been entirely overlooked by flower breeding programmes.

The large arsenal of breeding methodologies, laboratory techniques, plant transformation technologies, or methods of circumventing reproductive barriers is available for use, when necessary. In the future, additional technologies will become available for novel transformation possibilities. None the less, domestication and breeding of flower seed crops will continually need to overcome each crop's morphological, genetic or floral limitations and barriers to gene exchange. Knowledgeable use of available technologies, coupled with strategic planning (breeding objectives, crop ideotypes), will ensure that the consumer's palette remains abundant and colourful with successful products.

Acknowledgements

Journal Series Paper No. 031210122 of the Department of Horticultural Science. This research was funded in whole or in part by the Minnesota Agricultural Experiment Station.

References

Allard, R.W. 1960. *Principles of Plant Breeding*. John Wiley & Sons, New York.

Allard, R.W., S.K. Jain and P.L. Workman. 1968. The genetics of interbreeding populations. *Advances in Gen.* 14:55–131.

Allavena, A., A. Giovannini, T. Berio, A. Spena, M. Zottini, G. Accotto and A.M. Vaira. 2000. Genetic engineering of *Osteospermum* spp.: a case story. *Acta Horticulturae* 508:129–133.

Anderson, N.O. 1987. Reclassifications of the genus *Chrysanthemum* L. *HortSci.* 22(2):313.

Anderson, N.O. 2001. Commercial and research oriented plant breeding. Chicago Botanic Garden, New Ornamental Crops Research Symposium Program and Abstracts, 26–29 Sept., 2001, pp. 7–8.

Anderson, N.O. and P.D. Ascher. 1993. Male and female fertility of loosestrife (*Lythrum*) cultivars. *Jour. Amer. Soc. Hort. Sci.* 118(6): 851–858.

Anderson, N.O. and P.D. Ascher. 1996. Inheritance of pseudo-self compatibility in self-incompatible garden and greenhouse chrysanthemums, *Dendranthema grandiflora* Tzvelv. *Euphytica* 87:153–164.

Anderson, N.O. and P.D. Ascher. 2000. Fertility changes in inbred families of self-incompatible chrysanthemums (*Dendranthema grandiflora*). *Jour. Amer. Soc. Hort. Sci.* 125(5): 619–625.

Anderson, N.O. and P.D. Ascher. 2001. Selection of day-neutral, heat-delay-insensitive *Dendranthema* x *grandiflora* genotypes. *Jour. Amer. Soc. Hort. Sci.* 126(6):710–721.

Anderson, N.O. and P.D. Ascher. 2004. Yield components and ideotype breeding for daylength neutrality/heat delay insensitivity in garden chrysanthemums, *Dendranthema × grandiflora*. *Jour. Amer. Soc. Hort. Sci.* In press.

Anderson, N.O. and B. Dunn. 2003. Inbreeding depression in seed-propagated *Lilium* x-*formolongi* inbreds. *Acta Horticulturae* (ISHS) 624:43–49.

Anderson, N.O. and E. Gesick. 2004. Phenotypic markers for selection of winter hardy garden chrysanthemum (*Dendranthema × grandiflora* Tzvelv.) genotypes. *Scientia Horticulturae* 101: 153–167.

Anderson, N.O. and B.E. Liedl. 1987. SIGMAS, Version 2.0. *Plant Cell Incompatibility Newsletter* 19:1–3.

Anderson, N.O. and W. Peters. 2001. Breeding for winter hardy *Gaura. Perennial Plant Quarterly*, Spring 2001:45–52.

Anderson, N.O., B.E. Liedl, P.D. Ascher, R.E. Widmer and S.L. Desborough. 1988. Evaluating self incompatibility in *Chrysanthemum*: the influence of ovule number. *Sexual Plant Reproduction* 1:173–181.

Anderson, N.O., B.E. Liedl, P.D. Ascher and S.L. Desborough. 1989. Distinguishing between self incompatibility and other reproductive barriers in plants using male (MCC) and female (FCC) coefficient of crossability. *Sexual Plant Reproduction* 2:116–126.

Anderson, N.O., P.D. Ascher, R.E. Widmer and J.J. Luby. 1990. Rapid generation cycling of chrysanthemums, *Dendranthema grandiflora* Tzvelv. (*Chrysanthemum morifolium* Ramat.), using laboratory seed development and embryo rescue techniques. *Jour. Amer. Soc. Hort. Sci.* 115(2):329–336.

Anderson, N.O., P.D. Ascher and R.E. Widmer. 1992a. Inbreeding depression in garden and glasshouse chrysanthemums: germination and survivorship. *Euphytica* 62:155–169.

Anderson, N.O., P.D. Ascher, and R.E. Widmer. 1992b. Lethal equivalents and genetic load. *Plant Breeding Reviews* 10:93–127.

Anderson, N.O., P.D. Ascher and J.J. Luby. 1993. Variations of recombinant inbreeding: Circumventing inbreeding depression in chrysanthemum populations. *In*: K.K. Dhir and T.S. Sareen (eds). *Frontiers in Plant Science Research*. Bhagwati Enterprises, Delhi, India, pp. 1–16.

Anderson, N.O., P.D. Ascher and K. Haghighi. 1996. Congruity backcrossing as a means of creating genetic variability in self pollinated crops: seed morphology of *Phaseolus vulgaris* L. and *P. acutifolius* A. Gray hybrids. *Euphytica* 87:211–224.

Anderson, N.O., G. Pietsch and N. Gomez. 2003. Plant collection and genetic variation in Texas *Gaura* populations. *Perennial Plant Quarterly* Spring 2003:28–43.

Arus, P. 2000. Molecular markers for ornamental breeding. *Acta Horticulturae* 508:91–98.

Ascher, P.D. 1976. Self incompatibility systems in floricultural crops. *Acta Horticulturae* 63: 205–215.

Ascher, P.D. 1986. Breeding garden chrysanthemums. Ornamental Plant Breeding Workshop, International. Horticulture Congress, Davis, California. *HortSci.* 21:616 (Abstr.).

Ascher, P.D. and S. Peloquin. 1968. Pollen tube growth and incompatibility following intra- and inter-specific pollinations in *Lilium longiflorum. Amer. Jour. Bot.* 55:1230–1234.

Asen, S. 1976. Known factors responsible for the infinite flower colour variations. *Acta Horticulturae* 63:217–223.

Asen, S., K.H. Norris and R.N. Stewart. 1971. Effect of pH and concentration of the anthocyanin-flavonol co-pigment complex on the colour of 'Better Times' roses. *Jour. Amer. Soc. Hort. Sci.* 96:770–773.

Asen, S., R.N. Stewart and K.H. Norris. 1972. Co-pigmentation of anthocyanins in plant tissues and its effect on colour. *Phytochem.* 11: 1139–1144.

Bailey, D.W. 1971. Recombinant-inbred strains: an aid to finding identity, linkage, and function of histocompatibility and other genes. *Transplantation* 11(3):325–327.

Bailey, L.H. 1927. *The Standard Cyclopedia of Horticulture. New Edition.* Vols. I-III. The MacMillan Co., New York.

Ball Seed Co. 2002. *2002–2003 Product Catalog.* Ball Seed Company, W. Chicago, Illinois.

Boase, M.R., R. Miller and S.C. Deroles. 1997. Chrysanthemum systematics, genetics, and breeding. *Plant Breeding Rev.* 14:321–361.

Boucher, W. 1988. Calculation of the inbreeding coefficient. *Jour. Mathematical Biol.* 26:57–64.

Boyle, T.H. and R.L. Wicke. 1996. Responses of *Zinnia angustifolia* × *Z. elegans* backcross hybrids to three pathogens. *HortSci.* 31: 851–854.

Brits, G.J. 1995. Selection criteria for protea flowering pot plants. *Acta Horticulturae* 387:47–54.

Broertjes, C. and H. Verboom. 1974. Mutation breeding of *Alstroemeria. Euphytica* 23:39–44.

Brouillard, R. 1988. Flavonoids and flower colour. *In*: J.B. Harborne (ed.). *The Flavonoids: Advances in Research.* Chapman & Hall, London. pp. 525–538.

Bryant, C. 1982. High technology speeds flower breeding successes. *New Zealand Jour. Ag.* 145(3):40–41.

Buitendijk, J.H., N. Pinsonneaux, A.C. Van Donk, M.S. Ramanna and A.A.M. Van Lammeren. 1995. Embryo rescue by half-ovule culture for the production of interspecific hybrids in *Alstroemeria. Scientia Horticulturae* 64:65–75.

Burchi, G., A. Mercuri, C. Bianchini, R. Bregliano and T. Schiva. 2000. New interspecific hybrids of *Alstroemeria* obtained through *in vitro* embryo-rescue. *Acta Horticulturae* 508: 233–235.

Busmann-Loock, A., M. Dambroth, and U. Menge-Hartmann. 1992. Histological observations on interspecific crosses in the genus *Lupinus. Plant Breeding* 109:82–85.

Callaway, D.J. and M.B. Callaway (eds). 2000. *Breeding Ornamental Plants.* Timber Press, Portland, Oregon.

Campbell, R.B. 1988. Mating structure and the cost of deleterious mutation. 1. Postponing inbreeding. *Jour. Heredity* 79:179–183.

Carlson, W.H. and F. Johnson. 1985. The Bedding Plant Industry *In*: J.W. Mastalerz and E.J. Holcomb (eds). *Bedding Plants III: a Manual on the Culture of Bedding Plants as a Greenhouse Crop.* Pennsylvania Flower Growers. pp. 1–7.

Cloyd, R.A., D.F. Warnock, and K. Holmes. 2001. Technique for collecting thrips for use in insecticide efficacy trials. *HortSci.* 36(5):925–926.

Correns, C. 1912. Selbsterilität und Individualstoffe Festschrifte der mat.-nat. Gesell. zur 84. Versammrnl. Duet. Naturforsch. Ärtze, Münster, i. W.:1–32.

Correns, C. 1913. Selbsterilität und Individualstoffe. *Biol. Zentral.* 33:389–423.

Courtney-Gutterson, N., C. Napoli, C. Lemieux, A. Morgan, E. Firoozabady and K.E.P. Robinson. 1994. Modification of flower colour in florist's chrysanthemum: production of a white flowering variety through molecular genetics. *Biotech.* 12:268–271.

Craig, R. and L. Laughner. 1985. Breeding new cultivars. *In*: J.W. Mastalerz and E.J. Holcomb (eds). *Bedding Plants III: a Manual on the Culture of Bedding Plants as a Greenhouse Crop.* Pennsylvania Flower Growers. pp. 526–539.

Crowe, L.K. 1970. The polygenic control of outbreeding in *Borago officinalis. Heredity* 25: 111–118.

Cui, M.L., T. Handa, K. Takayanagi, and H. Kamada. 2000. Development of *Agrobacterium*-mediated transformation system for snapdragon (*Antirrhinum majus*). *Acta Horticulturae* 508:241. (Abstr.).

Cumming, R.W. 1964. *The Chrysanthemum Book*. D. Van Nostrand, Princeton, New Jersey.

Dana, M.N. and P.D. Ascher. 1985. Discriminating styles (DS) and pollen-mediated pseudo-self compatibility (PMPSC) in *Petunia hybrida* Hort. *Euphytica* 35:237–244.

Dana, M.N. and P.D. Ascher. 1986a. Sexually localized expression of pseudo-self compatibility (PSC) in *Petunia × hybrida* Hort. 1. Pollen inactivation. *Theor. App. Gen.* 71:573–577.

Dana, M.N. and P.D. Ascher. 1986b. Sexually localized expression of pseudo-self compatibility (PSC) in *Petunia × hybrida* Hort. 2. Stylar inactivation. *Theor. App. Gen.* 71: 578–584.

Darwin, C.R. 1876. *The Effect of Cross- and Self-fertilization in the Vegetable Kingdom*. John Murray, London.

Darwin, C.R. 1895. *The Different Forms of Flowers on Plants of the Same Species*. D. Appleton, New York.

Davies, D.R. 1960. The embryo culture of interspecific hybrids of *Hordeum*. *New Phytol.* 59:9–14.

De Jeu, M.J. 2000. *In vitro* techniques for ornamental breeding. *Acta Horticulturae* 508:55–60.

De Jeu, M.J. and F. Garriga Calderé. 1997. Retarded embryo growth and early degeneration of sporophytic tissue are associated with embryo abortion in the interspecific cross *Alstroemeria pelegrina × Alstroemeria aurea*. *Can. Jour. Bot.* 75:916–924.

De Jeu, M.J. and R. Jacobsen. 1995. Early post-fertilization ovule culture in *Alstroemeria* L. and barriers to interspecific hybridization. *Euphytica* 86:15–23.

De Jong, J. 2000. Genetic engineering for resistance, quality, and plant habit. *Acta Horticulturae* 508: 123–127.

Dekker, E.K., A.F.L.M. Derks, C.J. Asjes, M.E.C. Lemmers, J.F. Bol and S.A. Langeveld. 1993. Characterization of potyviruses from tulip and lily which cause flower-breaking. *Jour. Gen. Vir.* 74:881–887.

De Nettancourt, D. 1977. *Incompatibility in Angiosperms*. Springer-Verlag, New York.

Denward, T. 1963. The function of the incompatibility alleles in red clover (*Trifolium pratense* L.). II. Results of crosses within inbred families. *Hereditas* 49:203–236.

Deroles, S., M. Bradley, K. Davies, K. Schwinn, and D. Manson. 1995. Generation of novel patterns in *Lisianthus* flowers using an antisense

chalcone synthase gene. *Acta Horticulturae* 420:26–28.

de Vetten, N., F. Quattrocchio, J. Mol and R. Koes. 1997. The *an11* locus controlling flower pigmentation in *Petunia* encodes a novel WD-repeat protein conserved in yeast, plants, and animals. *Genes and Dev.* 11:1422–1434.

de Vlaming, P., A.W. Schram and H. Wiering. 1983. Genes affecting flower colour and pH of flower limb homogenates in *Petunia hybrida*. *Theor. App. Gen.* 66:2171–2178.

Dole, J.M. 2002. 2001 ASCFG national cut flower seed trials. *Cut Flower Quarterly* 14(1):1–13.

Dole, J.M. and H.F. Wilkins. 1999. *Floriculture: Principles and Species*. Prentice Hall, Upper Saddle River, New Jersey.

Donald, C.M. 1968. The breeding of crop ideotypes. *Euphytica* 17:385–403.

Drewlow, L.W., P.D. Ascher and R.E. Widmer. 1973. Rapid method of determining pollen incompatibility in *Chrysanthemum morifolium* Ramat. *Theor. App. Gen.* 43:1–5.

Duvick, D.N. 1984. Progress in conventional plant breeding. *In*: J.P. Gustafson (ed.). *Gene Manipulation in Plant Improvement*. 16th Stadler Genetic Symposium. Plenum Press, New York. pp. 17–31.

East, E.M. 1932. Studies on self sterility. IX. The behaviour of crosses between self-sterile and self-fertile plants. *Genetics* 17:175–202.

East, E.M. and A.J. Mangelsdorf. 1925. A new interpretation of the behavior of self-sterile plants. *Proc. Nat. Acad. Sci.* 11:166–171.

Elomaa, P., J. Honkanen, R. Puska, P. Seppanen, Y. Helariutta, M. Mehto, M. Kotilainen, L. Nevalainen and T.H. Teeri. 1993. *Agrobacterium*-mediated transfer of antisense chalcone synthase cDNA to *Gerbera hybrida* inhibits flower pigmentation. *Biotech.* 11:508–511.

Fausey, B, A. Cameron, R. Heins, and W. Carlson. 2000. Forcing perennials: Species: *Hosta* spp. *In*: Greenhouse Grower Magazine and Michigan State University (eds.). *Firing Up Perennials, The 2000 Edition*. G.G. Plus, Willoughby, Ohio. pp. 84–87.

Finical, L., A. Frane, A. Cameron, R.D. Heins, and W. Carlson. 2000. Forcing perennials: species: *Campanula* 'Birch Hybrid'. *In*: Greenhouse Grower Magazine and Michigan State University (eds). *Firing Up Perennials, The 2000 Edition*. G.G. Plus, Willoughby, Ohio. pp. 48–51.

Firmin, L., D. Courtois, V. Petiard, C. Ehret and K. Lerch. 1998. Evaluation of the natural variability in iron content and selection of *Iris* sp. for perfume production. *HortSci.* 33(6):1046–1047.

Flaschenriem, D.R. 1974. A new method of producing *Petunia hybrida* seed *in vitro*. MS thesis, Mankato State College, Mankato, Minnesota.

Flaschenriem, D.R. and P.D. Ascher. 1979. *S* allele discrimination in styles of *Petunia hybrida* bearing stylar-conditioned pseudo-self compatibility. *Theor. App. Gen.* 55:23–28.

Frane, A., S.-Y. Wang, R.D. Heins, W. Carlson and A. Cameron. 2000. Forcing perennials: species: *Delphinium grandiflorum* 'Blue Mirror'. *In*: Greenhouse Grower Magazine and Michigan State University (eds). *Firing Up Perennials, The 2000 Edition*. G.G. Plus, Willoughby, Ohio. pp. 60–63.

Franklin, E.C. 1972. Genetic load in loblolly pine. *American Naturalist* 106:262–265.

Fukai, S., J. De-Jong and W. Rademaker. 1995. *Agrobacterium*-mediated transformation of chrysanthemum. *Plant Cell Rep.* 1994, 14(1): 59–64.

Galatowitsch, S.M., N.O. Anderson and P.D. Ascher. 1999. Invasiveness in wetland plants in temperate North America. *Wetlands* 19(4):733–755.

Gepts, P. and F.W. Bliss. 1985. F_1 hybrid weakness in the common bean. *Jour. Heredity* 76:447–450.

Goldsmith Seed Co. 2002. *Goldsmith seeds from A to Z.* Goldsmith Seed Company, Gilroy, California.

Grant, V. 1981. *Plant Speciation*. 2nd edn. Columbia University Press, New York.

Griesbach, R.J. 1996. The inheritance of flower colour in *Petunia hybrida*. *Jour. Heredity* 87: 241–245.

Griesbach, R.J. 1998. The effect of the *Ph 6* gene on the colour of *Petunia hybrida* Vilm. Flowers. *Jour. Amer. Soc. Hort. Sci.* 123:647–650.

Haghighi, K. 1986. Methods of hybridization of two bean species: *Phaseolus vulgaris* and *P. acutifolius*. PhD dissertation, University of Minnesota, St Paul, Minnesota.

Haghighi, K.R. and P.D. Ascher. 1988. Fertile, intermediate hybrids between *Phaseolus vulgaris* and *P. acutifolius* from congruity backcrossing. *Sex. Plant Repro.* 1:51–58.

Harlan, J.R. and J.M.J. DeWet. 1975. The compilospecies concept. *Evolution* 17:497–501.

Haufler, C.H. 1987. Electrophoresis is modifying our concepts of evolution in homosporous pteridophytes. *Amer. Jour. Bot.* 74:953–966.

Heil, J.J. 2002. Carnation plant named 'Sunflor Triton'. *US Plant Patent No. 12,743*. US Patent and Trademark Office, Washington, DC.

Helsper, J.P.F.G., J.A. Davies, H.J. Bouwmeester, A.F. Krol and M.H. van Kampen. 1998. Circadian rhythmicity in emission of volatile compounds by flowers of the *Rosa hybrida* L. cv. Honesty. *Planta* 207:88–95.

Henny, R.J. and P.D. Ascher. 1976. The inheritance of pseudo-self compatibility (PSC) in *Nemesia strumosa* Benth. *Theor. App. Gen.* 48:185–195.

Hogenboom, N.G. 1973. A model for incongruity in intimate partner relationships. *Euphytica* 22: 219–233.

Honda, K. and K. Tsutsui. 1997. Production of interspecific hybrids in the genus *Delphinium* via ovule culture. *Euphytica* 96:331–337.

Hottes, A.C. 1937a. *The Book of Annuals*. 4th edn. A.T. De La Mare Company, New York.

Hottes, A.C. 1937b. *The Book of Perennials*. 5th edn. A.T. De La Mare Company, New York.

Hunter, N.T. 2000. *The Art of Floral Design*. 2nd edn. Delmar Publishing, Albany, New York.

Hvoslef-Eide, T. and N.I. Vik. 2000. Modern methods for breeding ornamentals. *In*: E. Strømme (ed.). *Advances in Floriculture Research*. Agricultural University of Norway. Report no. 6/2000. pp. 18–131.

Ishizaka, H. and J. Uematsu. 1992. Production of interspecific hybrids of *Cyclamen persicum* Mill. And *C. hederifolium* Aiton. by ovule culture. *Jap. Jour. Plant Breed.* 42:353–366.

Ishizaka, H., H. Yamada and K. Sasaki. 2002. Volatile compounds in the flowers of *Cyclamen persicum*, *C. purpurascens* and their hybrids. *Scientia Horticulturae* 94(1/2):125–135.

Janson, J., M.C. Reinders, J.M. Van Tuyl and C.J. Keijzer. 1993. Pollen tube growth in *Lilium longiflorum* following different pollination techniques and flower manipulations. *Acta Bot. Neerl.* 42:461–472.

Jinks, J.L. 1983. Biometrical genetics of heterosis. *In*: R. Frankel (ed.). *Heterosis*. Springer-Verlag, Berlin. pp.1–46.

Johannsen, W. 1926. *Elements der exakten Erblichkeitslehre*. 3rd edn. Fischer, Jena, Germany.

Kelley, K.M., B.K. Behe, J.A. Biernbaum and K.L. Poff. 2001. Consumer and professional chef perceptions of three edible-flower species. *HortSci.* 36(1):162–166.

Kempthorne, O. 1957. *An Introduction to Genetic Statistics*. John Wiley & Sons, New York.

Kim, H.J., K. Kim, N.S. Kim and D.S. Lee. 2000. Determination of floral fragrances of *Rosa hybrida* using solid-phase trapping-solvent extraction and gas chromatography-mass spectrometry. *Jour. Chrom.* 902(2):389–404.

Klekowski, E.J. 1988. *Mutation, Developmental Selection, and Plant Evolution*. Columbia University Press, New York.

Kölreuter, J.G. 1761–1766. Verloäufige Nachricht von einigen das Geschlecht der Pflanzen betreffenden Versuchen und Beobachtungen, nebst Fortsetzungen 1, 2, und 3. *Ostwald's Klassiker*, Nr. 41. Engelmann, Leipzig, Germany.

Kondo, T., K. Yoshida, A. Nakagawa, T. Kawai, H. Tamura and T. Goto. 1992. Structural basis of blue-colour development in flower petals from *Commelina communis*. *Nature* 358:515–518.

Kress, W.J. 1983. Self incompatibility in Central American *Heliconia*. *Evolution* 37:735–744.

Kuanprasert, N., A.R. Kuehnle and C.S. Tang. 1998. Floral fragrance compounds of some *Anthurium* (Araceae) species and hybrids. *Phytochem.* 49(2):521–528.

Kuehnle, A.R. 1996. *Anthurium* cut flower breeding and economics. University of Hawaii at Manoa, College of Tropical Agriculture & Human Resources. *Research Extension Series* 0271–9916 :165.

Kuehnle, A.R. and N. Sugii. 1992. Transformation of *Dendrobium* orchid using particle bombardment of protocorms. *Plant Cell Rep.* 11:484–488.

Kuehnle, A.R., F.C. Chen and J.M. Jaynes. 1993. Engineering bacterial blight resistance into *Anthurium*. *Proc. EUCARPIA Ornamentals*, San Remo, Italy. pp. 127–129.

Lande, R. and D.W. Schemske. 1985. The evolution of self-fertilization and inbreeding depression in plants. I. Genetic models. *Evolution* 39:24–40.

Langton, F.A. and K.E. Cockshull. 1979. Screening chrysanthemums for leaf number in long days. *Research Report 1978*, Glasshouse Crops Research Institute, Littlehampton, W. Sussex, UK. pp. 177–186.

Laws, N. 1999. Rose open houses in France. *FloraCulture International* 26:28–30.

Lawson, J. and W. Williams. 1976. Environmental and genetic effects of pseudocompatibility in *Brassica oleracea* in relation to the production of hybrid seed. *Jour. Hort. Sci.* 51:359–365.

Levin, D.A. and Z. Bulinska-Radomska. 1988. Effects of hybridization and inbreeding on fitness in *Phlox*. *Amer. Jour. Bot.* 75:1632–1639.

Lewis, D. 1949. Structure of the incompatibility gene. 2. Induced mutation rate. *Heredity* 3:339–355.

Lewis, D. and L.K. Crowe. 1954. Structure of the incompatibility gene. 4. Types of mutations in *Prunus avium* L. *Heredity* 8:357–363.

Liedl, B.E. and N.O. Anderson. 1986a. Incompatibility hermeneutics: the determination of cutoff values. *Plant Cell Incompatibility Newsletter* 18:6–12.

Liedl, B.E. and N.O. Anderson. 1986b. SIGMAS-1: a software programme for self-incompatibility genetic modeling. *Jour. Heredity* 77:480.

Liedl, B.E. and N.O. Anderson. 1993. Reproductive barriers: identification, uses, and circumvention. *Plant Breed. Rev.* 11:11–154.

Liedl, B.E. and N.O. Anderson. 1994. Statistical differentiation between self incompatibility and pseudo-self compatibility in *Petunia hybrida* Hort. using female and male coefficient of crossability. *Sex. Plant Repro.* 7:229–238.

Lin, H.S., M.J. De Jeu and E. Jacobsen. 2000. Development of a plant regeneration system applicable for gene transformation in the ornamental *Alstroemeria*. *Acta Horticulturae* 508:61–65.

MacKey, J. 1970. Significance of mating systems for chromosomes and gametes in polyploids. *Hereditas* 66:165–176.

Martin, F.W. 1972. Inheritance of self-incompatibility versus self-fertility in the sweet potato. *Incompatibility Newsletter* 1:17–19.

Mather, K. 1943. Polygenic inheritance and natural selection. *Biol. Rev.* 18:32–64.

Meerow, A. 2002. Amaryllis plant named 'Rio'. *US Plant Patent No. 12,633*. US Patent and Trademark Office, Washington, DC.

Meynet, J. 1978. Obtention of seed-propagated gerbera varieties. *In*: *Proc. EUCARPIA Meeting on Carnation and Gerbera*. Alassio, pp. 203–210.

Miller, N.F. 1954. A preliminary study of rose fragrance. *American Rose Annual*, pp. 79–89.

Mitiouchkina, T.Y. and S.V. Dolgov. 2000. Modification of chrysanthemum plant and flower architecture by *ROLC* gene from *Agrobacterium rhizogenes* introduction. *Acta Horticulturae* 508:163–169.

Mol, J., A. Stuitje, A. Gerats, A. Krol and R. van der Jorgensen. 1989. Saying it with genes: molecular flower breeding. *Trends in Biotech.* 7(6):148–153.

Mol, J., T.A. Holton and R.E. Koes. 1995. Floriculture: genetic engineering of commercial traits. *Trends in Biotech.* 13(9):350–355.

Mol, J., E. Grotewold and R. Koes. 1998. How genes paint flowers and seeds. *Trends in Plant Sci.* 3:212–217.

Mulford, F.L. 1937. Results of selfing twenty-four early blooming chrysanthemums. *Proc. Amer. Soc. Hort. Sci.* 35:818–821.

Murashige, T. and F. Skoog. 1962. A revised medium for rapid growth and bioassays with tobacco tissue culture. *Physiologia Plantarum* 15:473–497.

Mutschler, M. and E. Cobb. 1985. Crosses of *Lycopersicon pennellii* and *L. esculentum* using *L. pennellii* as the female parent. *Tomato Gen. Coop. Report* 35:14.

Nakano, M. and M. Mii. 1993. Somatic hybridization between *Dianthus chinensis* and *D. barbatus* through protoplast fusion. *Theor. App. Gen.* 86:1–5.

Niwa, T. 1931. Pollination and self incompatibility in chrysanthemum (Outlines). *Jap. Assoc. Adv. Sci. Rep.* 6:479–487.

Norstog, K. 1973. New synthetic medium for the culture of premature barley embryos. *In Vitro* 5(4):307–308.

Oka, N., H. Ohishi, T. Hatano, M. Hornberger, K. Sakata and N. Watanabe. 1999. Aroma formation during flower opening in *Rosa damascena* Mill. *Zeit. fur Naturforsch. Teil. C: Biosci.* 54: 889–895.

Oud, J.S.N., H. Schneiders, A.J. Kool and M.Q.J.M. van Grivsven. 1995. Breeding of transgenic *Petunia hybrida* varieties. *Euphytica* 84: 175–181.

Palloix, A., Y. Herve, R.B. Know and C. Dumas. 1985. Effect of carbon dioxide and relative humidity on self incompatibility in cauliflowers, *Brassica oleracea. Theor. App. Gen.* 70: 628–633.

PanAmerican Seed Co. 1999. *Product Information Guide.* W. Chicago, Illinois.

Pellegrineschi, A., J.P. Damon, N. Valtora, N. Paillard, and D. Tepfer. 1994. Improvement of ornamental characters, fragrance, and production in lemon-scented geranium through genetic transformation by *Agrobacterium rhizogenes. Biotech.* 12(1):64–68.

Peters, W.L. 2002. Statistical discrimination between pollen tube growth and seed set in establishing self incompatibility in *Gaura lindheimeri.* MS thesis, University of Minnesota, St Paul, Minnesota.

Poehlman, J.M. 1979. *Breeding Field Crops.* AVI, Westport, Connecticut.

Pouw, A.A. 2002. Miniature rose plant named 'Ruiovat'. *US Plant Patent No. 12,657.* US Patent and Trademark Office, Washington, DC.

Price, H.J. 1976. Evolution of DNA content in higher plants. *Bot. Rev.* 42:27–52.

Price, H.J. 1988. DNA content variation among higher plants. *Annals of the Missouri Botanical Garden* 75:1248–1257.

Quarles, W. 1998. Disease resistant roses. *Common Sense Pest Control Quarterly* 14(2):4–6.

Raver, A. 1991. Flameless citronella, or how plants may join the war on mosquitoes. *New York Times,* July 25, 1991:B1, B5.

Reimann-Philipp, R. 1969. *Die Zuchtung der Blumen.* Paul Parey, Berlin.

Reimann-Philipp, R. 1983. Heterosis in ornamentals. *In*: R. Frankel (ed.). *Heterosis. Monographs on Theor. App. Gen.* 6:234–259.

Reimann-Philipp, R. and G. Fuchs. 1971. F1-hybriden bei *Ageratum houstonianum*, ein

neues Verfahren für ihre Erzeugung. *Zeit. Zierpflanzenbau* 20:433–444.

Reiser, W. and C.M. Ziessler. 1989. Die Uberwindung postgamer Inkompatibilität bei Freesia-Hybriden. Tag,-Ber., Akad. Landwirtsch.-Wiss. DDR, Berlin 281:135–138.

Robacker, C.D. and P.D. Ascher. 1978. Restoration of pseudo-self compatibility (PSC) in derivatives of a high-PSC × no-PSC cross in *Nemesia strumosa* Benth. *Theor. App. Gen.* 53:135–141.

Robacker, C.D. and P.D. Ascher. 1982. Effect of selection for pseudo-self compatibility in advanced inbred generations of *Nemesia strumosa* Benth. *Euphytica* 31:591–601.

Roggen, H. 1982. Breaking self incompatibility in *Brassica oleracea* with high frequency alternating electric current. *Incompatibility Newsletter* 14:92–95.

Ronald, W.G. 1974. Genetic and high temperature control self compatibility in *Chrysanthemum morifolium* Ramat. PhD dissertation, University of Minnesota, St Paul, Minnesota.

Ronald, W.G. and P.D. Ascher. 1975. Effects of high temperature treatments on seed yield and self incompatibility in chrysanthemum. *Euphytica* 24:317–322.

Ruiz, T. and M.T.K. Arroyo. 1978. Plant reproductive ecology of a secondary deciduous tropical forest in Venezuela. *Biotropica* 10:221–230.

Sanders, M.E. and N.K. Ziebur. 1963. Artificial culture of embryos. *In*: P. Maheshwari (ed.). *Recent Advances in the Embryology of Angiosperms.* Int'l. Soc. Plant Morphol. University of Delhi, Delhi, India. pp. 297–325.

Satory, M. 1986. Tagneutrale, F1-Minichrysanthemum fuer Beet- und Topfkulture als 'Nach-Urlaubs-Geschaeft' bis in den Spaetherbst geiegnet. *Zierpflanzen bau Gartenbautechnik* 26:86.

Schemske, D.W. and R. Lande. 1985. The evolution of self-fertilization and inbreeding depression in plants. II. Empirical observations. *Evolution* 39:41–52.

Schmidt, D. and H.T. Erickson. 1981. Inheritance of several plant and floral characters in *Streptocarpus. Jour. Amer. Soc. Hort. Sci.* 106: 170–174.

Schoen, D.J. 1983. Relative fitness of selfed and outcrossed progeny in *Gilia achilleifolia* (Polemoniaceae). *Evolution* 37:292–301.

Schuster, W. 1970. Effect of continued inbreeding from the 1 to 18 on various characters of sunflowers. *Zeit. für Pflanzenzuch.* 64: 310–334.

Scott, E.L. 1957. *The Breeder's Handbook.* National Chrysanthemum Society Handbook No. 4. Dancey Printing Co., Bogota, New Jersey.

Snow, A.B. 2002. Begonia plant named '77WL peach'. *US Plant Patent No. 12,786*. US Patent and Trademark Office, Washington, DC.

Sommer, H. and H. Saedler. 1986. Structure of the chalcone synthase gene of *Antirrhinum majus*. *Mol. Gen. Genet.* 202:429–434.

Sorenson, F. 1969. Embryogenic genetic load in coastal Douglas fir, *Pseuotsuga menziesii var. menziesii. Amer. Naturalist* 103:389–398.

Souq, F., P. Coutos-Thevenot, H. Yean, G. Delband, Y. Maziere, J.P. Barbe and M. Boulay. 1996. Genetic transformation of roses, two examples: one on morphogenesis, the other on anthocyanin biosynthetic pathway. *Acta Horticulturae* 424:381–388.

Sparnaaij, L.D. 1979. Polyploidy in flower breeding. *HortSci.* 14(4):496–499.

Sparnaaij, L.D. 1991. Breeding for disease and insect resistance in flower crops. *In*: J. Harding, F. Singh and J.N.M. Mol (eds). *Genetics and Breeding of Ornamental Species*. Kluwer Academic Publishers, Dordrecht, the Netherlands. pp. 179–211.

Spelt, C., F. Quattrocchio, J. Mol and R. Koes. 2000. *Anthocyanin 1* of *Petunia* encodes a basic helix-loop-helix protein that directly activates transcription of structural anthocyanin genes. *Plant Cell* 12:1619–1631.

Stephenson, A.G. and R.I. Bertin. 1983. Male competition, female choice and sexual selection in plants. *In*: L.E. Real (ed.). *Pollination Biology*. Academic Press, New York. pp.109–149.

Stimart, D. and P.D. Ascher. 1974. Culture medium suitable for growing small excised lily embryos. *North American Lily Society Yearbook* 37:77–84.

Stout, A.B. 1916. Self- and cross-pollinations in *Cichorium intybus* with reference to sterility. *Memoirs of the New York Botanic Garden* 6:334–455.

Sullivan, J.G. and F.A. Bliss. 1983a. Expression of enhanced seed protein content in inbred backcross lines of common bean. *Jour. Amer. Soc. Hort. Sci.* 108:787–791.

Sullivan, J.G. and F.A. Bliss. 1983b. Genetic control of quantitative variation in phaseolin seed protein of common bean. *Jour. Amer. Soc. Hort. Sci.* 108:782–787.

Takahashi, H. 1973. Genetical and physiological analysis of pseudo-self compatibility in *Petunia hybrida. Jap. Jour. Gen.* 48:27–33.

Tanksley, S.D. and T.J. Orton. (eds) 1983. *Isozymes in Plant Genetics and Breeding. Part A*. Elsevier, The Netherlands.

Templeton, A.R. and B. Read. 1984. Factors eliminating inbreeding depression in a captive herd of Speke's Gazelle (*Gazella spekei*). *Zoo Biology* 3:177–199.

Tian, H.W. and S.D. Russell. 1997. Micromanipulation of male and female gametes of *Nicotiana tabacum*. II. Preliminary attempts for *in vitro* fertilization and egg cell culture. *Plant Cell Rep.* 16:657–661.

USDA 2002. *Floriculture Crops 2001 Summary*. United States Department of Agriculture, National Agricultural Statistics Service, Washington, DC.

Vainstein, A., E. Lewinsohn, E. Pichersky and D. Weiss. 2001. Floral fragrance: new inroads into an old commodity. *Plant Physiol.* 127(4): 1383–1389.

Van Creij, M.G.M. 1997. Interspecific hybridization in the genus *Tulipa* L. PhD dissertation, Wageningen Agricultural University, The Netherlands.

Van Tuyl, J.M. 1997. Interspecific hybridization of flower bulbs: a review. Proc. VIIth Int'l Symp. on flower bulbs, Vol. II, Herzliya, Israel. *Acta Horticulturae* 430:465–476.

Van Tuyl, J.M. and M.J. De Jeu. 1997. Methods for overcoming interspecific crossing barriers. *In*: V.K. Sawhney and K.R. Shivanna (eds). *Pollen Biotechnology for Crop Production and Improvement*. Cambridge University Press, New York. pp. 273–292.

Van Tuyl, J.M., M.P. Van Diën, M.G.M. Van Creij, T.C.M. Van Kleinwee, J. Franken and R.J. Bino. 1991. Application of *in vitro* pollination, ovary culture, ovule culture and embryo rescue for overcoming incongruity barriers in interspecific *Lilium* crosses. *Plant Sci.* 74: 115–126.

Van Tuyl, J.M., H. Meijer and M.P. Van Diën. 1992. The use or oryzalin as an alternative for colchicines *in vitro* chromosome doubling of *Lilium* and *Nerine. Acta Horticulturae* 325: 625–630.

Van Tuyl, J.M., M.G.M. Van Creij, W. Eikelboom, D.M.F.J. Kerckhoffs and B. Meijer. 1993. New genetic variation in the *Lilium* and *Tulipa* assortment by wide hybridization. *In*: T. Schiva and A. Mercuri (eds). *Proc XVIIth EUCARPIA Symposium*, San Remo, Italy. pp. 141–149.

Van Tuyl, J.M., A. van Dijken, H.S. Chi, K.B. Lim, S. Villemoes and B.C.E. van Kronenburg. 2000. Breakthroughs in interspecific hybridization of lily. *Acta Horticulturae* 508:83–88.

Vazquez-Thello, A., L.J. Yang, M. Hidaka and T. Uozumi. 1996. Inherited chilling tolerance in somatic hybrids of transgenic *Hibiscus rosa-sinensis* × transgenic *Lavatera thuriengiaca* selected by double-antibiotic resistance. *Plant Cell Rep.* 15:506–511.

Wan, Y., D.R. Duncan, A.L. Rayburn, J.F. Petolino and J.M. Widholm. 1991. The use of antimicrotubule herbicides for the production of doubled haploid plants from anther-derived maize callus. *Theor. App. Gen.* 81(2):205–211.

Wang, S-Y., R.D. Heins, W. Carlson and A. Cameron. 2000. Forcing perennials, species: *Hibiscus moscheutos* 'Disco Belle Mixed'. *In*: Greenhouse Grower Magazine and Michigan State University (eds). *Firing Up Perennials, The 2000 Edition.* G.G. Plus, Willoughby, Ohio. pp. 80–83.

Watanabe, K. 1977. Successful ovary culture and production of F_1 hybrids and androgenic haploids in Japanese *Chrysanthemum species. Jour. Heredity* 68:317–320.

Wellensiek, S.J. 1959. The effect of inbreeding in *Cyclamen. Euphytica* 8:125–130.

Wergin, W. 1936. Cyto-genetische Untersuchungen an *Petunia hybrida* Hort. *Zeitsch. Indukt. Abamm. Vererbungslehre* 71:120–155.

Werner, C.P., A.P. Setter, B.M. Smith, J. Kubba and M.F. Kearsey. 1989. Performance of recombinant inbred lines in Brussels sprouts (*Brassica oleracea var. gemmifera*). *Theor. App. Gen.* 77:527–534.

Westerdahl, B.B., D. Giraud, S. Etter, L.J. Riddle and C.A. Anderson. 1998. Problems associated with crop rotation for management of *Pratylenchus penetrans* on Easter lily. *Jour. Nematology* 30(4, suppl.):581–589.

Whitson, J., R. John and H.S. Williams. 1914. *Luther Burbank: His Methods and Discoveries and their Practical Application.* Vol. 2. Luther Burbank Press, New York.

Wiens, D. 1984. Ovule survivorship, brood size, life history, breeding systems, and reproductive success in plants. *Oecologia* 64:47–53.

Wilkins, H.F. 1985. Asexually propagated bedding plants. *In*: J.W. Mastalerz and E.J. Holcomb (eds). *Bedding Plants III: A Manual on the Culture of Bedding Plants as a Greenhouse Crop.* Pennsylvania Flower Growers. pp. 471–501.

Williams, P.H. and C.B. Hill. 1986. Rapid-cycling populations of *Brassica. Science* 232:1385–1389.

Winsor, J.A., L.E. Davis and A.G. Stephenson. 1987. The relationship between pollen load and fruit maturation and the effect of pollen load on offspring vigour in *Curcubita pepo. Amer Naturalist* 129:643–656.

Yepes, L.M., V. Mittak, S.Z. Pang, C. Gonsalves, J.L. Slightom and D. Gonsalves. 1995. Biolistic transformation of chrysanthemum with the nucleocapsid gene of tomato spotted wilt virus. *Plant Cell Rep.* 14:11, 694–698.

Zagorski, J.S., P.D. Ascher and R.E. Widmer. 1983. Multigenic self incompatibility in hexaploid *Chrysanthemum. Euphytica* 32:1–7.

Zlesak, D.C., C.A. Thill and N.O. Anderson. 2002. Trifluralin-mediated chromosome doubling of *Rosa chinensis minima* Voss seedlings. On-site programme, XXVIth International Horticultural Congress, Toronto, Canada, 11–17 August, 2002, p. 459.

6 Factors Affecting Flowering in Ornamental Plants

John Erwin

Department of Horticultural Science, University of Minnesota, 1970 Folwell Avenue, St Paul, MN 55108, USA

Introduction

Flowering, or the transition from leaf to flower production by a meristem, can be stimulated by internal or external cues. Internal or autonomous cues include flowering responses that result from factors such as plant age or size. In contrast, external cues include flowering responses that result from environmental stimuli such as day/night length, low temperature, fire and/or the presence of water. The development of internal cues to control flowering enables plants to regulate flowering when a plant is at an optimal size or age. The development of external cues allows for optimal timing of flowering during a year to ensure successful pollination and seed development prior to inclement conditions as well as for synchronized flowering within a population. Such synchrony is essential to enable species to have successful cross-pollination.

Steps in the flowering process as well as autonomous and external cues that result in flowering and how they can be applied by flower breeders and plant physiologists will be discussed in this chapter. Both basic and applied literature will be reviewed. The chapter will distinguish between factors that result in flowering as a result of induction versus the breaking of dormancy. In addition, detailed information on specific conditions that promote flowering of a number of herbaceous ornamental species will be presented to enable a grower and/or scientist to induce flowering at any desired time. Much of this specific information is very recent. In addition, data on classification of species into irradiance response groups (see below) has recently been introduced (Erwin and Warner, 2002; Mattson, 2002).

The Flowering Process and Terminology

The processes whereby events in a shoot meristem are altered in such a way as to produce flowers as opposed to leaves are collectively referred to as 'floral evocation'. The actual signal that results in evocation is often referred to as the 'floral induction' signal. The formation of flower buds after induction is referred to as 'flower initiation'. The process after flower initiation until anthesis is referred to as 'flower development'. 'Anthesis' refers to the shedding of pollen by the stamen. It must be noted that flower opening can occur prior to or after anthesis.

A meristem is 'competent' to flower when it can respond in the expected manner when given an appropriate developmental signal. A meristem is referred to as 'determined' if it follows the same developmental programme even after it is removed from a source of environmental or biochemical stimulus. In some cases the 'expression' of flowering can be delayed until a second developmental signal is received. For instance, some species require a succession of two different photoperiods for successful evocation. Similarly, some species require a cold temperature treatment followed by a specific photoperiod for successful evocation.

Often, flower induction and initiation have occurred but flower development is interrupted. Such a suspension in flower development is, in some cases, referred to as 'dormancy'. In most cases, a single environmental event or a series of environmental cues, must occur for dormancy to be broken. Dormancy is common for flowering in spring-flowering woody species and ephemerals.

The complexity of control mechanisms associated with flowering is enormous. It is amazing that flowering can occur at all, given the sensitivity of each step and the possibility for interruption during the flowering process, which is significant. If a single factor is promotive, it is possible for flowering to be inhibited if other conditions are not met. Yet, plants have built a significant redundancy into the flowering process to ensure that a number of cues can enable flowering to compensate for environmental fluctuations. The inherent redundancy in the flowering system, as well as ways in which several environmental cues can result in the same flower induction, is apparent in recent models.

Autonomous Regulation of Flowering

Organisms pass through a series of developmental phases during growth and maturation. Among animal species, developmental phase changes are ubiquitous throughout the organism. In contrast, among plants such phase changes take place only in the shoot apical meristem and flowering is possible only if a meristem is competent to flower, and it receives an inductive signal.

Whether a meristem is competent to flower is dependent on the phase which that meristem is in. The transition between phases in development is referred to as 'phase change'. Put simply, a plant passes through three phases: juvenile, adult vegetative (competent) and adult reproductive (determined) phases. The critical difference between the juvenile and adult phases is inherent in the ability of that meristem to successfully flower, which is observed only in the adult phases. The critical difference between the adult vegetative and adult reproductive phase is simply whether that meristem has or has not been evoked to flower or is determined.

Whether a plant is competent or determined with respect to flowering can be evaluated using grafting experiments (McDaniel et al., 1992). If a non-flowering scion is grafted on to an induced rootstock and that scion flowers, then the scion must have been competent to respond to the floral stimulus. In contrast, if the scion does not flower, it is not yet competent. If a scion is grafted on to a juvenile rootstock and the scion flowers regardless, it is likely to be determined. For instance, Betula verrucosa J.F. Ehrh. tissues derived from the base of the tree (juvenile) grafted on to a rootstock remain juvenile or vegetative, i.e. the scion was not competent to flower (Longman, 1976). In contrast, tissues collected from the top of the flowering tree (mature) were competent to flower after 2 years.

It must be emphasized that the transition from juvenile to adult phases is a continuous process and not discontinuous. For instance, the ability to flower is a process and is transitional. Lunaria biennis L. (Wellensiek, 1958), Brussels sprouts (Stokes and Verkerk, 1951) and beets (Wellensiek and Hakkaart, 1955) pass through a clear juvenile to adult phase transition as evidenced by the increasing ability of cold temperatures to induce flowering as the plant ages rather than a sudden transition.

Once a plant attains the adult phase, there is an increasing tendency to flower as it ages, regardless of inductive conditions. For instance, Euphorbia pulcherrima Willd. ex Klotzsch 'stock', or 'mother', plants will produce numerous successive populations of vegetative cuttings, i.e. can have an extended adult vegetative phase. However, the percentage of cuttings that spontaneously produce floral organs (cyathia) increases with each successive generation of cuttings. As the stock plant gets older, cuttings taken from it are more likely to flower even under non-inductive conditions (Siraj-Ali *et al.*, 1990; Carver and Tayama, 1992). Similarly, the ability of celery plants to flower in response to a low temperature treatment increases as a plant grows older as evidenced by a shorter time for flower initiation and days to anthesis after completion of a cold treatment (4 weeks at 8°C) (Pawar and Thompson, 1950).

The transition from juvenile to adult phases in woody species and some herbaceous species can be accompanied by changes in morphology, phyllotaxy, thorniness, rooting capacity and/or leaf retention, i.e. a phenotypic change. For instance, leaf arrangement in *Antirrhinum majus* L. changes from opposite during the juvenile phase to alternate during the adult phase. Leaf morphology changes dramatically in the aquatic plant *Hippuris vulgaris* (common marestail) as the plant transitions from a juvenile to an adult phase (Goliber and Feldman, 1989). *Hedera helix* L. rooting occurs readily when tissue is in the juvenile phase but is nearly 'non-existent' in adult phases (Poethig, 1990).

Because there is a temporal order to the phase changes during plant development, there can be a spatial gradient in phases along a shoot. With some species in which the juvenile phase is relatively short there may simply be a few juvenile structures at the base of the plant. In contrast, on species in which there is a prolonged juvenile period, a significant portion of the plant may be composed of juvenile structures. Meristematic regions then switch to adult phases and, as a result, the adult or reproductive structures are often located on the peripheral portions of a plant. This can be commonly observed in *Fagus sylvatica* L. and *Quercus* spp. where juvenility is associated with leaf retention during the winter and can be observed on the base of the tree.

Because many herbaceous or annual species do not exhibit a morphological or phenotypic change after transitioning from juvenile to adult phases, we often identify the phase change from juvenile to adult phase based on another developmental marker: the number of leaves that have unfolded since germination, as leaf number is a direct measure of a plant's developmental age and is temperature dependent. For this reason, identification of competence to flower based on leaf number is commonly used as a general phase change indicator in commercial floriculture production.

Once the adult phase is achieved, it is relatively stable. For instance, the juvenile or adult phase is maintained through vegetative propagation as well as grafting (Longman, 1976). For instance, *H. helix* cuttings taken from juvenile tissues will remain juvenile after propagation and vice versa (Poethig, 1990). However, in rare cases, non-optimal conditions and/or stress can cause rejuvenation, or reversion to the juvenile phase. For instance, low irradiance conditions can result in reversion from an adult to a juvenile phase. However, the incidence of such reversion decreases as the length of time since the transition from juvenile to adult phase increases. It is noted also that sexual or apomictic reproduction will result in regeneration of the juvenile phase (Hackett, 1980).

A number of factors affect the transition between phases in plants. Some of those factors that impact the transition from the juvenile phase to an adult phase are outlined below. In particular, the impact of species, meristem size and environmental factors are briefly mentioned.

- **Species**. Species vary considerably in the length of the juvenile phase (Tables 6.1 and 6.2). In general, the 'longer-lived' a species is, the longer the juvenile phase. For instance, some nut-trees can have a 25-year juvenile period compared with a 4-month juvenile period (14 leaves) for some herbaceous perennial species or a 2-week (3–4 leaf) juvenile phase for some annuals. In some cases, as for *Raphanus sativus* L., there is initially no juvenile phase, as imbibed seed can be vernalized (Engelen-Eigles and Erwin, 1997). Variation in woody plant juvenile phase lengths are shown in Table 6.1.

  In contrast to woody plant species, many herbaceous species' juvenile phase lengths are identified based on leaf

Table 6.1. Juvenile phase length across woody plant species.

Species	Juvenile period length
Rose (*Rosa × hybrida*)	20–30 days
Grape (*Vitis* spp.)	1 year
Orchid spp.	4–7 years
Apple (*Malus* spp.)	4–8 years
Orange (*Citrus* spp.)	5–8 years
English ivy (*Hedera helix*)	5–10 years
Redwood (*Sequoia sempervirens*)	5–15 years
Sycamore maple (*Acer pseudoplatanus*)	15–20 years
English oak (*Quercus robur*)	25–30 years
European beech (*Fagus sylvatica*)	30–40 years

Adapted from Clark, 1983; Goh and Arditti, 1985.

Table 6.2. Juvenile period lengths as identified by leaf number.

Species	Leaf number at which plants become competent to flower
Aquilegia 'McKana's Giant'	12 nodes
Aquilegia 'Fairyland'	15 nodes
Calceolaria herbeohybrida	5 nodes
Callistephus chinensis	4 nodes
Coreopsis grandiflora 'Sunray'	8 nodes
Gaillardia × grandiflora 'Goblin'	16 nodes
Heuchera sanguinea 'Bressingham'	19 nodes
Lavandula angustifolia 'Munstead'	18 nodes
Rudbeckia fulgida 'Goldstrum'	10 nodes

Adapted from Yuan, 1995; Whitman, 1995; Sheldron and Weiler, 1982.

number. However, the developmental time of the juvenile period differs with herbaceous species as with woody plants as shown in Table 6.2 (Cameron *et al.*, 1996).

- **Meristem size**. The association between leaf number and phase transition may be related to other morphological characteristics such as meristem size. There is an association between meristem size and transition out of the juvenile phase where a plant may progress from juvenile to adult phases after the meristem increases beyond a minimum diameter meristem size which increases as leaf number increases. For instance, *Chrysanthemum × morifolium* Ramat. flower primordia are not initiated until after a minimum apex size has been attained (Cockshull, 1985).

Similarly, Singer and McDaniel (1986) showed that tobacco plants produced 37 leaves from terminal buds prior to flowering. After numerous grafting experiments across tobacco varieties, they concluded that the number of nodes a meristem produces prior to flowering was a function of the strength of the floral stimulus and the competence of the meristem to respond to that flowering signal. Pomologists have also noted that the passage of a tree out of the juvenile period may be associated with the attainment of a specific size (Visser and DeVries, 1970). However, more recent work has shown that *H. helix* phase change is independent of meristem size (Hackett and Srinivasani, 1983/1985).

However, this assumption of an association between meristem size and phase transition excludes any involvement of carbohydrate status and we know that carbohydrate status impacts leaf number below the flower or juvenile period phase length (see Table 6.4).

- **Environmental conditions**. Environmental conditions that retard growth can delay the transition from juvenile to adult phase. For instance, as noted before, exposure to low irradiance conditions can delay the transition from the juvenile to adult phase or cause reversion from the adult back to the juvenile phase. Such an association has led to a questioning of the role of carbohydrate availability in the transition from juvenile to adult phase. Such assumptions are borne out in that increasing irradiance can decrease the length of the juvenile phase developmentally in numerous herbaceous annuals as evidenced by a decrease in the leaf number below the first flower (see Irradiance section below and Table 6.4). For example, juvenile period length can be decreased as irradiance increases in *Pelargonium × hortorum* L.H. Bailey (Armitage and Tsujita, 1979). Similarly, numerous species' juvenile period length can be reduced by addition of supplemental lighting (Warner and Erwin, 2001b; Erwin and Warner, 2002; Table 6.4) or by providing conditions that promote growth (Poethig, 1990).

Physiological basis for transition from juvenile to adult phase

The physiological basis for the transition from juvenile to the adult phase is not well understood. Application of plant growth regulators can hasten or delay the transition from the juvenile to the adult phase or cause a reversion from the adult to the juvenile phase. For instance, application of GAs to *Cupressus arizonica* Greene caused male cone formation (indicator of transition to adult phase) when plants were only 2 months old (Pharis and King, 1985). Similarly, other treatments that result in accumulation of endogenous GAs in conifers can reduce the juvenile phase length.

In contrast to conifers, application of GAs to *H. helix* can cause a reversion from an adult to the juvenile phase (Hackett and Srinvasani, 1985). Application of Amo 1618 (GA-synthesis inhibitor) was not able to cause a phase change from juvenile to adult in *H. helix* (Frydman and Wareing, 1974). However, ancymidol (GA-synthesis inhibitor) prevented the spontaneous reversion to the juvenile state in low light in *H. helix* (Rogler and Hackett, 1975). Similar inhibition of transition from juvenile to adult phases following GA application has been noted on *Citrus* (Cooper and Peynado, 1958), deciduous fruit trees (Luckwill, 1970) and *Ipomoea caerula* L. (Njoku, 1958).

External Regulation of Flowering

There are four primary external environmental cues that affect flowering in plants:

photoperiod, temperature, irradiance and stress (fire, water). In addition, lack of stress or supra-optimal levels of nutrients or saturating water levels can also impact progression towards flowering. Of these environmental cues, photoperiod, temperature and the presence of water, in particular, allow plants to synchronize flowering with the seasons. Fire can be an indicator of seasonal transition but is less precise. In general, horticultural manipulation of flowering involves manipulation of photoperiod, temperature and/or irradiance and will be discussed in more detail below.

When photoperiod impacts flowering, a plant is said to be 'photoperiodic'. Photoperiodism is primarily associated with longer-lived species that survive for at least a single growing season. Development of such species can often depend on photoperiod to synchronize flowering to ensure that flowering occurs at a specific time of year in order to allow successful seed set and maturation along with cross-pollination.

In contrast to photoperiodism, prolonged exposure to temperatures between 0°C and 16°C can also synchronize flowering. For example, flowering in biennial and perennial species will often have a cool temperature requirement (2°C–4°C) for successful flower induction (often 6–12 weeks). Such low temperature induction of flowering is referred to as 'vernalization'. Presumably, a vernalization requirement ensures that a plant will not flower until after winter has occurred in temperate climates.

Further, some species exhibit a dual requirement for successful flowering, i.e. they may require a prescribed sequence of inductive conditions such as SD–LD (short day–long day) or LD–SD. In some cases, as with many perennial species in temperate climates, a plant may have a vernalization requirement followed by a LD requirement. Lastly, daylength and vernalization can act synergistically as in the case of *Lilium longiflorum* or *Apium graveolens* L. 'Florida' (Pressman and Negbi, 1980).

A number of other environmental factors such as irradiance, light quality and/or water stress can impact flowering and/or interact with vernalization and/or photoperiodic induction. These factors appear to be associated more with whether conditions are optimal for flowering under inductive conditions (growth) than with the inductive process itself. For instance, low irradiance can decrease flowering of the day-neutral *Rosa* × *hybrida* L. by increasing the occurrence of 'blind' shoots, i.e. aborted flowers (Nell and Rasmussen, 1979). In contrast, high irradiance can substitute for a photoperiod requirement entirely in *Hibiscus moscheutos* L. (Warner and Erwin, 2003b). Additionally, high temperatures can inhibit flowering induced by photoperiod or vernalization. The following sections elaborate on the primary environmental cues for flowering. In addition, the physiological bases for each response are discussed.

Photoperiodism

Photoperiodism refers to the ability of an organism to detect daylength (or more correctly night length). Photoperiodic events can be observed in animals and plants. In animals, hibernation, egg laying and migration are all controlled by photoperiod. In plants, photoperiodism can regulate growth processes as different as flowering, tuber formation, onset of dormancy and rhizome formation.

Photoperiodism in flowering was initially discovered at the US Department of Agriculture laboratories in Beltsville, Maryland by Wightman Garner and Henry Allard in the 1920s. Garner and Allard (1920) noted that a mutant variety of tobacco (Maryland Mammoth) did not flower in the field during the summer but did flower in the greenhouse during the winter. They subsequently covered outdoor field-grown plants during the summer with black cloth to artificially shorten the daylength; this treatment resulted in flowering plants. They concluded that day length controlled when plants flower.

Photoperiodism was not applied in the commercial floriculture industry until the 1930s–1950s when Gus Poesch noted that street lights outside a chrysanthemum facility inhibited flowering (G. Poesch, personal communication). The recognition of the importance of daylength in that chrysanthemum facility resulted in the application of photoperiod manipulation in greenhouse production and resulted in a year-round flowering chrysanthemum industry as well as extended flowering seasons for many other potted plants and cut flowers (Poesch, 1931). Such application of photoperiod to bedding plant production is in its infancy and is a current primary focus of our research programme at the University of Minnesota.

Photoperiodic response groups

Photoperiodic flowering responses can be divided into several different response groups: short-day plants, long-day plants, day-neutral plants, intermediate-day plants and ambiphotoperiodic plants (Thomas and Vince-Prue, 1997). What characterizes each group is outlined below:

- *Short-day plants* – those species in which night length must be greater than some critical length. Generally species from low latitudes (i.e. coffee, cotton and rice) or species which flower in the late summer, such as chrysanthemum (Cockshull, 1984).
- *Long-day plants* – those species in which night length must be less than some critical length. Generally species from high latitudes, such as temperate grasses (Deitzer, 1984).
- *Day-neutral plants* – flower induction is not affected by night length. Usually species with a wide latitudinal distribution, such as potato and tomato (Halevy, 1984).
- *Intermediate-day plants* – those species that require a night length between 12 and 14 hours.
- *Ambiphotoperiodic plants* – those species in which flower induction occurs under long or short nights but not intermediate nights.

Subdivisions within response groups

Within short- and long-day plant groups, there are plants that have a facultative (quantitative) or obligate (qualitative) response. Species that exhibit a 'facultative', or quantitative, response will flower under any photoperiod; however, flowering is hastened under the described photoperiod. In contrast, species that exhibit an obligate, or qualitative, response will not flower unless they receive the described photoperiod. Examples of common herbaceous annual species and their corresponding photoperiodic classifications are shown in Table 6.4.

Species within genera can vary widely in their photoperiodic requirement. For instance, Warner and Erwin (2003b) showed that all photoperiodic response groups were evident in a group of *Hibiscus* spp. studied. In addition, early work by Cumming (1969) showed that *Chenopodium rubrum* L. strains collected from different latitudes possessed a wide range of photoperiodic requirements for flowering.

Photoperiodism in floriculture crops

Many herbaceous species grown for spring crop production in temperate climates are photoperiodic. Shillo (1976) showed flowering of *Limonium sinuata* Mill. was delayed 6 weeks and leaf number below the first flower was increased when plants were grown under ambient light (10–12 h photoperiod) versus ambient light plus a 4-h night interruption with incandescent lights; *L. sinuata* was found to be a facultative long-day plant. *Eschscholzia californica* Cham. flowered only when grown under 9 h ambient light plus a 4-h night interruption (3–$4 \ \mu mol/m^2/s$ from incandescent lamps); *E. californica* was classified as an obligate long-day plant by Lyons and Booze-Daniels (1986). Keatinge *et al.* (1998) noted that *Dolichos lablab* L. flowered after 69 and 172 days when grown under short day (11.5 h) and long day (14.5 h) photoperiods, respectively (leaf number below the first flower not reported); *D. lablab* was classified as a facultative short-day plant. Armitage and Garner (1999) showed *Catananche caerulea* L. Per. was an obligate long-day plant, flowering after 164 days when grown under long days, but did not flower when grown under short days. Recent screening experiments have resulted in the characterization of >50

commercial floriculture or herb species into photoperiodic response groups (Erwin and Warner, 2002; Mattson, 2002).

Flowering response to photoperiod can vary with cultivar. For example, Ross and Murfet (1985) describe three flowering classes of *Lathyrus odoratus* L.: a day-neutral (winter) flowering group, and long-day groups exhibiting a facultative (spring) or obligate (summer) long-day response. Such variation in photoperiodic responses among cultivars is also obvious in *Salvia splendens* and *Petunia × hybrida* cultivars (J. Erwin, personal observation). In particular, introduction of new germplasm (often wild in origin) can often increase the photoperiod requirement of nearly day-neutral existing cultivars. Perhaps the most obvious recent example is the increase in long-day requirement within *P. × hybrida* with the introduction of the 'Wave' and 'Fantasy' series of petunias in recent years (J. Erwin, personal observation; Erwin *et al.*, 1997).

Length of inductive conditions required for flower induction

The length of time for complete induction varies with plant age, temperature and irradiance. The tendency to flower increases as a plant ages regardless of whether it is in inductive conditions or not. Evidence of this is seen when cuttings harvested from poinsettias (*Euphorbia pulcherrima*) have a greater and greater proportion that exhibit 'splitting', a condition where buds initiate but do not develop fully (Siraj-Ali *et al.*, 1990; Carver and Tayama, 1992) as the stock or mother plant age increases. Similarly, there is an inverse relationship between plant age and the number of short days required for complete flower induction with *Pharbitis nil* Chois.

In addition to the natural tendency of a plant to eventually flower, the number of inductive cycles required for complete induction will often vary with the age of a plant. For instance, in *Lolium temulentum* (darnel ryegrass) time for complete induction decreases from 4 LD cycles on plants with 2–3 leaves to 1 LD cycle on plants with 6–7 leaves (Taiz and Zeiger, 1998). Similarly, *P. × hybrida* 'Purple

Wave' required 24 LD to flower when 10 days old and 18 LD to flower when 24 days old (Mattson, 2002). These observations suggest that the effect of photoperiod may be to accelerate a process that is already occurring at a slow pace even under non-inductive conditions.

Although there is no exact number of inductive cycles, often some estimate of time to completely induce a plant to flower is important from a commercial standpoint. For instance, an understanding of the time required to induce a floriculture crop to flower can identify how long a plant must stay under conditions with supplemental lighting or short days with a blackout curtain before it can be moved to non-inductive conditions. To this end, we have made some general observations and estimates. Many species appear to require approximately 21 days for complete induction when grown at 16°C–20°C. For example, Begonia × hiemalis complete flower induction requires 2–3 weeks (Karlsson and Heins, 1992). In some cases, flower induction and initiation can be achieved in 14 days; however, flower development may not be successful. Similarly, P. × hybrida 'Purple Wave' flowering will occur when plants are exposed to 20 LD (Mattson, 2002; 20°C, natural daylight conditions, St Paul, Minnesota).

The length of time required for complete flower induction can vary with temperature and irradiance. In general, time required for complete induction and initiation increases as temperature decreases and as irradiance decreases (Zrebiec and Tayama, 1990; Mattson, 2002).

Interaction of photoperiodic induction with temperature

Photoperiod can interact with temperature to affect flowering. It is common for some plant species to be photoperiodic at higher temperatures but exhibit a day-neutral response to photoperiod at lower temperatures. For instance, Begonia × hiemalis Fotsch is an obligate short-day plant at temperatures greater than 24°C (Sandved, 1969). In contrast B. × hiemalis is a facultative short-day plant (SDP) at temperatures >24°C but a day-neutral

plant (DNP) at temperatures lower than 24°C. Similarly, Calceolaria herbeohybrida Cav. Stout (facultative long-day plant (LDP)) flower induction will occur under SD conditions (Poesch, 1931). Campanula isophylla Moretti critical day-length is 14 h with temperatures of 15°C–21°C; however, it increases to 15 h when temperatures exceed 21°C (Heide, 1965). In most cases, this is due to an increase in the critical photoperiod required for flower induction (Heide and Runger, 1985).

It is also common with many SDP to exhibit delayed flowering under inductive SD conditions when night temperature exceeds 22°C. This phenomenon is commonly referred to as 'heat delay' in the commercial floriculture industry. High night temperature inhibition of flowering has been studied extensively with E. pulcherrima and C. × morifolium (SDP). In addition to these species, Gomphrena globosa L. exhibits a heat delay when night temperature is 25°C (Warner et al., 1997). In general, there is evidence that as night temperature increases above 22°C the length of an inductive long night for successful flowering increases. Therefore, heat delay can be overcome in some cases by providing a longer night to SDP grown with night temperatures higher than 22°C.

Physiological basis for photoperiodism

Plants perceive daylength by measuring the night length. Flowering short-day plants is night length, regardless of the daylength. Similarly, LDP also measure night length and only flower when the length of the night is less than some critical length regardless of the length of the photoperiod.

The importance of the night length in determination of photoperiodic responses with respect to flowering was demonstrated by early experiments by Hamner and Bonner (1938) and Hamner (1940) where flowering of Xanthium strumarium and Glycine max (L.) Merrill occurred only when night length exceeded 8.5 and 10 h, respectively. Night break lighting of X. strumarium or P. nil (both SDP) of as little as a few minutes prevented flowering even when the total night length was sufficient to promote flowering.

Similarly, night break lighting can result in stimulation of flowering in LDP. In general, the length of time of a night break required to inhibit flowering in SDP is considerably less than the time required to promote flowering in LDP.

When a night break occurs this greatly impacts the effectiveness of that break in either promoting or inhibiting flowering. In general, SDP and LDP are most sensitive to receiving a night interruption 8 h after the onset of darkness (Salisbury, 1963; Papenfuss and Salisbury, 1967; Vince-Prue, 1975). Harder and Bode (1943) showed that inhibition of *Kalanchoe blossfeldiana* flowering by light was most effective when the night interruption occurred between 5 and 7 h after the onset of darkness. In contrast, Salisbury and Bonner (1956) showed that *X. strumarium* was most sensitive to night interruption lighting 8 h after the onset of darkness. This observation is the basis for the commercial practice of interrupting the night from 10 p.m. to 2 a.m. with low-intensity lighting. Such an interruption is long enough to achieve a desired promotive or inhibitory effect and is delivered at the critical time of the night.

The observation of night interruption lighting led to the development of an action spectrum for photoperiodism. Different wavelengths of light varied in their ability to inhibit flowering of the SDP *X. strumarium* (Hendricks and Borthwick, 1954). Red light (640–660 nm) was shown to be most effective in inhibiting flowering. Perhaps most significant was the observation that the night-break action spectrum for inhibition of flowering in SDP (*Glycine* and *Xanthium*) was similar to that for stimulation of flowering in LDP (*Hordeum vulgare* and *Hordeum niger*), indicating a similar pigment as a photoreceptor (Borthwick *et al.*, 1948).

Interestingly, species can differ in the 'mixture' of light that is optimal for stimulation or inhibition of flowering. Whereas many SDP are most sensitive to red light (660 nm), stimulation of flowering in some LDP by a night interruption is increased when there is a mix of both red and far red light. It is for this reason that induction of flowering in the LDP *Fuchsia × hybrida* requires a light source that contains both red and far red light (incandescent lamps and non-fluorescent lamps).

The observation that light delivered in the middle of the night inhibited flowering of tobacco, a SDP, led to the discovery of phytochrome. Borthwick *et al.* (1952) showed that red light was the primary colour of light which inhibited flowering when delivered as a night break. In contrast, a subsequent exposure to far red light restored the flowering response and suggested a photoconversion process (Downs, 1956). Action spectra of the inhibition of flowering and an action spectrum for reversal of inhibition identify a peak at 660 nm and 720–740 nm, respectively (Saji *et al.*, 1983). There is also evidence for the involvement of a blue light receptor, although these data are not as clear as with phytochrome involvement (Millar *et al.*, 1995; Hicks *et al.*, 1996; Zagotta *et al.*, 1996).

Subsequent to this work was the significant finding that the red-light inhibition of flowering was reversible by far red light (720 nm). Hendricks *et al.* (1956) demonstrated that red/far red light effects were reversible. This reversibility was likely a result of two forms of a pigment: a red and far red light absorbing form. The pigment was named phytochrome (Butler *et al.*, 1959).

The variation in sensitivity of plants during the night to a night-interruption lighting is not well understood. Bunning (1948) suggested that plants exhibit diurnal cycles between a photophile and skotophile phase. The photophile phase was light-sensitive and the skotophile phase was not. Carr (1952) subsequently showed that the effectiveness of night interruption lighting in inhibiting flowering followed a periodic 24-h rhythm when the night was extended. Important to such a time-keeping mechanism is the observation that this rhythmicity is mostly temperature independent (Bunning, 1963; Takimoto and Hamner, 1964).

It is important to note that the effectiveness of a night interruption lighting in inhibiting flowering of SDP is affected by temperature. The effectiveness of a red-light night interruption decreases as temperature increases from 18.5°C to 25°C (Takimoto and Hamner, 1964). In addition, the length of night required to stimulate flowering in SDP is altered by temperature. The degree to which temperature affects this required night length for flower induction in SDP is species-specific (Salisbury and Ross, 1969). In addition, those plants that are sensitive to high temperatures in the night are most sensitive 8 h after the onset of darkness.

Photoperiodic stimulus

The photoperiodic stimulus is perceived in the leaves. When a single leaf of *X. strumarium* is exposed to short-day conditions when the rest of the plant is grown under long-day conditions, the plant will flower (Hamner and Bonner, 1938). Similarly, in a classic study by Zeevaart (1969), a single excised leaf of *Perilla crispa* (L.) Britt. exposed to short-day conditions was capable of repeatedly causing plants grown under long-day conditions to flower after being grafted to those plants. Interestingly, stimulation of flowering occurred even with a day-neutral plant when a single leaf of *G. max* 'Agate' was grafted on to the short-day *G. max* cultivar 'Biloxi'. Flowering was induced in Biloxi even when grown under long-day conditions (Heinze *et al.*, 1942). Also, leaves that were induced indirectly via grafting experiments can themselves also be donors and result in flowering of non-induced plants (Lona, 1946; Zeevaart and Lang, 1962; Wellensiek, 1966). Studies by Handro (1977) demonstrated that only *Streptocarpus nobilis* L. tissues derived from leaf tissue were capable of flowering *in vitro* under inductive conditions.

The degree to which an induced leaf can induce a non-induced donor plant to flower can vary with species and the degree to which the leaf has been induced. *X. strumarium* leaves will remain induced for a few days and are competent in inducing plants to which they are grafted (Carr, 1959). Imamura (1953) showed that a *P. nil* leaf was also induced for a few days following a single inductive night but could remain in an induced state for many days when the leaves received numerous SD compared with one. Similarly, *P. crispa* leaves could be inductive for a few days or many months depending on the number of inductive short days received

prior to removal from the donor plant (Zeevaart, 1957).

Grafting between species in a different photoperiodic class showed the stimulus to be similar across species. For instance, grafting the LDP *Hyoscyamus niger* L. to the SDP *Nicotiana tabacum* L. caused the SDP to flower (Melchers and Lang, 1941). Although more difficult, pairing a flowering SDP to a LDP has induced flowering in the LDP (Lang and Melchers, 1943; Zeevaart, 1957). In addition, flowering of parasites such as dodder can be stimulated when it is parasitizing flowering plants from any photoperiodic class (Frattianne, 1965). Together, the grafting experiments suggest the existence of a mobile flowering promoter which is produced by induced leaves. The term used to describe this promoter is 'florigen'. Although considerable effort has been spent trying to identify the nature of florigen, it remains unknown.

In contrast to the theory that photoperiodism is associated with a floral promoter, there is evidence that photoperiod can affect the synthesis of a floral inhibitor. Removal of all leaves of the LDP *H. niger* resulted in flowering, suggesting that the leaves produced an inhibitor and photoperiod (LD) acted to depress the synthesis of that inhibitor during LD (Lang and Melchers, 1943). Guttridge (1959) was able to show a similar induction of flowering following leaf removal on the SDP strawberry. Leaves from a photoperiodic tobacco (non-inductive conditions) are capable of inhibiting flowering of a day-neutral tobacco cultivar (Lang *et al.*, 1977).

Wellensiek (1959) identified an association between the day and degree of inhibition in SDP. Supportive of the flowering inhibitor concept is the observation that reduced irradiance during the day (Krumweide, 1960) or low temperatures (de Zeeuw, 1957; Wellensiek, 1959; Ogawa, 1960) can promote flowering under non-inductive photoperiods. However, this contradicts more recent work when higher irradiance levels overcome a LD requirement in *H. moscheutos* when grown under SD conditions (Warner and Erwin, 2003b).

Regardless of whether there is a chemical promoter/inhibitor, this substance is believed to be transferred in the phloem. Imamura and Takimoto (1955) estimated the timing of the flowering stimulus by developing a two-branched *P. nil* system and identifying the node at which flowering was induced. Subsequent experiments showed that translocation of the compound from the leaf to a receptive bud is comparable to or somewhat slower than that for the translocation of sugars (2–4 mm/h) (Evans and Wardlaw, 1966; King *et al.*, 1968).

Of all the known plant growth regulators, there is evidence that a promoter/inhibitor of flowering is associated with gibberellins. A 100 ppm application of GA_3 to a facultative SDP, *Cosmos*, was able to substitute for SD when plants were grown under an 18 h photoperiod (Wittwer and Bukovac, 1959). *L. temulentum* plants are committed to flower after exposure to a single long day (plants will not flower under SD). Excised terminal meristems exposed to a single long day will flower in tissue culture in the presence of GAs. Short-day-exposed meristems in tissue culture will not flower in the presence of GAs. Therefore, long days are required for *determination* of *L. temulentum* L. and GAs are required for the *expression* of the determined state (McDaniels and Harnett, 1996).

Vernalization

Vernalization is the low temperature induction of flowering in an imbibed seed or a growing plant (Chouard, 1960; Taiz and Zeiger, 1998). As with photoperiodism, plants must advance through the juvenile stage into an adult stage before they have the capacity to perceive a vernalization treatment. The length of time required for this varies. Winter annuals have the capacity to perceive a vernalization treatment as an imbibed seed. *R. sativus* similarly has the capacity to perceive a vernalization treatment as an imbibed seed.

Plants can be vernalized with a wide range of temperatures. In general, vernalization can occur at temperatures ranging from −6°C to 14°C (Chouard, 1960). There is an optimal temperature range of 6°C–10°C for most species (Lang, 1959; Chouard, 1960). Still other studies show that plants can be vernalized at temperatures between 5°C and 17°C. However, the length of time that plants must be exposed to cool temperatures increases as the temperature deviates from the common optimum of 4°C.

The time required for complete vernalization varies with species as well. For instance, *Lunaria biennis* L. requires 9 weeks (Wellensiek, 1958) and *Apium graveolens* L. (celery) and *Secale cereale* L. (winter rye) require 6 weeks (Friend, 1965; Ramin and Atherton, 1991). *Alstroemeria* L. hybrids commonly require 6 weeks at 5°C (Healy and Wilkins, 1981, 1982). In contrast, vernalization time can be as short as 6–8 days for *R. sativus* 'Chinese Jumbo Radish Scarlet' (Engelen-Eigles and Erwin, 1997; Erwin *et al.*, 2002). Some species exhibit sensitivity to vernalization immediately after seed is imbibed; a subsequent period of temperature insensitivity is followed by a period of increasing sensitivity to low temperature induction (Napp-Zinn, 1969).

The length of time required for vernalization is related to whether a species has an obligate or facultative vernalization requirement. Plants that require a short vernalization period (less than 30 days) such as *Brassica campestris pekinensis* L. (Suge, 1984) and *Arabidopsis thaliana* L. (Martinez-Zapater and Somerville, 1990; Bagnall, 1993) often have a facultative vernalization requirement and photoperiod is often the primary flower induction stimulus. In contrast, plants that require a long vernalization period often have an obligate vernalization requirement. A noted exception to this is *R. sativus* 'Chinese Jumbo Radish Scarlet' (Engelen-Eigles and Erwin, 1997).

Perception of vernalization

The site of perception of vernalization is the shoot tip. Studies conducted where different portions of the plant were cooled relative to the rest of the plant indicate that the shoot tip solely is capable of perception (Curtis and Chang, 1930; Metzger, 1988, 1996). Similarly, Gregory and de Ropp (1938) showed that excised winter rye embryos, or even a fragment thereof, can still be effectively vernalized in tissue culture. However, Wellensiek (1961, 1962) reported that young expanding leaves of *L. biennis* were capable of being vernalized. The commonality between these two tissues is the presence of active cell division. Therefore, active cell division is required for vernalization as isolated cells in tissue culture (originating from different locations on a mother plant) are capable of perceiving vernalization as well (Metzger *et al.*, 1992).

Once tissue has been effectively vernalized, all growth that develops from that tissue is vernalized (Schwabe, 1954). Regrowth of perennials following flowering results in a 'reversion' to a non-vernalized state in those buds (Schwabe, 1954). Unlike photoperiodism, the vernalization stimulus is not graft transmissible.

Devernalization

The period between the completion of a vernalization treatment and flower initiation can be divided into two phases. Phase I is a period immediately after vernalization when flower induction can be reversed/eliminated by exposure to warm temperatures, low irradiance and/or SD conditions. The reversal of vernalization by environmental conditions is referred to as 'devernalization' (Lang, 1965). Phase II is that period after Phase I when flower induction is stable and cannot be reversed.

Devernalization refers to the reversal of the vernalization process (Purvis and Gregory, 1952). Devernalization results from an interaction between the degree of vernalization and environmental conditions immediately after the vernalization treatment, i.e. during Phase I (see Table 6.3). The most common agent of devernalization is high temperature (25°C–30°C). However, devernalization by short-day conditions and/or low irradiance during or after vernalization has also been demonstrated in some species (Thomas and Vince-Prue, 1997). Devernalization treatments are most effective immediately after vernalization, if a plant is not completely vernalized (Table 6.1), when light intensity during and/or after

vernalization is low and/or plants are exposed to short-day conditions (Thomas and Vince-Prue, 1997). Interestingly, exposure of *Beta* to SD caused devernalization at any stage of development, even when stem elongation had initiated. *Oenothera biennis* L. and *Cieranthus allionii* L. can also be devernalized by transfer to short-day conditions (Wellensiek, 1965). In fact, nearly all cold-requiring plants can, hypothetically, be devernalized. For instance, Petkus winter rye (Purvis and Gregory, 1952), *C. allionii* (Barendse, 1964), carrot (Hiller and Kelly, 1979), cauliflower (Fujime and Hirose, 1980), kohlrabi (Wiebe *et al.*, 1992), celery (Boojiu and Meurs, 1993), *L. longiflorum* Thunb. (Miller and Kiplinger, 1966) and cineraria (Yeh *et al.*, 1997) can all be devernalized. Devernalization can be a commercial problem in Easter lily, cineraria and Regal geranium production (J. Erwin, personal observation). The first 5 days after vernalization appear to be the most critical period, i.e. non-optimal temperatures or irradiance after this period have little impact on induction (J. Erwin, personal observation).

Photoperiod interaction with vernalization

Photoperiod interacts with vernalization to affect flower induction (Yui and Yoshikawa, 1991; Thomas and Vince-Prue, 1997; Yeh *et al.*, 1997). In some cases, as with winter cereals, short-day conditions substitute for vernalization entirely. For instance, *Coreopsis grandiflora* Hogg ex Sweet. Per. is a SDP–LDP (dual photoperiod requirement); however, the short-day requirement can be substituted for a vernalization treatment (Ketellapper and Barbaro, 1966). In contrast, long-day conditions are additive with vernalization in hastening flower initiation of some species, including *L. longiflorum* (Wilkins, *et al.*, 1968). In addition, thermoinduction affects photoperiodic requirements after vernalization (Thomas and Vince-Prue, 1997). Most cold-requiring plants require long-day conditions after cooling. The number of long days required/the critical daylength required for flowering after vernalization decreases as the length of the vernalization treatment increases (Lang, 1965).

Irradiance and light quality interaction with irradiance

Irradiance during vernalization is critical with some species. For instance, *Apium graveolens* L. var. Dulce. does not perceive vernalizing temperatures unless plants are exposed to light during the vernalization process (Ramin and Atherton, 1994). Similarly, *Dianthus carophyllus* L. and *Ajuga reptans* L. flowering was hastened as the irradiance during vernalization increased (Eltzroth and Link, 1970).

Recent research demonstrates that a low red:far red light ratio is antagonistic to vernalization of *R. sativus* (Engelen-Eigles, 1996) and *L. longiflorum* (J. Erwin and G. Engelen-Eigles, unpublished data). In contrast, a high red:far red ratio appears to hasten vernalization. For instance, *L. longiflorum* plants were completely induced by a vernalization treatment in 4 weeks when bulbs were exposed to light with a high red:far red ratio but were not completely induced until after 8 weeks when bulbs were exposed to light with a low red:far red ratio.

Stage of sensitivity to vernalization

The time when a plant becomes sensitive to a vernalization treatment, or the length of the juvenile period, varies with species. Some species can be vernalized as imbibed seed. For instance, beet, *Oenothera* (Thomas and Vince-Prue, 1997) and *R. sativus* (Engelen-Eigles and Erwin, 1997) can be vernalized as imbibed seed. In contrast, most plants must reach a specific stage of development before the meristem is capable of responding to vernalizing temperatures. For instance, *H. niger* must grow for 10 days and *L. biennis* must grow for 7 weeks before plants will respond to a vernalization treatment (Thomas and Vince-Prue, 1997). Carrot cv. 'Chantenay Red Cored' required 8–12 leaves before plants were capable of responding to vernalizing temperatures (Atherton *et al.*, 1990). Similarly, many herbaceous perennial species have a minimum leaf requirement (often over 10) before plants can successfully flower in response to a vernalization treatment (Lopes and Weiler, 1977; Sheldron and

Weiler, 1982a,b; Iverson and Weiler, 1994; Armitage *et al.*, 1996; Whitman *et al.*, 1996). In addition to variation among species, variation in sensitivity among cultivars of a given species in leaf number required to achieve a mature state and in the required length of the vernalization treatment to induce flowering also exists (Yui and Yoshikawa, 1991; Wurr *et al.*, 1994; Engelen-Eigles and Erwin, 1997).

Stability of the vernalization process

The vernalization process, once completed, can be very stable (Table 6.3) (Yeh *et al.*, 1997). For instance, once *Hyoscyamus* is vernalized (requires both vernalization and long days for flowering), the vernalized state can persist for several months under short-day conditions: flowering occurs when plants are returned to long-day conditions (Thomas and Vince-Prue, 1997). *L. longiflorum* bulbs exposed to vernalizing temperatures retained that information even when several warm temperature periods occurred prior to the completion of a vernalization treatment (Erwin and Engelen-Eigles, 1998). No loss of the vernalization stimulus was evident on henbane after 190 days. In fact, there was no measurable loss in degree of vernalization on henbane until 300 days had passed (Thomas and Vince-Prue, 1997). Similarly, some cereal seeds can be moistened, vernalized and then re-dried and maintained for months without loss of the vernalization status (Purvis and Gregory, 1952).

Physiological basis for vernalization

Early work on the physiology of vernalization focused on the identification of a specific, cold-induced, gibberellin responsible for flowering (Lang, 1965). Gibberellins are

Table 6.3. Progressive stabilization of vernalization with increasing duration of exposure to cold in winter rye 'Petkus' (adapted from Thomas and Vince-Prue, 1984).

Cold treatment (weeks)	2	3	4	5	6	8
Plants remaining vernalized after 2 days at 25°C (devernalizing) (%)	0	42	44	75	84	97

known to interact with vernalization to affect flowering of a number of species (Metzger and Dusbabek, 1991). Application of GAs to certain biennial species, including wild-type *A. thaliana* grown under unfavourable conditions for flowering, promotes flower formation and bolting (Lang, 1957; Zeevaart, 1983). Conversely, application of 2-chloroethyltrimethylammonium chloride (CCC–Cycocel), a GA synthesis inhibitor, delayed flowering in several wild-type *A. thaliana* lines, suggesting that the presence of GA is essential for direct flower induction (Napp-Zinn, 1985).

Although exogenous applications of specific gibberellins can overcome a cold-requirement for flowering of some species, these applications are successful on some species and not others (Ketellapper and Barbaro, 1966). It has been proposed that the vernalization process is associated with a progressive demethylation of DNA. Demethylation of DNA is believed to remove a block in gene expression that ultimately leads to floral initiation. Evidence for this theory is based on effects of exogenous 'demethylators' on flower induction and changes in the degree of demethylation of DNA during cooling. For example, application of 5-azacytidine, a demethylator, promoted flowering of non-vernalized cold-requiring *Thlaspi* and *A. thaliana* (Metzger, 1996). In addition, vernalizing temperatures result in demethylation of plant DNA (Burn *et al.*, 1993). Transformed *A. thaliana* that have reduced levels of DNA methylation because of the presence of a methyltransferase (METI) antisense gene flowered earlier than untransformed control plants (Finnegan *et al.*, 1998). However, a 70% reduction of methylation occurred in antisense plants where the promotion of flowering was comparable to that of a vernalization treatment that resulted in only a 15% decrease in DNA methylation. These data suggest that other methyltransferases may be affected by vernalization, since METI affected only one.

Irradiance induction of flowering

In general, increased irradiance reduces the length of the juvenile period with many plant species. In contrast, low irradiance can extend the length of the juvenile period with some species. Irradiance reduction of the juvenile period is the basis for the common practice of lighting seed geraniums for earlier flower induction where the general 'rule-of-thumb' is that every day seedlings are illuminated with supplemental light results in a single day decrease in time to flower.

In addition to photoperiod, irradiance can affect earliness of flowering of many floriculture species (Armitage and Tsujita, 1979; Armitage *et al.*, 1981; Zhang *et al.*, 1996; Erwin *et al.*, 1997; Dole and Wilkins, 1999). For instance, *Achillea millefolium* L. 'Summer Pastels' grown under a 16-h photoperiod in growth chambers flowered after 57, 45 and 37 days when grown under 100, 200 or 300 $\mu mol/m^2/s$, respectively (plant temperature and leaf number below the first flower not reported) (Zhang *et al.*, 1996). An increase in mean dry weight gain per day (MDWG) from 0.32 to 1.02 g/day occurred as irradiance increased from 100 to 300 $\mu mol/m^2/s$ in *A. millefolium* (Zhang *et al.*, 1996). Armitage and Tsujita (1979) grew four *Pelargonium × hortorum* Bailey cultivars under natural daylight conditions (Feb–Mar 1977, Guelph, Ontario, Canada) plus supplemental HPS and LPS lighting (27 or 54 $\mu mol/m^2/s$). HPS irradiance (27 $\mu mol/m^2/s$) hastened flowering for some cultivars but not others, suggesting that *P × hortorum* cultivars' flowering responses to supplemental HPS lighting varied. *Asclepias curassavica* L. node number below the inflorescence decreased from 21 to 16 nodes when plants were grown under continuous lighting versus night interruption lighting (Nordwig, 1999). However, it is often unclear whether irradiance effects are a result of a hastening of flowering strictly through infrared heating from supplemental lighting or from a reduction in juvenile period length as determined through leaf number data.

More recent studies on irradiance effects on flowering have sought to distinguish between thermal hastening of flowering versus developmental hastening of flowering. Erwin *et al.* (1997) showed supplemental lighting hastened flowering developmentally in *Viola × wittrockiana* Gams. and *P. × hybrida* Vilm. For example, *P × hybrida* 'Fantasy Pink

Morn' flowered in as little as 28 days after germination when grown at constant 24°C under 8–9 h ambient daylight conditions (St Paul, Minnesota) plus continuous 100 µmol/m²/s HPS lighting. Similarly, Adams *et al.* (1999) showed *P. hybrida* flowering was hastened by long days, but that decreased daily light integrals (DLI) lengthened the time to flowering. Pearson *et al.* (1993) also noted that increasing DLI shortened the time to anthesis for *D. grandiflora* Ramat. Further, the reduction in time to flower was related to DLI rather than to maximum light intensity. Erwin and Warner (2002) reported that 11 species flowered earlier developmentally when plants were grown with ambient light (10–14 h photoperiod, Sep 1999–May 2000, St Paul, Minnesota) plus 25–50 µmol/m²/s HPS lighting for up to 18 h daily. Hedley (1974) showed that *A. majus* 'Orchid Rocket' flowering was hastened developmentally (leaf number below the flower decreased from 80 to 34 leaves) under long-day conditions as irradiance increased from 115 to 500 µmol/m²/s.

We have 'coined' the term 'facultative irradiance response' to describe a developmental hastening of flowering by addition of supplemental lighting. Species that exhibit a facultative irradiance response (FI) will show a decrease in leaf number below the first flower as irradiance increases (Erwin and Warner, 2002). The increased irradiance is capable of reducing the length of the juvenile period. In contrast, we refer to plants that do not have hastened flowering, developmentally, in response to increased irradiance as 'irradiance indifferent' (II) (Erwin and Warner, 2002). With these species, increased irradiance does not reduce leaf number below the first flower, i.e. juvenile period length is not reduced developmentally with additional lighting.

Some species exhibiting a facultative irradiance response include: *Convolvulus*, *Dianthus*, *Gazania*, *Lavatera*, *Limnanthes*, *Linaria*, *Nemophila*, *Nicotiana*, oregano, *Silene*, snapdragon, petunia, pansy, seed geranium. Some species with a neutral irradiance response include: *Ageratum*, *Celosia*, *Cleome*, *Cosmos*, *Gomphrena*, *Statice*, *Lobelia*, *Mimulus*, *Nigella*, *Salvia*, *Tithonia*, *Zinnia*.

The majority of species studied here did not flower earlier developmentally with increasing irradiance (see Table 6.4; Warner and Erwin, 2002; Mattson and Erwin, 2004b). Erickson *et al.* (1980) found a linear relationship between DLI and the DTF for *P.* × *hortorum* until a threshold level between 6.89 and 9.01 µmol/m²/day was reached (leaf number not reported). Fausey *et al.* (2001) noted that 100% flowering of *Digitalis purpurea* L. 'Foxy' was only reached with DLI greater than 11 µmol/m²/day. Supplemental irradiance (at 30, 60 and 90 µmol/ m²/s) on *Gerbera jamesonii* H. Bolus hastened flowering by up to 23 days in the winter, but only up to 11 days during the spring (leaf number below the first flower not reported) (Gagnon and Dansereau, 1990). Thus, the impact of supplemental irradiance on flowering can be dependent on ambient light conditions. In our experiment, average DLIs (Table 6.2) under all lighting treatments were greater than the threshold levels found by Erickson *et al.* (1980) and Fausey *et al.* (2001). It is likely that the light saturation point for impact on flower induction was reached under the lowest irradiance treatments for many of the species during the relatively high natural light conditions of spring in Minnesota. The five species identified as having a FI response represent plants that probably have a higher threshold of light than other species.

The hastening of flowering with exposure to supplemental irradiance is likely to be due to two components: increased plant temperature from infrared and long-wave radiation from the HPS lamps and additional light available for photosynthesis. More research is needed to determine if FI responses seen on some species are a result of increased photosynthesis or a photomorphogenetic high irradiance response (Beggs *et al.*, 1980). In our experiment, ADT at plant level was 1–2°C greater under the highest lighting treatments (data not presented). Faust and Heins (1998) reported that supplemental irradiance increased shoot-tip temperatures of *Catharanthus roseus* L. by 1.2, 1.5 and 1.7°C with HPS light additions of 50, 75 and 100 µmol/m²/s, respectively. Pietsch *et al.* (1995) concluded that increased development rates of *C. roseus* under supplemental light were a result of

Table 6.4. Photoperiodic and irradiance classifications are based on mean leaf number below the first open flower. Data presented below were primarily from the following references: Armitage, 1996; Dole and Wilkins, 1999; Erwin and Warner, 2002; Motum and Goodwin, 1987a,b; Nordwig, 1999; Mattson and Erwin 2004/5; Seeley, 1985; and Zanin and Erwin, 2003). Photoperiod classifications: FSDP (facultative short-day plant); FLDP (facultative long-day plant); OSDP (obligate short-day plant) OLDP (obligate long-day plant); DNP (day neutral plant). Irradiance classifications: 'FI' (facultative irradiance response: supplemental irradiance hastened induction developmentally); 'II' (irradiance indifferent response; increasing irradiance did not hasten flowering developmentally) (Erwin and Warner, 2002; Mattson and Erwin, 2003a,b). A question mark identifies an uncertain photoperiodic classification.

Species	Photoperiod	Irradiance
Ageratum houstonianum L. 'Blue Danube'	FLDP	II
Alcea rosea	?LDP	
Amaranthus hybridus L. 'Pygmy Torch'	DNP	II
Ammi majus L.	OLDP	II
Anethum graveolens L. 'Mammoth'	OLDP	II
Anigozanthos flavidus	FLDP	
Anigozanthos manglesii	FSDP	
Anigozanthos pulcherrimus Hook.	DNP	
Anigozanthos rufus Labill.	DNP	
Anisodontea × *hypomandarum* K. Presl.	FLDP	
Antirrhinum majus L.	FLDP	FI
Asclepias curassavica L.	DNP	FI
Asclepias tuberosa L.	OLDP	
Asperula arvensis L. 'Blue Mist'	OLDP	II
Begonia × *hiemalis* Fotsch	O/FSDP	
Begonia tuberhybrida	OLDP	
Begonia semperflorens	DNP	FI
Bougainvillea spp.	FSDP	FI
Calceolaria herbeohybrida	FLDP	
Calendula officinalis 'Calypso Orange'	FLDP	II
Callistephus chinensis L.	FLDP	
Catananche caerulea L. Per. 'Blue'	OLDP	FI
Carpanthea pomeridiana L. 'Golden Carpet'	DNP	II
Celosia plumose L. 'Flamingo Feather Purple'	OSDP	II
Centaurea cyanus L. 'Blue Boy'	OLDP	II
Centranthus macrosiphon Boiss.	DNP	FI
Cleome hasslerana Chodat 'Pink Queen'	FLDP	II
Cleome hasslerana Chodat 'Rose Queen'	DNP	FI
Clerodendrum thomsoniae	DNP	
Clerodendrum × *speciosum*	DNP	
Coleus spp.	?SDP	
Cobaea scandens Cav.	DNP	II
Convolvulus tricolor L. 'Blue Enchantment'	DNP	FI
Cosmos bipinnatus Cav. Ann. 'Diablo'	FSDP	II
C. bipinnatus Cav. Ann. 'Sensation White'	FSDP	FI
Cosmos sulphureus Cav.	OSDP	
Collinsia heterophylla Buist	FLDP	II
Crossandra infundibuliformis L.	DNP	
Cucumis sativus H.	DNP	
Cyclamen persicum Mill.	DNP	FI
Dendranthema × *grandiflorum*	FSDP	
Dianthus barbatus L.	DNP	
Dianthus chinensis L. 'Ideal Cherry Picotee'	FLDP	II
Dimorphotheca sinuata DC. 'Mixed Colors'	DNP	II
Dolichos lablab L.	OSDP	II

Table 6.4. *Continued.*

Species	Photoperiod	Irradiance
Eschscholzia californica Cham. 'Sundew'	FLDP	II
Euphorbia pulcherrima Willd. ex Klotzsch.	OSDP	
Exacum affine Balf. F.	DNP	
Fuchsia × *hybrida*	OLDP	
Fuchsia 'Gartenmeister'	DNP	
Gazania rigens L. 'Daybreak Red Stripe'	OLDP	FI
Gomphrena globosa L. 'Bicolor Rose'	FSDP	II
Gypsophila spp.	?LDP	
Hatoria gaertneri Reg.	OSDP	
Helianthus annus L. 'Vanilla Ice'	FLDP	II
Helipterum roseum Hook.	OLDP	II
Hibiscus cisplatinus	DNP	
Hibiscus laevis	OLDP	
Hibiscus moscheutos	OLDP	FI
Hibiscus radiatus	OSDP	
Hibiscus rosa-sinensis L.	DNP	
Hibiscus trionum	FLDP	
Impatiens balsamina	DNP	
Impatiens hawkeri Bull.	DNP	
Impatiens wallerana Hook.f.	DNP	
Ipomoea × *multifida* Shinn. 'Scarlet'	FSDP	II
Ipomopsis rubra Wherry 'Hummingbird Mix'	OLDP	II
Kalanchoe blossfeldiana Poelln.	OSDP	
Lathyrus odoratus L. 'Royal White'	OLDP	FI
Lavatera trimestris L. 'Silver Cup'	OLDP	FI
Legousia speculum-veneris Chaix	OLDP	II
Leonotis menthaefolia R. Br.	DNP	
Leptosiphon hybrida	OLDP	II
Lilium spp.	FLDP	
Limnanthes douglasii R. Br.	OLD	FI
Limonium sinuata Mill. 'Fortress Deep Rose'	FLDP	II
L. sinuata Mill. 'Heavenly Blue'	FLDP	II
Linaria maroccana Hook. f.	FLDP	FI
Linum perenne L.	OLDP	FI
Lobelia erinus L. 'Crystal Palace'	OLDP	II
Lobularia maritima	DNP	
Lycopersicon esculentum Mill.	DNP	
Matthiola longipetala Venten.'Starlight Scentsation'	DNP	II
Mimulus × *hybridus* L. 'Magic'	OLDP	II
Mina lobata Cerv.	OSDP	II
Mirabilis jalapa L.	OLDP	II
Nemophila maculate Benth. 'Pennie Black'	DNP	FI
Nemophila menziesii Hook. & Arn.	DNP	II
Nicotiana alata Link & Otto 'Domino White'	DNP	FI
Nigella damascene L. 'Miss Jekyll'	OLDP	II
Oenothera pallida Lindl. 'Wedding Bells'	OLDP	II
Origanum vulgare L.	DNP	FI
Oxypetalum caerulea D. Don 'Blue Star'	DNP	FI
Pelargonium × *domesticum* L.H. Bail.	FLDP	
Pelargonium × *hortorum* L.H. Bail.	DNP	FI
Pelargonium peltatum L.	DNP	
Perilla frutescens	?SDP	

continued

Table 6.4. *Continued.*

Species	Photoperiod	Irradiance
Petunia × *hybrida*	FLDP	
P. × *hybrida* 'Purple Wave'	OLDP	
Phacelia campanularia A. Gray.	DNP	II
Phacelia tanacetifolia Benth.	FLDP	II
Pharbitis nil	FSDP	
Polemonium viscosum Nutt.	OLDP	II
Primula malacoides Franch	OSDP	
Primula obconica Hance	DNP	
Primula × *polyantha*	DNP	FI
Rhododendron spp.	OSDP	
Rosa × *hybrida* spp.	DNP	FI
Saintpaulia ionantha Wendl.	DNP	FI
Salpiglossus sinuata	?LDP	
Salvia farinacea 'Strata'	FLDP	FI
Salvia splendens F. Sellow 'Vista Red'	FLDP	II
Sanvitalia procumbens Lam.	FSDP	II
Scabiosa caucasia	?LDP	
Schlumbergera truncata Haw.	OSDP	II
Silene armeria L. 'Elektra'	OLDP	FI
Sinningia speciosa Lodd.	DNP	
Solidago L. spp.	SDP	
Streptocarpus × *hybridus* Voss.	DNP	FI
Streptocarpus nobilis Clarke	FSDP	
Tagetes erecta L.	FSDP	
Tagates patula L.	DNP	
Tagetes tenuifolia Cav.	FSDP	
Thunbergia alata Bojer.	DNP	II
Tithonia rotundifolia Mill. 'Fiesta Del Sol'	FLDP	II
T. rotundifolia Mill. 'Sundance'	FSDP	FI
Verbascum phoeniceum L.	DNP	II
Verbena × *hybrida*	?LDP	
Viguiera multiflora S.F. Blake	FLDP	II
Viola tricolor L.	F/O LDP	II
Viola × *wittrockiana* Gams.	FLDP	FI
Zea mays H.	DNP	
Zinnia angustifolia Kurth.	DNP	
Zinnia elegans Jacq. 'Exquisite Pink'	FSDP	II
Z. elegans Jacq. 'Peter Pan Scarlet'	FSDP	II

increased shoot-tip temperatures. Graper and Healy (1991) conducted an experiment to separate the components of HPS lighting into the effects of thermal radiation vs. photosynthetic photon flux (PPF) using a circulating water bath. They concluded that light plays a greater role in development of *P.* × *hybrida* seedlings via greater photosynthesis than the increase in plant temperature associated with supplemental HPS lighting. Leaf number below the first flower was not reported.

Clearly, controlled environment chamber experiments where lighting conditions can be precisely controlled are necessary to more accurately evaluate the effect of irradiance on herbaceous ornamental plant species flower induction. Determining the threshold levels of irradiance required for optimal growth and impact on flower induction of herbaceous ornamental plant species would be useful information for growers deciding whether to purchase supplemental lighting.

Stress induction of flowering

Short-term plant stress will cause a shortening of the juvenile phase in some species. Perhaps the best example of this phenomenon is the induction of early flowering in celosia after a short-term water stress (J. Erwin, personal observation). Similarly, it appears as though short-term heat stress can cause earlier flowering on newer cultivars of chrysanthemums. High or low temperatures can also affect a plant's photoperiodic response. For instance, at temperatures below 17°C, cineraria is a day-neutral plant. At temperatures above 17°C, cineraria is a short-day plant. In contrast, when temperatures exceed 23°C, flowering of many species can be delayed. Some species are sensitive to high night temperatures (*E. pulcherrima*, *G. globosa*, *Tagetes erecta*), some species are sensitive to high day temperatures (*Fuchsia* × *hybrida*, *A. majus*), and some species are sensitive to high day or night temperatures (*I. hawkeri*, *C. morifolium*, *Schlumbergera truncata*, many *Pelargonium* spp.) (Erwin and Warner, 1999a).

Ethylene

Exposure to ethylene is capable of inducing flowering in some plant species. Most notably, ethylene exposure induces flowering in Bromeliads. Clark and Kerns (1942) showed that application of auxin to pineapple could stimulate early flowering. Subsequent research by Burg and Burg (1966) demonstrated that the induction of flowering was associated with auxin-induced stimulation of ethylene. Commercially, pineapple is now stimulated to flower by applying ethephon, which releases ethylene on plants. Application of ethephon also stimulates early flowering in mango (Chacko *et al.*, 1976).

Flower Development

Once flower induction and initiation have occurred, successful flower development must follow to have successful flowering.

Successful flower development can require specific environmental conditions. For instance, some species require a specific photoperiod (which can be different from that for flower induction) for successful flower development. Additionally, species may have a minimum irradiance required for flower development as was noted with *Rosa* spp. previously (Nell and Rasmussen, 1979). Temperature also interacts with flowering as described below. In contrast, other species may have no environmental requirements for successful flower development other than a lack of stress.

Photoperiod requirement for flower development

Flower development can have a photoperiodic requirement. For instance, *D. grandiflora* flower development has a shorter critical photoperiod than that for flower induction. *D. grandiflora*, therefore, has evolved in such a way as to naturally induce flowers as daylength starts to shorten after 21 June and then has successive flower development as the daylength continues to shorten into autumn and the vernal equinox. Therefore, if a grower induced *D. grandiflora* in the spring and then exposed plants to a photoperiod greater than the critical photoperiod for flower development, development would cease and a 'crown bud' would develop.

Species differ in whether they have a specific photoperiod requirement for flower bud development. For instance *C.* × *morifolium* flower development has a photoperiod requirement that is shorter than the photoperiod requirement for flower induction and initiation. Therefore, if chrysanthemum (SDP) plants are induced during spring and placed under long-day conditions following induction, flower development would be inhibited, which results in 'crown bud' formation (Dole and Wilkins, 1999). Similarly, if a LDP is induced very early in spring and placed back under short-day conditions after induction in a lighted environment too early, flower bud development can be arrested.

Temperature requirement for flower development

There is an optimal temperature for flower development that is probably species, photo-period and irradiance dependent. Whether temperature effects are due to impacts on the latter stages of flower initiation or whether the effect is on actual flower bud abortion is not clear. It is likely that, depending on the timing of non-optimal temperature conditions, both conditions are occurring. Additionally, whether non-optimal tempera-ture conditions constitute a 'stress' is a matter of debate. Here, non-optimal temperature conditions are separated from temperature stress as discussed below.

High temperature inhibition of flowering has been noted for some time. For instance, the node number at which the first flower cluster appears in *L. esculentum* increases dra-matically at 27°C versus 10°C–16°C (Calvert, 1957). In general, many plant species have temperature optima for flowering that are between 18°C and 22°C. As temperature increases or decreases from this optimum, flower number decreases. There is evidence which anecdotally suggests that these temper-ature optima are based on differences in flower bud abortion during development. Such a conclusion is based on the observation that temperature significantly affects flower number on mature day-neutral plants where flower induction is continuous. For example, *Pelargonium* spp. flower number per inflores-cence decreases as average daily temperature increases. Flower number per inflorescence on *Pelargonium peltatum* L. 'Nicole' decreased from 9 to 3.8 flowers per inflorescence as average daily temperature increased from 12°C to 29°C (Erwin, 1999). Similarly, *Pelargonium × domesticum* L.H. Bailey flower number per inflorescence decreases as average daily temperature decreases and flowering is inhibited altogether at average daily temperatures above 17°C (Erwin and Engelen, 1992). *P. × hortorum* 'Veronica' flower number decreased from 52 to 15 flowers per inflorescence as average daily temperature increased from 12°C to 29°C (Erwin and Heins, 1992). *Fuchsia hybrida* L. 'Dollar Princess' flower number per node increased from 2.3 to 6 flowers as average daily temperature increased from 12°C to 15°C then decreased to 2.3 flowers per node as average daily temperature further increased to 25°C (Erwin and Kovanda, 1990).

Flower development of different species is sensitive to temperature and can vary diurnally. *Schlumbergera truncata* Lem. and *Euphorbia pulcherrima* flower development is most sensitive to night temperature. For instance, flower number was greatest on *S. truncata* 'Madisto' when night temperature was 20°C (Erwin *et al.*, 1990). In contrast, *Antirrhinum majus* flower development appears to be most sensitive to day tempera-ture. *Impatiens hawkeri* Bull., *D. grandiflora*, and *F. hybrida* flower development is most sensitive to day and night temperature. *Pelargonium* spp. flower development appears to be most sensitive to average daily temperature (Erwin and Warner, 1999a).

Impact of stress on flower development

High temperatures can decrease flowering. Whether such temperature exposures consti-tute a 'stress' is a matter for debate. However, temperature exposure that results in an extended period of reduced growth should be considered a stress.

Prolonged exposure to high temperatures reduces flowering of numerous flower crops. Exposure of *A. majus*, *Calendula officinalis*, *Impatiens walleriana*, *Mimilus × hybridus* Hort. ex Siebert & Voss, and *Torenia fournieri* Linden ex E. Forum to high temperature reduced flowering of all species compared with cooler temperatures across a variety of irradiance treatments (Warner and Erwin, 2001b). The basis for reduced flower number is likely to be due, in part, to reduced photosynthesis. Exposure of *I. hawkeri* and *V. × wittrockiana* to high temperatures (35°C) for as little as 2 h reduced photosynthetic rates for at least 3 days for some cultivars compared with 'un-stressed' plants (Warner and Erwin, 2002). Photosynthetic recovery varied between *I. hawkeri* and *V. × wittrockiana* with *V. wittrockiana* recovering more quickly.

Variation in high temperature tolerance can be seen across cultivars within a species as well. For instance, prolonged exposure of *V. × wittrockiana* to high temperatures (30°C) compared with a 'non-stress' environment (20°C) reduced flowering of all cultivars studied (Warner and Erwin, 2003a). However, there was cultivar variation with flowering being reduced 23% for 'Crystal Bowl Purple' and 79% for 'Majestic Giants Red and Yellow'. Similarly, high temperature exposure (32°C/28°C DT/NT) affected flower number per cluster on *L. esculentum* cultivars differently (Warner and Erwin, 2003a). Lastly, Strope (1999) showed that *I. hawkeri* cultivars varied in their ability to continue to flower under high temperature conditions (30°C).

In addition to reduced flower number, yield in some crops can be reduced by high temperature induced floral sterility, resulting in reduced seed set, fruit size and yield. Such high temperature reduction in yield is a common problem in commercial horticulture production of peppers as well as many crops worldwide.

High temperature sensitivity of flowering may be due to reduced overall carbon assimilation under high temperatures (Ranney and Peet, 1994), increased respiration (Berry and Bjorkman, 1980), or to a reduced ability of floral organs to recruit photoassimilates compared with vegetative tissues (Dinar and Rudich, 1985). Photosynthesis is a metabolic process that is susceptible to high temperature stress (Larcher, 1995). For example, photosynthetic rates of five birch taxa declined sharply when temperature exceeded 30°C (Ranney and Peet, 1994). This may be in part due to a reduction in chloroplast number as chloroplast biogenesis is reduced at temperatures greater than 32°C in barley (Smillie *et al.*, 1978). Respiration rates continue to increase at temperatures above the optimum temperature for photosynthesis resulting in a net carbon loss (Berry and Bjorkman, 1980). Lastly, reduced concentrations of plant growth hormones in reproductive tissues following high temperature exposure may well reduce the ability of those tissues to attract assimilates (Kuo and Tsai, 1984).

We are currently evaluating the physiological basis for high temperature inhibition of flower development (R. Warner, PhD thesis). Promising data are emerging showing variation in *A. thaliana* heat sensitivity which may provide a basis for gaining new insight into this significant floriculture crop issue.

Dormancy

Dormancy will be discussed only briefly since it is not directly related to flower evocation, only to continued flower development in a limited number of floriculture crops. Most notably, some bulb crops and woody plant species can initiate flowers, but then have flower development stopped at a specific stage until specific environmental conditions are met. Note below that not all bulb crops express dormancy.

Bulb crops can vary considerably as to when flowers are initiated during the year and during the life cycle of the species. Species have been classified into the following five response groups by Hartsema (1961):

1. Flowers initiated during the spring or early summer of the year preceding that in which they reach anthesis and before the bulbs are 'lifted' (*Narcissus, Galanthus, Leucojum*).

2. Flowers initiated following the previous growing period, so that the bulbs have initiated flowers by replanting time in the autumn (*Tulipa, Hyacinthus, Iris reticulata*).

3. Flowers initiated after replanting, at the low temperatures of winter or early spring (bulbous *Iris*, but not *I. reticulata*).

4. Flowers initiated more than a year before anthesis (*Nerine*).

5. Flower initiation alternates with leaf formation through the whole growing period (*Hippeastrum, Zephyranthes*).

The physiological basis for flowering in each of these groupings can differ significantly. For instance, there is evidence that group 3 has a vernalization requirement, whereas groups 1 and 2 have a dormancy requirement. By dormancy requirement, flower development can be arrested until environmental or time constraints are fulfilled. For example, ephemerals, i.e. *Tulipa × hybrida* or *Narcissus pseudonarcissus*, are not photoperiodic as the length of a growth cycle is relatively short. Such plants will only successfully flower after these requirements are met. These plants are said to be 'dormant'. Temperature control of dormancy is common in temperate perennial species and water control of dormancy is common in locations where water can be limiting, i.e. a desert region. For instance, water stress can induce dormancy in *Achimenes* hybrids (Zimmer and Junker, 1985).

In general, flower bud dormancy can be broken on many woody plant species by exposing tissue to conditions similar to vernalization. Between 6 and 12 weeks of temperatures between 2°C and 6°C is often optimal. In some cases, GA_3 application can overcome dormancy and enhance shoot emergence from tubers (Lurie *et al.*, 1992). Application of GA_3 to overcome flower bud dormancy is common commercially in florist's Azalea.

Conclusions

At this point, I will take some latitude and emphasize the need for traditional physiological research on floriculture crops. It seems that a complete understanding of what factors result in flowering in floriculture crops is essential to the success of any breeding or physiology project related to that crop. It is amazing how much research has been conducted on crops in which we have little or no understanding of the flowering mechanisms in that crop! How valid are the conclusions?

Our recent work on photoperiodism in annual crops was started by questions from Terry Smith (Smith Garden, Bellingham, Washington) when he simply wanted to know how to flower petunias from 1 February to 1 March. What grew out of that question was a realization that our understanding of how photoperiod and temperature affect flowering of many of the commercially significant bedding plant crops is in its infancy. There is clearly a need to renew fundamental research on the flowering physiology of flower crops to meet the needs for scheduling, timing and marketing that are upon the floriculture industry.

A fundamental understanding of flowering physiology of floriculture crops will provide the foundation from which we can utilize molecular biology to improve our crops.

References

Adams, S.R., S. Pearson, P. Hadley and W.M. Patefield. 1999. The effects of temperature and light integral on the phases of photoperiod sensitivity in *Petunia* × *hybrida*. *Ann. Bot.* 83: 263–269.

Armitage, A.M. 1996. Forcing perennials in the greenhouse. *GrowerTalks* 60(3): 86, 88, 93, 94, 96, 97.

Armitage, A.M. and J.M. Garner. 1999. Photoperiod and cooling duration influence growth and flowering of six herbaceous perennials. *Jour. Hort. Sci. Biotech.* 74(2):170–174.

Armitage, A.M. and M.J. Tsujita. 1979. The effect of supplemental light source, illumination and quantum flux density on the flowering of seed-propagated geraniums. *Jour. Hort. Sci.* 54(3): 195–198.

Armitage, A.M., W.H. Carlson and J.A. Flore. 1981. The effect of temperature and quantum flux density on the morphology, physiology, and flowering of hybrid geraniums. *Jour. Amer. Soc. Hort. Sci.* 106(5):643–647.

Armitage, A.M., L. Copeland, P. Gross and M. Green. 1996. Cold storage and moisture regime influence flowering of *Oxalis adenophylla* and *Ipheion uniflorum*. *HortSci.* 31:1154–1155.

Atherton, J.G., J. Craigon and E.A. Basher. 1990. Flowering and bolting in carrot. I. Juvenility, cardinal temperatures and thermal times for vernalization. *Jour. Hort. Sci.* 65:423–429.

Bagnall, D.J. 1993. Light quality and vernalization interact in controlling late flowering in *Arabidopsis* ecotypes and mutants. *Ann. Bot.* 71:75–83.

Barendse, G.W. 1964. Vernalization of *Cieranthus allionii* Hort. *Meded. Landbou., Wageningen*, 64: 1–64.

Beggs, C.J., M.G. Holmes, M. Jabben and E. Schaefer. 1980. Action spectra for the inhibition of hypocotyl growth by continuous irradiation in light- and dark-grown *Sinapsis alba* L. seedlings. *Plant Physiol.* 66:615–618.

Berry, J. and O. Bjorkman. 1980. Photosynthetic response and adaptation to temperature in higher plants. *Ann. Rev. Plant Physiol.* 31: 491–543.

Booiju, R. and E.J.J. Meurs. 1993. Flower induction and initiation in celeriac (*Apium graveolens* L. var. *rapaceum* (Mill.) DC.): effects of temperature and plant age. *Scient. Hort.* 55: 227–238.

Borthwick, H.A., S.B. Hendricks and M.W. Parker. 1948. Action spectrum for the photoperiodic control of floral initiation of a long day plant, Wintex barley (*Hordeum vulgare*). *Bot. Gaz.* 110: 103–118.

Borthwick, H.A., S.B. Hendricks and N.W. Parker. 1952. The reaction controlling floral initiation. *Proc. Natl. Acad. Sci., USA* 38:929–934.

Bunning, E. 1948. Die entwicklungsphysiologische Bedeutung der endogenen Tagesrhythmik bei den Pflanzen. *In:* A.E. Murneek, and R.O. Whyte (eds) *Vernalization and Photoperiodism.* Chronica Botanica, Waltham, Massachusetts.

Bunning, E. 1963. *Die Physiologische Uhr.* Springer-Verlag, Berlin, Germany. pp. 153.

Burg, S.P. and E.A. Burg. 1966. Auxin-induced ethylene formation: its relation to flowering in pineapple. *Sci.* 152:1269.

Burn, J.E., D.J. Bagnall, J.D. Metzger, E.S. Dennis and W.J. Peacock. 1993. DNA methylation, vernalization, and the initiation of flowering. *Proc. Natl. Acad. Sci. USA* 90:287–291.

Butler, K.H., H. Norris, W. Siegelman and S.B. Hendricks. 1959. Detection assay and purification of the pigment controlling photoresponsive development of plants. *Proc. Natl. Acad. Sci. USA* 45:1703–1708.

Calvert, A. 1957. Effect of early environment on the development of flowering in tomato. *Jour. Hort. Sci.* 32:9–17.

Cameron, A., M. Yuan, R. Heins and W. Carlson. 1996. Juvenility: your perennial crop's age affects flowering. *GrowerTalks* 60(8):30–32, 34.

Carr, D.J. 1952. A critical experiment on Bunning's theory of photoperiodism. *Z. Naturforsch.* 76:570.

Carr, D.J. 1959. Translocation between leaf and meristem in the flowering response of short-day plants. *9th Int. Bot. Congr.* 2:11.

Carver, S.A. and H.K. Tayama. 1992. Number of stock plant shoot nodes influences splitting of 'Lilo' poinsettia. *HortTech.* 2(2):206–207.

Chacko, E.K., R.R. Kohli, R. Dore Swamy and G.S. Randhawa. 1976. Growth regulators and flowering in juvenile mango (*Mangifera indica* L.) seedlings. *Acta Horticulturae* 56:173–181.

Chouard, P. 1960. Vernalization and its relations to dormancy. *Ann. Rev. Plant Physiol.* 11, 191–238.

Clark, H.E. and K.R. Kerns. 1942. Control of flowering with phytohormones. *Science* 95:536–537.

Clark, J.R. 1983. Age-related changes in trees. *Jour. Arboriculture* 9:201–205.

Cockshull, K.E. 1984. The photoperiodic induction of flowering in short-day plants. *In:* D. Vince-Prue, B. Thomas and K.E. Cockschull (eds), *Light and the Flowering Process.* Academic Press, London. pp. 33–50.

Cockshull, K.E. 1985. *Chrysanthemum morifolium. In:* A.H. Halevy, (ed.). *Handbook of Flowering, Vol. 2.* CRC Press, Boca Raton, Florida. pp. 238–257.

Cooper, W.C. and A. Peynado. 1958. Effect of gibberellic acid on growth and dormancy of Citrus. *Proc. Amer. Soc. Hort. Sci.* 72: 284–289.

Crosthwaite, S.K. and G.I. Jenkins. 1993. The role of leaves in the perception of vernalizing temperatures in sugar beet. *Ann. Bot.* 69:123–127.

Cumming, B.G. 1969. *Chenopodium rubrum* L. and related species. *In:* L.T. Evans (ed.), *The Induction of Flowering.* MacMillan, Melbourne, Australia. pp. 156–185.

Curtis, O.F. and C.K. Chang. 1930. The relative effectiveness of temperature of the crown as contrasted with that of the rest of the plant upon flowering of celery plants. *Amer. Jour. Bot.* 17:1047–1048.

Deitzer, G.F. 1984. Photoperiodic induction in long-day plants. *In:* D. Vince-Prue, B. Thomas, and K.E. Cockschull (eds). *Light and the Flowering Process.* Academic Press, London. pp. 51–64.

de Zeeuw. 1957. Flowering of *Xanthium* under long-day conditions. *Nature* 180:588.

Dinar, M. and J. Rudich. 1985. Effect of heat stress on assimilate partitioning in tomato. *Ann. Bot.* 56:239–248.

Dole, J.M. and H.F. Wilkins. 1999. *Floriculture: Principles and Species.* Prentice Hall, Upper Saddle River, New Jersey.

Downs, R.J. 1956. Photoreversibility of flower initiation. *Plant Physiol.* 31:279–284.

Eltzroth, D.E. and C.B. Link. 1970. The influence of light during vernalization on the flowering response of *Ajuga* and *Dianthus. Jour. Amer. Soc. Hort. Sci.* 95:95–98.

Engelen-Eigles, G. 1996. Irradiance and light quality interact to affect vernalization in *Raphanus sativus* L.. MS thesis. Department of Horticultural Science, University of Minnesota, St Paul, Minnesota.

Engelen-Eigles, G. and J.E. Erwin. 1997. A new model plant for vernalization studies. *Scient. Hort.* 70:197–202.

Erickson, V.L., A. Armitage, W.H. Carlson and R.M. Miranda. 1980. The effect of cumulative photosynthetically active radiation on the growth and flowering of the seedling geranium, *Pelargonium × hortorum. HortSci.* 15(6):815–817.

Erwin, J.E. 1991. Cool temperatures are still critical on regals. *Minn. Comm. Flow. Grow. Bull.* 40(3): 3–4.

Erwin, J.E. 1999. Ivy geranium production. *Ohio Flor. Bull.* 831:1, 15–20.

Erwin, J.E. and G. Engelen. 1992. Regal geranium production. *Minn. Comm. Flow. Grow. Bull.* 41(6):1–9.

Erwin, J.E. and G. Engelen-Eigles. 1998. Influence of simulated shipping and rooting temperature and production year on Easter lily (*Lilium longiflorum* Thunb.) development. *Jour. Amer. Soc. Hort. Sci.* 123:230–233.

Erwin, J.E. and R.D. Heins. 1992. Environmental effects on geranium development. *Minn. Comm. Flow. Grow. Bull.* 41(1):1–9.

Erwin, J.E., R.D. Heins, R.D. Berghage and B. Kovanda. 1990. Temperature effects *Schlumbergera truncata* 'Madisto' flower initiation. *Acta Horticulturae* 272:97–101.

Erwin, J. and B. Kovanda. 1990. Fuchsia production. *Minn. Comm. Flow. Grow. Bull.* 39(3):1–3.

Erwin, J.E, and R. Warner. 1999a. Temperature. *In:* C.A. Buck, S.A. Carver, M.L. Gaston, P.S. Konjoian, L.A. Kunkle and M.F. Wilt (eds). *Tips on Growing Bedding Plants*, 4th edn. OFA Services, Columbus, Ohio. pp. 69–82.

Erwin, J.E. and R. Warner. 2002. Determination of photoperiodic response group and effect of supplemental irradiance on flowering of several bedding plant species. *Acta Horticulturae* 580:95–100.

Erwin, J.E., R. Warner, T. Smith and R. Wagner. 1997. Photoperiod and temperature interact to affect *Petunia × hybrida* Vilm. development. *HortScience* 32:501.

Erwin, J.E., R.M. Warner and A.G. Smith. 2002. Vernalization, photoperiod and GA3 interact to affect flowering of Japanese radish (*Raphanus sativus* Chinese radish Jumbo Scarlet). *Physiol. Plant* 155:298–302.

Evans, L.T. and I.F. Wardlaw. 1966. Independent translocation of ^{14}C-labelled assimilates and of the floral stimulus in *Lolium temulentum*. *Planta* 68:310–326.

Fausey, B.A., A.C. Cameron and R.D. Heins. 2001. Daily light integral, photoperiod, and vernalization affect flowering of *Digitalis purpurea* L. 'Foxy'. *HortSci.* 36(3):565.

Faust, J., and R.D. Heins. 1998. Modeling shoot-tip temperature in the greenhouse environment. *Jour. Amer. Soc. Hort. Sci.* 123:208–214.

Finnegan, E.J., R.K. Genger, K. Kovac, W.J. Peacock and E.S. Dennis. 1998. DNA methylation and the promotion of flowering by vernalization. *Proc. Natl. Acad. Sci. USA* 95:5824–5829.

Frattianne, D.G. 1965. The interelationship between the flowering of dodder and the flowering of some long and short day plants. *Amer. Jour. Bot.* 52:556–562.

Friend, D.J.C. 1965. Interaction of red and far red radiations with the vernalization process in winter rye. *Can. Jour. Bot.* 43:161–170.

Frydman, V.M. and P.F. Wareing. 1974. Phase change in *Hedera helix* L. III. The effects of gibberellins, abscisic acid and growth retardants on juvenile and adult ivy. *Jour. Expt. Bot.* 25:420–429.

Fujime, Y. and T. Hirose. 1980. Studies on thermal conditions of curd formation and development in cauliflower and broccoli. *Jour. Jap. Soc. Hort. Sci.* 49:217–227.

Gagnon, S. and B. Dansereau. 1990. Influence of light and photoperiod on growth and development of gerbera. *Acta Horticulturae* 272:145–151.

Garner, W.W. and H.A. Allard. 1920. Effect of relative length of day and night and other factors of the environment on growth and reproduction of plants. *Jour. Agr. Res.* 18: 553–607.

Goh, C.J. and J. Arditti. 1985. Orchidaceae. *In:* A.H. Halevy (ed.). *Handbook of Flowering,. Vol. I.* CRC Press, Boca Raton, Florida. pp. 309–336.

Goliber, T.E., and L.J. Feldman. 1989. Osmotic stress, endogenous abscisic acid and the control of leaf morphology in *Hippuris vulgaris* L. *Plant, Cell Environ.* 12:163–172.

Graper, D.F. and W. Healy. 1991. High pressure sodium irradiation and infrared radiation accelerate *Petunia* seedling growth. *Jour. Amer. Soc. Hort. Sci.* 116(3):435–438.

Gregory, F.G. and R.S. de Ropp. 1938. Vernalization of excised embryos. *Nature* 142:481–482.

Guttridge, C.D. 1959. Further evidence for a growth-promoting and flower-inhibiting hormone in strawberry. *Ann. Bot.* 23:612–621.

Hackett, W.P. 1980. Control of phase change in woody plants. *In: Control of Shoot Growth in Trees.* IUFRO, Fredericton, New Brunswick, Canada. pp. 257–272.

Hackett, W.P. and C. Srinivasani. 1983/5. *Hedera helix* and *Hedera canatiensis. In:* A.H. Halevy (ed.). *Handbook of Flowering, Vol. 3.* CRC Press, Boca Raton, Florida. pp. 89–97.

Halevy, A.H. 1984. Light and autonomous induction. *In:* D. Vince-Prue, B. Thomas and K.E. Cockschull (eds). *Light and the Flowering Process.* Academic Press, London. pp. 65–74.

Hamner, K.C. 1940. Interrelation of light and darkness in photoperiodic induction. *Bot. Gaz.* 101:658–687.

Hamner, K.C. and J. Bonner. 1938. Photoperiodism in relation to hormones as factors in floral initiation and development. *Bot. Gaz.* 100: 388–431.

Handro, W. 1977. Photoperiodic induction of flowering on different explanted tissues from *Streptocarpus nobilis* cultured *in vitro. Bol. Botan. Univ. Sao Paulo* 5:21–26.

Harder, R. and O Bode. 1943. Wirkung von zwischenbelichtungen wahrend der dunkel-periode auf Kalanchoe. *Planta* 33:469–504.

Hartsema, A.M. 1961. Influence of temperatures on flower formation and flowering of bulbous and tuberous plants. *In*: W. Ruhland (ed.). *Encyclopedia of Plant Physiol. Vol. 16.* pp. 123–161.

Healy, W.E. and H.F. Wilkins. 1981. Interaction of soil temperature, air temperature and photo-period on growth and flowering of *Alstroemeria* 'Regina'. *HortSci.* 16:459.

Healy, W.E. and H.F. Wilkins. 1982. The interaction of temperature on flowering of *Alstroemeria* 'Regina'. *Jour. Amer. Soc. Hort. Sci.* 107: 248–251.

Hedley, C.L. 1974. Response to light intensity and daylength of two contrasting flower varieties of *Antirrhinum majus. Jour. Hort. Sci.* 49:105–112.

Heide, O.M. 1965. *Campanula isophylla* som langdag-splante. *Gartneryrket* (Oslo) 55:210–212.

Heide, O.M. and W. Runger. 1985. Begonia. *In*: A.H. Halevy (ed.). *Handbook of Flowering, Vol. 2.* CRC Press, Boca Raton, Florida. pp. 4–23.

Heinze, P.H., M.W. Parker and H.A. Borthwick. 1942. Floral initiation in Biloxi soybean as influenced by grafting. *Bot. Gaz.* 103: 518–530.

Hendricks, S.B. and H.A. Borthwick. 1954. Photo-periodism in plants. *Proc. 1st Int'l. Photobiol. Congress Amsterdam* pp. 23–35.

Hendricks, S.B., H.A. Borthwick and R.J. Downs. 1956. Pigment conversion in the formative responses of plants. *Proc. Natl. Acad. Sci. USA* 42:19–26.

Hicks, K.A., A.J. Millar, I.A. Carre, D.E. Somers, M. Straume, D.R. Meeks-Wagner and S.A. Kay. 1996. Conditional circadian dysfunction of the *Arabidopsis* early-flowering 3 mutant. *Sci.* 274: 790–792.

Hiller, L.K. and W.C. Kelly. 1979. The effect of post-vernalization temperature on seedstalk elongation and flowering in carrots. *Jour. Amer. Soc. Hort. Sci.* 104:253–257.

Imamura, S. 1953. Photoperiodic initiation of flower primordia in Japanese morning glory, *Pharbitis nil. Proc. Jap. Acad.* 29:368–373.

Imamura, S. and A. Takimoto. 1955. Transmission rate of photoperiodic stimulus in *Pharbitis nil. Bot. Mag.* 68:260–266.

Iverson, R. and Weiler, T. 1994. Strategies to force flowering of six herbaceous garden perennials. *HortTech.* 4:61–65.

Karlsson, M.G. and R.D. Heins. 1992. Begonias. *In*: R.A. Larson (ed.). *Introduction to Floriculture*, 2nd edn. Academic Press, San Diego, California. pp. 409–427.

Keatinge, J.D.H., A. Qi, T.R. Wheeler, R.H. Ellis and R.J. Summerfield. 1998. Effects of temperature and photoperiod on phenology as a guide to the selection of annual legume cover and green manure crops for hillside farming systems. *Field Crops Research* 57(2): 139–152.

Ketellapper, H.J. and Barbaro, A. 1966. The role of photoperiod, vernalization and gibberellic acid in floral induction in *Coreopsis grandiflora* Nutt. *Phyton* (Buenos Aires) 23:33–41.

King, R.W., L.T. Evans, and I.F. Wardlaw. 1968. Translocation of the floral stimulus in *Pharbitis nil* in relation to that of assimilates. *Z. Pflanzenphysiol.* 59:377–388.

Krumweide, D. 1960. Uber de wirkung von starkund schwachlichtkombinationen auf das bluhen von *Kalanchoe blossfeldiana. Biol. Zentralbl.* 79:258–278.

Kuo, C.G. and C.T. Tsai. 1984. Alteration by high temperature of auxin and gibberellin concentrations in the floral buds, flowers and young fruit of tomato. *HortSci.* 19:870–872.

Lang, A. 1957. The effect of gibberellin upon flower formation. *Proc. Natl. Acad. Sci. USA* 43:699–504.

Lang, A. 1959. Physiology of flowering. *Ann. Rev. Plant Physiol.* 3:265–306.

Lang, A. 1965. Physiology of flower initiation. *In*: H. Ruhland (ed.). *Encyclopedia of Plant Physiology.* Springer-Verlag, Berlin, Germany. pp. 1380–1536.

Lang, A. and G. Melchers. 1943. Die photoperiod-ische reaktion von Hyoscyamus niger. *Planta* 33:653–702.

Lang, A., M.K. Chailakhyan and I.A. Frolova. 1977. Promotion and inhibition of flower formation in a day-neutral plant in grafts with a short-day plant and a long-day plant. *Proc. Natl. Acad. Sci. USA* 74:2412–2416.

Larcher, W. 1995. *Physiological Plant Ecology.* Springer-Verlag, Berlin, Germany. p. 506.

Lona, F. 1946. Sui fenomeni di induzione, post-effeto e localizzazione fotoperiodica. *Nuovo Giorn. Bot. Ital.* 53:548–575.

Longman, K.A. 1976. Some experimental approaches to the problem of phase change in forest trees. *Acta Horticulturae* 56:81–90.

Lopes, L.C. and Weiler, T.C. 1977. Light and temper-ature effects on the growth and flowering of

Dicentra spectabilis (L.) Lem. *Jour. Amer. Soc. Hort. Sci.* 102:388–390.

Luckwill, L.C. 1970. Control of growth and fruitfulness of apple trees. *In:* L.C. Luckwill and C.V. Cutting (eds). *Physiology of Tree Crops.* Academic Press, New York, pp. 237–254.

Lurie, G., A.A. Watad and A. Borochov. 1992. *Aconitum*: effect of tuber size, daylength and GA$_3$ on growth, flowering and tuber production. *Acta Horticulturae* 325:113–117.

Lyons, R.E. and J.N. Booze-Daniels. 1986. Characteristics of the photoperiodic response of California poppy. *Jour. Amer. Soc. Hort. Sci.* 111(4):593–596.

Martinez-Zapater, J.M. and C.R. Somerville. 1990. Effect of light quality and vernalization on late-flowering mutants of *Arabidopsis thaliana*. *Plant Physiol.* 92:770–776.

Mattson, N. 2002. Photoperiod, irradiance, and light quality affect flowering of herbaceous annuals. MS thesis. Department of Horticultural Science. University of Minnesota, St Paul, Minnesota.

Mattson, N., and J.E. Erwin. 2004/5. Impact of photoperiod and irradiance on flowering of several herbaceous ornamentals I. *Scientia Hort.*

McDaniel, C.N., S.R. Singer and S.M.E. Smith. 1992. Developmental stated associated with the floral transmission. *Dev. Biol.* 153:59–69.

McDaniels, C.N., and L.K. Harnett. 1996. Flowering as metamorphosis: two sequential signals regulate floral initiation in *Lolium temulentum*. *Dev.* 122:3661–3668.

Melchers, G. and A. Lang. 1941. Weitere untersuchungen zur frage der bluhhormone. *Biol. Zentralbl.* 61:16–39.

Metzger, J.J. 1988. Localization of the site of perception of thermo-inductive temperatures in *Thlaspi arvense* L. *Plant Physiol.* 88:424–428.

Metzger, J.D. 1996. A physiological comparison of vernalization and dormancy chilling requirement. *In:* G.A. Lang (ed.). *Plant Dormancy: Physiology, Biochemistry and Molecular Biology.* CAB International, Wallingford, UK. pp. 147–156.

Metzger, J.D. and K. Dusbabek. 1991. Determination of the cellular mechanisms regulating thermo-induced stem growth in *Thlaspi arvense* L. *Plant Physiol.* 97:630–637.

Metzger, J.D., E.S. Dennis and W.J. Peacock. 1992. Tissue specificity of thermoinductive processes: *Arabidopsis* roots respond to vernalization. *Plant Physiol.* S99:52.

Millar, A.J., I.A. Carre, C.A. Strayer, N.H. Chua and S.A. Kay. 1995. Circadian clock mutants in *Arabidopsis* identified by luciferase imaging. *Sci.* 267:1161–1163.

Miller, R.O. and D.C. Kiplinger. 1966. Reversal of vernalization in northwest Easter lilies. *Proc. Amer. Soc. Hort. Sci.* 88:646–650.

Motum, G.J. and P.B. Goodwin. 1987a. Floral initiation in kangaroo paw (*Anigozanthos* spp.): a scanning electron microscope study. *Scient. Hort.* 32:115–122.

Motum, G.J. and P.B. Goodwin. 1987b. The control of flowering in kangaroo paw (*Anigozanthos* spp.). *Scient. Hort.* 32:123–133.

Napp-Zinn, K. 1969. *Arabidopsis thaliana* (L.) Heynh. *In:* Evans L.T. (ed.). *The Induction of Flowering.* Macmillan Press, Melbourne, Australia. pp. 291–304.

Napp-Zinn, K. 1985. *Arabidopsis thaliana. In:* A.H. Halevy (ed.). *Handbook of Flowering, Vol. I.* CRC Press, Boca Raton, Florida. pp. 492–503.

Nell, T.A. and H.P. Rasmussen. 1979. Blindness in roses: effects of high intensity light and blind shoot prediction severity. *Jour. Amer. Soc. Hort. Sci.* 104:21–25.

Nordwig, G. 1999. Evaluation of floral inductive requirements and commercial potential of *Asclepias* species. MS thesis. Department of Horticultural Science, University of Minnesota, St Paul, Minnesota.

Njoku, E. 1958. Effect of gibberellic acid on leaf form. *Nature* 182:1097–1098.

Ogawa, Y. 1960. Uber die auslosung der blutenbildung von *Pharbitis nil* durch nieder temperatur. *Bot. Mag.* 73:334–335.

Papenfuss, H.D. and F.B. Salisbury. 1967. Aspects of clock resetting in flowering of *Xanthium*. *Plant Physiol.* 42:1562–1568.

Pawar, S.S., and H.C. Thompson. 1950. The effect of age and size of plant at the time of exposure of low temperature on reproductive growth of celery. *Proc. Amer. Soc. Hort. Sci.* 55:367–371.

Pearson, S., P. Hadley and A.E. Wheldon. 1993. A reanalysis of the effects of temperature and irradiance on time to flowering in chrysanthemum (*Dendranthema grandiflora*). *Jour. Hort. Sci.* 68(1):89–97.

Pharis, R.P., and R.W. King. 1985. Gibberellins and reproductive development in seed plants. *Ann. Rev. Plant Physiol.* 36: 517–568.

Pietsch, G.M., W.H. Carlson, R.D. Heins and J.E. Faust. 1995. The effect of day and night temperature and irradiance on development of *Catharanthus roseus* (L.) 'Grape Cooler'. *Jour. Amer. Soc. Hort. Sci.* 120(5):877–881.

Poesch, G.H. 1931. Forcing plants with artificial light. *Proc. Amer. Soc. Hort. Sci.* 28:402–406.

Poethig, R.S. 1990. Phase change and the regulation of shoot morphogenesis in plants. *Sci.* 250: 923–930.

Pressman, E. and M. Negbi. 1980. The effect of daylength on the response of celery to vernalization. *Jour. Exp. Bot.* 31:1291–1296.

Purvis, O.N. and F.G. Gregory. 1952. Studies in vernalization of cereals. XII. The reversibility by high temperature of the vernalized condition in Petkus winter rye. *Ann. Bot.* 1:569–592.

Ramin, A.A. and J.G. Atherton. 1991. Manipulation of bolting and flowering in celery (*Apium graveolens* L. var. dulce) II. Juvenility. *Jour. Hort. Sci.* 66(6):709–717.

Ramin, A.A. and J.G. Atherton 1994. Manipulation of bolting and flowering in celery (*Apium graveolens* L. var. dulce). III. Effects of photoperiod and irradiance. *Jour. Hort. Sci.* 69:861–868.

Ranney, T.G. and M.M. Peet. 1994. Heat tolerance of five taxa of birch (*Betula*): physiological responses to supraoptimal leaf temperatures. *Jour. Amer. Soc. Hort. Sci.* 119:243–248.

Rogler, C.E. and W.P. Hackett. 1975. Phase change in *Hedera helix*: stabilization of the mature form with abscisic acid and growth retardants. *Physiol. Plant.* 34:148–152.

Ross, J.J. and I.C. Murfet. 1985. Flowering and branching in *Lathyrus odoratus* L.: environmental and genetic effects. *Ann. Bot.* 55:715–726.

Salisbury, F.B. 1963. Biological timing and hormone synthesis in flowering of *Xanthium*. *Planta* 49:518–524.

Salisbury, F.B. and H.A. Bonner. 1956. The reactions of the photoinductive dark period. *Plant Physiol.* 31:141–147.

Salisbury, F.B. and C. Ross. 1969. *Plant Physiology*, Wadsworth Publishing Co., Belmont, California. pp. 608.

Saji, H., D. Vince-Prue and M. Furuya. 1983. Studies on the photoreceptors for the promotion and inhibition of flowering in dark-grown seedlings of *Pharbitis nil* Choisy. *Plant Cell Physiol.* 67:1183–1189.

Sandved, G. 1969. Flowering in *Begonia* × *hiemalis* Fotsch. As affected by day length and temperature. *Acta Horticulturae* 14:61–66.

Schwabe, W.W. 1954. Acceleration of flowering in non-vernalized chrysanthemums by the removal of apical sections of the stem. *Nature* 174:1022.

Seeley, J.G. 1985. Finishing bedding plants, effects of environmental factors – temperature, light, carbon dioxide, growth regulators. *In:* J.W. Mastalerz and E.J. Holcomb (eds). *Bedding Plants III*. Pennsylvania Flower Growers, USA. pp. 212–244.

Sheldron, K.G. and T.C. Weiler 1982a. Regulation of growth and flowering in Basket of Gold,

Autinia saxatilis (L.) Desv. *HortScience* 17: 338–340.

Sheldron, K.G. and T.C. Weiler, 1982b. Regulation of growth and flowering of *Aquilegia* x *hybrida* Sims. *Jour. Amer. Soc. Hort. Sci.* 107:878–882.

Shillo, R. 1976. Control of flower initiation and development of statice (*Limonium sinuatum*) by temperature and daylength. *Acta Horticulturae* 64:197–203.

Singer, S.R. and C.N. McDaniel. 1986. Floral determination in the terminal and axillary buds of *Nicotiana tabacum* L. *Dev. Biol.* 118:587–592.

Siraj-Ali, Y.S., H.K. Tayama, T.L. Prince and S.A. Carver. 1990. The relationship between maturity level and splitting of poinsettia. *HortSci.* 25:1616–1618.

Smillie, R.M., C. Critchley, J.M. Bain and R. Nott. 1978. Effect of growth temperature on chloroplast structure and activity in barley. *Plant Physiol.* 62:191–196.

Stokes, P. and K. Verkerk. 1951. Flower formation in Brussels sprouts. *Meded. Landbouwhogesch. Wageningen* 50:141–160.

Strope, K. 1999. The effects of heat stress on flowering and vegetative growth of New Guinea impatiens (*Impatiens hawkeri* Bull.). MS thesis. Dept. of Horticultral Science, University of Minnesota, St Paul, Minnesota.

Suge, H. 1984. Re-examination on the role of vernalization and photoperiod in the flowering of *Brassica* crops under controlled environment. *Jap. Jour. Plant Breed.* 34:171–180.

Taiz, L. and E. Zeiger (eds). 1998. *Plant Physiology*, 2nd edn. Sinauer Associates, Inc. Sunderland, Massachusetts.

Takimoto, A. and K.C. Hamner. 1964. Effect of temperature and preconditioning on photoperiodic response of Pharbitis. *Plant Physiol.* 39:1024–1030.

Thomas, B. and D. Vince-Prue 1984. Juvenility, photoperiodism and vernalization. *In:* M.B. Wilkins (ed.). *Advanced Plant Physiology*. Pitman, London. pp. 408–439.

Thomas, B. and D. Vince-Prue. 1997. *Photoperiodism in Plants*, 2nd edn. Academic Press, New York. pp. 1–26.

Vince-Prue, D. 1975. *Photoperiodism in Plants*. McGraw-Hill, London.

Visser, T. and D.P. DeVries. 1970. Precocity and productivity of propagated apple and pear seedlings as dependent on the juvenile period. *Euphytica* 19:141–144.

Warner, R.M. and J.E. Erwin. 2001a. Effect of high-temperature stress on flower number per inflorescence of 11 *Lycopersicon esculentum* Mill. genotypes. Poster presented at the 98th International Conference of the American

Society of Horticultural Science, Sacramento, CA, July 22–25. *HortSci.* 36:508.

Warner, R.M. and J.E. Erwin. 2001b. Impact of high-temperature and irradiance on development of five bedding plant species. Poster presented at the 98th International Conference of the American Society of Horticultural Science, Sacramento, California, 22–25 July. *HortSci.* 36:550.

Warner, R.M. and J.E. Erwin. 2002. Photosynthetic responses of heat-tolerant and heat-sensitive cultivars of *Impatiens hawkeri* and *Viola × wittrockiana* to high temperature exposures. *Acta Horticulturae* 580:215–219.

Warner, R.M. and J.E. Erwin. 2003a. High temperature developmentally reduces growth and flowering of twelve pansy cultivars. IHC Program, Toronto, Canada.

Warner, R.M. and J.E. Erwin. 2003b. Effect of photoperiod and daily light integral on flowering of five *Hibiscus* spp. *Scient. Hort.* 97:341–351.

Warner, R., J.E. Erwin and R. Wagner. 1997. Photoperiod and temperature interact to affect *Gomphrena globosa* L. and *Salvia farinacea* Benth. development. *HortSci.* 32:501.

Wellensiek, S.J. 1958. Vernalization and age in *Lunaria biennis*. *Kon. Ned. Akad. Wet., Proc. Ser. C.* (Amsterdam). 61:561–571.

Wellensiek, S.J. 1959. The inhibitory action of light on the floral induction of *Perilla*. *Proc. K. Ned. Acad. Wet. Amst.* 62:195–203.

Wellensiek, S.J. 1961. Temperature. *In: Encyclopedia of Plant Physiology, Vol. 16.* Springer Verlag, Berlin, Germany. pp. 1–23.

Wellensiek, S.J. 1962. Phytotronics. Plant Science Symposium, Campbell Soup Co., Camden, New Jersey, pp. 149–162.

Wellensiek, S.J. 1964. Dividing cells as a prerequisite for vernalization. *Plant Physiol.* 39:832–835.

Wellensiek, S.J. 1965. Recent developments in vernalization. *Acta Bot. Neerl.* 14:308–314.

Wellensiek, S.J. 1966. The flower forming stimulus in *Silene armeria*. *Z. Pflanzenphysiol.* 55:1.

Wellensiek, S.J. and F.A. Hakkaart. 1955. Vernalization and age. *Proc. K. Ned. Akad. Wet. Amst. Proc. Sec. Sci.* C58:16–21.

Whitman, C.M. 1995. Influence of photoperiod and temperature on flowering of *Campanula carpatica* 'Blue Chips', *Coreopsis grandiflora* 'Early Sunrise', *Coreopsis verticillata* 'Moonbeam', *Rudbeckia fulgida* 'Goldstrum', and *Lavandula angustifolia* 'Munstead'. MS thesis, Dept. of Horticulture, Michigan State University, East Lansing, Michigan.

Whitman, C.M., R.D. Heins, A.C. Cameron and W.H. Carlson. 1996. Cold treatments, photoperiod, and forcing temperature influence flowering of *Lavandula angustifolia*. *HortSci.* 31:1150–1153.

Wiebe, H.J., R. Habegger and H.P. Liebig. 1992. Quantification of vernalization and devernalization effects for kohlrabi (*Brassica oleracea* convar. *acephala* var. *gonglyodes* L.). *Scient. Hort.* 50:11–20.

Wilkins, H.F., W.E. Waters and R.E. Widmer. 1968. University of Minnesota's Easter Lily Research Report: Paper No. II. An insurance policy: lighting lilies at shoot emergence will overcome inadequate bulb precooling. *Minnesota State Florists' Bulletin*, Dec:10–12.

Wittwer, S.J. and M.J. Bukovac. 1959. Effects of gibberellin on photoperiodic response of some higher plants. *In: Photoperiodism and Related Phenomenon in Plants and Animals.* R.B. Withrow. (ed.). American Association for the Advancement of Science, Washington, DC. pp. 373–380.

Wurr, D.C.E., J.R. Fellows, K. Phelps and A.J. Reader 1994. Testing a vernalization model on field-grown crops of four cauliflower cultivars. *Jour. Hort. Sci.* 69:251–255.

Yeh, D.M., J.G. Atherton and J. Craigon 1997. Manipulation of flowering in cineraria. IV. Devernalization. *Jour Hort. Sci.* 72:545–551.

Yuan, M. 1995. Effect of juvenility, temperature and cultural practices on flowering of *Coreopsis*, *Gaillardia*, *Heuchera*, *Leucanthemum* and *Rudbeckia*. MS thesis, Dept. of Horticulture, Michigan State University, East Lansing, Michigan. USA.

Yui, S. and H. Yoshikawa 1991. Bolting resistence breeding of Chinese cabbage. 1. Flower induction of late bolting variety without chilling treatment. *Euphytica* 52:171–176.

Zagotta, M.T., K.A. Hicks, C.I. Jacobs, J.C. Young, R.P. Hangarter and D.R. Meeks-Wagner. 1996. The *Arabidopsis* ELF3 gene regulates vegetative photomorphogenesis and the photoperiodic induction of flowering. *Plant Jour.* 10:691–702.

Zanin, P. and J.E. Erwin. 2004/5. Factors affecting flowering of *Anisodontea × hypomandarum* K. Presl. and *Leonotis menthaefolia* (Pers.) R. Br. *HortSci.* (in press).

Zeevaart, J.A.D. 1957. Studies on flowering by means of grafting. II. Photoperiodic treatment of detached *Perilla* and *Xanthium* leaves. *Proc. K. Ned. Acad. Wet. Amst.* 6D:332–337.

Zeevaart, J.A.D. 1969. Perilla. In: Evans, L.T. (ed.). *The Induction of Flowering: Some Case Histories.* MacMillan, Melbourne, Australia, pp. 116–144.

Zeevaart, J.A.D. 1983. Gibberellins and flowering. *In:* A. Crozier (ed.). *The Biochemistry and Physiology of Gibberellins.* Springer-Verlag, New York. pp. 333–374.

Zeevaart, J.A.D. and A. Lang. 1962. Physiology of flowering. *Sci.* 137:723–731.

Zhang, D., A.M. Armitage, J.M. Affolter and M.A. Dirr. 1996. Environmental control of flowering and growth of *Achillea millefolium* L. 'Summer Pastels'. *HortSci.* 31(3):364–365.

Zimmer, K. and K. Junker. 1985. *Achimenes. In:* A.H. Halevy (ed.). *Handbook of Flowering, Vol. I.* CRC Press, Boca Raton, Florida. pp 391–392.

Zrebiec, V. and H.K. Tayama. 1990. Short-day photoperiod duration influences splitting of poinsettia. *HortSci.* 25:1663.

7 Seed Development and Structure in Floral Crops

Deborah J. Lionakis Meyer

Seed Laboratory, Plant Pest Diagnostics Center, California Department of Food and Agriculture, 3294 Meadowview Road, Sacramento, CA 95832-1448, USA

Introduction

A basic knowledge of plant anatomy and morphology is critical to understanding seed development and structure in flowering plants. Technically a mature seed consists of an embryo, and in some cases additional nutritive tissue, surrounded by a protective covering. The term 'seed' is often applied to plant propagules that are not true seeds, but may consist of one or more seeds surrounded by fruiting structures and floral parts. Some authors (Booth and Haferkamp, 1995) prefer the term 'diaspore' (from the Greek word *diaspora*, meaning scattering or sowing), while others (Meyer, 2001; AOSA, 2003) prefer the inclusive term 'seed unit' to account for seeds, fruits, fruits with attached structures, etc. Although seed-bearing plants include both the gymnosperms (cone-bearing plants) and angiosperms (flower-bearing plants), this chapter will be limited to seed characteristics of angiosperms, with emphasis on floral crops.

Flowering Plants

Angiosperms or flowering plants are a diverse group including such things as grasses, herbs, vines, shrubs and trees. The name angiosperm is derived from the Greek words *angeion* and *sperma*, meaning vessel and seed, respectively. In contrast to gymnosperms (naked-seeded plants), in which the seeds are usually borne on ovulate cone scales, in angiosperms the seeds are protected in a 'vessel' called the carpel, which matures into the fruit. A carpel is a specialized, longitudinally folded, female spore-bearing leaf or megasporophyll. The carpel contains one to many ovules that develop into seeds following fertilization. A flower may have one or more separate or fused carpels, depending on the species.

Currently there are approximately 250,000 known species of flowering plants comprising the largest and most diverse group within the plant kingdom (Stern, 1997; Judd *et al.*, 1999). Angiosperms have traditionally been divided into two groups, the monocotyledons (monocots) and the dicotyledons (dicots). Represented by nearly 65,000 species (Raven *et al.*, 1999), monocots include such floral crops as alliums, bromeliads, grasses, irises, lilies, orchids, sedges, spiderworts and tulips, to name only a few. The orchid family (Orchidaceae) is the largest family of flowering plants, with approximately 24,000 species (Raven *et al.*, 1999). Dicots encompass over 300 plant families ranging from delicate herbaceous plants to enormous trees. Separation of the two groups has traditionally been based on the number of cotyledons (i.e. monocots with one, dicots with two), numbers of floral parts, type of leaf venation, vascular bundle arrangement and type of root system. Recent morphological and molecular studies have led some taxonomists to subdivide the angiosperms into four groups: monocots, eudicotyledons (eudicots), magnoliids and paleoherbs. Although the magnoliids and paleoherbs have traditionally been grouped with the eudicots under the dicotyledons because their embryos have two cotyledons, they differ in several ways. One major difference between these groups is that eudicots have triaperturate pollen (i.e. three pores or furrows), while the magnoliids and paleoherbs have monoaperturate pollen (i.e. a single pore or furrow), similar to the pollen found in monocots. Details of this recent work are beyond the scope of this chapter; however, an excellent summary of the recent systematic literature on this subject is given in Judd *et al.* (1999).

Angiosperm Life Cycle

A typical angiosperm life cycle requires two processes, fertilization and meiosis, resulting in an alternation of gametophyte and sporophyte generations. During fertilization, two haploid (1N) nuclei fuse to form a diploid (2N) zygote. Meiosis is a type of cell division in which a diploid (2N) nucleus divides to produce four haploid (1N) nuclei. All diploid cells belong to the sporophyte generation and all haploid cells belong to the gametophyte generation. The life cycles of all sexually reproducing plants have both a sporophyte and a gametophyte plant form and these forms are noticeably different. In some types of plants (e.g. mosses, liverworts and hornworts), the gametophyte plant form is dominant, while in others (e.g. ferns, clubmosses, quillworts, gymnosperms and angiosperms) the sporophyte plant form is dominant (Plate 7.1). For flowering plants, the portion of the plant visible to the naked eye is the mature sporophyte. The microscopic pollen grain and embryo sac represent the mature male and female gametophytes, respectively. The embryo contained within the seed is the young sporophyte of a new generation.

Flowers

Flowers are the sexual reproductive structures in angiosperms. A flower is a highly modified stem bearing modified leaves (floral parts). The enlarged part at the base of the flower to which the floral parts are attached is the floral axis or receptacle. Typically, a flower consists of four whorls of floral parts, two sterile and two fertile (Fig. 7.1). The lowermost sterile whorl usually consists of three or more leaf-like sepals, collectively called the calyx. The next sterile whorl usually consists of three or more, often showy, petals and is collectively referred to as the corolla. If the calyx and corolla are similar in size, shape and colour they are referred to as tepals. The calyx and corolla together form the perianth. In angiosperms, the microsporophyll (male spore-bearing leaf) is called the stamen. The typical stamen consists of an anther attached to a filament. The stamens form the first fertile whorl of floral parts, collectively called the androecium. The second fertile whorl of floral parts may consist of one or more separate or fused carpels, each carpel or group of fused carpels is called a pistil. A typical pistil consists of an ovary containing one or more ovules, a stigma that receives the pollen, and a style connecting the stigma to the ovary. Collectively, the carpels are called the gynoecium. Flowers that have all four whorls of floral parts are considered complete, while those lacking one or more whorls of floral parts are considered incomplete. A flower that has both male and female reproductive structures is said to be perfect. An imperfect flower has only male or only female reproductive structures; such flowers may be respectively referred to as staminate or pistillate.

The ripened ovary will become the fruit. Within the ovary, the portion of the ovary bearing the ovules is called the placenta. The cavity within the ovary in which the ovules are located is called the locule. An ovary may have one or more locules. Four examples of ovule arrangement within the ovary, or placentation, are shown in Plate 7.2. In an ovary with parietal placentation, as in *Helianthemum* Mill., *Lathyrus* L., *Lupinus* L. (Judd *et al.*, 1999), *Passiflora* L. and *Schlumbergera*

7.1

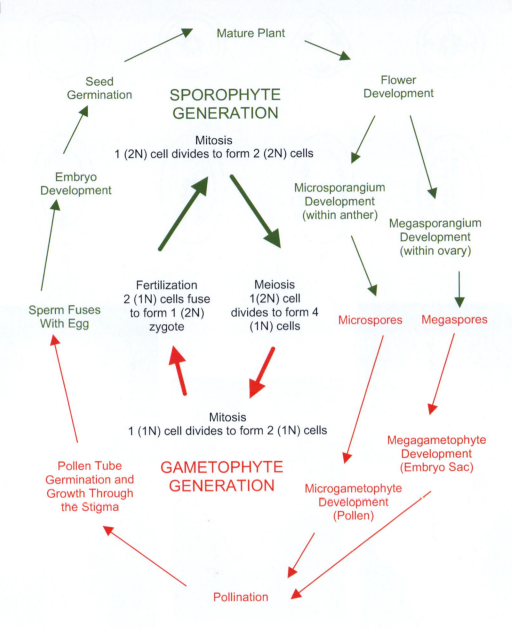

Plate 7.1. Generalized sporophyte dominant angiosperm life cycle.

7.2

Plate 7.2. Ovary placentation. **A–D.** Diagrammatic representations of four types of ovary placentation (top row cross-section and bottom row longitudinal section): **A.** axile; **B.** parietal; **C.** free central; **D.** basal. (A–D redrawn from Meyer, 2001.) **E** and **F.** Cross-section and longitudinal section, respectively, showing parietal placentation, *Schlumbergera* Lemarie. **G.** Longitudinal section showing free central placentation, *Cyclamen* L. **H.** Cross-section showing axile placentation, *Hemerocallis* L. **I.** longitudinal section showing axile placentation, *Alstroemeria* L.

7.3

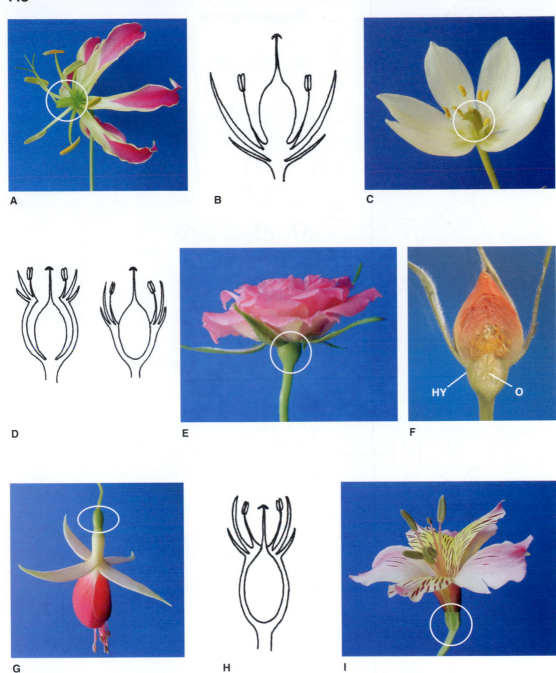

Plate 7.3. Ovary positions. Hypogynous, ovary superior: **A.** *Gloriosa superba* L. **B.** Diagram of hypogynous flower. **C.** *Ornithogalum* L. Perigynous: **D.** Diagram of perigynous flower with ovary surrounded by a cup-shaped hypanthium of fused floral parts (left) or surrounded by an extension of the hypanthium (right). **E.** *Rosa* L. **F.** Longitudinal section of rose flower showing hypanthium surrounding numerous ovaries. HY = hypanthium, O = ovaries. Epigynous, ovary inferior: **G.** *Fuchsia* L. **H.** Diagram of epigynous flower. **I.** *Alstroemeria* L. Drawings from Meyer 2001.

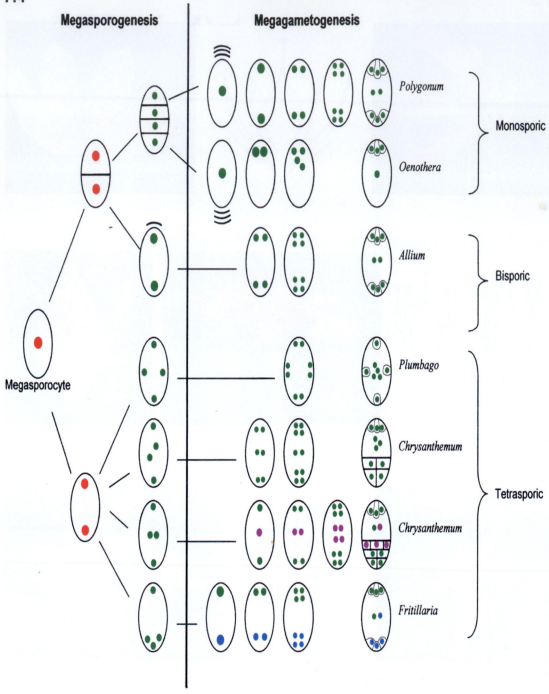

Plate 7.4. Comparison of seven types of megasporogenesis and megagametogenesis in floral genera. Micropyle oriented toward top of page. Red nuclei = 2N, green nuclei = 1N, purple fusion nuclei = 2N, blue fusion nuclei = 3N. Modified and redrawn from Foster and Gifford, 1974; Willemse and van Went, 1984.

Plate 7.5. Simple, dry dehiscent fruits. **A.** Unopened circumscissile capsule, *Portulaca* L. **B.** Circumscissile capsule opened along transverse suture, *Portulaca*. **C.** Poricidal capsule, *Papaver* L. **D.** Unopened loculicidal capsule, *Viola* L. **E.** Partially opened loculicidal capsule, *Viola*. **F.** Fully opened loculicidal capsule, *Viola*. **G.** Unopened silique (left) and silique with one side removed to expose the seed and septum (right), *Lobularia maritima* (L.) Desv. **H.** Legume, *Lathyrus odoratus* L. **I.** Craspedium with one-seeded segments separated transversely but still remaining attached to the replum, *Mimosa pudica* L.

Plate 7.6. Simple, dry indehiscent fruits. *Gomphrena globosa* L.: **A.** Inflorescence. **B.** Flower subtended by two purple bracts. **C.** Flower, showing hair-covered perianth. **D.** Perianth separated and ring of fused anthers (androecium) partially removed to expose utricle. **E.** Utricle. *Eriogonum umbellatum* Torrey: **F.** Flower with perianth segment partially removed to expose achene derived from a superior ovary. **G.** Achene (left) and seed (right). *Scabiosa* L.: **H.** Inflorescence. **I.** Fruit surrounded by involucel (fused bracts) and topped with bristle-like calyx. **J.** Longitudinal section of involucel exposing the achene derived from an inferior ovary (cypsela).

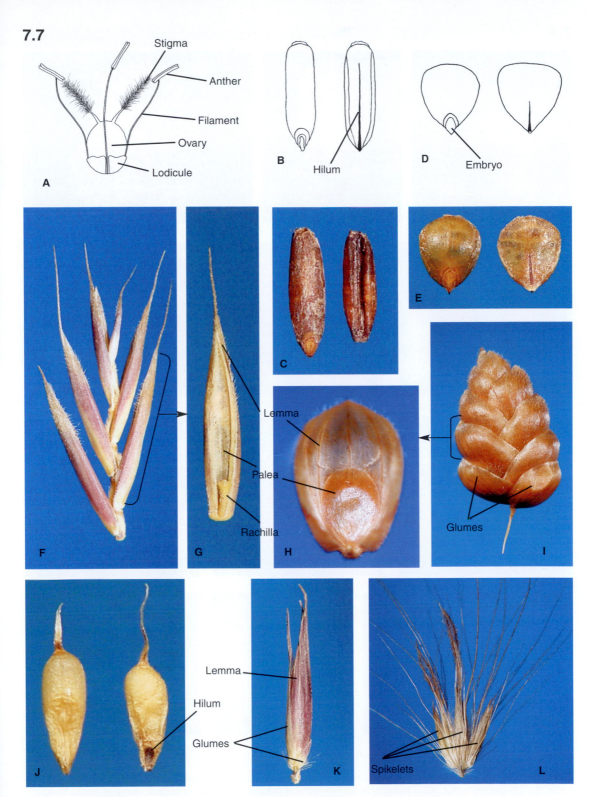

7.7

Plate 7.7. Caryopses and surrounding structures. **A.** Generalized grass flower. (A from Meyer, 2001.) *Festuca* sp.: **B** and **C.** Caryopses, dorsal (left) and ventral (right) views. *Briza maxima* L.: **D** and **E.** Caryopses, dorsal (left) and ventral (right) views. (B and D from Meyer, 2000.) *F. ovina* L.: **F.** Cluster of florets (glumes removed). **G.** Floret. *B. maxima:* **H.** Floret. **I.** Spikelet. *Pennisetum setaceum* (Forssk.) Chiov.: **J.** Caryopses, dorsal (left) and ventral (right) views. **K.** Spikelet. **L.** Cluster of spikelets with subtending bristles.

Plate 7.8. Schizocarpic nutlets: **A.** Immature fruit with four nutlets, *Lavandula angustifolia* Mill. (Lamiaceae). CA = calyx, IM = immature nutlet, RE = receptacle. **B.** Side view of mature fruit with four nutlets, *Verbena stricta* Vent. (Verbenaceae). NU = nutlet. **C.** Top view of mature fruit with one aborted nutlet and three mature nutlets, *Cynoglossum amabile* Stapf & J.R. Drumm. (Boraginaceae). AN = aborted nutlet. Lamiaceae nutlets: **D.** *L. angustifolia*. **E.** *Moluccella laevis* L. **F.** *Salvia splendens* Sellow ex Schult. **G.** *S. coccinea* Buch'hoz ex Etl. Boraginaceae nutlets: **H.** *Borago officinalis* L. **I.** *Echium vulgare* L. **J.** *Heliotropium arborescens* L. Verbenaceae nutlets: **K.** *Glandularia gooddingii* (Brig.) Solbrig. **L.** *Verbena rigida* Spreng.

7.9

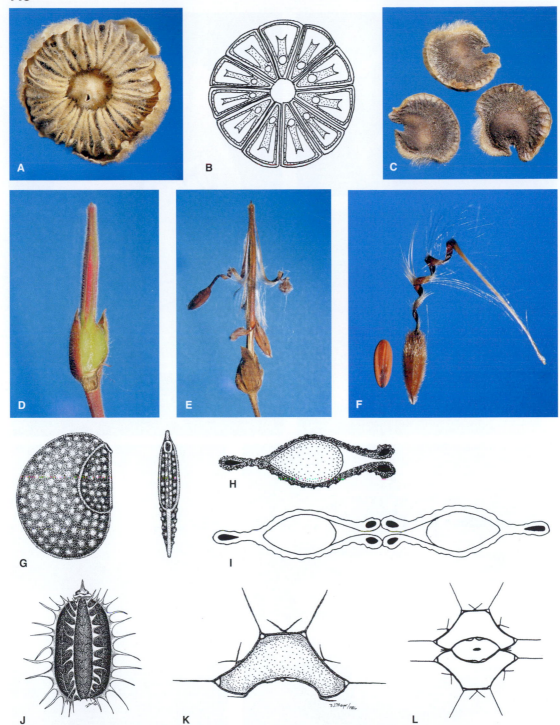

Plate 7.9. Simple, schizocarpic fruits. *Alcea rosea* L.: **A.** Top view of mature ovary. **B.** Cross-section of mature ovary showing numerous segments each containing one seed. (B from Meyer, 2001.) **C.** Single seeded fruit segments. *Pelargonium* L'Hér.: **D.** Immature fruit showing globose basal segments and elongated upper sterile portion. **E.** Mature fruit with segments separating from central column. **F.** Seed (left) and fruit segment (right). *Trachymene coerulae* Graham: **G.** Side view (left) and ventral view (right) of mericarp. **H.** Cross-section of mericarp. **I.** Cross-section of fruit. *Daucus carota* L.: **J.** Dorsal view of mericarp. **K.** Cross-section of mericarp. **L.** Cross-section of fruit. (G–L from Meyer, 1999.)

Plate 7.10. Berry with succulent pericarp: **A.** *Asparagus densiflorus* (Kunth) Jessop 'Sprengeri'. **B.** Longitudinal section of *A. densiflorus* berry showing succulent pericarp and single large seed. **C.** Cross-section of *Fuchsia* L. berry showing axile placentation with four locules. **D.** *Fuchsia* berry. Berry with leathery pericarp: **E.** Longitudinal section of dried *Lagenaria* Ser. berry.

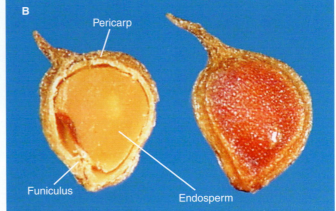

Pericarp

Funiculus

Endosperm

Ovaries

Plate 7.11. Aggregate fruits. **A.** Achenecetum, *Ranunculus* L. **B.** Longitudinal section of *Ranunculus* achene (left) and external view of achene (right). **C.** Achenecetum, *Ranunculus muricatus* L. **D.** *Helleborus* L. flower with stamens removed showing five superior ovaries that will become a follicetum. **E.** Follicetum, *Nigella orientalis* L.

7.12

Plate 7.12. Accessory fruits. Diclesia: **A.** Achene with persistent calyx, *Centranthus ruber* (L.) DC.; **B.** persistent hardened perianth and **C** longitudinal section of fruit, *Mirabilis jalapa* L.; **D.** persistent calyx (left), achene (middle) and seed (right), *Armeria* Willd.; **E.** persistent calyx and **F** longitudinal section of calyx exposing berry, *Physalis alkekengi* L. Hip: **G.** External view and **H** longitudinal section showing fleshy hypanthium surrounding achenes, *Rosa* L. Pseudocarp: **I.** Exterior covered with achenes and **J** longitudinal section exposing fleshy receptacle, achenes embedded on perimeter, *Fragaria* L. Capitulum: **K.** Inflorescence (pseudocephalium) an aggregation of single-flowered capitula and **L** achene and floral parts of the capitulum that fall together as a unit (some phyllaries and subtending bristles removed for photograph), *Echinops* L.

Plate 7.13. Examples of seed surface textures. *Continued on next page.*

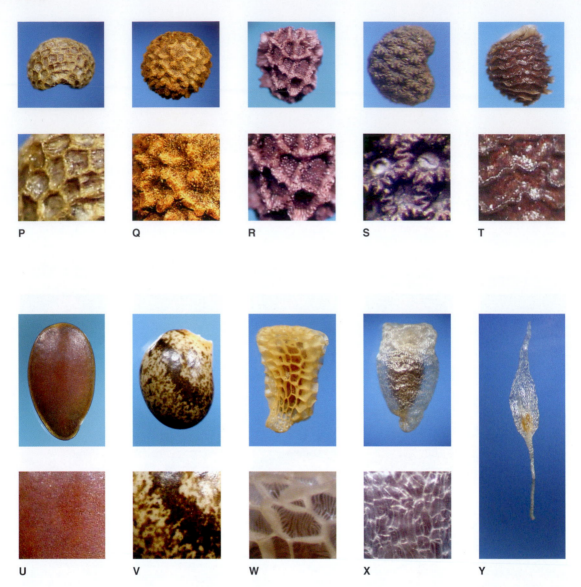

Plate 7.13. (*Continued*) Examples of seed surface textures. **A.** *Gypsophila paniculata* L. **B.** *Gypsophila elegans* M. Bieb. **C.** *Hypoestes phyllostachya* Baker. **D.** *Eschscholzia lobbii* E. Greene. **E.** *Impatiens walleriana* Hook. f. **F.** *Echinocereus engelmanii* (Englm.) Lemarie. **G.** *Phlox drummondii* Hook. **H.** *Linaria maroccana* Hook. f. **I.** *Passiflora caerulea* L. **J.** *Phacelia campanularia* A. Gray. **K.** *Browallia speciosa* Hook. **L.** *Nicotiana alata* Link & Otto. **M.** *Nicotiana glauca* Graham. **N.** Petunia Juss. **O.** *Eschscholzia californica* Cham. **P.** *Papaver orientale* L. **Q.** *Thunbergia alata* Bojer ex Sims. **R.** *Antirrhinum* L. **S.** *Schizanthus pinnatus* Ruíz & Pavón. **T.** *Consolida ajacis* (L.) Schur. **U.** *Linum grandiflora* Desf. **V.** *Lupinus polyphyllus* Lindl. **W.** *Castilleja* Mutis ex L.f. **X.** *Delphinium variegatum* Torr. & A. Gray. **Y.** *Epidendrum ibaguense* Kunth.

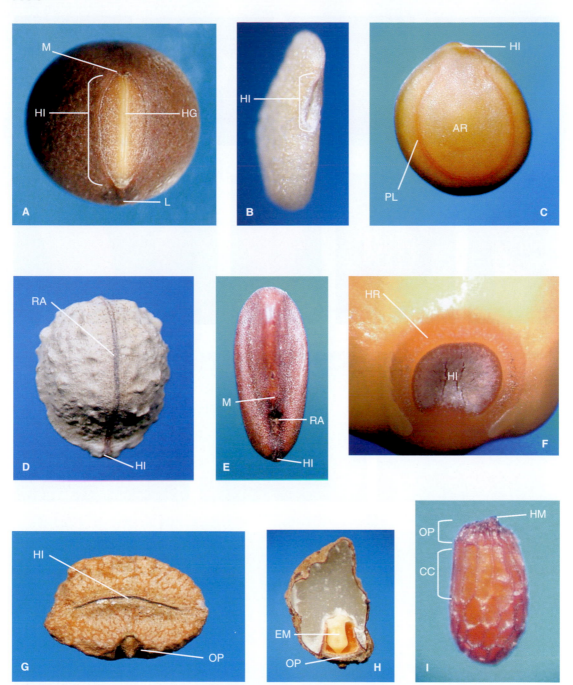

Plate 7.14. Examples of external seed features. **A.** Hilum, *Lathyrus odoratus* L. **B.** Slit-like hilum, *Physalis alkekengi* L. **C.** Pleurogram and areole, *Mimosa pudica* L. **D.** Small hilum and longitudinal raphe, *Euphorbia marginata* Pursh. **E.** Small hilum, short raphe and micropyle, *Pelargonium* L'Hér. **F.** Emarginate *Ipomoea*-type hilum with hilar ridge, *Ipomoea alba* L. **G.** Lateral operculum and linear hilum, *Commelina communis* L. **H.** Longitudinal section of seed exposing embryo and operculum, *C. communis.* **I.** Apical operculum, *Begonia* L. AR = areola, CC = collar cells, EM = embryo, HG = hilar groove, HI = hilum, HM = hilum & micropyle, HR = hilar ridge, L = lens, M = micropyle, OP = operculum, PL = pleurogram.

7.15

Plate 7.15. Specialized external seed structures. Different types of arils: **A.** *Strelitzia reginae* Banks ex Dryander (aril consisting of two tufts of multi-cellular hair-like threads, one an outgrowth of the funiculus and the other an outgrowth of the outer integument, cf. Serrato-Valenti *et al*., 1991). **B.** *Portulaca oleracea* L. **C.** *Dicentra formosa* (Andrews) Walp. (elaiosome developed primarily by cell enlargement and division of the epidermal cells, cf. Werker, 1997). **D.** *Opuntia* Mill. (bony aril). **E.** *Nemophila menziesii* Hook. & Arn. (elaiosome derived from epidermal cells of the chalaza, cf. Chuang and Constance, 1992). **F.** *Ricinus communis* L. (caruncle). **G.** *Viola cornuta* L. (elaiosome). **H.** *Passiflora* L., seed (left), fleshy pink aril surrounding seed with attached white funiculus (right). Seeds with hairs: **I.** *Epilobium canum* (Green) Raven (apical plume). **J.** *Crossandra nilotica* Oliver (fringed scales). **K.** *Asclepias* L. (apical plume). Winged seeds: **L.** *Asclepias* L. **M.** *Nemesia* Vent. **N.** *Gladiolus* L. **O.** *Lunaria annua* L.

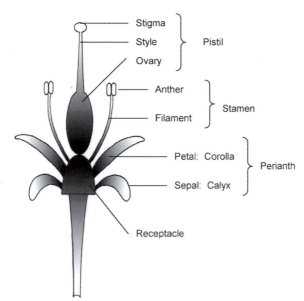

Fig. 7.1. Parts of a generalized flower.

Lemárie, the ovules are borne on the ovary wall. In a partitioned ovary with several locules, such as is found in *Antirrhinum* L., *Fuchsia* L. (Esau, 1977), *Ipomoea* L., *Petunia* Juss., *Phlox* L. (Judd *et al.*, 1999), *Alstroemeria* L. and *Hemerocallis* L., the ovules are borne on a central column of tissue and the placentation type is axile. In *Dodecatheon* L. (Esau, 1977), *Exacum* L., *Gentiana* L. (Mabberley, 1997), *Primula* L. (Judd *et al.*, 1999) and *Cyclamen* L., the ovary is not divided (unilocular) and the ovules are borne on a central column of tissue. This type of ovule arrangement is called free central placentation. In an ovary with basal placentation, as found in the Asteraceae (Fahn, 1982), a single ovule is borne at the base of the unilocular ovary.

When the sepals, petals and stamens are attached below the ovary, the flower is hypogynous and the ovary is in a superior position (Plate 7.3, A–C). When these structures are attached above the ovary, the flower is epigynous and the ovary is in an inferior position (Plate 7.3, D–F). In some flowers, the ovary is surrounded by a cup-shaped hypanthium derived either from the fusion of the sepals, petals, and stamens, or an extension of the receptacle to which the floral parts are attached. These flowers are called perigynous (Plate 7.3, G–I). Although not part of the true seed, the ovary and other floral parts that surround the ovary may remain attached to the mature seeds and become part of the dispersed plant propagule.

Ovule

In seed-bearing plants, the megagametophyte or female gametophyte is retained within the megasporangium, which is called the nucellus and is usually surrounded by one or two layers of tissue called integuments. This entire structure is called an ovule. The ripened ovule will eventually become the seed. The number of ovules within the ovary varies among species. Ovule development begins with cell divisions below the surface of the placenta. As shown in Fig. 7.2, the ovule primordium appears as a conical projection as the nucellus enlarges and the integuments begin to surround the nucellus, eventually leaving only a small opening called the micropyle (Fahn, 1982; Bouman, 1984). In approximately 50% of angiosperm families, nucelli are rather large, increasing in size through meristematic activity, and the ovules containing such nucelli are called crassinucellate (Rangan and Rangaswamy, 1999). In the remaining half of the angiosperm families, the nucelli are thin, comprising one to three cell layers, and ovules with these nucelli are referred to as tenuinucellate. Most tenuinucellate dicot

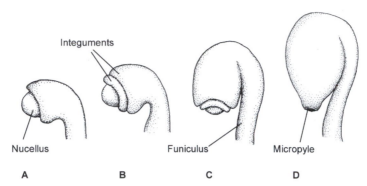

Fig. 7.2. Diagrammatic illustration of ovule development: A – integument beginning to surround the nucellus; B – further integument development; C – elongation of funiculus and ovule; D – integuments surround nucellus except for micropylar opening (based on Fahn, 1982, and Bouman, 1984).

ovules are unitegmic (i.e. have only one integument). The nucellus appears to play a role in nutrient transfer between the mother plant and the developing embryo sac (Rangan and Rangaswamy, 1999). By the time the embryo sac is completely developed the nucellus usually degenerates, or in some instances it may persist in the mature seed as a storage tissue called perisperm (e.g. Amaranthaceae, Cactaceae, Caryophyllaceae, Nyctaginaceae, Portulacaceae).

As the ovule matures, the histological structure of the integuments may change. In many plants the inner epidermis of the inner (or single) integument develops into the integumental tapetum or endothelium (Fahn, 1982; Werker, 1997). Cells in this layer usually have large nuclei and dense cytoplasm containing numerous organelles. In species where the nucellus is almost completely absorbed during megagametogenesis, this layer comes in direct contact with the embryo sac. Presence of an integumental tapetum has been reported in 65 dicot and seven monocot families, and is typically found in the more evolutionarily advanced families. In most cases this layer is absorbed during seed maturation, but can still be found in mature seeds of the Polemoniaceae, Linaceae, Scrophulariaceae, Solanaceae and others. During seed development, the integumental tapetum is involved in the transfer and accumulation of nutrients, in enzyme secretion into integumentary tissue, and serves as a barrier to restrict growth of the embryo and endosperm (Werker, 1997).

The funiculus connects the ovule to the placenta within the ovary. Although usually cylindrical, the funiculus can vary in shape, length and curvature or can be completely lacking, in which case the ovules are sessile, as in *Dicentra* Benth. (Werker, 1997). The anatomy, vasculature and function of the funiculus vary considerably. However, in early ovule and seed development, the funiculus transfers nutrients and water from the mother plant to the developing ovule and seed, and directs the pollen tube to the micropyle (Werker, 1997). The area where the integuments and nucellus fuse with the funiculus is called the chalaza.

Ovules are classified by the shape and position of the different ovular parts (Fig. 7.3). Anatropous and orthotropous (atropous) ovules have straight nucelli and campylotropous and amphitropous ovules have curved nucelli. Some families may have examples of more than one ovule type (Mabberley, 1997; Judd *et al.*, 1999). In the species that have been studied, anatropous ovules occur most often (Bouman, 1984). During ontogeny of an anatropous ovule, intercalary growth of the funiculus causes the ovule to rotate 180°, shifting the micropyle toward the future hilum. Bouman (1984) reported that orthotropous ovules occur in about 20 families (e.g. Araceae, Cistaceae and Polygonaceae). The funiculus, chalaza, nucellus and micropyle are in an upright

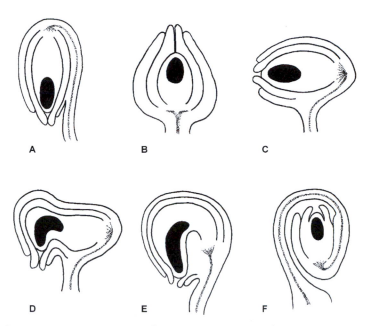

Fig. 7.3. Ovule types: A – anatropous; B – orthotropous (atropous); C – hemianatropous; D – campylotropous; E – ammphitropous; F – circinotropous. Embryo sac shown in black, course of vascular strand stippled (based on Foster and Gifford, 1974, and Fahn, 1982).

linear arrangement in an orthotropous ovule. Ovules in which the nucellus is perpendicular to the funiculus are called hemianatropous. In circinotropous ovules (e.g. Cactaceae, Plumbaginaceae), rotation exceeds 180°, causing the funiculus to nearly encircle the ovule (Bouman, 1984). Examples of families in which the ovules contain a curved embryo sac include: Aizoaceae, Amaranthaceae, Brassicaceae, Caryophyllaceae, Fabaceae, Malvaceae, Nyctaginaceae and Portulacaceae (Judd *et al.*, 1999). Names used to describe such ovules include campylotropous (slightly curved, kidney-shaped embryo sac) and amphitropous (strongly curved, horseshoe-shaped embryo sac). In non-orthotropous type ovules, the funiculus is partially or wholly adnate to the integument forming the raphe (Werker, 1997). The length of the raphe depends on the ovule type.

Embryo sac development

There are three major types of megaspore development (megasporogenesis): monosporic (from a single megaspore nucleus), bisporic (from two megaspore nuclei) and tetrasporic (from four megaspore nuclei) leading to various types of embryo sac development or megagametogenesis (Plate 7.4). Of the angiosperm species studied so far, the monosporic type megasporogenesis is most common (Willemse and van Went, 1984). Typical monosporic megasporogenesis leads to development of an embryo sac consisting of seven cells: the egg cell, two synergids, the central cell and three antipodal cells. This type of embryo sac is referred to as the 'normal' or '*Polygonum*' type (Fig. 7.4).

Development of the normal or *Polygonum* type embryo sac begins within the nucellus, as a single megasporocyte (2N) that develops and divides meiotically to produce a linear tetrad of megaspores, three of which degenerate, leaving one functional megaspore (1N). The functional megaspore enlarges and undergoes three mitotic divisions yielding eight nuclei. The nuclei are arranged in two groups of four, one group at the micropylar end of the cell and one group at the opposite end. One nucleus from each group migrates to the centre of the cell and both are referred to as polar nuclei. Cell walls form around the remaining six nuclei. At the end opposite

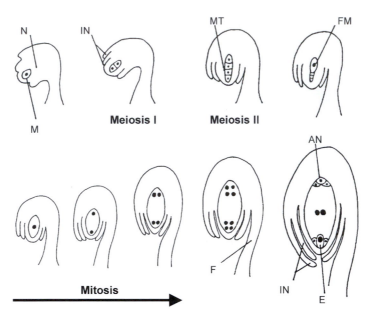

Meiosis I   Meiosis II

Mitosis

Fig. 7.4. Development of an angiosperm ovule and formation of a *Polygonum* type embryo sac.
AN = antipodals, E = egg, F = funiculus, FM = functional megaspore, IN = integuments, M = megasporocyte, MT = megaspore tetrad, N = nucellus (from Meyer, 2001; redrawn from Gifford and Foster, 1989).

the micropyle, the three cells are called antipodals, and at the micropylar end, the egg apparatus consists of an egg cell flanked by two synergids. The region between the antipodals and egg apparatus is called the central cell and contains the polar nuclei. This eight-nucleate, seven-celled structure is the mature female gametophyte or embryo sac (Fig. 7.4).

In the normal or *Polygonum* type embryo sac, the antipodals and the cells of the egg apparatus are haploid. In the large and highly vacuolate central cell the two haploid polar nuclei eventually fuse to form a diploid fusion nucleus (Willemse and van Went, 1984), or secondary nucleus (Fahn, 1982), or primary endosperm nucleus (Esau, 1977). Timing of this fusion can vary among species, occurring either before or during fertilization (Maze and Bohm, 1973; Newcomb, 1973; Esau, 1977; Fahn, 1982). As shown in Plate 7.4, the synergids and egg cell are haploid; however, the ploidy level of the central cell and antipodal cells varies among different embryo sac types (Willemse and van Went, 1984). The number of cells constituting embryo sacs in other angiosperm species may vary because of the lack of one or more synergids (no

synergids are formed in *Plumbago* L.), and an increase or decrease in the number of antipodal cells. The number of antipodal cells varies among species, for example, zero in *Oenothera* L. (Willemse and van Went, 1984) two in *Helianthus* L. and most Asteraceae (Newcomb, 1973), seven in *Chrysanthemum* L. (Willemse and van Went, 1984) and up to more than 100 cells in certain grasses (Maze and Bohm, 1973). The antipodals are usually the smallest of the cells of the embryo sac and in many plants the antipodal cells degenerate during maturation of the embryo sac, while in some species they persist through embryo and endosperm formation (Willemse and van Went, 1984).

Fertilization

In angiosperms pollination involves the transfer of pollen grains (young male gametophyte) from the mature anther to the sticky surface of a receptive stigma of the appropriate species. In most cases this transfer is achieved via external agents (wind currents, gravity, water, insects, birds, mammals, etc.). If the pollen grain finds a suitable stigma the male gametophyte will continue to develop into a mature male gametophyte as follows: the pollen tube emerges from an aperture in the pollen grain wall and grows through the stigma and style; the tube nucleus is usually located near the tip of the pollen tube and the two male gametes are positioned behind the tube nucleus however, this order may be reversed in some species (Fahn, 1982); after reaching the ovary, the pollen tube may enter an ovule by several possible routes, most commonly through the micropyle and the nucellus (Esau, 1977) (Fig. 7.5).

The fertilization processes described below may vary slightly among species (van Went and Willemse, 1984); however, in general, when contact is made with the embryo sac, the pollen tube ruptures releasing the two male gametes and tube nucleus into one of the two synergid cells (degenerate synergid) through the thickened micropylar end of the degenerate synergid called the filiform apparatus. The male gametes travel through the degenerate synergid and one is transferred to the egg and the other to the central cell. The synergids and antipodal cells degenerate. One male gamete and the egg fuse to form the zygote (2N). The other male gamete fuses with the polar nuclei of the central cell to become the primary endosperm nucleus. This process of gametic fusion in angiosperms is called double fertilization. The primary endosperm nucleus divides repeatedly to produce the endosperm. Ploidy of the endosperm is dependent upon the number and ploidy of the nuclei that fuse in the central cell.

Fertilization in angiosperms usually initiates the following processes: development of the zygote into an embryo; development of endosperm, a food reserve; development of seed coat from integuments; absorption or disintegration of nucellar tissue, although some plants retain this tissue as a food reserve called perisperm; maturation of the ovary to form a fruit; and, in some species, stimulation of accessory floral parts (e.g. receptacle, sepals, petals) to increase growth for incorporation into the fruiting structure.

Seeds may sometimes develop in the absence of fertilization. Apomixis is a type of reproduction in which an embryo develops

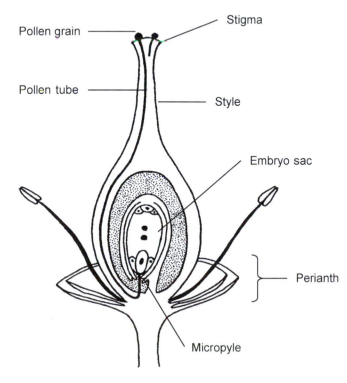

Pollen grain

Pollen tube

Stigma

Style

Embryo sac

Perianth

Micropyle

Fig. 7.5. Longitudinal section through flower showing one possible route of pollen tube growth through the style and ovary wall, entering the ovule via the micropyle. Sepals and petals (perianth) partially removed (revised from Meyer, 2001).

within a seed without the formation and fusion of gametes. In this type of asexual reproduction, an embryo may develop in one of two basic ways, depending on the species. In the first, the embryo may develop from a cell within an embryo sac following meiosis, in which case the embryo and subsequent plant will be haploid and sterile. Secondly, the embryo may develop from a diploid cell of an ovule, either from a sporogenous cell, other cells of the nucellus or from the integuments. Embryos produced from diploid cells of the mother plant are genetically identical to that plant. Over 300 species of angiosperms, representing more than 35 families, reproduce apomictically (Raven *et al.*, 1999). Apomictic embryos derived from nucellar tissue have been reported for 29 angiosperm families (e.g. Araceae, Cactaceae, Euphorbiaceae, Fabaceae, Liliaceae, Onagraceae, Orchidaceae, Poaceae, Rosaceae) (Rangan and Rangaswamy, 1999). Families in which apomictic embryos are derived from sporogenous cells include Asteraceae and Poaceae (Nogler, 1984).

Embryo development

During embryo development (embryogenesis), the structural axis of the new sporophyte body, apical meristems and patterns of tissue systems are established. This development may occur within a few hours following fertilization or may be delayed by days, weeks or months (Maheshwari, 1950; Fahn, 1982; Matsumoto *et al.*, 1998). Embryo development can be divided into three general phases: (i) establishment of the embryonic axis, shoot and root apices, embryonic tissue and organ systems; (ii) embryo maturation, including accumulation of storage reserves; and (iii) developmental arrest, embryo desiccation and metabolic quiescence (West and Harada, 1993).

Variations in cell wall orientation and the sequence of cell divisions in the first phase of embryo development are the basis for classification of embryo types. Traditionally, six embryo types have been recognized (Johansen, 1950) and these are briefly described in Table 7.1. Within each of these embryo types, slight variations occur, leading

some authors to suggest more elaborate classification schemes. For a detailed summary on classification of embryo types see Johansen (1950) and Natesh and Rau (1984). A seventh type of embryo development (Table 7.1) was reported in *Paeonia* L., in which the first division of the zygote and several subsequent divisions lack cell wall formation, yielding a free-nuclear proembryo (Raghavan, 1986; Johri, 2001). A large vacuole occupies the central part of the proembryo and the nuclei are peripheral. Eventually, cell wall formation takes place followed by differentiation into a typical dicot embryo. In Orchidaceae, the cell divisions are irregular and arrested so early that it is not possible to assign the embryos to a type (Davis, 1966; Johri, 2001).

Histological analysis indicates that cotyledon and epicotyl initials are established in the early globular stage of embryo development (West and Harada, 1993). In general, establishment of the epicotyl and cotyledon(s) follows one of two pathways: (i) in dicots, early cell divisions in the apical region of the globular embryo result in the establishment of four sectors, of which two sectors differentiate into the two cotyledons and the two remaining sectors become the epicotylar meristem; and (ii) in monocots, the apical region of the proembryo is bisected, the cotyledonary and epicotylar initials each occupy one-half of the apex (Swamy and Krishnamurthy, 1977). In both monocots and dicots, the cells of the cotyledons divide more rapidly than those of the epicotylar meristem (Swamy and Krishnamurthy, 1977; Lakshmanan, 1978; West and Harada, 1993). In the case of monocots, this differential growth rate gives the impression that the shoot apex originates from a lateral position, and this point has been the topic of much debate (Swamy and Krishnamurthy, 1977; Lakshmanan, 1978; Fahn, 1982; Natesh and Rau, 1984; Gifford and Foster, 1989; Matsumoto *et al.*, 1998).

Embryo development of *Dianthus chinensis* L., China pinks (Fig. 7.6), as described by Buell (1952), follows the caryophyllad type embryo development, except for the formation of two large basal cells rather than a single cell typical of this embryo type. The zygote divides transversely to produce

Table 7.1. Classification of seven basic embryo types, important character states, and flower seed families in which the embryo type has been reported (based on Johansen, 1950; Raghavan, 1986; Johri, 2001).

Embryo type	Plane of zygotic division	Plane of terminal cell division	Basal cell contribution to embryo proper	Families with reported occurrence
Onagrad (Crucifer)	Transverse	Longitudinal	Minimal	Onagraceae, Brassicaceae, Lamiaceae (*Mentha*), Scrophulariaceae (*Veronica*), Liliaceae (*Lilium*)
Asterad	Transverse	Longitudinal	Substantial	Asteraceae, Geraniaceae, Lamiaceae, Liliaceae, Poaceae, Polygonaceae, Rosaceae
Solanad	Transverse	Transverse	None	Solanaceae, Linaceae, Papaveraceae, Rubiaceae
Caryophyllad	Transverse	Transverse	None	Caryophyllaceae, Crassulaceae, Fumariaceae, Portulacaceae, Alismataceae, Araceae, Ruppiaceae, Zannichelliaceae
Chenopodiad	Transverse	Transverse	Substantial	Chenopodiaceae, Amaranthaceae, Polemoniaceae
Piperad	Longitudinal	Longitudinal		Balanophoraceae, Dipsacaceae (*Scabiosa*), Piperaceae
Paeonia	No cell wall formation	No cell wall formation		Paeoniaceae (*Paeonia*)

two unequal cells; a large basal cell near the micropylar end of the embryo sac and a smaller apical cell. The apical cell divides transversely to produce a second basal cell. The two basal cells enlarge but do not divide again. The proembryo consists of these three cells. Further transverse divisions of the apical cell produce a linear embryo. Longitudinal and further transverse divisions produce a spherical embryo. By the time the embryo becomes spherical, cells within the embryo begin to differentiate to form the precursors of tissue systems (i.e. dermal, cortical and vascular). Just prior to cotyledon formation, the embryo apex becomes flattened as cell divisions form the cotyledon primordia, giving the embryo a heart-shaped appearance. At the same time, cell divisions between the cotyledon primordia give rise to the epicotyl primordium, and cell divisions near the suspensor give rise to the root cap initials and root meristem. With further cell division and elongation of the cotyledons and hypocotyl the embryo becomes torpedo-shaped. During the torpedo stage the suspensor and basal cells begin to disintegrate and are absorbed by the endosperm. The mature embryo lies straight and fills a large portion of the seed.

During the early stages of embryo development, the endosperm is free nuclear, and becomes cellular during the late spherical stage. As the embryo matures, the central portion of the endosperm is digested and only a single layer of endosperm cells remains in the mature seed. The ovule continues to enlarge to accommodate the growing embryo. Much of the nucellus in the micropylar and chalazal regions is digested by the growing endosperm. The remaining lateral cells of the nucellus continue to enlarge as the seed matures and perisperm starch is accumulated in two flat lateral wings of the seed. The mature seed is dorso-ventrally flattened, with the embryo outline visible on the dorsal surface (Fig. 7.6).

Studies have shown that the proembryos of dicots and monocots are very similar in development up to the globular stage (Esau, 1977). In the non-grass monocot, *Allium cepa* L., the young embryo is club-shaped, becoming spherical with a thin suspensor (Fig. 7.7, A). The cotyledon develops upward from the spherical-shaped embryo. As the cotyledon

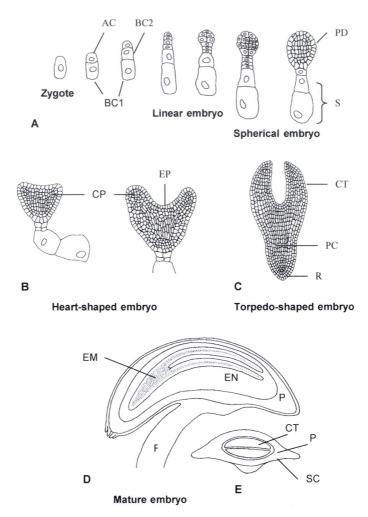

Fig. 7.6. Embryo development in *Dianthus chinensis* L., china pinks. A – division of zygote into terminal and basal cell with further divisions to form the linear and spherical embryo stages; B – flattening of embryo into the heart-shaped stage, development of cotyledon and epicotyl primordia; C – torpedo-shaped embryo; D – longitudinal section of mature embryo surrounded by endosperm and perisperm within the seed; E – cross-section of mature seed. (AC = apical cell, BC1 = first basal cell, BC2 = second basal cell, CT = cotyledon, CP = cotyledon primordia, EM = embryo, EN = endosperm, EP = epicotyl primordium, F = funiculus, P = perisperm, PC = procambium, PD = protoderm, R = root apex, S = suspensor, SC = seed coat.) (Modified and redrawn, based on Buell, 1952.)

elongates, a notch forms on one side. This notch deepens as a marginal sheath-like extension of the cotyledon forms. Within the notch surrounded by the sheath, the shoot apex develops. During germination, the first leaf will emerge from this enclosure through a small slit above the sheath. In *Commelina* L., the shoot apex is surrounded by a cotyledonary sheath and the bulk of the cotyledon

over-tops the shoot apex at a 90° angle (Fig. 7.7, B) (Maheshwari and Baldev, 1958; Lakshmanan, 1978).

In grasses, following fertilization, numerous cell divisions produce a club-shaped embryo (Fig. 7.8, A and B) (Esau, 1977). The upper enlarged part gives rise to the main body of the embryo and the lower part is the suspensor. The shield-shaped cotyledon,

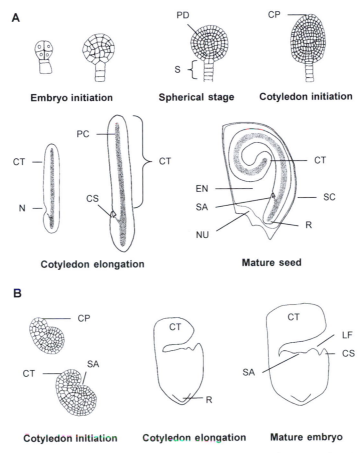

Fig. 7.7. Embryo development: A - *Allium cepa* L., longitudinal sections showing embryo initiation, spherical stage, cotyledon initiation, cotyledon elongation and mature seed; B – *Commelina benghalensis* L., longitudinal sections showing cotyledon initiation, cotyledon elongation and mature embryo. (CS = cotyledon sheath, CT = cotyledon, CP = cotyledon primordium, EN = endosperm, LF = leaf, N = notch, NU = nucellus, PC = procambium, PD = protoderm, R = root apex, S = suspensor, SA = shoot apex, SC = seed coat.) (A – from Meyer, 2001, based on Esau, 1977; B – modified and redrawn from Lakshmanan, 1978.)

or scutellum as it is called in grasses, becomes elongated and the shoot apex, surrounded by the coleoptile, becomes visible on one side (Fig. 7.8, C and D) (Esau, 1977). At the lower end of the embryo axis, the radicle and root cap develop above the suspensor (Fig. 7.8, E) (Avery, 1930). The coleorhiza tissue becomes separated from the radicle as the embryo matures. For comparison, see embryos of *Deschampsia cespitosa* (L.) P. Beauv., *Phalaris arundinacea* L. and *Cortaderia selloana* (Schult.) Aschers. & Graebn. shown in Fig. 7.8, F–H.

Fruits

Many plant propagules are not true seeds, but are, in fact, fruits or fruit segments containing seeds. In many cases, it is difficult to recognize the difference between a true seed and a seed-like fruit. Strictly defined, a fruit is a ripened ovary. However, a broader definition more commonly used recognizes the fruit as a derivative of the gynoecium and any other extracarpellary parts the gynoecium may be united with during maturation (Esau, 1977). The following

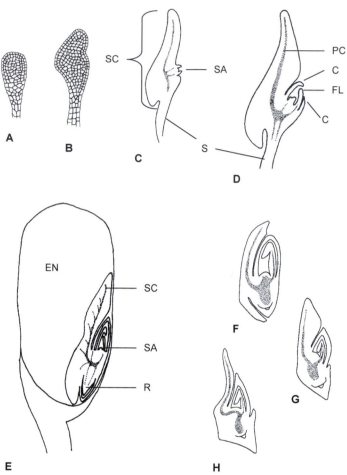

Fig. 7.8. Embryo development in *Zea mays* L. Longitudinal sections of: A – 5-day old club-shaped embryo; B – 10-day old embryo showing scutellum initiation; C – 15-day old embryo; D – scutellar elongation and differentiation of shoot apex; E – mature embryo within caryopsis. Longitudinal sections of grass embryos: F – *Cortaderia selloana* (Schult.) Aschers. & Graebn.; G – *Deschampsia cespitosa* (L.) P. Beauv.; H – *Phalaris arundinacea* L. (C = coleoptile, EN = endosperm, FL = first leaf, PC = procambium, R = root apex, S = suspensor, SA = shoot apex, SC = scutellum.) (A–E from Meyer, 2001, adapted from Esau, 1977 and Avery, 1930; F–H adapted from Reeder, 1957.)

discussion of fruits is based on Radford (1986) as these terms are in general use in most botany textbooks. Classification of fruits into the general grouping of simple, aggregate, multiple and accessory is based primarily on origin (i.e. number and location of carpels), texture (i.e. dry or fleshy), ability to spontaneously open at maturity (dehiscence) and fusion with extracarpellary parts. Placement of fruits into these groups is arbitrary, since a satisfactory method for fruit classification has yet to be developed (Radford, 1986;

Judd *et al.*, 1999). Spjut, (1994) gives an extensive historical review of the problems associated with fruit classification and reintroduces carpological terminology that has been forgotten over the centuries.

A simple fruit is derived from a single pistil consisting of one carpel or several fused carpels. An aggregate fruit is derived from a number of separate carpels of one gynoecium. A multiple fruit is derived from a combination of gynoecia from many flowers within an inflorescence. Fruits that incorporate

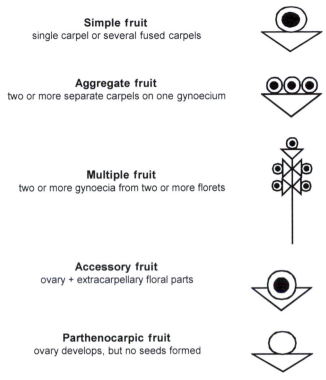

Simple fruit
single carpel or several fused carpels

Aggregate fruit
two or more separate carpels on one gynoecium

Multiple fruit
two or more gynoecia from two or more florets

Accessory fruit
ovary + extracarpellary floral parts

Parthenocarpic fruit
ovary develops, but no seeds formed

Fig. 7.9. General fruit types: simple, aggregate, multiple, accessory and parthenocarpic. Each clear circle represents a pistil containing an ovule, solid circles represent seeds; each triangle represents a floral receptacle (from Meyer, 2001).

extracarpellary parts (calyx, receptacle, bracts, etc.) are called accessory fruits. It is generally accepted that fruits develop after fertilization, but this is not always the case. Parthenocarpic fruits are produced in plants such as bananas and certain varieties of grapes and oranges in which the ovary develops without fertilization and the development of seeds. Not all seedless fruits are parthenocarpic. The seedless condition may be artificially induced by hormonal treatment, by hybridization of plants with incompatible chromosome numbers, or by the lack of seed development following fertilization. General fruit types are shown in Fig. 7.9.

In simple fruits, the mature fruit wall or pericarp may be either hard and dry or soft and fleshy. The pericarp may consist of three regions: outermost is the exocarp, central is the mesocarp, and innermost is the endocarp, as found in *Olea europaea* L. (Fig. 7.10). Hard, dry fruits may be further classified as

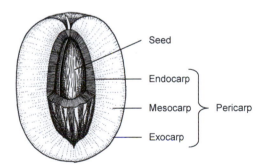

Fig. 7.10. Parts of a mature *Olea europaea* L. (olive) fruit (from Meyer, 2001).

dehiscent or indehiscent based on their ability to split open and shed seeds upon maturity.

Dry, dehiscent fruits

Simple, dry, dehiscent fruits include various types of capsules derived from one or more carpels (e.g. *Agapanthus* L'Hér., *Dietes* Salisb. ex Klatt, *Papaver* L., *Portulaca* L., *Viola* L.). At maturity capsules may open by various methods (slits, pores, splitting along transverse or longitudinal sutures, etc.) to release the seeds. Examples of other dry, dehiscent, simple fruits include: legume – derived from one carpel that splits open along two sutures, as in *Lathyrus*, *Lupinus* and *Phaseolus* L.; follicle – consisting of a single carpel that splits open along one suture, as in *Asclepias* L.; silique – derived from two or more carpels that open along two sutures and has a persistent marginal replum (rim) and septum (partition), as in *Lobularia maritima* (L.) Desv.; craspedium – consisting of a single carpel that disarticulates transversely into one-seeded segments and longitudinally from the marginal replum, as in *Mimosa pudica* L. A few of the many possible forms of simple, dry, indehiscent fruits are shown in Plate 7.5.

Dry, indehiscent fruits

A one-seeded, dry, indehiscent fruit in which the seed is attached to the fruit wall at only one point is called an achene. The seed coat is usually very thin or in some cases may disintegrate during development. Achenes may be derived from either a superior (e.g. *Eriogonum* Michx., *Ranunculus* L.) or an inferior ovary (e.g. Asteraceae). Achenes derived from an inferior ovary to which accessory structures remain attached (i.e. pappus in Asteraceae, connate involucel of fused bracts and a bristle-like calyx in Dipsacaceae) have been referred to as cypselas by some authors. Examples of various achenes are shown in Fig. 7.11 and Plate 7.6. A utricle is a one-seeded, dry, indehiscent fruit with a thin bladder-like pericarp that does not tightly adhere to the seed (Plate 7.6). At maturity, utricles often remain surrounded by floral structures and subtending bracts, as in *Gomphrena* L. and *Atriplex* L. For further discussion on extracarpellary structures, refer to the section on accessory fruits.

The caryopsis is the typical fruit of the grass family (Poaceae). A caryopsis is a one-seeded, dry, indehiscent fruit derived from a superior ovary. Size, shape and anatomy of caryopses vary considerably within the

family and are useful taxonomic characters (Sendulsky *et al.*, 1986). The ovule is attached to the ovary wall by a short funiculus along the ventral suture or at its base. The pericarp is thin and fused to the seed coat in the hilar region only. The area of pericarp and seed coat fusion is reflected in the size and shape of the hilum. In grasses, the term 'hilum' refers to the outer marking (usually circular to linear) on the pericarp, on the side of the caryopsis opposite the embryo that reflects the inner attachment of the ovule funiculus to the ovary wall. The remainder of the pericarp is closely adnate over the seed coat surface. Some authors consider the pericarp and seed coat to be completely fused as the result of cellular destruction and compression during the growth of the embryo and endosperm within the ovary (see Sendulsky *et al.*, 1986, and Spjut, 1994, for further details). In grasses, the fruit may remain surrounded by accessory structures (e.g. lemma, palea, glumes) at the time of dispersal, and the term 'floret' applies to more than the individual flower. In most grasses, the flower consists of a pistil with a superior ovary, stamens and two scale-like structures called lodicules. Two bracts surround the grass flower; the upper is the palea and the lower is the lemma. This entire structure is called a floret. One or more florets may be grouped together on an axis called the rachilla and two additional bracts called glumes subtend the floret or group of florets. This entire structure is called a spikelet. Additional bristle-like structures may subtend the spikelet. Spikelets are usually clustered into larger groups of various arrangements to form the inflorescence. Examples of ornamental grasses include *Briza maxima* L., *Festuca ovina* L., and *Pennisetum setaceum* (Forssk.) Chiov. (Plate 7.7).

Schizocarpic fruits

A schizocarpic fruit is a simple fruit derived from a compound pistil consisting of two or more carpels. At maturity, the carpellary components of the gynoecium separate and, in some species, the carpels will separate into half-carpels. Each fruit segment simulates a fruit derived from a simple pistil and functions as a separate, typically one-seeded,

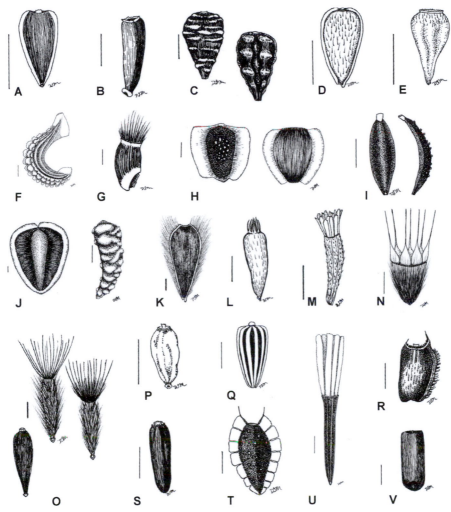

Fig. 7.11. Simple dry indehiscent fruit, achenes (cypselas) of the Asteraceae: A – *Achillea millefolium* L. (common yarrow); B – *Ageratum* L. (floss flower); C – *Arctotis fastuosa* Jacq. (Cape daisy); D – *Bellis perennis* L. (English daisy); E – *Brachycome iberidifolia* Benth. (Swan River daisy); F – *Calendula officinalis* L. (calendula); G – *Centaurea cyanus* L. (bachelor's-button); H – *Coreopsis lanceolata* L. (perennial coreopsis); I – *C. tinctoria* Nutt. (calliopsis); J – *Dimorphotheca* Moench., flake (left), stick (right) (African daisy); K – *Encelia californica* Nutt.; L – *Erigeron speciosus* (Lindl.) DC. (aspen daisy); M - *Eriophyllum confertiflorum* (DC.) A. Gray (golden yarrow); N – *Gaillardia pulchella* Foug. (Indian blanket-flower); O – *Layia platyglossa* (Fisch. & C.A. Mey.) A. Gray (tidytips daisy); P – *Leontopodium alpinum* Cass. (edelweiss); Q – *Leucanthemum vulgare* Lam. (Shasta daisy); R – *Ratibida columnifera* (Nutt.) Wooton & Standl. (prairie coneflower); S – *Rudbeckia hirta* L. (black-eyed-Susan); T – *Sanvitalia procumbens* Lam. (sanvitalia); U – *Tagetes* L. (marigold); V – *Xerochrysum bracteatum* (Vent.) Tzvelev (helichrysum, strawflower). Scale line = 1mm. A–E, G, I, K–M, O–T, and V from Meyer (1999a).

dispersal unit. In Lamiaceae, Boraginaceae and Verbenaceae, the ovary is superior and the schizocarpic fruit separates into four dry indehiscent nutlets at maturity, each with a hard pericarp and distinctive marking at the abscission point (Plate 7.8). In certain members of the Malvaceae (e.g. *Alcea* L., *Lavatera* L. and *Sphaeralcea* A. St. Hil.), the ovary is superior and composed of numerous carpels that separate into wedge-shaped segments at maturity. Each segment functions as a single-seeded achene (Plate 7.9, A–C). In *Geranium* L. and *Pelargonium* L'Hér. (Geraniaceae), the upper sterile portion of the five-lobed superior ovary elongates greatly during fruit development. At maturity, the five single-seeded globose bases of the fruit separate as the elongated upper portion of the ovary separates into five awn-like segments that curve away from the central column (Plate 7.9, D–F). In the Apiaceae, the schizocarpic fruit is derived from an inferior ovary consisting of two carpels. At maturity, the fruit splits down a septum (commissure) into two one-seeded mericarps. The style base usually remains attached at the apex of the mericarp. Integumentary tissue is almost entirely destroyed during fruit development. The pericarp usually consists of two or more distinct layers, some of which may be closely appressed to the seed. The mericarps of some genera may remain suspended from a common carpophore. The fruit may be nearly circular or compressed (either dorsiventrally or laterally) in cross-section and the surface may bear ornamentations of various kinds (e.g. longitudinal ridges, spine-like projections, hooks, hairs or tubercles) or may be smooth (Plate 7.9 G–L).

Succulent fruits

A berry is an indehiscent fruit with a succulent pericarp that contains one or more seeds. The flesh may be succulent and homogeneous throughout, as in *Asparagus densiflorus* (Kunth) Jessop, *Fuchsia*, *Physalis alkekengi* L., and members of the Cactaceae, or the outer part may be hard, firm or leathery, as in *Lagenaria* Ser. and *Cucurbita* L. (Plate 7.10). In a drupe, the pericarp is divided into three layers: a leathery exocarp, a fleshy mesocarp and a hard endocarp. The endocarp usually surrounds the seed after the fleshy part of the fruit disintegrates (e.g. *O. europaea* and *Prunus* L.).

Aggregate fruits

An aggregate fruit develops from several separate pistils within a single flower (Plate 7.11). The individual units of the aggregate

fruit may be any of the basic fruit types, for example, the fruit of *Ranunculus* consists of an aggregation of achenes (achenecetum); in *Helleborus* L. and *Nigella* L. the fruit is an aggregation of follicles (follicetum); and in *Rubus* L. the fruit consists of an aggregation of drupes (drupecetum).

Accessory fruits

In accessory fruits the basic type fruit remains associated with extracarpellary structures (Plate 7.12). A diclesium is a simple indehiscent fruit enclosed by a persistent calyx (e.g. *Armeria* Willd., *Centranthus ruber* (L.) DC., *Limonium sinuatum* (L.) Mill., *P. alkekengi* and *Plumbago*) or perianth (e.g. *Mirabilis jalapa* L.). In *Rosa* L., the fruit is called a hip and consists of an aggregation of achenes surrounded by an urn-shaped receptacle and hypanthium. A pseudocarp is an aggregation of achenes embedded in a fleshy receptacle (e.g. *Fragaria* L.). In *Echinops* L., the inflorescence consists of an aggregation of one-flowered capitula (heads), and is called a pseudocephalium. Each capitulum falls separately from the inflorescence and the mature fruit (achene) remains surrounded by persistent phyllaries and subtending bristles. Persistent bracts surround the mature fruit (usually a utricle) in some members of the Amaranthaceae (e.g. *Gomphrena*) and Chenopodiaceae (e.g. *Atriplex*).

Seeds

A true seed is a mature ovule containing an embryo and stored nutrients surrounded by a protective seed coat or testa derived from the integument(s). In most species, desiccation is the last stage of seed development (West and Harada, 1993). During desiccation, the embryo stops growing, becomes quiescent, and the entire seed loses water and becomes dry. However, not all seeds tolerate desiccation. Seeds of some species, mostly aquatic or tropical representatives from 20 families, lose viability if their moisture content drops below 12–31% (Werker, 1997). These seeds are called recalcitrant. In viviparous seeds, the embryo continues to grow, never undergoing quiescence or desiccation, and the seedling emerges from the seed while still attached to the mother plant. Regardless of the seed type, the embryo always represents the new sporophyte generation.

Shape

Seed shape is more or less species-specific, determined genetically, and is relatively constant for most species, but can be variable within other species (Werker, 1997). The shape of seeds may be further moulded by the space available for growth within the fruit, embryo shape, or position of the funicular attachment. Common seed shapes include ellipsoid, globose, lenticular, oblong, ovoid, pyramidal, reniform and secteroid, etc. (Kozlowski and Gunn, 1972). Cross-sectional seed shapes may be described as circular (1:1), compressed (2:1), flattened (3:1 or more) (Gunn, 1981, 1991), triangular, rectangular, reniform, etc.

Seed coat

The seed coat develops primarily from the integumental tissue; however, chalazal, raphal and nucellar tissues may also contribute to the seed coat (Werker, 1997). All these tissues are of maternal origin. The seed coat can consist of few to several layers of various cell types that may include: tannin, crystal (e.g. *Anthurium* Schott. and *Calla* L.), cork, mucilage, sclerenchyma, collenchyma, parenchyma, aerenchyma and chlorenchyma (e.g. *Amaryllis* L., *Gladiolus* L., *Lilium* L.). The presence of stomata in the seed coat has been reported in 30 families: e.g. Cannaceae – *Canna* L., Capparidaceae – *Cleome* L., Euphorbiaceae – *Ricinus* L., Geraniaceae – *Geranium*, Liliaceae – *Lilium*, *Fritillaria* L., Malvaceae – *Hibiscus* L., Papaveraceae – *Argemone* L., *Eschscholzia* Cham. and Violaceae – *Viola* (Corner, 1976; Boesewinkel and Bouman, 1984).

Seed coat development is triggered at the time of fertilization, but the precise point at which the integuments become the seed coat is not distinct. Following fertilization, functions of the integument(s) or seed coat may include: transport and conversion of amino acids and sugars (reserve materials) from the pericarp into the embryo sac; temporary storage for reserve materials used in embryo and seed coat development; provide enzymes for digestion of seed coat reserve materials; gas exchange; protection against embryo desiccation and mechanical injury; supply growth substances for embryo sac and maternal organs; and aid in photosynthesis in some species (Werker, 1997). When mature, seed coat functions may include: protection of the mature embryo against environmental factors; regulation of water uptake; aid in seed dispersal; and serve as reserve tissue for the germinating seed (Werker, 1997).

Seeds from about half of the angiosperm families are bitegmic (Bouman, 1984); derived from ovules that have two integuments (Table 7.2). The seed coat of a bitegmic seed consists of the testa or outer integument and the tegmen or inner integument (Corner, 1976). This condition is generally believed to occur in the more primitive families (Bouman, 1984). About one-quarter of all angiosperm families are solely unitegmic (i.e. derived from ovules that have one integument), 5% of the families have both bitegmic and unitegmic representatives, and the number of integuments has not been determined for the remaining 20% of angiosperm families (Davis, 1966). Many parasitic species in the Olacaceae, Santalaceae, Viscaceae, Loranthaceae and Balanophoraceae lack integuments (ategmic) and therefore have no seed coats. In some species, the seed coat remains relatively undeveloped, consisting of one layer of cells (e.g. some species in Orchidaceae). In species where the dispersal unit is an indehiscent fruit (e.g. Asteraceae, Lamiaceae, Boraginaceae and Verbenaceae), the pericarp serves as the major source of protection for the embryo and the seed coat is usually either underdeveloped or wholly or partially destroyed during seed and fruit maturation (Werker, 1997). Seventeen families of monocots and dicots have species in which the seed coat remains fleshy and juicy at maturity (e.g. Annonaceae, Cucurbitaceae, Euphorbiaceae, Liliaceae, Magnoliaceae, Meliaceae, Paeoniaceae; cf. Werker, 1997). This type of seed coat is called a sarcotesta. The seed coat for some species in the

Table 7.2. Typical number of integuments found in ornamental flower families (based on Mabberley, 1997; Judd *et al.*, 1999).

Unitegmic	Bitegmic	Uni and Bitegmic
Dicots	Dicots	Dicots
Acanthaceae, Apiaceae, Apocynaceae, Asclepiadaceae, Asteraceae, Bignoniaceae, Boraginaceae, Campanulaceae, Convolvulaceae, Gentianaceae, Gesneriaceae, Hydrophyllaceae, Lamiaceae, Loasaceae, Polemoniaceae, Rubiaceae, Scrophulariaceae, Solanaceae, Valerianaceae, Verbenaceae	Aizoaceae, Amaranthaceae, Balsaminaceae, Begoniaceae, Brassicaceae, Cactaceae, Capparidaceae, Caryopphyllaceae, Chenopodiaceae, Cistaceae, Crassulaceae, Cucurbitaceae, Dipsacaceae, Euphorbiaceae, Geraniaceae, Linaceae, Lythraceae, Malvaceae, Onagraceae, Passifloraceae, Portulacaceae, Plumbaginaceae, Papaveraceae, Resedaceae, Rutaceae, Sapindaceae, Tropaeolaceae, Violaceae	Fabaceae, Nyctaginaceae, Primulaceae, Polygonaceae, Ranunculaceae, Rosaceae, Saxifragaceae
Monocots	Monocots	Monocots
N/A	Araceae, Cannaceae, Commelinaceae, Iridaceae, Juncaceae, Musaceae	Amaryllidaceae, Liliaceae, Orchidaceae, Poaceae

Balsaminaceae, Tropaeolaceae, Rutaceae, Sapindaceae and Cannaceae is derived from a proliferation of chalazal tissue, which encloses the embryo and endosperm, while the integumental tissue remains relatively undeveloped. These seeds are called pachychalazal seeds. The presence of pachychalazal seeds has been reported in species from 16 dicot families, and is generally associated with the following characters: bitegmic ovules, crassinucellate ovules, nuclear endosperm, woody plants and tropical habitats (von Teichman and van Wyk, 1991).

Seed coat ornamentation

Surface ornamentation is an important diagnostic feature of the seed coat (Plate 7.13). The main contributor to surface ornamentation is the outermost cell layer of the seed coat and this becomes most evident in mature dry seeds (Werker, 1997). Topographic features result from a combination of epidermal cell size, shape and arrangement, waxy secretions (rodlets, filaments, flakes, plates, etc.), subepidermal wrinkles, reticulations and ruminations, and the presence of stomata and trichomes (Boesewinkel and Bouman, 1984). General arrangement of cells of different sizes and shapes contributes to the overall pattern of the seed coat. Primary sculpturing of the seed coat is reflective of the shape of the outer layer of epidermal cells and occasionally internal cell layers. Curvature of the outer periclinal wall (i.e. outer epidermal cell wall parallel to the surface of the epidermis) may be domed, conical, papillate or elongated into a unicellular trichome (hair). If the outer periclinal wall is thin it may collapse and take on the shape of the inner periclinal wall, giving the appearance of a depression. The anticlinal cell walls (i.e. vertical boundaries perpendicular to the seed coat surface) may be straight, curved, lobed or undulated, either regularly or irregularly (Chuang and Heckard, 1983; de Lange and Bouman, 1999; Koul *et al.*, 2000). Thickness of the anticlinal walls, as well as the condition of the middle lamella (obscure, channelled, raised, etc.) can contribute to the ornamentation and may have diagnostic value (Axelius, 1992; Watanabe *et al.*, 1999). Secondary sculpturing or fine relief of the periclinal wall may also serve as a diagnostic feature (Hufford, 1995; Koul *et al.*, 2000). The surface of the outer periclinal wall can be smooth, striate, reticulate or micropapillate. Possible sources of these structures include secondary wall thickenings in reticulate or helical patterns, cuticular patterns and cellular inclusions such as crystals (Werker, 1997). In *Castilleja* Mutis ex L.f. and *Orthocarpus* Nutt. the anticlinal walls and the inner periclinal wall develop lignified thickenings forming various patterns of diagnostic value (Chuang and Heckard, 1983). Further depth on the anatomy (i.e. cellular composition of the various cell layers) of the seed coat is beyond the scope of this chapter but may be found in Corner (1976), Boesewinkel and Bouman (1984), Kumar and Singh (1990) and Werker (1997).

Specialized structures of the seed coat

The seed coat may have several specialized structures. The micropyle is a small gap in the integuments through which the pollen tube usually passes to enter the embryo sac, which may remain visible as a small opening on the mature seed surface. The micropyle may be variously shaped (e.g. round, deltoid or slit-like). When the seed coat is bitegmic, the opening in the inner and outer integuments, called exostome and endostome, respectively, may not be aligned along the same line, resulting in a zigzag micropyle, as is found in the Fabaceae, Lythraceae, Malvaceae and Saxifragaceae (Werker, 1997). The hilum is a scar formed by the abscission of the seed from the funiculus. Hila vary in size and shape depending on the species (Plate 7.14). Some hila are inconspicuous while others occupy a large portion of the seed surface. In a mature seed, the severed vascular bundle of the hilum may serve as a route for evaporation during seed desiccation and may also serve as the route for water entry upon imbibition (Werker, 1997). The raphe is the zone where the funiculus is partly or entirely fused to the ovule (Werker, 1997) and sometimes appears as a raised ridge (Fahn, 1982) or distinct line on the seed surface (Plate 7.14, D and E). A pleurogram, also known as a linea fissura or linea sutura, is a specialized structure found on the seeds of some members of the

Fabaceae (i.e. Caesalpinioideae and Mimosoideae) and some members of the Cucurbitaceae. The pleurogram (Plate 7.14, C) is located on both sides of the seed and resembles an oval or a U-shaped depression with the open end directed toward the hilum (Gunn, 1981; Werker, 1997). The region surrounded by the pleurogram is called the areola (Gunn, 1981). The lens is a rounded protuberance (Plate 7.14, A), usually located on the side of the hilum opposite the micropyle in some members of the Fabaceae. The terms 'chalaza' and 'strophiole' have been incorrectly applied to this structure by some authors (Gunn, 1970). An operculum is a plug or lid in the micropylar region that abscises during germination (Plate 7.14, G–I). Opercula have been reported in 25 monocot (e.g. Araceae, Commelinaceae) and 20 dicot (e.g. Begoniaceae, Cactaceae) families, with the origin, structure, shape and opening mechanisms of opercula varying among families (Werker, 1997). In *Begonia* L., the extremely small seeds (typically 300–600 µm long) have a transverse ring of collar cells bordering the operculum at the micropylar-hilar end of the seed (Plate 7.14, I). During germination the middle lamella between the collar cells and operculum rupture, allowing the seedling to emerge (Werker, 1997; de Lange and Bouman, 1999).

Other specialized, external seed structures may include hairs, wings, arils, caruncles and elaiosomes (Plate 7.15). Hairs (trichomes) and scales are usually derived from the seed coat, but may also originate from the funiculus. They may be unicellular or multicellular, simple or branched, and vary in shape, size, colour and function. Examples of seeds with hairs include *Asclepias* L., *Crossandra* Salisb., *Epilobium* L. and *Hibiscus* L. Wings are extensions of one or more layers of the seed coat and sometimes the funiculus (Werker, 1997) and can vary in size, shape and thickness. Examples of winged seeds include *Asclepias*, *Gladiolus* L., *Lilium* L., *Lunaria* L., *Nemesia* Vent., *Oenothera* L. and *Penstemon* Schmidel. 'Aril' is a general term for a pulpy outgrowth of any part of the ovule or funiculus. Arils vary in shape, size, colour and origin. There is a considerable difference in opinion as to the precise use of the term 'aril' and other related terms, such as 'caruncle' and 'elaiosome' (Boesewinkel and Bouman, 1984; Lisci *et al.*, 1996). Seeds of *Passiflora* have fleshy arils that envelop the entire seed. The large, bright orange, multi-cellular hair-like threads of the arils on seeds of *Strelitzia reginae* Banks ex Dryander attract birds, and may also play a role in water absorption (Serrato-Valenti *et al.*, 1991). A caruncle is a hardened aril in the Euphorbiaceae. An elaiosome is a specialized aril, usually white or yellow in colour, containing oils that attract ants. Elaiosomes can also develop on fruit tissue, as in the Asteraceae, Boraginaceae and Lamiaceae (Pemberton and Irving, 1990; Lisci *et al.*, 1996).

Embryo structures

A mature seed usually contains an embryo and non-embryonic nutrient storage tissues.

A mature embryo in angiosperms consists of an axis bearing either one or two cotyledons (Fig. 7.12). The portion above the cotyledon(s) is called the epicotyl and the portion below the cotyledon(s) is the hypocotyl. The embryonic root is called the radicle, consisting of the root apical meristem surrounded by the root cap.

In grasses, the shoot apex is surrounded by the coleoptile; the root apex and root cap are surrounded by the coleorhiza; and the cotyledon is a specialized absorptive organ called the scutellum (Fig. 7.12, B). Some grass species have a scale-like appendage opposite the scutellum called the epiblast. Traditionally, grass embryos have been classified based on four factors: elongation of the vascular system, presence or absence of the epiblast, presence or absence of a notch between the scutellum and coleorhiza, and position of the margins of the embryonic leaves. Based

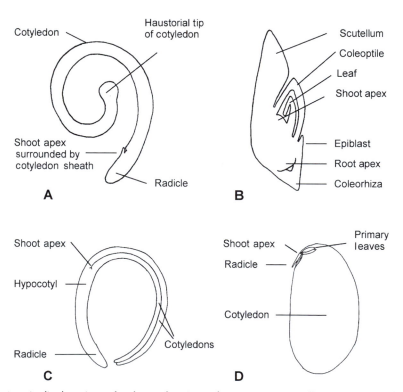

Fig. 7.12. Longitudinal sections of embryos showing embryonic parts: A – *Allium cepa* L., cotyledon with haustorial tip, cotyledon sheath surrounding shoot apex; B – *Deschampsia cespitosa* (L.) P. Beauv. with large scutellum and small epiblast; C – *Gomphrena globosa* L. with thin cotyledons; D – *Phaseolus coccineus* L. with thick, fleshy cotyledons, one removed to expose the relatively small embryonic axis.

on these criteria, six general categories were established (Reeder, 1957): festucoid, panicoid, chloridoid-eragrostoid, bambusoid, oryzoid-olyriod and arundnoid-danthonoid. Although much work has been subsequently published on grass embryo structure, these same basic categories are still employed.

Maturation rates of the various embryonic parts are species-dependent (Werker, 1997). In many species, the embryo is fully mature at the time of dispersal from the mother plant. In some species, certain portions of the embryo may be fully developed while others remain undifferentiated. Members of at least 20 angiosperm families (e.g. Papaveraceae, Orchidaceae and Ranunculaceae) are reported to disperse seeds containing quiescent undifferentiated embryos (i.e. devoid of organs), which differentiate only near the onset of germination (Natesh and Rau, 1984).

Cotyledons

As the names indicate, monocot embryos have one cotyledon and dicot embryos have two; however, there are exceptions. Underdeveloped monocot and dicot embryos lack cotyledons and are referred to as acotyledonous. Some dicots have only one cotyledon and are called monocotylar or moncotyledonous dicots (e.g. some members of Apiaceae, Gesneriaceae, Primulaceae – *Cyclamen* L. and Ranunculaceae) while some dicots appear to have only one cotyledon because the two are fused together (pseudomonocotyledonous) (Haccius and Lakshmanan, 1967). Occasionally, an embryo will develop more than the normal two cotyledons, a condition called schizocotyly (e.g. *Opuntia* Mill. and *Phaseolus*), while some embryos, as in gymnosperms, normally produce more than two cotyledons, a condition called polycotyledony.

Cotyledons vary in size, shape and function (Booth and Haferkamp, 1995; Danilova, et al., 1995; Werker, 1997). The cotyledons may be thick and fleshy and serve as a nutrient source for the embryo. In seeds with thick fleshy cotyledons, the endosperm is often almost completely absorbed by the growing embryo. This type of cotyledon is typically hemispherical (*Lathyrus odoratus* L.) to reniform (*Phaseolus coccineus* L.) in shape. Some embryos have thin cotyledons that may be ligulate to broad and leaf-like, with margins that are entire, bifid or lobed. Seeds with this type of embryo may or may not contain other nutrient storage tissues. This type of cotyledon usually serves as a photosynthetic organ following seed germination. In some non-grass monocots (e.g. *Asparagus* L. and *Commelina*) the entire cotyledon or the upper portion serves as an absorption organ (haustorium) deriving nutrients from other tissues, while the lower portion of the cotyledon serves to push and protect the shoot apex during seedling emergence (Esau, 1977; Lakshmanan, 1978). In the case of *Allium* L., the tip of the swollen cotyledon remains within the seed coat to serve as a haustorium, while the remainder of the cotyledon elongates and becomes photosynthetic (Esau, 1977). In most grasses, the cotyledon (scutellum) is shield-shaped and remains closely adpressed to the endosperm, from which it absorbs nutrients to transport to the embryonic axis (Esau, 1977).

Non-embryonic storage tissues

Nutrient storage tissues in angiosperm seeds, other than cotyledonary tissue, may include endosperm, perisperm or chalazosperm (Fig. 7.13). These tissues may be completely consumed, or nearly so (a thin layer of cells may persist), during embryo development and appear to be absent in the mature seed (Table 7.3). Mature seeds that appear to lack non-embryonic storage tissues are called exalbuminous. Seeds in which the nutrient storage tissues persist at maturity are called albuminous. Some authors reserve the terms 'albuminous' and 'exalbuminous' for those mature seeds containing or not containing endosperm, respectively, while other authors utilize the terms to include endosperm, perisperm and chalazosperm, since it is difficult to distinguish between these types of tissues in mature seed (Martin, 1946; Werker, 1997).

Endosperm is the most common nutritive tissue for the embryo during development and germination (Table 7.3). It is derived from

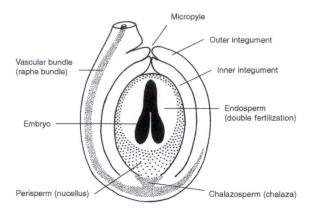

Fig. 7.13. Diagrammatic representation of an ovule with three types of nutritive tissues (i.e. endosperm, perisperm and chalazosperm) and tissue origin (from Meyer, 2001).

Table 7.3. Various nutritive storage tissues in mature seeds (+ found in mature seed; – not found in mature seed) (based on Boesewinkel and Bouman, 1984; Mabberley, 1997; Werker, 1997; Judd *et al.*, 1999).

Family	Perisperm	Chalazosperm	Endosperm
Monocotyledons			
Araceae			+/–
Cannaceae	+	+	+
Commelinaceae, Iridaceae, Juncaceae, Liliaceae	+	–	–
Musaceae	+	–	+
Orchidaceae	–	–	–
Poaceae	–	–	+
Dicotyledons			
Aizoaceae, Amaranthaceae	+	–	+/–
Capparidaceae, Caryophyllaceae, Chenopodiaceae, Nyctaginaceae	+	–	–/very little
Portulacaceae	+	–	–
Cactaceae	+/–	–	–
Apiaceae, Convolvulaceae, Euphorbiaceae, Polygonaceae, Rubiaceae, Scrophulariaceae, Solanaceae, Apocynaceae, Asclepiadaceae, Campanulaceae, Cistaceae, Crassulaceae, Dipsacaceae, Gentianaceae, Hydrophyllaceae, Malvaceae, Papaveraceae, Passifloraceae, Polemoniaceae, Primulaceae, Ranunculaceae, Saxifragaceae, Violaceae	–	–	+
Boraginaceae, Plumbaginaceae, Gesneriaceae, Loasaceae, Lythraceae, Rutaceae	–	–	+/–
Asteraceae, Balsaminaceae, Begoniaceae, Brassicaceae, Cucurbitaceae, Fabaceae, Geraniaceae, Lamiaceae, Linaceae, Resedaceae	–	–	–/very little
Acanthaceae, Bignoniaceae, Onagraceae, Rosaceae, Sapindaceae, Tropaeolaceae, Valerianaceae, Verbenaceae	–	–	–

the fusion of the second male gamete with the polar nuclei of the central cell of the embryo sac. Ploidy of endosperm is dependent upon the number of nuclei that fuse in the central cell. There are three basic types of endosperm development found in floral crops (Fahn, 1982; Johri, 2001): *Cellular*, in which cell walls are formed immediately following nuclear division; *Nuclear*, in which cell wall formation is postponed, sometimes indefinitely; and *Helobial*, where nuclear and cellular divisions are carried out at different rates at either end of the embryo sac. A fourth type of endosperm development (*Composite*) has been described but is not of significance to floral crops (Johri, 2001). Rates of occurrence of the various types of endosperm within angiosperm families are given in Davis (1966).

Perisperm is derived from the nucellus; it is maternal tissue and therefore diploid (2N), and serves a similar function as endosperm. Perisperm is found in both monocots and dicots (Table 7.3). In some cases, perisperm is the only nutritive tissue found in the mature seed, but it is usually present in combination with endosperm (Werker, 1997). In rare cases, nutrient storage tissue may also be derived from the chalaza and is called chalazosperm. The chalaza is maternal tissue located at the point where the funicular vascular bundle reaches the nucellus.

The nutrient contents of these storage tissues vary among species, and may also vary within a seed, as different regions within the storage tissue may contain differing amounts of proteins, lipids and carbohydrates (Werker, 1997). For more detailed discussion on the development and usage of these nutrient storage tissues, particularly endosperm, see Vijayaraghavan and Prabhakar (1984), Erdelská (1986), Werker (1997), Floyd and Friedman (2000) and Olsen (2001).

Embryo placement

The following discussion of embryo placement within the seed follows Martin (1946). In mature albuminous seeds, the embryo shares the seed volume with the nutritive tissues. The embryo may be variously positioned relative to, or embedded in, the nutritive tissue. In exalbuminous seeds, where all or almost all the nutritive tissues have been consumed by the developing embryo, the mature embryo usually fills the entire seed volume. In most species, the radicle points toward the micropyle. Placement of the embryo within the seed may be either peripheral or axile (Fig. 7.14). As the name indicates, peripheral embryos lie along the perimeter of the seed, occupying one-quarter to three-quarters of the seed volume, and the

nutritive tissue is usually abundant. Axile embryos may occupy less than one-quarter of the seed volume to the entire seed volume, are centrally located, and the nutritive tissue may be abundant or lacking. Embryos within these two groups are further subdivided into 12 categories (Table 7.4). Extremely small embryos found in medium to large seeds are referred to as rudimentary by some authors. Rudimentary embryos are globular to oval-oblong shaped, and the cotyledons are usually underdeveloped and obscure, but may be evident and appear like miniature versions of linear or spatulate types. This group of embryos does not clearly fit into either peripheral or axile embryo placements.

Seed and fruit dispersal

In the natural world, seeds and fruits (diaspores) have developed the ability to disperse from the mother plant, sometimes for great distances. These diaspores may have various appendages, mucilaginous properties or nutrient-rich properties that aid in dispersal. Wind is a very effective dispersal agent (anemochory) and wind-borne diaspores often have wings (e.g. *Briza maxima*, *Coreopsis lanceolatus* L., *Dimorphotheca* Moench, *Gladiolus* and *Nemesia* Vent.), plumes (e.g. *Armeria*, *Asclepias*, *Centranthus ruber* and *Epilobium*), pappi (e.g. *Gaillardia pulchella* Foug., *Layia platylossa* (Fisch. & C.A. Mey.) A. Gray) and net-shaped, air-filled, balloon-like seed coats (e.g. *Castilleja* Mutis ex L. f., *Delphinium variegatum* Torr. & A. Gray and members of the Orchidaceae) that give the falling diaspores air resistance, enabling them to travel great distances from the mother plant. Zoochory is diaspore dispersal by animals. Diaspores may have appendages or coverings that are consumed by animals; the diaspores usually pass through the digestive tract unharmed and can later germinate (e.g. *Strelitzia reginae*). Some animals will cache diaspores for later consumption. Diaspores with appendages (e.g. *Cynoglossum amabile* Stapf. & J.R. Drumm., *Daucus carota* L., *Pelargonium* and *Ranunculus muricatus* L.) and mucilaginous coverings (e.g. *Salvia* L.

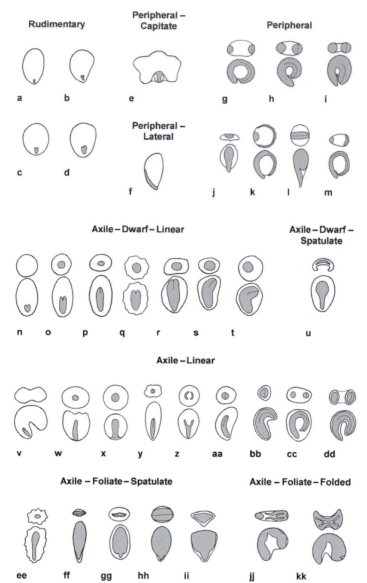

Fig. 7.14. Embryo placement. Rudimentary: a – *Aquilegia canadensis* L.; b – *Ranunculus californicus* Benth.; c – *Anemone caroliniana* Walt.; d – *Ranunculus sceleratus* L. Peripheral – Capitate: e – *Tradescantia virginiana* L. Peripheral – Lateral: f – *Briza* L. Peripheral: g – *Gypsophila paniculata* L.; h – *Portulaca oleracea* L.; i – *Cylindropuntia leptocaulis* (DC.) F.M. Kunth; j – *Dianthus armeria* L.; k – *Mirabilis multiflora* (Torr.) A. Gray; l – *Eriogonum latifolium* Sm.; m – *Gomphrena globosa* L. Axile – Dwarf – Linear: n - *Gentiana cruciata* L.; o – *Campanula rotundifolia* L.; p – *Gentiana sceptrum* Griseb.; q – *Eustoma exaltatum* (L.) Salisb. ex G. Don; r – *Primula eximia* Greene; s – *Nicotiana glauca* Graham; t – *Petunia axillaris* (Lam.) Britton *et al.*, Axile – Dwarf – Spatulate: u – *Collinsia heterophylla* G. Buist ex Graham. Axile – Linear: v – *Dicentra chrysantha* (Hook. & Arn.) Walp.; w – *Gladiolus gandavensis* Van Houtte; x – *Canna* L.; y – *Eryngium articulatum* Hook.; z – *Eschscholzia californica* Cham.; aa – *Papaver somniferum* L.; bb – *Reseda odorata* L.; cc – *Salpiglossa sinuata* Ruiz & Pav.; dd – *Cleome hassleriana* Chodat. Axile – Foliate – Spatulate: ee – *Nemophila menziesii* Hook. & Arn.; ff – *Limonium californicum* (Boiss.) A. Heller; gg – *Ricinus communis* L.; hh – *Scabiosa caucasica* M. Bieb.; ii – *Moluccella laevis* L. Axile – Foliate – Folded: jj – *Alcea rosea* L.; kk – *Hibiscus moscheutos* L. (Redrawn and modified, based on Martin, 1946.)

Table 7.4. Twelve embryo types as described by Martin (1946).

Embryo type	Description
Rudimentary Division	
Rudimentary	Seeds medium to large; embryo small, globular to oval-oblong, cotyledons usually obscure, but sometimes evident; endosperm present; monocots and dicots.
Peripheral Division	
Broad	Seeds usually medium to large; embryo globular, broader than long; endosperm present; monocots and dicots.
Capitate	Seeds usually medium to large; embryo expanded above, capitate or turbinate; endosperm present; monocots only.
Lateral	Seeds usually medium to large; embryo basal-lateral or lateral, evident from the exterior, size variable; endosperm present; Poaceae only.
Peripheral	Embryo usually elongate and large, occupying one-quarter to three-quarters or more of the periphery of the seed, often curved; cotyledons narrow or expanded; storage tissues usually central, sometimes lateral; dicots only, but in some, one cotyledon is lacking.
Division Axile – Subdivision Linear	
Linear	Seeds normally not minute; embryo generally several times longer than broad, straight, curved or coiled; cotyledons not expanded; monocots and dicots.
Division Axile – Subdivision Miniature	
Dwarf	Seeds small, generally 0.3 mm to 2 mm long exclusive of testa, often nearly as broad as long; embryo variable in relative size, small to total, generally stocky, usually oval to elliptic or oblong, cotyledons inclined to be poorly developed; endosperm +/– present.
Micro	Seeds minute, usually less then 0.2 mm long exclusive of testa, generally globular and consisting of relatively few cells, *c.* 50 to 150 within the testa; embryo minute to total; endosperm +/– present.
Division Axile – Subdivision Foliate	
Spatulate	Seeds generally medium to large; embryo large, erect, generally one-quarter to total, central, cotyledons variable, thin to thick slightly expanded to broad; stalk may be slightly invested by cotyledons; endosperm +/– present; dicots only.
Bent	Seeds generally medium to large; embryo large, generally one-quarter to total, central, bent in jackknife form; cotyledons expanded to broad, generally thick; endosperm +/– present; dicots only.
Folded	Seeds generally medium to large; embryo large, generally one-quarter to total, central; cotyledons usually thin, extensively expanded and folded in various ways; endosperm +/– present; dicots only.
Investing	Seeds generally medium to large; embryo large, erect, generally total, central; thick cotyledons overlapping and encasing the somewhat dwarfed stalk for at least half its length; endosperm wanting or limited; dicots only.

and *Solenostemon scutellariodes* (L.) Codd.) can attach to fur, feathers, clothing, etc. and be carried great distances. Some diaspores have special oily appendages called elaiosomes that attract ants. In ant dispersal (myrmecochory) the ants carry the diaspores to their nest, consume the elaiosomes, leaving the diaspores relatively undamaged and able to germinate (Pemberton and Irving, 1990). Some plants disperse their seeds ballistically, either by explosive fruit dehiscence or by explosive ejection of the seeds after the fruit is open (e.g. *Viola cornuta* L.). Some species may have diaspores with more than one body form (e.g. *Catananche* L., *Coreopsis* L., *Dimorphotheca*) and thus may be dispersed by different means (Ruiz de Clavijo, 1995). For an extensive review on diaspore dispersal see Willson and Traveset (2000). In the world of cultivated flower seed production these appendages, mucilaginous properties, fleshy coverings and dispersal mechanisms are of little consequence and may even present problems for harvesting, cleaning, conditioning and mechanical planting (Harrington, 1977; Hall, 1998).

Conclusions

The seed is the beginning of life for most flowering plants. The bulk of knowledge we have about seed development and structure has come from the study of agronomic crops as the result of our need to better provide food for the masses. While there is no less emphasis today on the need to improve our understanding of seed development in agronomic crops, there is an increasing demand for knowledge about other types of seed as well. Beautiful flowers in the garden, along the roadside, and in our homes provide aesthetically pleasing surroundings for humans. In the quest to beautify our surroundings there are infinite possibilities for research related to understanding the mysteries of seed development in floral crops. With more than 250,000 species of flowering plants, humans have only scratched the surface of knowledge about these fascinating organisms. The information provided in this chapter is a very brief glimpse into the complex world of seed development and structure in floral crops.

Acknowledgements

I would like to thank Ms Evelyn Ramos, Ms Connie Weiner, Dr Marian Stephenson and Mr Jim Effenberger for generously providing fresh floral material for photographing. Thank you also to Mr Effenberger for his assistance with digitizing all the illustrations included in this chapter and for assistance with designing Plate 7.4. I am grateful to Dr Miller McDonald, Dr Jeri Langham and Mr Greg Meyer for their suggestions for improving the manuscript.

References

AOSA. 2003. *Rules for Testing Seeds.* Association of Official Seed Analysts. 166 pp.

Avery, G.S. 1930. Comparative anatomy and morphology of embryos and seedlings of maize, oats, and wheat. *Bot. Gaz.* 89:1–39.

Axelius, B. 1992. Testa patterns in some species of *Physalis* L. and some other genera in the tribe Solaneae (Solanaceae). *Int'l Jour. Plant Sci.* 153(3): 488–502.

Boesewinkel, F.D. and F. Bouman. 1984. The seed: structure. *In*: B.M. Johri (ed.). *Embryology of Angiosperms.* Springer-Verlag, Berlin, Germany. pp. 567–610.

Booth, D.T. and M.R. Haferkamp. 1995. Morphology and seedling establishment. *In*: D.J. Bedunah and R.E. Sosebee (eds). *Wild Plants: Physiological Ecology and Developmental Morphology.* Society for Range Management. pp. 239–290.

Bouman, F. 1984. The ovule. *In*: B.M. Johri (ed.). *Embryology of Angiosperms.* Springer-Verlag, Berlin, Germany. pp. 123–157.

Buell, K.M. 1952. Developmental morphology in *Dianthus*. I. Structure of the pistil and seed development. *Amer. Jour. Bot.* 39: 194–210.

Chuang, T.I. and L. Constance. 1992. Seeds and systematics in Hydrophyllaceae: tribe Hydrophylleae. *Amer. Jour. Bot.* 79:257–264.

Chuang, T.I. and L.R. Heckard. 1983. Systematic significance of seed-surface features in *Orthocarpus* (Scrophulariaceae – subtribe Castillejinae). *Amer. Jour. Bot.* 70(6): 877–890.

Corner, E.J.H. 1976. *The Seeds of Dicotyledons.* Cambridge University Press, Cambridge, UK. Vol. 1. 311 pp.

Danilova, M.F., E.N. Nemirovich-Danchenko, G.A. Komar and M.M. Lodkina. 1995. The seed structure of monocotyledons. *In*: Rudall, P.J., P.J. Cribb, D.F. Cutler and C.J. Humphries (eds). *Monocotyledons: Systematics and Evolution.* Royal Botanic Gardens, Kew, UK. pp. 461–472.

Davis, G.L. 1966. *Systematic Embryology of the Angiosperms.* John Wiley & Sons, New York. 528 pp.

de Lange, A. and F. Bouman. 1999. *Seed Micromorphology of Neotropical Begonias. Smithsonian Contributions to Botany.* No. 90. Smithsonian Institution Press, Washington, DC. 49 pp.

Erdelská, O. 1986. Cytolysis of the endosperm in different types of correlation in endosperm and embryo development. *Acta Bot. Neerl.* 35(4): 437–441.

Esau, K. 1977. *Anatomy of Seed Plants.* 2nd edn. John Wiley & Sons, New York. 550 pp.

Fahn, A. 1982. *Plant Anatomy.* 3rd edn. Pergamon Press, Washington, DC. 544 pp.

Floyd, D.K. and W.F. Friedman. 2000. Evolution of endosperm development patterns among basal flowering plants. *Int'l Jour. Plant Sci.* 161(6 suppl.): S57–S81.

Foster, A.S. and E.M. Gifford, Jr. 1974. *Comparative Morphology of Vascular Plants.* 2nd. edn. W.H. Freeman & Co., San Francisco, 751 pp.

Gifford, E.M., Jr. and A.S. Foster. 1989. *Morphology and Evolution of Vascular Plants.* 3rd edn. W.H. Freeman & Co., New York. 626 pp.

Gunn, C.R. 1970. A key and diagrams for the seeds of one hundred species of *Vicia* (Leguminosae). *Proc. Int'l Seed Test. Assoc.* 35(3):773–790.

Gunn, C.R. 1981. Seeds of Leguminosae. *In*: R.J. Polhill and P.H. Raven (eds). *Advances in Legume Systematics. Part 2.* Royal Botanic Gardens, Kew, UK. pp. 913–925.

Gunn, C.R. 1991. Fruits and seeds of genera in the subfamily Caesalpinioideae (Fabaceae). US Department of Agriculture, *Technical Bulletin* No. 1755. 408 pp.

Haccius, B. and K.K. Lakshmanan. 1967. Experimental studies on monocotyledonous dicotyledons: phenylboric acid-induced 'dicotyledonous' embryos in *Cyclamen persicum. Phytomorphology* 17:488–494.

Hall, J. 1998. A walk on the wild side: quality assurance problems unique to the wildflower seed trade. *Seed Technol.* 20:162–175.

Harrington, J.F. 1977. Cleaning vegetable and flower seeds. *Seed Sci. and Technol.* 5: 225–231.

Hufford, L. 1995. Seed morphology of Hydrangeaceae and its phylogenetic implications. *Int'l Jour. Plant Sci.* 156(4):555–580.

Johansen, D.A. 1950. *Plant Embryology.* Waltham, Massachusetts.

Johri, B.M. 2001. Angiosperm embryology in the 21st century – reflections. *Phytomorphology –* Golden Jubilee Issue:217–224.

Judd, W.S., C.S. Campbell, E.A. Kellogg and P.F. Stevens. 1999. *Plant Systematics: A Phylogenetic Approach.* Sinauer Associates, Inc., Sunderland, Massachusetts. 464 pp.

Koul, K.K., R. Nagpal and S.N. Raina. 2000. Seed coat microsculpturing in *Brassica* and allied genera (subtribes Brassicinae, Raphaninae, Moricandiinae). *Ann. Bot.* 86:385–397.

Kozlowski, T.T. and C.R. Gunn. 1972. Importance and characteristics of seeds. *In*: T.T. Kozlowski (ed.). *Seed Biology: Importance, Development and Germination.* Vol. I. Academic Press, New York. pp. 1–8.

Kumar, P. and D. Singh. 1990. Development and structure of seed-coat in *Hibiscus* L. *Phytomorphology* 40:170–188.

Lakshmanan, K.K. 1978. Studies on the development of *Commelina benghalensis.* III. Cotyledon. *Phytomorphology* 28(3):253–261.

Lisci, M., M. Bianchini and E. Pacini. 1996. Structure and function of the elaiosome in some angiosperm species. *Flora* 191:131–141.

Mabberley, D.J. 1997. *The Plant-Book: a Portable Dictionary of the Vascular Plants.* 2nd edn. Cambridge University Press, Cambridge, UK. 858 pp.

Maheshwari, P. 1950. *An Introduction to the Embryology of Angiosperms.* McGraw-Hill Book Co., Inc., New York. 453 pp.

Maheshwari, S.C. and B. Baldev. 1958. A contribution to the morphology and embryology of *Commelina forskalaei* Vahl. *Phytomorphology* 8:277–298.

Martin, A.C. 1946. The comparative internal morphology of seeds. *Amer. Midland Nat.* 36(3):513–660.

Matsumoto, T.D., A.R. Kuehnle and D.T. Webb. 1998. Zygotic embryogenesis in *Anthurium* (Araceae). *Amer. Jour. Bot.* 85(11): 1560–1568.

Maze, J. and L.R. Bohm. 1973. Comparative embryology of *Stipa elmeri* (Gramineae). *Can. Jour. Bot.* 51:235–247.

Meyer, D.J.L. 1999a. Seed unit identification in the Apiaceae (Umbelliferae). California Dept. of Food and Agriculture, Sacramento, California. 19 pp.

Meyer, D.J.L. 1999b. Seed unit identification in the Asteraceae. California Dept. of Food and Agriculture, Sacramento, California. 15 pp.

Meyer, D.J.L. 2000. The grass caryopsis. California Dept. of Food and Agriculture, Sacramento, California. 12 pp.

Meyer, D.J.L. 2001. Basic botany for seed testing. *In*: M. McDonald, T. Gutormson and B. Turnipseed (eds). *Seed Technologist Training Manual.* Society of Commercial Seed Technologists. pp. 2.1–2.38.

Natesh, W. and M.A. Rau. 1984. The embryo. *In*: B.M. Johri (ed.). *Embryology of Angiosperms.* Springer-Verlag, Berlin, Germany. pp. 377–443.

Newcomb, W. 1973. The development of the embryo sac of sunflower, *Helianthus annuus*, before fertilization. *Can. Jour. Bot.* 51:863–878.

Nogler, G.A. 1984. Gametophytic apomixis. *In*: B.M. Johri (ed.). *Embryology of Angiosperms.* Springer-Verlag, Berlin, Germany. pp. 475–518.

Olsen, O. 2001. Endosperm development: cellularization and cell fate specification. *Ann. Rev. Plant Physiol. Plant Mol. Biol.* 52:233–267.

Pemberton, R.W. and D.W. Irving. 1990. Elaiosomes on weed seeds and the potential for myrmecochory in naturalized plants. *Weed Sci.* 38:615–619.

Radford, A.E. 1986. *Fundamentals of Plant Systematics.* Harper & Row, Publishers, Inc., New York. 498 pp.

Raghavan, V. 1986. *Embryogenesis in Angiosperms: A Developmental and Experimental Study.* Cambridge University Press, Cambridge, UK. 302 pp.

Rangan, T.S. and N.S. Rangaswamy. 1999. Nucellus – a unique embryologic system. *Phytomorphology* 49(4):337–376.

Raven, P.H., R.F. Evert and S.E. Eichhorn. 1999. *Biology of Plants.* 6th edn. W.H. Freeman & Co./Worth Publishers, New York. 944 pp.

Reeder, J.R. 1957. The embryo in grass systematics. *Amer. Jour. Bot.* 44:756–768.

Ruiz de Clavijo, E. 1995. The ecological significance of fruit heteromorphism in the amphicarpic species *Catananche lutea* (Asteraceae). *Int'l Jour. Plant Sci.* 156(6):824–833.

Sendulsky, T., T.S. Filgueiras and A.G. Burman. 1986. Fruits, embryos and seedlings. *In*: T.R. Soderstrom, K.W. Hilu, C.S. Campbell and M.E. Barkworth (eds). *Grass Systematics and Evolution.* Smithsonian Institution Press, Washington, DC. pp. 31–36

Serrato-Valenti, G., L. Cornara, P. Modenesi and P. Profumo. 1991. The aril of the *Strelitzia reginae* Banks seed: structure and histochemistry. *Ann. Bot.* 67:475–478.

Stern, K.R. 1997. *Introductory Plant Biology.* 7th edn. McGraw Hill. 570 pp.

Spjut, R.W. 1994. A systematic treatment of fruit types. *Memoirs of the New York Botanical Garden* 70:1–182.

Swamy, B.G. and K.V. Krishnamurthy. 1977. Certain conceptual aspects of meristems. II. Epiphysis and shoot apex. *Phytomorphology* 27(1):1–8.

Watanabe, H., T. Ando, E. Hishino, H. Kokubun, T. Tsukamoto, G. Hashimoto and E. Marchesi. 1999. Three groups of species in *Petunia sensu* Jussieu (Solanaceae) inferred from the intact seed morphology. *Amer. Jour. Bot* 86(2): 302–305.

Werker, E. 1997. *Seed Anatomy.* Gebruder Borntraeger, Berlin, Germany. 424 pp.

West, M.A.L. and J.J. Harada. 1993. Embryogenesis in higher plants: an overview. *Plant Cell* 5: 1361–1369.

Willemse, M.T.M. and J.L. van Went. 1984. The female gametophyte. *In*: B.M. Johri (ed.). *Embryology of Angiosperms.* Springer-Verlag, Berlin, Germany. pp. 159–196.

Willson, M.F. and A. Traveset. 2000. The ecology of seed dispersal. *In*: M. Fenner (ed.). *Seeds: The Ecology of Regeneration in Plant Communities.* 2nd edn. CAB International, Wallingford, UK. pp. 85–110

van Went, J.L. and M.T.M. Willemse. 1984. Fertilization.. *In*: B.M. Johri (ed.). *Embryology of Angiosperms.* Springer-Verlag, Berlin, Germany. pp. 273–317

Vijayaraghavan, M.R. and K. Prabhakar. 1984. The endosperm. *In*: B.M. Johri (ed.). *Embryology of Angiosperms.* Springer-Verlag, Berlin, Germany. pp. 319–376.

von Teichman, I. and A.E. van Wyk. 1991. Trends in the evolution of dicotyledonous seeds based on character associations, with special reference to pachychalazy and recalcitrance. *Bot. Jour. Linn. Soc.* 105: 211–237.

8   Flower Seed Physiology and Plug Germination

Miller B. McDonald

*Seed Biology Program, Department of Horticulture and Crop Science,
Ohio State University, 2021 Coffey Road, Columbus, OH 43210-1086, USA*

Introduction

Before describing the factors that govern seed germination, it is important first to understand the determinants of flower seed quality. These include genetic purity, mechanical purity, germination, vigour and dormancy.

Genetic purity

Genetic purity confirms that the variety developed by the breeder is the same variety received by the grower. Most flower crops are marketed as hybrids, which are the progeny of two inbred parents. Genetic purity of hybrid crops verifies that the hybridization process has been successful and that the seed lot is not contaminated with undesirable selfs or outcrosses. This assures uniform appearance of the crop.

Mechanical purity

Mechanical purity identifies the important kinds of seeds present in the seed lot and includes percentages by weight of pure seed, other crop seed, weed seed and inert matter. The ideal is 100% pure desired seed. The reality is that during seed production, inevitable physical contamination by other undesired seeds, sand and plant materials occurs, and these are not always completely eliminated from the seed lot during cleaning. In most cases, this physical contamination can be visibly detected and corrected. However, if the seed is pelleted, other undesirable seeds/materials may also be pelleted and hidden from the consumer until germination occurs.

Germination

Germination testing is a procedure to estimate the percentage emergence expected of the seed lot under *favourable* conditions since visible examination of the seed cannot reveal its capacity to produce a seedling. In general, germination conditions include the type of growing medium, moisture level, duration of the test and temperature requirements. In all cases, germination tests are conducted under optimum conditions. As a result, the germination value identifies the optimum level of seed performance expected of the seed lot.

Vigour

Under stress, seed germination values often overestimate the emergence of seeds in the greenhouse. To provide a more accurate appraisal of seed quality related to greenhouse performance, the concept of seed vigour was established. Seed vigour is those seed properties that determine the potential for rapid, uniform emergence and development of normal seedlings *under a wide range of environmental conditions*. Most plug producers are keenly interested in this value because it better portrays how the seed will perform in the greenhouse. The seed industry is actively developing and identifying vigour tests that better assess seed quality for a variety of flower crops (see chapter 16).

From the perspective of a plug grower whose principal concern is complete, uniform germination, other factors influence seed quality. Many of these, such as harvesting and handling, are out of the control of the grower but are extremely important to the seed industry. Others, such as dormancy and proper storage conditions, can be encountered and managed by the grower.

Dormancy

Dormancy is a condition where a viable seed is unable to germinate due to a physical and/or physiological barrier. The degree of dormancy encountered can vary from year to year, seed lot to seed lot, and from seed to seed in the lot. In most cases, when it occurs, it is transitory and diminishes with time after harvest. For example, verbena seeds require storage at cool temperatures for 3–5 months after harvest to improve germination. In other cases, action must be taken to alleviate the dormancy-imposing condition. For example, geranium has a hard, impermeable seed coat that does not allow the entry of water until the seed coat has been abraded (a process called scarification). In still other cases, a physiological barrier to germination is encountered that must be removed. An example is the number of flower crops that require light to enhance germination (see Table 8.1). In most cases, seed companies know which seed lots have a high incidence of dormancy and take appropriate action prior to shipping the seed to the grower so that optimum performance is attained.

Flower Seed Physiology

A seed is a reproductive unit. Its primary function is to ensure the development of a normal seedling, ultimately leading to successful stand establishment. Thus, an understanding of the physiology of seed germination and seedling establishment is important in a consideration of components that influence flower production. Many factors affect germination and seedling establishment from normal development and maturation of the seed to its harvest, storage, and final planting. Excellent detailed considerations of these events are given in this volume and have been presented elsewhere (Murray, 1984a,b; Bewley and Black, 1985; Mayer and Poljakoff-Mayber, 1989; Copeland and McDonald, 2001) and will not be considered here. Further, there is relatively little information on flower seed germination, but many principles found in other crop seeds are relevant to flower seeds. This chapter begins with the assumption that a viable and vigorous seed is available to the flower grower and is planted using traditional practices. Because flower seeds differ in their anatomy, chemistry and physiological processes regulating germination, it is difficult to describe one common physiological mechanism that governs seed germination. However, the process of seed germination begins with imbibition.

Imbibition

The early stages of water uptake are a crucial period for seed germination. Seeds are sensitive to rapid imbibition, chilling and anoxia; common events that occur under uncontrolled planting conditions. Imbibition is an essential process initiating seed germination. It is the first key event that moves the seed from a dry, quiescent, dormant organism to a resumption of embryo growth. Consequently, an orderly transition of increased hydration, enzyme activation, storage product breakdown and resumption of seedling development must occur. Imbibition is not merely an uncontrolled physical event; it is now recognized that chemical conformational events, seed coat effects, and seed quality factors govern the directed flow of water into the seed (Leopold and Vertucci, 1989). Thus, any consideration of flower seed germination physiology and its resultant impact on seedling establishment should initially focus on water uptake.

Flower seeds typically possess extremely low water potentials attributed to their osmotic and matric characteristics (Vertucci, 1989). These potentials may be as low as −400 MPa (McDonald, 1994) and are a consequence of the relationship of water with components of the seed. Water in a seed basically exists in three forms (Fig. 8.1), dependent on its hydrational status (Leopold and Vertucci, 1986). Below 8% moisture (region 1), water is 'chemi-sorbed' to macromolecules through ionic bonding, has limited mobility, and acts as a ligand rather than a solvent. Between 9% and 24% moisture (region 2), water is weakly bound to macromolecules and begins to have solvent properties, and diffusion gradually becomes evident as water takes on the properties of a bulk solution. Above 24% moisture (region 3), water is bound with negligible energy, and its properties are similar to bulk water. The macromolecular surface of the seed is fully wetted at 35% moisture, since freezing damage occurs at that moisture content (Leopold and Vertucci, 1986).

As flower seeds take up water, there is an enormous increase in tissue swelling that is likely to be due to an unfolding of proteins and association of water with the matric forces of the cell walls. The greatest relative change in volume occurs between 4% and 8% moisture. Between 8% and 21% moisture content, seed swelling on an absolute volume basis is greatest, and beyond 21% moisture content, very little increase in seed volume is observed (Vertucci and Leopold, 1986). Different seeds and different parts of the seed imbibe water differently. Seeds high in protein content absorb more water than seeds high in lipid content. Similarly, starch stored in the endosperm is far more hydrophobic than high levels of protein found in the embryo, which often has a greater water content at the same relative humidity. Seed coat morphology also plays a major role in determining water permeability into the seed. Many flower species with hard seed coats (impermeable to water) have small elongated pores and a high density of waxy materials embedded in the testa epidermis (Calero et al., 1981; Tully et al., 1981; Yaklich et al., 1986; Mugnisjah et al., 1987). The incidence of hard-seededness is both genetically and environmentally controlled and is greatest when seed maturation occurs under high temperature, high humidity conditions (Potts et al., 1978; Dassou and Kueneman, 1984) or under dry conditions (Hill et al., 1986). Seed coats are also extremely hydrophilic and able to absorb as much as 3.8 times their weight in water (McDonald et al., 1988). This water-holding capability assists the seed in avoiding imbibitional damage from the rapid inrush of water into tissues with very negative water potentials, culminating in tissue damage (Powell and Matthews, 1978; Tully et al., 1981). Any damage to the integrity of the seed coat can influence the rate of water uptake (McDonald, 1985; Evans et al., 1990), increase the incidence of imbibitional chilling injury (Tully et al., 1981), and decrease seedling emergence (Luedders and Burris, 1979).

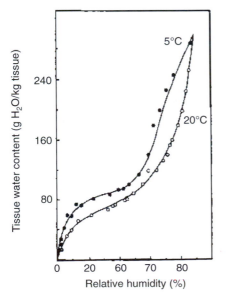

Fig. 8.1. Moisture isotherms for pellets of soybean seed tissue at two different temperatures (from Leopold and Vertucci, 1986).

Temperature influences the rate of imbibition, with low temperatures resulting in slow imbibition (Murphy and Noland, 1982). One factor responsible for the temperature effect may be an increase in water viscosity at low temperatures and/or the ease with which tissues are wetted (Vertucci and Leopold, 1983). The rate of water entry into seeds is also enhanced by high temperatures and occurs at the same rate in both dead and living seeds (Leopold, 1983).

Breakdown of storage materials

The process of seed germination requires energy. Following water uptake, many long-chain macromolecules are hydrolysed to simple forms, translocated and incorporated into energy-producing biochemical pathways. The rapidity with which these events occur (within 2 h) led to early confusion regarding the temporal requirements for the transcription/translation process that must occur for the orderly sequence of biochemical germination events. Specific messenger ribonucleic acids (mRNAs) that code for the synthesis of proteins in the early stages of germination appear to be conserved in the dry seed. Thus, the early synthesis of enzymes following imbibition is 'preprogrammed' by stored mRNAs. After hydration of the dry seed is complete, and adequate time has elapsed, RNA synthesis that directs later stages of germination follows. The specific physiological pathways involved for the breakdown of storage materials are dependent on the type of seed and storage form available, which typically falls into three groups: carbohydrates, lipids and proteins.

Carbohydrates

Starch and soluble sugars are the major constituents of the total non-structural carbohydrates found in flower seeds. During germination, the degradation of starch to free sugars and dextrins occurs initially by starch phosphorylase. These are converted in the cotyledons of dicot seeds into sucrose, which is translocated to the axis. There the sucrose is hydrolysed by invertase to glucose and fructose. Much later, α- and β-amylase activities increase and provide even greater quantities of sucrose. Since β-amylase does not interact with starch grains, it is likely that α-amylase initiates the hydrolysis of the starch grain, followed by β-amylase reacting with the resultant starch fragments (Adams *et al.*, 1981). The glucose moieties are converted to sucrose by sucrose-phosphate synthase and sucrose-phosphate phosphatase, and the sucrose is readily translocated to the embryonic axis and used as an energy source.

In monocot flower seeds possessing a caryoposis, the initial production of α-amylase occurs in the region adjacent to the scutellum, where it is released into the endosperm. Later, the enzyme is synthesized in the aleurone layer and secreted into the endosperm to provide a complete hydrolysis of the starch reserves. The β-amylase also is present in the endosperm in an inactive form bound by disulphide linkages to protein. It becomes activated following initial hydrolysis of the starch grain by α-amylase. In both cases, the major product of α- and β-amylase activity is the disaccharide maltose, which is further hydrolysed to glucose by α-glucosidase, an enzyme present in both the embryo and aleurone, where it is released into the endosperm during germination. Generally, the maltose and glucose products of starch hydrolysis are translocated to the scutellum, where they are converted to sucrose, which is then mobilized to the embryo for growth. The small quantities of carbohydrates present in the embryo are responsible for the early metabolic reactions in the axis until the substantial energy reserves from the endosperm and scutellum become available.

Lipids

Triglycerides are the major storage lipids in flower seeds. Their initial hydrolysis is by lipases located in oil bodies that hydrolyse fatty acid ester bonds from glycerol, producing one glycerol molecule and three fatty acids. Glycerol is then used as an energy source in glycolysis and ultimately the citric acid cycle. The fatty acids are predominantly broken down by β-oxidation, which sequentially removes two C atoms at a time from the fatty acid to form acetyl-CoA, which enters the citric acid cycle directly to produce energy. As seedling growth progresses, the fatty acids are also hydrolysed by β-oxidation, but the resultant acetyl-CoA is shunted into the glyoxylate cycle for the production of carbohydrates.

Proteins

Classically, proteins are defined into four categories based on their solubility in various solvents. Albumins (12–100 kDa) are found in nearly all flower seeds, soluble in water, easily coagulated (denatured by heat), and yield mostly amino acids. Globulins (18–360 kDa) are soluble in 0.5–1.0 M NaCl and contain high levels of glutamine, asparagine and arginine with low levels of S-containing amino acids such as methionine and cysteine. Prolamins (16–60 kDa) are soluble in 700 ml/l (70%) ethanol and are enriched in glutamine and proline. Glutenins (30–50 kDa) are insoluble in water and soluble in alkaline solvents. Most proteins are associated with protein bodies that are digested by proteolytic enzymes during the germination process. Initially, there is a hydrolysis of the protein into subunits followed sequentially by hydrolytic release into amino acids and small peptides. Among the principal enzymes involved are the endopeptidases that cleave peptide linkages possessing α-carbonyl groups of aspartic and glutamic acids, releasing the acyl portion of acidic amino acids. Other proteases such as exopeptidases complete the peptide digestion. The types of proteins found in flower seeds are highly species-dependent and more needs to be done in their characterization.

Seedling establishment

Flower seed germination is described as being hypogeal or epigeal, depending on the position of storage tissues in the soil. If the storage tissue, such as the endosperm in grasses, remains below the soil, it is hypogeal germination. Since the cotyledons of many dicot flower seeds are moved above the soil surface during seedling establishment, this is identified as epigeal germination. In most flower seeds, the root is the first seedling structure to emerge, followed by the shoot structure (mesocotyl/coleoptile in grasses and hypcotyl/cotyledons in most dicots). Root growth of germinating seedlings is inhibited less by low water potentials compared with shoot growth (Creelman et al., 1990). To account for this response, it has been proposed that abscisic acid (ABA) accumulates at low water potentials and thereby preferentially maintains primary root growth while inhibiting shoot growth (Sharp et al., 1990). This response may be advantageous since a reduction in stem (leaf) growth and increased root growth improves overall plant water status. In addition, it assures greater exploration of the soil by seedling roots to enhance water uptake.

The gravitropic response of seedling roots has been examined and is mediated by auxin (indole-3-acetic acid) (Fig. 8.2). Auxin is synthesized in the root apical meristem (Feldman, 1981), moves forward through the stele to the root cap, and is preferentially redistributed toward the lower side of the cap. At that point, the redistributed auxin moves through the root cortical cells to the zone of elongation. There, the auxin on the lower side of the root inhibits cell growth and causes a downward curvature (Evans et al., 1986). Thus, while the meristem is the site of auxin synthesis, the root cap functions in redistributing auxin asymmetrically so that roots grow downward (Young et al., 1990).

Other factors influence the gravitropic response in seedling roots. Red light, but not blue light, also induces root curvature, implicating a role of phytochrome (Feldman and Briggs, 1987). It also has been shown that certain cations have major effects on root growth. Hasenstein and Evans (1988) demonstrated that Al inhibited auxin loading into the xylem transport stream from the root cap, while Ca and Mg strongly increased basipetal transport of auxin, leading to root curvature greater than the control. This physiological mechanism may explain why seedling roots grow toward Ca and Mg and away from Al cations.

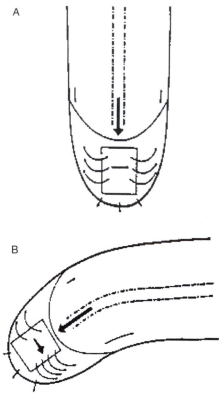

Fig. 8.2. Proposed patterns of auxin movement in the apical region of vertically and horizontally orientated primary roots of maize. In a vertically orientated root, auxin is redistributed symmetrically in the cap and transported in equal amounts toward the elongating zone (A). After gravistimulation, auxin entering the cap is transported preferentially downward and then back toward the lower side of the elongating zone. The increased amount of auxin on the lower side leads to growth inhibition and downward curvature (B). Chemical interference with the auxin transport system within the root cap, e.g. by unilateral application of Al^{3+} or Ca^{2+}, also results in root curvature (from Hasenstein and Evans, 1988).

With respect to seedling shoot growth, studies have demonstrated that grasses rely on the elongation of the mesocotyl and coleoptile before leaves can unfold above the soil surface. While both tissues elongate, it is the mesocotyl that is primarily responsible for the greatest elongation. Two studies suggest that auxin governs this elongation. Iino and Carr (1982) found that mesocotyl elongation ceased after 1 h when the coleoptile and primary leaves were removed. They concluded that mesocotyl elongation is regulated by auxin produced at the coleoptile tip and then diffused into the mesocotyl. Other studies have shown that the gravitropic response of grass shoots is rapid (1.5 min) and asymmetrical (higher concentration of auxin in the lower half of the mesocotyl) in the mesocotyl cortex, which results in observable curvature within 2 min for the coleoptile and 3 min for the mesocotyl (Bandurski et al., 1984). These data further support the hypothesis that auxin is synthesized in the coleoptile tip and moves basipetally to the mesocotyl. Phytochrome has also been implicated in governing mesocotyl growth. Far red light stimulates mesocotyl elongation, while red light inhibits this reversible response (Yahalom et al., 1987).

In dicot seeds, immediately following radicle protrusion, there is rapid growth of the hypocotyl. Cavalieri and Boyer (1982) demonstrated that water potentials decreased from the root to the hypocotyl crook and radially from the stele to the cortex. They also showed that water potential in the hypocotyl elongating zone was not uniform and was most negative immediately following the hypocotyl crook (Fig. 8.3). A hormonal mechanism relying on a balance between gibberellic acid (GA_1) and ABA has been proposed as influencing hypocotyl elongation rates. Applied ABA inhibited hypocotyl elongation in well-watered seedlings (Bensen et al., 1988) and transfer of these seedlings to a low-water-potential vermiculite medium reduced hypocotyl elongation rates and increased ABA levels. With rewatering, there was a rapid decrease in ABA content with concurrent increases in GA_1 levels (Bensen et al., 1990). These hormonal effects reflect the notion that changes in growth rates result from a complex interplay between physical pressure for expansion (turgor) and the yield threshold of the cell walls, which is, in turn, influenced by hormones.

Plug Germination

Today's plug producer possesses modern germination chambers specifically constructed to accommodate plug trays in volume. These chambers maintain precise high relative humidity, light (if needed) and temperature conditions so that complete and rapid germination is consistently attained. Variable emergence rates result in non-uniform plug trays and loss of revenue to the grower (Fig. 8.4). When determining successful germination in plugs, the grower must consider

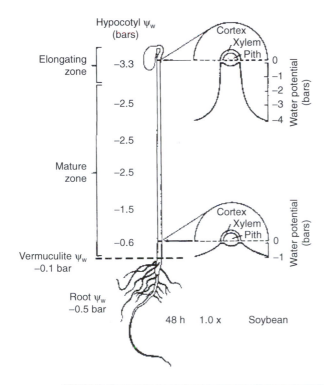

Fig. 8.3. Summary of the location and extent of growth-induced water potential in a soybean hypocotyl adequately supplied with water for 48 h after transplanting. Mature zone contains cells slowly increasing in diameter as well as basal cells not enlarging. Radial water potential gradients are shown in section views. Longitudinal gradient is shown alongside hypocotyl (from Cavalieri and Boyer, 1982).

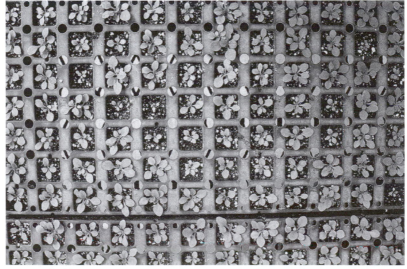

Fig. 8.4. Variable emergence rates result in non-uniform plug trays that require more production time and often result in unsatisfied customers.

scheduling, plug media and seed planting equipment.

Scheduling

Because germination conditions are carefully controlled during plug production, the scheduling of planting approximates the time when plugs attain their correct size and are in optimum shape for shipping and marketing. For example, planting dates are often established by first knowing the expected marketing date and then counting back the number of days to flower or sale required for the crop under given environmental conditions. This number is only approximate, however, since such tasks as sowing, transplanting, refilling empty cells and shipping are not always conducted as planned. Inclement weather and unfavourable market conditions also affect actual shipping dates. When germination and early emergence of plugs needs to be delayed, movement of plugs into cold rooms at 5°C–10°C is helpful.

Plug media

The medium used to germinate seeds is a critical aspect of successful germination. Attention should be given to water-holding capacity and air porosity. The ideal medium must provide sufficient water so that rapid germination occurs while simultaneously providing adequate oxygen in the root zone to allow respiration for optimum root growth. Most commercial and individually prepared media mixes are a combination of peat moss and aggregates such as perlite or vermiculite to enhance drainage and increase aeration. Commercial media can come in differing concentrations of the aggregates for growing moist or dry crops (24% instead of 18% perlite for vinca and petunia, as examples) to keep them drier or wetter. Growers blending their own germination media frequently use a 1:1 ratio, by volume, of sphagnum peat moss and perlite or vermiculite. Soil-based media are not recommended for plug production because of the need for sterilization

and the lack of uniformity in soil sources. The type of plug mix used may also vary depending on the number of cells per tray, e.g. 128s, 288s and 512s. As cells become smaller, they are more prone to flooding after irrigation and have more rapid dry-out at the seed zone. Thus, growers should water plugs four or five times a day to ensure enough water for germination and still permit dry-out of the medium for restoration of needed air space for optimal root growth. In all cases, when handling media, compaction during shipping, tray filling or stacking must be avoided to maintain vital air space while still enhancing seed to medium contact. For this reason, commercial media should be thoroughly mixed prior to filling trays. Since peat moss expands on hydration, air space can also be enhanced by wetting the media prior to tray filling. The media should be tested for pH and electrical conductivity (EC) of soluble salts before seeding to avoid reduced germination and potential burning of seedling roots. The ideal pH is between 5.5 and 6.5. Pulverized limestone can adjust the acidic pH characteristic of peat-based plug mixes. The EC value should be less than 1.0 mmhos/cm at a 2:1 dilution to minimize burn damage of sensitive primary roots, particularly for vinca, snapdragon, impatiens and begonia.

Seed planting equipment

Seeds can be either broadcast or planted in rows in flats by hand or planted directly into plugs using specialized equipment. Generally, broadcast planting is considered inefficient and unpredictable, while planting in rows often culminates in twisting and intertwining of seedling shoots/roots as well as stretching when excessive crowding is encountered. Direct seeding of flower crops into plugs has gained increasing acceptance by growers. This is accomplished using vacuum seeders that place one seed per cell in each tray. Almost all seeders rely on the same general operational principles. Individual seeds are picked up in a seed tray by vacuum through manifolds (openings) in a specialized plate, needle or drum. The seeds are held in place, positioned over the plug tray cells, the vacuum released, and the singulated seeds dropped into each individual cell on top of the planting medium. The size of the manifolds can be adjusted to pick up either large or small seeds. Pelleting seeds benefits accurate planting of small seeds such as begonia and petunia by making them larger and heavier. In addition, many pelleted seeds have a contrasting colour added to the pellet to make visualization of seed placement and number easier in the plug (Fig. 8.5). Other seeds, such as marigold, that have ridges and tails can be detailed and pelleted to make the seed surface smoother for vacuum pick-up. Three types of vacuum seeder are in common use: manual, needle and drum.

Manual

Manual seeders vary from flat trays to hand-held needles spaced to accommodate rows in a plug flat. Flat trays possess a front and back face with a cavity for air passage between which has precisely spaced perforations to match the number of cells in the flat. Seeds are placed on the tray, the vacuum turned on, and a single seed retained on each hole. The tray is placed over the substratum, the vacuum turned off and the seeds deposited on to the plug medium. This method can also be used on seeding flats for accurate seed numbers and placement uniformity. Hand-held needles allow small seed samples to be precisely placed using a vacuum to retain the seeds and then deposit them into the plug flat. Manual seeding is the most inexpensive approach to automated seeding. Because of the manual approach, this method works well for small numbers of seeds. However, it is a laborious and time-consuming operation for high-volume seeding.

Needle

Needle seeders are the most versatile automated seeder. They have hollow needles arranged in rows spaced according to the number of cells widthwise in the plug flat (Fig. 8.5a). The needles are usually made of steel, although rubber tips are available for singulation of flower crop seeds with rough surfaces. Needle seeders function by establishing a vacuum through the row of needle tips. The needle tips are placed into the seed tray, where each tip picks up a seed. The tips containing the seeds are raised from the seed tray, lowered over a wide mouth drop tube, the vacuum turned off, and seeds released where they slide down the individual drop tubes for accurate placement in the plug cell. Needle seeders can accurately plant large and odd-shaped seeds such as marigold, dahlia and zinnia as well as small seeds such as petunia and begonia when properly calibrated and operated. This versatility makes them popular in seeding operations. However, their rate of seeding is not as fast as drum seeders.

Drum

Drum seeders are primarily used in highly automated, high-volume seeding operations. They consist of a hollow steel drum with precisely spaced perforations to conform to a plug tray (Fig. 8.5b). The vacuum is turned on and the drum rotates in a seed tray where seeds are picked up. As the drum continues to turn over the tray, the vacuum is turned off and the seeds released into individual cells. The advantage of drum seeders is the high volume of seeding, where over 1000 trays can be seeded in 1 h.

Flower seed germination stages

Growth stages

Flower crop plug production has been described using four growth stages, each requiring different management strategies for optimum seedling growth. The process of germination and the initiation of seedling growth includes stages 1 and 2. Stage 1 is germination as signified by radicle emergence from the seed. This process requires high levels of moisture for imbibition and optimum temperature and sufficient oxygen to begin the germination process. Stage 2 is when the radicle has penetrated the soil and hypocotyl elongation and leaf expansion occurs so that the seedling is observed in the plug. This

Fig. 8.5.   Examples of automated seeders: (a) needle seeder; (b) drum seeder.

stage generally requires less water and more root aeration and cooler temperatures for optimum seedling growth. While starter fertilizers can be used in plug production, they have more important benefits in stages 3 and 4, since most seeds rely on storage reserves in the endosperm and/or cotyledons for the necessary energy to initiate the earliest stages of germination. Stages 1 and 2 can be considered the most critical stages in plug production since any failure in seedling emergence leads to loss in grower revenue. It is important for growers to understand the requirements for seed germination.

Speed and uniformity of seed germination are critical for timing and shipping of plugs. These factors can be controlled by ensuring optimal environmental conditions during seed germination which vary according to flower crop. The successful plug grower is aware of the differences in germination requirements. Table 8.1 presents specific germination requirements for important flower crops. The environmental factors most easily controlled during seed germination are moisture, temperature and light.

MOISTURE.   After sowing of seeds, the medium should be irrigated thoroughly for rapid germination. Avoid overwatering. Watering is accomplished through specialized irrigation tunnels or misting operations that provide uniform irrigation without washing seeds from the medium. Many factors contribute to the availability of water for successful germination. Among these are medium composition, plug cell size, and frequency and amount of irrigation.

MEDIUM COMPOSITION. Identifying the appropriate medium for optimum seedling growth begins with the selection and proper mixing of media components. Water-holding capacity is a measure of how much water a substance can hold against the pull of gravity. The finer the particles in the plug mix, the greater the water-holding capacity. Conversely, as water-holding capacity is increased, which assists rapid and complete germination, aeration of the medium is reduced, which limits optimum root growth. Growers need to identify the best plug medium composition that represents a compromise between water-holding capacity for germination and porosity to support seedling growth. Filling plugs with medium is also important. The medium should not be older than 1 day so that the moisture content stays constant and the medium in the cells does not dry out. After the

plugs are filled, they should be rolled, compacted and dibbled to make a depression for the seed. This assists in maintaining medium uniformity for seedling growth and the dibble ensures proper seed placement at planting.

PLUG CELL SIZE. Plug cell size affects the amount and frequency of irrigation. Smaller plugs require less water to fully saturate the medium and they dry out more rapidly than larger plugs. Thus, they should be watered more frequently to ensure that sufficient water is available for germination. After germination, less water can be used, allowing sufficient aeration for root growth.

FREQUENCY AND AMOUNT OF IRRIGATION. How often and how much irrigation is required for germination? The process of water uptake for seed germination happens in three phases (Fig. 8.6). Rapid water uptake (imbibition) occurs initially, followed by a plateau and then rapid water uptake again after germination (radicle protrusion). Thus, to initiate germination, the medium should be fully saturated to the point of drip, thus minimizing nutrients leaching from the plug.

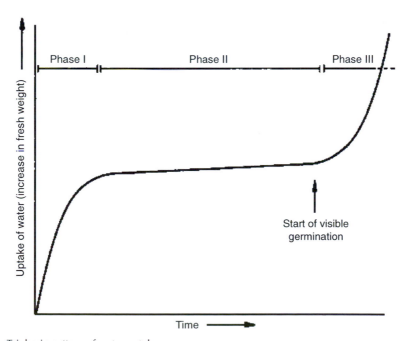

Fig. 8.6.   Triphasic pattern of water uptake.

As the plug begins to dry, additional mistings become necessary to maintain sufficient water around the seed. In the greenhouse, particularly for small plugs, this may mean four to five mistings per day depending on the ambient RH and temperature. Another approach is to thoroughly water the plug flats and then cover them with polythene film to ensure that evaporation is minimized. To avoid these labour-intensive processes, and to enhance uniformity of emergence, many growers have specific germination rooms maintained at high relative humidities to minimize plug dry-out and provide precise temperatures for optimum germination.

After germination occurs, the flats are removed from the germination room to avoid stretching of seedlings. While these germination rooms are expensive, germination occurs more rapidly, uniformly, and stronger seedlings are produced, often justifying this initial investment. Maintaining optimum moisture levels for seed germination is important for successful plug production. Flats that become too dry produce fewer seedlings and irregular germination.

TEMPERATURE.   Optimum temperatures of the media for seed germination vary according to flower crop (Table 8.1). To obtain the

Table 8.1. Minimum germination standards, laboratory and greenhouse count day, laboratory and greenhouse germination temperature, media moisture status at stages 1 and 2 (D = dry, M = moist, W = wet), and greenhouse cover (Y = yes, N = no) for germination of ornamental and vegetable crops. (Table courtesy of Val Maxwell, Greiling Farms.)

Class	Germ std.	Lab. count day	Grhse. count day	Lab. temp.	Grhse. temp.	Soil moist. 1	2	Grhse. cover Y/N
Abelmoschus	75	7	NO	73	72–75	M	M	Y
Achillea	75	7	NO	67	65–70	M	M	N
Aethionema	75	NO	21	67	65–70	M	M	N
Agastache	70	10	NO	67	65–70	M	M	N
Ageratum	88	7	NO	77	76–80	M	M	N
Alchemilla	85	NO	14	67	65–70	M	M	N
Alyssum	75	7	NO	77	76–80	M	M	N
Amaranthus	85	7	NO	73	72–75	M	M	N
Amsonia	75	14	NO	67	65–70	M	M	N
Anaphalis	65	21	NO	67	65–70	M	M	Y
Anchusa	75	NO	28	67	65–70	M	M	Y
Anemone	65	28	NO	67	65–70	M	M	N
Anthemis	85	7	NO	67	65–70	M	M	N
Aquilegia	85	14	NO	73	72–75	M	M	N
Arabis	75	10	NO	67	65–70	M	M	N
Arenaria	75	14	NO	67	65–70	M	M	N
Armeria	80	14	NO	67	65–70	M	M	Y
Asclepias	65	21	NO	73	72–75	M	M	Y
Asparagus sprengeri	75	28	NO	77	76–80	M	M	Y
Aster	80	10	NO	67	65–70	M	M	Y
Astilbe	75	21	NO	67	65–70	M	M	N
Aubrieta	70	14	NO	67	65–70	M	M	N
Balsam	75	10	NO	73	72–75	W	M	N
Baptista	75	14	NO	73	72–75	W	W	Y
Basil, ornamental	80	7	NO	67	65–70	M	M	N
Begonia, fibrous	88	NO	21	77	76–80	W	W	N
Begonia, tuberous nonstop	85	21	NO	77	76–80	W	W	N
Bellis	80	7	NO	73	72–75	M	M	N
Bergenia	65	14	NO	73	72–75	M	M	N
Boltonia	75	14	NO	73	72–75	M	M	N

Table 8.1. *Continued.*

Class	Germ std.	Lab. count day	Grhse. count day	Lab. temp.	Grhse. temp.	Soil moist. 1	Soil moist. 2	Grhse. cover Y/N
Brachycome	80	14	NO	67	65–70	M	M	N
Briza	75	14	NO	67	65–70	M	M	N
Browallia	85	10	NO	73	72–75	M	M	N
Calceolaria	80	14	NO	73	72–75	M	M	N
Calendula	80	10	NO	67	65–70	M	D	Y
Campanula	80	21	NO	67	65–70	M	M	N
Canna	60	28	NO	73	72–75	M	D	Y
Carnation	88	7	NO	67	65–70	M	M	Y
Caryopteris	88	10	NO	67	65–70	M	M	Y
Catananche	75	10	NO	67	65–70	M	M	Y
Celosia	80	7	NO	73	72–75	W	M	Y
Centauria (cornflower)	75	14	NO	67	65–70	M	M	Y
Centranthus	75	14	NO	67	65–70	M	M	N
Cerastium	75	10	NO	67	65–70	M	M	N
Chrysanthemum	80	14	NO	67	65–70	M	M	N
Chrysothemis	75	14	NO	67	65–70	M	M	N
Cineraria	88	10	NO	73	72–75	M	M	N
Cleome	75	NO	21	67/77	67/77	M	M	Y
Coleus	80	7	NO	73	72–75	M	M	N
Coreopsis	70	NO	14	73	72–75	M	M	Y
Coronilla	75	14	NO	73	72–75	M	M	N
Cortaderia	75	14	NO	67	65–70	M	M	N
Corydalis	65	21	NO	67	65–70	M	M	Y
Cosmos	80	10	NO	67	65–70	M	M	Y
Crossandra	75	NO	28	67/77	67/77	M	D	Y
Cuphea	80	14	NO	73	72–75	M	M	N
Cyclamen	75	28	NO	67	65–70	W	W	Y
Dahlia	88	NO	10	67	65–70	D	D	Y
Daisy, Dahlberg	75	14	NO	67	65–70	M	M	Y
Daisy, Shasta	80	14	NO	67	65–70	M	M	N
Delphinium	75	14	NO	67	65–70	M	M	N
Dianthus	88	7	NO	73	72–75	M	M	Y
Digitalis	75	10	NO	67	65–70	M	M	N
Doronicum (Leopard's bane)	85	NO	21	73	72–75	M	M	N
Dracaena spikes	65	NO	28	77	76–80	M	M	Y
Dusty miller	75	NO	14	73	72–75	M	M	N
Echinacea	80	14	NO	67	65–70	M	M	N
Echinops	88	10	NO	67	65–70	M	M	N
Edelweiss (*Leontopodium*)	75	14	NO	67	65–70	M	M	N
Erigeron	75	NO	14	67	65–70	M	M	N
Euphorbia	70	14	NO	67/77	67/77	M	M	Y
Exacum	88	14	NO	73	72–75	M	M	Y
Felicia	88	14	NO	67	65–70	M	M	N
Festuca	75	14	NO	67	65–70	M	M	N
Flowering cabbage	88	7	NO	67	65–70	M	M	Y
Flowering kale	85	7	NO	67	65–70	M	M	Y
Fuchsia	80	NO	14, 28	73	72–75	M	M	N
Gaillardia	80	10	NO	67	65–70	M	M	Y
Gazania	70	10	NO	67	65–70	M	M	Y
Geranium	88	NO	10	73	72–75	W	M	Y

continued

Table 8.1. *Continued.*

Class	Germ std.	Lab. count day	Grhse. count day	Lab. temp.	Grhse. temp.	Soil moist. 1	Soil moist. 2	Grhse. cover Y/N
Gerbera	80	NO	21	73	72–75	M	M	N
Geum	88	NO	14	67/77	67/77	M	M	N
Godetia	75	10	NO	67	65–70	M	M	N
Gomphrena	75	NO	10	73	72–75	M	D	N
Grass, ornamental	75	14	NO	67	65–70	M	M	N
Gypsophila	80	14	NO	67	65–70	M	M	N
Helenium	75	21	NO	67	65–70	M	M	N
Helichrysum	75	10	NO	67	65–70	M	M	N
Heliopsis	75	14	NO	67	65–70	M	M	N
Heliotrope	75	21	NO	67	65–70	M	M	N
Heuchera	75	14	NO	67	65–70	M	M	N
Hibiscus	80	10	NO	73	72–75	M	M	Y
Hollyhock	65	14	NO	67	65–70	M	M	Y
Hypoestes	85	NO	14	73	72–75	M	M	N
Iberis	85	14	NO	67	65–70	M	M	N
Impatiens	88	NO	10, 24	77	76–80	W	M	N
Impatiens, New Guinea	65	NO	14, 28	77	76–80	W	M	Y
Knautia	75	21	NO	67	65–70	M	M	Y
Kniphofia	65	28	NO	67	65–70	M	M	Y
Lagerstroemia	75	NO	21	73	72–75	M	M	Y
Lagurus	75	14	NO	67	65–70	M	M	N
Larkspur	75	21	NO	67	65–70	M	M	N
Lavender	80	14	NO	67	65–70	M	M	Y
Lewisia	65	14	NO	73	72–75	M	M	N
Liatris	85	14	NO	67	65–70	M	M	Y
Limonium	65	14	NO	67	65–70	M	M	Y
Linaria	75	14	NO	67	65–70	M	M	N
Linum	85	14	NO	67	65–70	M	M	N
Lisianthus	80	NO	21	73	72–75	M	M	N
Lobelia	85	14	NO	77	76–80	M	M	N
Lotus	80	14	NO	67	65–70	M	M	N
Lupin	75	21	NO	67	65–70	M	M	Y
Lychnis(*Viscaria*)	85	14	NO	67	65–70	M	M	N
Lysimachia	75	21	NO	67	65–70	M	M	N
Malva	75	14	NO	67	65–70	M	M	N
Marigold	88	NO	7	73	72–75	M	D	Y
Matricaria	60	7	NO	67	65–70	M	M	N
Melampodium	70	NO	10	67	65–70	M	M	Y
Mimulus	80	7	NO	67	65–70	M	M	N
Myosotis	75	14	NO	67	65–70	M	M	N
Nasturtium	75	14	NO	67	65–70	M	M	Y
Nemesia	75	10	NO	67	65–70	M	M	Y/N
Nepeta	75	NO	14	67	65–70	M	M	N
Nicotiana	80	10	NO	73	72–75	M	M	N
Nierembergia	80	NO	10	67	65–70	M	M	N
Nolana	75	21	NO	67	65–70	M	M	N
Oenothera (Evening primrose)	75	28	NO	67	65–70	M	M	N
Onion	88	7	NO	67	65–70	M	M	Y
Pansy	88	7	NO	67	65–70	M	M	Y
Penstemon	75	NO	21	67	65–70	M	M	Y
Pentas	88	14	NO	73	72–75	M	M	N

Table 8.1. *Continued.*

Class	Germ std.	Lab. count day	Grhse. count day	Lab. temp.	Grhse. temp.	Soil moist. 1	2	Grhse. cover Y/N
Pepper, ornamental	80	14	NO	73	72–75	M	M	Y
Perovskia	65	28	NO	67	65–70	M	M	N
Petunia	88	NO	10, 24	77	76–80	M	M	N
Phalaris	75	14	NO	67	65–70	M	M	N
Philodendron	80	14	NO	73	72–75	M	M	Y
Phlox	75	NO	21	67	65–70	M	D	Y
Physalis	75	NO	21	67	65–70	M	M	N
Physostegia	75	NO	21	67	65–70	M	M	N
Platycodon	75	NO	14	67	65–70	M	M	N
Plumbago	60	14	NO	73	72–75	M	M	Y
Polemonium	70	10	NO	67	65–70	M	M	N
Poppy	75	14	NO	67	65–70	M	M	N
Portulaca	75	10	NO	77	76–80	M	D	N
Potentilla	85	14	NO	67	65–70	M	M	N
Primula	75	28	NO	67	65–70	W	M	N
Prunella	80	10	NO	67	65–70	M	M	N
Pyrethrum	75	14	NO	67	65–70	M	M	N
Radermachera	75	10	NO	73	72–75	M	M	Y
Ranunculus	85	21	NO	67	65–70	M	M	N
Rudbeckia	80	14	NO	73	72–75	M	M	N
Sagina	75	14	NO	67	65–70	M	M	N
Salvia splendens	88	10	NO	77	76–80	W	M	Y
Salvia, perennial	85	14	NO	77	76–80	W	M	Y
Salvia farinacea	80	10	NO	77	76–80	W	M	Y
Saponaria	75	14	NO	67	65–70	M	M	N
Saxifraga	88	21	NO	67	65–70	M	M	N
Scabiosa	75	28	NO	67	65–70	M	M	Y
Schizanthus	80	10	NO	67	65–70	M	M	N
Sedum	75	14	NO	67/77	67/77	M	M	N
Silene	65	14	NO	67	65–70	M	M	Y
Snapdragon	75	14	NO	73	72–75	M	D	N
Solidago	75	21	NO	67	65–70	M	M	N
Stachys	65	14	NO	67	65–70	M	M	N
Statice	75	10	NO	67	65–70	M	M	N
Stokesia (Stokes aster)	75	21	NO	67	65–70	M	M	N
Strawberry	80	14	NO	67	65–70	M	M	Y
Sunflower	80	10	NO	73	72–75	M	M	Y
Thunbergia	88	10	NO	73	72–75	M	M	Y
Torenia	75	10	NO	67	65–70	M	M	N
Trollius	75	21	NO	67	65–70	M	M	N
Verbascum	65	21	NO	67	65–70	M	M	N
Verbena novalis	70	NO	14	77	76–80	D	D	Y
Verbena Quartz	90	NO	14	77	76–80	D	D	Y
Verbena Romance	75	NO	14	77	76–80	D	D	Y
Veronica	80	10	NO	67	65–70	M	M	N
Vinca	75	NO	14	77	76–80	W	M	N
Viola	88	7	NO	67	65–70	W	M	Y
Zinnia	80	7	NO	73	72–75	M	M	Y
Zinnia linearis	75	7	NO	73	72–75	M	M	N

continued

Table 8.1. *Continued.*

Class	Germ std.	Lab. count day	Grhse. count day	Lab. temp.	Grhse. temp.	Soil moist. 1	Soil moist. 2	Grhse. cover Y/N
Herb, basil	70	7	NO	67	65–70	M	M	N
Herb, catnip	65	7	NO	67	65–70	M	M	N
Herb, chives	70	14	NO	67	65–70	M	M	N
Herb, coriander	70	10	NO	67	65–70	M	M	Y
Herb, dill	70	10	NO	67	65–70	M	M	N
Herb, fennel	70	10	NO	67	65–70	M	M	Y
Herb, lemon balm	70	10	NO	67	65–70	M	M	N
Herb, lavender	70	21	NO	67	65–70	M	M	N
Herb, marjoram	70	7	NO	67	65–70	M	M	N
Herb, mint	70	NO	21	73	72–75	M	M	N
Herb, oregano	70	7	NO	67	65–70	M	M	Y
Herb, parsley	70	14	NO	67	65–70	M	M	Y
Herb, rosemary	35	14	NO	67	65–70	M	M	N
Herb, sage	70	10	NO	67	65–70	M	M	Y
Herb, savory	60	14	NO	67	65–70	M	M	N
Herb, thyme	70	7	NO	67	65–70	M	M	N
Herb, woodruff (*Galium odoratum*)	70	21	NO	67	65–70	M	M	Y

fastest and most uniform germination, the seeds must be exposed to optimum medium temperatures so that essential biochemical and physiological processes required for germination occur. Generally, these temperatures range from 15°C to 25°C. Germinating seeds in the greenhouse or in flats under polythene can rapidly produce variable temperatures that influence the rate of germination, culminating in less uniform performance. Germination rooms minimize this problem.

LIGHT.   Seeds differ in their responsiveness to light (Table 8.1). In some cases, light is essential for germination. In others, the absence of light promotes germination. In still other cases, seeds germinate regardless of the light regime. A general rule is that the smaller the seed, the more likely light is necessary for germination. If light is necessary, seeds must be fully imbibed for the optimum response. Dry seeds do not respond to light. For best light response, imbibed seeds should be uncovered without any topcoating at the time of exposure. The type and amount of light also affects the uniformity of germination. Light that is heavy in red is preferred. Incandescent bulbs alone or in combination with fluorescent lamps enriched with the red spectrum are best. The intensity of the light also enhances germination for light requiring species. No less than 100 foot-candles is recommended. Frequent measurements of light levels and uniformity should be conducted in germination rooms.

Conclusions

Flower growers must understand the components of high quality seeds. Each seed should germinate rapidly and uniformly within the seed lot when provided an optimum environment. Growers must know the underlying physiology of flower seeds to best understand how to properly handle them. This, in conjunction with a knowledge of the conditions required for optimal germination in plugs, can assure the greatest success in stand establishment. When seeds do not perform, the problem often lies with inappropriate handling and failure to provide optimum germination conditions. This chapter has identified the principles of germination and the environmental conditions favourable for germination and seedling growth of most flower crops. Purchasing the best quality seeds should not be viewed as a cost, but an investment in a successful flower operation.

References

Adams, C.A., T.H. Broman, S.W. Norby and R.W. Rinne. 1981. Occurrence of multiple forms of α-amylase and absence of starch phosphorylase in soybean seeds. *Ann. Bot.* 48:895–903.

Bandurski, R.S., A. Schulze, P. Dayanandan and P.B. Kaufman. 1984. Reponse to gravity by *Zea mays* seedlings. I. Time course of the response. *Plant Physiol.* 74:294–289.

Bensen, R.J., J.S. Boyer and J.E. Mullet. 1988. Water deficit induced changes in abscisic acid, growth, polysomes, and translatable RNA in etiolated soybean hypocotyls. *Plant Physiol.* 88: 289–294.

Bensen, R.J., R.D. Beal, J.E. Mullet and P.W. Morgan. 1990. Detection of endogenous gibberellins and their relationship to hypocotyl elongation in soybean seedlings. *Plant Physiol.* 88:289–294.

Bewley, J.D. and M. Black. 1994. *Seeds: Physiology and Development.* Plenum Press, New York.

Calero, E., S.H. West and K. Hinson. 1981. Water absorption of soybean seeds and associated causal factors. *Crop Sci.* 21:926–933.

Cavalieri, A.J. and J.S. Boyer. 1982. Water potentials induced by growth in soybean hypocotyls. *Plant Physiol.* 69:492–497.

Copeland, L.O. and M.B. McDonald. 2001. *Principles of Seed Science and Technology.* Kluwer Press, New York.

Creelman, R.A., H.S. Mason, R.J. Bensen, J.S. Boyer and J.E. Mullet. 1990. Water deficit and abscisic acid cause differential inhibition of shoot versus root growth in soybean seedlings: analysis of growth, sugar accumulation, and gene expression. *Plant Physiol.* 92:205–215.

Dassou, S. and E.A. Keuneman. 1984. Screening methodology for resistance to field weathering of soybean seed. *Crop Sci.* 24:774–779.

Evans, M.L., R. Moore and K.H. Hasenstein. 1986. How roots respond to gravity. *Sci. Amer.* 255:112–119.

Evans, M.D., R.G. Holmes and M.B. McDonald. 1990. Impact damage to soybean seed as affected by surface hardness and seed orientation. *Trans. Amer. Soc. Agr. Eng.* 33: 234–240.

Feldman, L.J. 1981. Effect of auxin on acropetal auxin transport in roots of corn. *Plant Physiol.* 67:278–281.

Feldman, L.J. and W.R. Briggs. 1987. Light regulated gravitropism in seedling roots of maize. *Plant Physiol.* 83:241–244.

Hasenstein, K.H. and M.L. Evans. 1988. Effects of cations on hormone transport in primary roots of *Zea mays. Plant Physiol.* 86:890–894.

Hill, H.H.J., S.H. West and K. Hinson. 1986. Effect of water stress during seed fill on impermeable seed expression in soybean. *Crop Sci.* 26: 807–813.

Iino, M. and D.J. Carr. 1982. Sources of free IAA in the mesocotyl of etiolated maize seedlings. *Plant Physiol.* 69:1109–1113.

Leopold, A.C. 1983. Volumetric components of seed imbibition. *Plant Physiol.* 73:677–680.

Leopold, A.C. and C.W. Vertucci. 1986. Physical attributes of desiccated seeds. *In*: A.C. Leopold (ed.). *Membranes, Metabolism, and Dry Organs.* Comstock, Cornell University Press, Ithaca, New York. pp. 22–34.

Leopold, A.C. and C.W. Vertucci. 1989. Moisture as a regulator of physiological reaction in seeds. *In*: P.C. Stanwood and M.B. McDonald (eds). *Seed Moisture.* CSSA Special Publication 14. ASA, Madison, Wisconsin. pp. 51–68.

Luedders, V.A. and J.S. Burris. 1979. Effect of broken seed coats on field emergence of soybeans. *Agron. Jour.* 71:877–879.

Mayer, A.M. and A. Poljakoff-Mayber. 1989. *The Germination of Seeds.* Pergamon Press, New York.

McDonald, M.B. 1985. Physical seed quality of soybean. *Seed Sci. and Technol.* 13:601–628.

McDonald, M.B. 1994. Seed germination and seedling establishment. *In*: K.J. Boote, J.M. Bennett, T.R. Sinclair and G.M. Paulsen (eds). *Physiology and Determination of Crop Yield.* ASA, Madison, Wisconsin. pp. 37–60.

McDonald, M.B., C.W. Vertucci and E.E. Roos. 1988. Soybean seed imbibition: water absorption by seed parts. *Crop Sci.* 28:993–997.

Mugnisjah, W.Q., I. Shimano and S. Matsumoto. 1987. Studies of the vigour of soybean seeds. II. Varietal differences in seed coat colour and swelling components of seed during moisture imbibition. *Jour. Fac. Agric., Kyushu Univ.* 31: 227–234.

Murphy, J.B. and T.L. Noland. 1982. Temperature effects on seed imbibition and leakage mediated by viscosity and membranes. *Plant Physiol.* 69:428–431.

Murray, D.R. 1984a. *Seed Physiology. I. Development.* Academic Press, New York.

Murray, D.R. 1984b. *Seed Physiology. II. Germination and Reserve Mobilization.* Academic Press, New York.

Potts, H.C., J. Duangpatra, W.G. Hairston and J.C. Delouche. 1978. Some influences of hard-seededness on soybean seed quality. *Crop Sci.* 18:221–224.

Powell, A.A. and S. Matthews. 1978. The damaging effect of water on dry pea embryos during imbibition. *Jour. Exp. Bot.* 29:1215–1229.

Sharp, R.E., T.C. Hsiao and W.K. Silk. 1990. Growth of the maize primary root at low water potentials. II. Role of growth and deposition of hexose and potassium in osmotic adjustment. *Plant Physiol.* 93:1337–1347.

Tully, R.E., M.E. Musgrave and A.C. Leopold. 1981. The seed coat as a control of imbibitional chilling injury. *Crop Sci.* 21:312–317.

Vertucci, C.W. 1989. The kinetics of seed imbibition: controlling factors and relevance to seedling vigour. *In*: P.C. Standwood and M.B. McDonald (eds). *Seed Moisture.* CSSA Special Publication 14. ASA, Madison, Wisconsin. pp. 93–115.

Vertucci, C.W. and A.C. Leopold. 1983. Dynamics of imbibition in soybean embryos. *Plant Physiol.* 72:190–193.

Yahalom, A.B., L. Epel, Z. Glinka, I.R. MacDonald and D.C. Gordon. 1987. A kinetic analysis of phytochrome controlled mesocotyl growth in *Zea mays* seedlings. *Plant Physiol.* 84: 390–395.

Yaklich, R.W., E.L. Vigil and W.P. Wergin. 1986. Pore development and seed coat permeability in soybean. *Crop Sci.* 26:616–624.

Young, L.M., M.L. Evans and R. Hertel. 1990. Correlations between gravitropic curvature and auxin movement across gravistimulated roots of *Zea mays*. *Plant Physiol.* J92:792–797.

9 Seed Dormancy in Wild Flowers

Carol C. Baskin[1,2] and Jerry M. Baskin[1]
*[1]Department of Biology, University of Kentucky, Lexington,
KY 40506-0225, USA; [2]Department of Agronomy, University of Kentucky,
Lexington, KY 40546-0321, USA*

Introduction

The dream of those individuals who wish to grow wild flowers from seeds is to be able to sow seeds and a few days later find that each has produced a healthy seedling. With the aid of highly controlled greenhouses, it is possible to provide seeds with an array of seemingly ideal conditions for germination, but all too often this does not guarantee success. Another approach to obtaining high germination percentages is to sow seeds out-of-doors in nurseries. However, depending on the species and time of sowing, natural temperature regimes may not result in immediate germination and/or high germination percentages.

Failure of seeds to germinate in greenhouses and/or nurseries usually causes people to re-examine the seed lot to make sure the seeds are viable. If seeds are viable but do not germinate when sown in the greenhouse and/or nursery, it is concluded they are dormant. Now, what should one do? People who work with seeds know about various dormancy-breaking treatments such as cold stratification, warm stratification, after-ripening and scarification that can be used to promote germination. Consequently, these treatments are used on dormant seeds, but all too often the results are very disappointing. Close examination of these cases frequently reveals that the incorrect dormancy-breaking treatment or sequence of treatments has been used.

There are five major classes of dormancy, and for some of them levels (i.e. subdivisions of the class) have been distinguished. Further, specific environmental conditions are required to overcome each class and level of dormancy. Obviously, if the incorrect environmental conditions are applied to seeds, they are not going to come out of dormancy and will not germinate when sown in the greenhouse and/or nursery. Thus, an important part of growing flowers from dormant seeds is to be able to identify the classes and levels of dormancy and know how to break each of them. Three objectives of this chapter are to: (i) distinguish between dormant, conditionally dormant and non-dormant seeds; (ii) describe the five classes of seed dormancy and present a dichotomous key for distinguishing them; and (iii) discuss the environmental conditions required to break the various classes and levels of dormancy.

People are constantly trying to propagate species from seeds and introduce them into the flower/plant market, but this may be especially difficult if no information is available on seed germination of the species under consideration. Trying to decide what treatments seeds should be given in an attempt to break dormancy can be a frustrating and time-consuming endeavour. For example, should seeds be given a warm and/or a cold stratification treatment(s)? Also, the number of seeds available for studies may be limited. To help facilitate successful determination of the appropriate dormancy-breaking treatment(s), we have developed the 'move-along experiment' (C. Baskin and Baskin, 2004). Thus, a fourth objective of this chapter is to explain how the 'move-along experiment' can be used to discover the dormancy-breaking requirements of water-permeable seeds.

Non-dormant, Dormant and Conditionally Dormant Seeds

In our studies, seeds are sown on a moist substrate and placed in light and darkness in five incubators, each set on a different daily alternating temperature regime. The five regimes simulate temperatures throughout the growing season in south-central eastern USA. If seeds from broadly different climatic regions are studied, temperature regimes are adjusted accordingly. An individual seed can be placed only at one test condition, and we do not know how that individual seed would respond at another set of conditions. Thus, seed collections are made from many individual plants in the field, and testing for germination involves taking samples from the population of seeds and placing them over a range of conditions. The assumption is that the germination responses obtained at the various test conditions are representative of the population.

If seeds germinate to 80–100% over a range of conditions and this range does not increase after seeds have been given a dormancy-breaking treatment, we conclude they are non-dormant. On the other hand, if no seeds germinate at any condition after 30 days, we conclude they are dormant. If seeds germinate to high percentages at some conditions and not others, they may be non-dormant or conditionally dormant (= relative dormancy). The seeds could be non-dormant, but there is only a very narrow range of conditions over which they will germinate. If seeds are conditionally dormant,

the range of conditions over which they germinate will increase during a dormancy-breaking treatment. That is, with exposure (or additional exposure) to dormancy-breaking conditions, the seeds will germinate over a wider range of conditions (J. Baskin and Baskin, 1985a). According to Gosling (1988), if a seed lot is in conditional dormancy, dormancy-breaking treatments will cause an increase in: (i) the temperature range over which seeds will germinate; (ii) germination percentages; and (iii) rate (speed) of germination. A further criterion, applicable to some species, varieties, ecotypes or seed lots, is that while conditionally dormant seeds may require light to germinate, non-dormant seeds may not (see Table 4.3 in C. Baskin and Baskin, 1998).

Classes of Seed Dormancy

The hierarchy of classification of seed dormancy includes (from highest to lowest tier) classes, levels and types (J. Baskin and Baskin, 2003b). Thus, classes may contain levels, and levels may contain types. We recognize five classes of seed dormancy. Two of the classes, physiological dormancy and morphophysiological dormancy, are subdivided into three and eight levels, respectively, and the non-deep level of physiological dormancy is subdivided into five types.

Morphological dormancy

Germination does not occur immediately (i.e. within a week or less) after seeds are sown because the embryo is undifferentiated or differentiated but underdeveloped (small). Consequently, in undifferentiated embryos, 'differentiation' and growth must precede the appearance of a small plant. In seeds with a differentiated, underdeveloped embryo, the embryo must grow to a species-specific critical length, after which the radicle emerges. Seeds with undifferentiated embryos are less than 0.2 mm in length (Martin, 1946), lack a cotyledon(s) and a radicle, and the embryo is a mass of only 2–100 cells. These are called micro seeds! Micro seeds occur in various angiosperm families (some or all genera), including Balanophoraceae, Bruniaceae, Burmanniaceae, Corsiaceae, Cytinaceae, Gentianaceae, Hydnoraceae (placed in Aristolochiaceae by Nickrent *et al.*, 2002), Lennoaceae (placed in Boraginaceae by APG, 1998), Lentibulariaceae, Mitrastemonaceae, Monotropaceae, Orchidaceae, Orobanchaceae, Pyrolaceae, Rafflesiaceae and Triuridaceae. We know of no gymnosperms in which the embryo is undifferentiated in the mature seed. During germination of micro seeds, cells in the embryo undergo mitosis to form a body called the protocorm, tubercle or haustorium, depending on the species, from which a shoot and sometimes roots are produced. Tissue culture techniques are frequently required to produce seedlings from micro seeds, e.g. Orchidaceae, but they will not be discussed here.

Seeds with differentiated, underdeveloped (small) embryos are known to occur in 72 families of angiosperms and gymnosperms, and some of the commonly known angiosperms are Amaryllidaceae, Annonaceae, Apiaceae, Aquifoliaceae, Araceae, Aristolochiaceae, Berberidaceae, Caprifoliaceae, Fumariaceae, Haemodoraceae, Illiciaceae, Liliaceae, Magnoliaceae, Paeoniaceae, Papaveraceae and Ranunculaceae (C. Baskin and Baskin, 1998). Gymnosperm families with differentiated, underdeveloped embryos include Cycadaceae, Ginkoaceae, Podocarpaceae and Taxaceae. The embryo is small to very small in relation to the amount/size of the endosperm in angiosperms or the female gametophyte tissue in gymnosperms, and it may be rudimentary (as broad as long), linear (longer than broad) or spatulate (longer than broad, with the cotyledon end wider than the radicle) in shape (Fig. 9.1). Growth (elongation) of the embryo must take place before the radicle emerges from the seed; thus, germination is delayed until the embryo reaches the critical length for germination. A recent phylogenetic study by Forbis *et al.* (2002) supported the long-held opinion that the differentiated, underdeveloped embryo, and thus morphological dormancy, is primitive among seed plants.

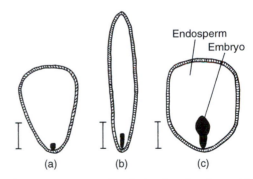

Fig. 9.1. Drawings of underdeveloped embryos. (a) *Paris quadrifolia*, rudimentary; (b) *Chaerophyllum hirsutum*, linear; (c) *Lonicera mackii*, spatulate. Bars are 1 mm in length.

Physiological dormancy

Germination does not occur because there in a 'physiological problem' in the embryo that prevents it from growing and overcoming the mechanical constraint of the seed coat and other covering tissues (if present). After the embryo becomes fully non-dormant, it has the growth potential to push through the seed coat and any other tissues that surround it. Physiological dormancy (PD) is common in all the vegetation/climatic zones on earth (*sensu* Walter, 1979), occurs in numerous families and is found in all 12 known types of seeds (C. Baskin and Baskin, 1998): bent, broad, capitate, dwarf, folded, investing, lateral, linear, micro, peripheral, rudimentary and spatulate (Martin, 1946). Three levels of PD (non-deep, intermediate and deep) have been recognized (Nikolaeva, 1969), and they are based on ability of the excised embryos to grow, temperature requirements for dormancy break and response to plant growth regulators (hormones) known as gibberellins (GA).

Embryos excised from seeds with non-deep or intermediate PD grow normally and produce healthy seedlings. Thus, germination in these seeds is prevented by a low growth (turgor) potential of the embryo, i.e. the embryo does not have enough 'push power' to break through the endosperm and/or seed (fruit) coat (Nikolaeva, 1977; C. Baskin and Baskin, 1998). However, embryos excised from seeds with deep PD either do not grow or they give rise to dwarfed plants. Thus, there is a physiological problem in the embryo, in addition to lack of enough growth potential to overcome the mechanical constraint of the seed coat and other covering tissues (if present).

Depending on the species, seeds with non-deep PD respond to GA, and they require either a cold (about 0°C to 10°C and moist) or warm (≥15°C and moist) stratification treatment for dormancy break, i.e. simulated winter and summer conditions in the temperate region, respectively. Also, it should be noted that seeds with non-deep PD will after-ripen during dry storage at room temperature. Seeds with intermediate PD may or may not respond to GA, and they require a cold stratification treatment for dormancy break. However, a warm stratification pretreatment may decrease the amount of cold stratification required for dormancy break, depending on the species. Seeds with deep PD require long periods of cold stratification for dormancy break, and they do not respond to GA.

Morphophysiological dormancy

This description is limited to seeds with differentiated, underdeveloped embryos, but there is some evidence of physiological dormancy in micro seeds, e.g. some temperate-zone Orchidaceae require cold stratification to break dormancy, indicating the presence of PD (Ballard, 1987; Ichihashi, 1989). Morphophysiological dormancy (MPD) means that a seed has an underdeveloped embryo, and the embryo has PD. Thus, prior to radicle emergence, the PD component of dormancy has to be broken, and the embryo must grow to some species-specific critical length (morphological component of dormancy). Depending on the species, PD is broken before embryo growth begins (J. Baskin and Baskin, 1990), or it occurs while the embryo is growing (C. Baskin and Baskin, 1994). Eight levels of MPD have been distinguished (Table 9.1) based on: (i) the temperature regime(s) that seeds must be subjected to before they complete germination, i.e. emergence of the radicle as well as of cotyledon(s); (ii) temperature requirement for embryo growth *per se*; and (iii) the response of seeds to GA.

Many woodland wild flowers of temperate eastern North America have MPD, and these species occur in the Apiaceae, Araceae, Araliaceae, Aristolochiaceae, Berberidaceae, Fumariaceae, Hydrophyllaceae, Liliaceae, Papaveraceae and Ranunculaceae. Further, corresponding species in genera of these families in western North America, eastern Asia and Europe also have MPD, e.g. *Aristolochia*,

Table 9.1. Characteristics of the eight levels of morphophysiological dormancy (MPD) and one example of a species with each level. Summer = 25/15°C, autumn = 20/10°C, winter = 5°C, spring = 20/10°C.

Level of MPD	First treatment	Time of embryo growth	Second treatment	Time of germination	Species
Simple					
Non-deep[a]	summer	autumn	none	autumn	*Chaerophyllum tainturieri*[b]
	winter	spring	none	spring	*Thalictrum mirabile*[c]
Intermediate[a]	summer	autumn	winter	spring	*Dendropanax japonicum*[d]
Deep	summer	autumn	winter	spring	*Jeffersonia diphylla*[e]
Deep epicotyl	summer	autumn	winter	spring	*Asarum canadense*[f]
Deep double	winter	spring	winter	spring	*Arisaema dracontium*[g]
Complex					
Non-deep[a]	summer	winter	none	spring	*Erythronium albidum*[h]
Intermediate[a]	winter	winter	none	spring	*Stylophorum diphyllum*[i]
Deep	winter	winter	none	spring	*Osmorhiza occidentalis*[j]

[a]GA promotes germination; [b]J. Baskin and Baskin, 1990; [c]Walck *et al.*, 1999; [d]Grushvitzky, 1967; [e]J. Baskin and Baskin, 1989; [f]J. Baskin and Baskin, 1986c; [g]Pickett, 1913; [h]J. Baskin and Baskin, 1985c; [i]J. Baskin and Baskin, 1984b; [j]C. Baskin *et al.*, 1995b.

Erythronium, Jeffersonia, Osmorhiza and *Panax* (C. Baskin and Baskin, 1998; Walck *et al.*, 2002; C. Adams, unpublished). Examples of species with each level of MPD are discussed below.

Physical dormancy

Germination is prevented because the seed coat (sometimes the fruit coat or part thereof, e.g. *Rhus*) is impermeable to water; this is called physical dormancy. Impermeability is due to the presence of one or more palisade layers of lignified cells, or macrosclereids (Fig. 9.2). Physical dormancy occurs in the plant families (*sensu* APG, 1998) Anacardiaceae, Bixaceae, Cannaceae, Cistaceae, Cochlospermaceae, Convolvulaceae (including Cuscutaceae), Curcurbitaceae, Dipterocarpaceae (subfamilies Montoideae and Pakaraimoideae, but not Dipterocarpoideae), Fabaceae (subfamilies Caesalpinioideae, Mimosoideae and Papilionoideae), Geraniaceae, Malvaceae (including Bombacacaceae, Sterculiaceae and Tiliaceae), Nelumbonaceae, Rhamnaceae, Sapindaceae and Sarcolaenaceae (see J. Baskin *et al.*, 2000). However, it should be noted that not all members of all of these families have impermeable seed/fruit coats. For example, in the Anacardiaceae, it appears that physical dormancy occurs only in the *Rhus* complex (Miller *et al.*, 2001) and is not present in most of the about 75 other genera in the family (Baskin and

Baskin, unpublished). Further, seeds of some tropical members of these families (e.g. Fabaceae) may be recalcitrant, i.e. lose viability if the moisture content falls below a certain species-specific critical level (Pritchard *et al.*, 1995). In 13 families, the impermeable palisade layers are in the seed coat, but in the Anacardiaceae and Nelumbonaceae they are in the fruit wall (pericarp). In 12 of the 15 families, a specialized structure ('water gap' or 'plug') has been identified in the impermeable seed or fruit coat. The three families in which a specialized water-gap structure has not been described are the Curcurbitaceae, Rhamnaceae and Sapindaceae (J. Baskin *et al.*, 2000); the anatomy of the seed coat in these three families needs to be studied from this perspective. There are several kinds of water plugs that differ in developmental origin and anatomy, and they serve as environmental 'signal detectors'. That is, in response to environmental cues, especially temperature, the water plug is dislodged or disrupted, thereby creating an entry point for water (J. Baskin *et al.*, 2000).

Combinational dormancy

Germination is prevented both by an impermeable seed (or fruit) coat and PD of the embryo; thus, a combination of physical dormancy and PD. Combinational dormancy occurs in a few herbaceous members of the Fabaceae (e.g. *Ornithopus, Stylosanthes* and

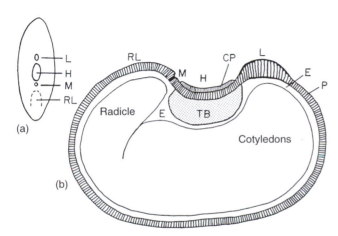

(a)

(b)

Fig. 9.2. An impermeable seed of a Papilionoid legume. (a) surface view; (b) sagittal section. CP = counter palisade; E = other layers of the seed coat and endosperm; H = hilum; L = lens; M = micropyle; P = palisade (impermeable layer); RL = radical lobe.

Trifolium), Cucurbitaceae (*Sicyos*), Geraniaceae (*Geranium*) and Malvaceae (*Malva*), and it is found in some woody members of the Fabaceae (*Cercis*), Rhamnaceae (*Ceanothus*), Sapindaceae (*Koelreuteria*), Anacardiaceae (*Rhus* and *Cotinus*) and Malvaceae (*Tilia*) (C. Baskin and Baskin, 1998). It should be noted that the occurrence of seeds with combinational dormancy in a genus does not necessarily mean that this is true for all members of the genus. For example, seeds of *Ceanothus* species from the montane zone in California have combinational dormancy, but those of coastal species have PY (Quick, 1935). Also, *Rhus aromatica* in subgenus *Lobadium* has combinational dormancy, whereas *Rhus glabra* in subgenus *Rhus* has physical dormancy (Li *et al.*, 1999a). On the other hand, it appears that all species of *Cercis* have combinational dormancy (C. Baskin and Baskin, 1998).

A Dichotomous Key for Classes of Seed Dormancy

After fresh seeds have been determined to be dormant, it is very helpful to know what kind of dormancy one is attempting to break. Thus, a key to distinguish the five classes of

seed dormancy has been developed using information on: (i) the developmental state/size of the embryo; (ii) whether the seed/fruit is permeable to water; and (iii) whether the seed germinates within about 30 days (Table 9.2).

To examine the embryo in seeds, first allow the seeds to imbibe water for 24 h at room temperatures and then cut open the seeds using a dissecting microscope. If the seed does not have an endosperm, it obviously has a fully developed embryo. However, if: (i) endosperm is present, and (ii) the species belongs to some family other than the Balanophoraceae, Bruniaceae, Burmanniaceae, Corsiaceae, Cytinaceae, Gentianaceae, Hydnoraceae, Lennoaceae, Lentibulariaceae, Mitrastemonaceae, Monotropaceae, Orchidaceae, Orobanchaceae, Pyrolaceae, Rafflesiaceae or Triuridaceae, the length of the embryo in relation to the total length of the endosperm needs to be determined. For example, if the embryo is *c.* ≤50% of the length of the endosperm, the embryo is underdeveloped and will be either rudimentary, linear or spatulate (Fig. 9.1). An outstanding source of information on seed anatomy, especially the shape and size of the embryo in relation to the endosperm, is Martin's (1946) paper. The paper contains a drawing (or description) of the embryo and endosperm of seeds of 1287

Table 9.2. A dichotomous key to distinguish non-dormancy and the five classes of seed dormancy. It is assumed that studies begin with freshly matured seeds and that seeds are incubated at temperatures appropriate for germination, e.g. daily alternating temperature regimes of 20/10°C and/or 25/15°C. (Modified from J. Baskin and Baskin, 2003b.)

1. Embryo differentiated and fully developed. 2
  2. Seeds[a] imbibe water 3
    3. Seeds[a] germinate in about 30 days . . . . Non-dormant
    3. Seeds[a] do not germinate in about 30 days . . . . Physiological dormancy
  2. Seeds[a] do not imbibe water 4
    4. Scarified seeds[a] germinate in 30 days or less. . . . . Physical dormancy
    4. Scarified seeds[a] do not germinate in 30 days . . . . Combinational dormancy
1. Embryo undifferentiated or if differentiated it is underdeveloped 5
  5. Embryo not differentiated . . . . Specialized type of morphological or morphophysiological dormancy
  5. Embryo differentiated but underdeveloped 6
    6. After seeds[a] are placed on a moist substrate, the embryo grows, and seeds germinate in about 30 days. . . . . Morphological dormancy
    6. After seeds[a] are placed on a moist substrate, the embryo does not grow, and seeds do not germinate in about 30 days . . . . . Morphophysiological dormancy

[a]Natural dispersal/germination unit may be a seed covered by one or more layers of the pericarp.

genera of seed plants, including gymnosperms and angiosperms.

Imbibed seeds of many species that lack physical dormancy increase in weight by only *c.* 25–35% during imbibition, and they may feel 'hard' if you try to pinch them e.g. *Phytolacca americana* (C. Baskin and Baskin, unpublished). Thus, the only way to be sure that some seeds have imbibed is to weigh them before and after they have been placed on a moist substrate for 24 h or longer. Seeds with impermeable seed (or fruit) coats will not show any increase in weight; however, it is not uncommon for a few seeds in a sample to imbibe water. However, these imbibed seeds are usually easy to recognize by their increase in size and by the fact that they are 'soft' when pinched. After seeds with impermeable seed coats are scarified, their weight may increase by 100% or more (e.g. J. Baskin and Baskin, 1997).

Breaking Dormancy in Permeable Seeds with Fully Developed Embryos

Deep physiological dormancy

This level of PD has been documented in only a few families, including the Aceraceae, Balsaminaceae, Celastraceae and Rosaceae, and, except for Balsaminaceae, it is restricted to seeds of shrubs and trees (C. Baskin and Baskin, 1998; G. Pendley, unpublished). Deep PD may have originated from ancestors with non-dormant seeds (C. Baskin and Baskin, 1998) in response to climatic cooling in the mid to late Eocene and early Oligocene that resulted in warm summers alternating with cool winters (Prothero, 1994). By the late Eocene, many modern families (and even genera including *Acer*, *Celastrus*, *Prunus* and *Rosa*) of angiosperms had already evolved (Graham, 1999). Dormancy-break takes place only during cold stratification, after which seeds require relatively low temperatures for germination. Thus, dormancy-break occurs during winter, and seeds have the capacity to germinate in late winter to early spring; they may even germinate at the cold stratification temperature regime (e.g.

1°C or 5°C). The length of the cold stratification period required to break dormancy varies with species and usually ranges from about 10 to 14 weeks, with the record being 18 weeks for seeds of the herbaceous species *Impatiens parviflora* (Nikolaeva, 1969).

Intermediate physiological dormancy

This level of PD has been documented in seeds of trees, shrubs, vines and herbaceous species in various plant families, including Aceraceae, Berberidaceae, Betulaceae, Brassicaceae, Cucurbitaceae, Empetraceae, Fagaceae, Limnanthaceae, Oleaceae, Polygonaceae, Portulacaceae, Rosaceae, Scrophulariaceae and Vitaceae (see references in J. Baskin *et al.*, 1988; C. Baskin *et al.*, 1993b, 2002b; C. Baskin and Baskin, 1995, 1998).

Dormancy is broken by cold stratification, but in many species the length of the cold stratification period required to break dormancy decreases with an increase in the length of a warm stratification pretreatment (e.g. J. Baskin *et al.*, 1988; C. Baskin and Baskin, 1995), or in some cases following a period of dry storage at room temperatures (e.g. Ransom, 1935). Germination of non-dormant seeds of many species is promoted by low temperatures, including those suitable for cold stratification (J. Baskin *et al.*, 1988; C. Baskin and Baskin, 1995).

Non-deep physiological dormancy

This level of PD occurs in numerous plant families (Table 9.3) and is also found in seeds of trees, shrubs, vines and herbaceous species. PD is the most important class of dormancy in all the vegetation/climatic zones on earth, except tropical deciduous forests and matorral, where PD and physical dormancy are equally important (J. Baskin and Baskin, 2003a). Based on 5250 species of seed plants, 69.6% of the species have dormant seeds, and 45.1% have PD (J. Baskin and Baskin, 2003a). We estimate that ≥90% of the physiologically dormant seeds in the world have non-deep PD. Thus, based on a

Table 9.3. Presence of various types of non-deep physiological dormancy in angiosperm families. Unless otherwise indicated, information is from C. Baskin and Baskin (1998).

Family	Type of non-deep PD 1	2	3	4	5	Family	Type of non-deep PD 1	2	3	4	5
Acanthaceae		x				Limnanthaceae	x				
Agavaceae	x					Lobeliaceae		x			
Aizoaceae		x				Lythraceae		x			
Amaranthaceae		x				Melastomataceae		x			
Apiaceae	x					Molluginaceae		x			
Asclepiadaceae		x				Onagraceae		x			
Asteraceae	x	x	x			Penthoraceae		x			
Betulaceae		x[a,b]				Phytolaccaceae		x			
Boraginaceae	x	x				Plantaginaceae	x				
Brassicaceae	x	x				Papaveraceae	x				
Campanulaceae	x	x				Poaceae	x	x		x?	
Caryophyllaceae	x	x				Polemoniaceae	x				x[a,h]
Chenopodiaceae		x				Polygonaceae		x			x[a,h]
Crassulaceae	x					Portulacaceae	x	x			
Cyperaceae		x	x[a,c]			Primulaceae			x[i]		
Diapensiaceae		x[a,d]				Ranunculaceae	x				
Dipsacaceae		x				Rosaceae	x[a,j]	x			
Droseraceae		x[a,e]				Rubiaceae		x			
Ericaceae		x[a,f]				Saxifragaceae	x[a,k]				
Euphorbiaceae		x				Scrophulariaceae	x	x	x[a,l]		
Fumariaceae	x					Solanaceae		x[a,m]			
Gentianaceae		x			x[g]	Valerianaceae	x				
Hydrophyllaceae	x					Verbenaceae				x	
Juncaceae		x				Violaceae	x				
Lamiaceae	x	x				Xyridaceae		x			
Liliaceae	x	x	x								

[a]Family represented in the southeastern USA but study was done on seeds collected outside this region; [b]McVean, 1955; [c]Schütz, 2000; [d]Densmore, 1997; [e]C. Baskin *et al.*, 2001b; [f]C. Baskin *et al.*, 2000; [g]Figure 9.3d; [h]C. Baskin *et al.*, 1993c; [i]C. Baskin *et al.*, 1996; [j]Roberts and Neilson, 1982; [k]Pemadasa and Lovell, 1975; [l]C. Baskin *et al.*, 1998; [m]Roberts and Boddrell, 1983.

sample of more than 5000 species, it appears that 41% ($1.00 \times 0.45 \times 0.90$) of seeds of all species on earth have non-deep PD.

Types of non-deep PD

Five patterns of temperature requirements for germination as non-deep PD is broken have been identified (Vegis, 1964; C. Baskin and Baskin, 1998), and these are the types of non-deep PD (Fig. 9.3). In *Type 1*, seeds germinate only at low temperatures in the initial stages of dormancy break, but as dormancy loss continues, the maximum temperature at which seeds germinate increases. In *Type 2*, seeds germinate only at high temperatures in the initial stages of dormancy break, but as

dormancy loss continues, the minimum temperature at which seeds germinate decreases. In *Type 3*, seeds germinate only at intermediate temperatures in the initial stages of dormancy break, but as dormancy loss continues, the minimum and maximum temperature at which seeds germinate decreases and increases, respectively. In *Type 4*, seeds germinate initially only at high temperatures and there is no decrease in the minimum temperature at which they will germinate, even when the dormancy-breaking treatment is prolonged. In *Type 5*, seeds germinate initially only at low temperatures and there is no increase in the maximum temperature at which they will germinate, even when the dormancy-breaking treatment is prolonged.

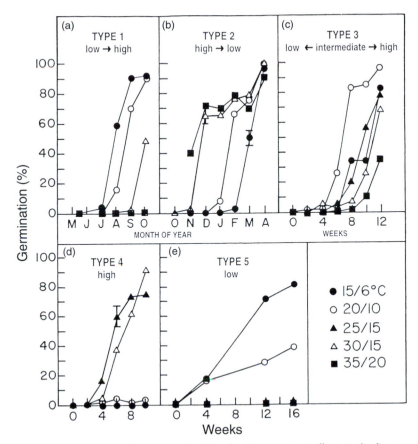

Fig. 9.3. Germination percentages (mean ± SE, if ≥5%) of various species to illustrate the five types of non-deep physiological dormancy (Baskin and Baskin, unpublished). For each species, seeds were tested over a range of alternating temperature regimes in light after various periods of exposure to either summer or winter temperatures, depending on conditions required to break dormancy in that species. (a) Type 1, *Valerianella olitoria* seeds stored dry at room temperatures during summer; (b) Type 2, *Polygonum pensylvanicum* seeds exposed to winter temperatures; (c) Type 3, *Aster ptarmacoides* seeds exposed to winter temperatures; (d) *Callicarpa americana* seeds exposed to summer temperatures; and (e) *Gentiana quinquefolia* seeds exposed to winter temperatures.

Type 1 has been documented in various families of angiosperms (Table 9.3) and is very common in temperate-zone winter annuals (C. Baskin and Baskin, 1988). It also occurs in perennials whose seeds mature in spring and germinate in autumn (C. Baskin and Baskin, 1988; C. Baskin *et al.*, 1994b). In both life cycle groups of species, seeds come out of dormancy during summer, and the high temperatures of summer are required for dormancy break. Thus, without exposure to high summer temperatures, the maximum temperature at which seeds germinate does not increase enough for seeds to germinate at autumn

temperatures (e.g. 25/15°C) in autumn (J. Baskin and Baskin, 1986a). Under natural temperature regimes in temperate regions seeds can germinate at the maximum temperature possible for the species, variety or ecotype by autumn. Thus, since temperatures are beginning to decline in autumn, there is an overlap between the maximum temperature at which seeds can germinate and temperatures occurring in the habitat, and seeds germinate when soil moisture becomes non-limiting.

Type 1 has also been documented in seeds of a few perennials in eastern temperate North

America whose seeds come out of dormancy during winter and germinate in spring, including *Diarrhena americana, Mertensia virginica, Northoscordum bivalve* and *Viola egglestonii* (C. Baskin and Baskin, 1988). The discovery of perennials whose maximum temperature for germination increases during winter raises questions concerning the evolutionary origin and world biogeography of Type 1.

Type 2 has been documented in many angiosperm families (Table 9.3) and is very common in temperate-zone summer annuals and spring-germinating perennials (C. Baskin and Baskin, 1988). Seeds come out of dormancy during winter, and the low temperatures of winter are required for dormancy break. Thus, without exposure to low winter temperatures, the minimum temperature at which seeds germinate does not decrease enough for seeds to germinate at spring temperatures (e.g. 15/6°C, 20/10°C) in spring (J. Baskin and Baskin, 1987). Under natural temperature regimes in temperate regions seeds can germinate at the minimum temperatures possible for the species, variety or ecotype by spring or early summer, and at this time temperatures are increasing. Germination occurs in late winter, spring or early summer as soon as there is an overlap between the minimum temperature at which seeds can germinate and temperatures in the habitat.

Type 2 also occurs in seeds of a few species that come out of dormancy during summer and germinate in autumn, including *Arctotheca calendula* (Chaharsoghi and Jacobs, 1998), *Barbarea vulgaris, Dianthus armeria, Heterotheca subaxillaris* (ray achenes only), *Lobelia appendiculata* var. *gattingerii, Lychnis alba* and *Scutellaria parvula* (C. Baskin and Baskin, 1988). Thus, Type 2 is not restricted to spring-germinating species, and it may be broken by low or high temperatures, depending on the species.

Type 3 has been documented in the Asteraceae, Cyperaceae, Primulaceae and Scrophulariacae (Table 9.3). Among the species whose seeds have Type 3 there is diversity in life cycle type and time of germination. The perennials *Chaptalia nutans* and *Hymenopappus scabiosaeus* germinate in autumn (C. Baskin et al., 1994b), the perennials *Aster ptarmicoides* (C. Baskin and Baskin, 1988) and *Echinacea*

angustifolia var. *angustifolia* (C. Baskin et al., 1992) germinate in spring, the winter annuals *Krigia oppositifolia* (C. Baskin et al., 1991) and *Hottonia inflata* (C. Baskin et al., 1996) germinate in autumn, and the summer annual *Agalinis fasciculata* (C. Baskin et al., 1998) germinates in spring. Thus, in some species with Type 3, high summer temperatures break seed dormancy, and the seeds effectively are like those of a winter annual with Type 1; thus, they germinate in autumn. However, in other species with Type 3, low winter temperatures break seed dormancy, and the seeds effectively are like those of a summer annual or spring-germinating perennial; thus, they germinate in spring. Obviously, much remains to be learned about the evolutionary origin of Type 3 and its relationship to Types 1 and 2.

Type 4 has been documented in seeds of *Callicarpa americana* (Verbenaceae) (Fig. 9.3d). Also, based on data published by Lodge and Whalley (1981), it appears that Type 4 is present in seeds of *Aristida ramosa* (Poaceae) growing in the dry woodland–savanna–grassland vegetation region of New South Wales, Australia. It is expected that seeds of many species in tropical or subtropical regions with an annual dry season might have Type 4, but this remains to be determined.

Type 5 has been reported in seeds of two species, *Eriastrum diffusum* (Polemoniaceae) and *Eriogonum abertianum* (Polygonaceae), from the deserts of the southwestern USA (C. Baskin et al., 1993c). However, Type 5 is not restricted to desert species and has been found in seeds of *Gentiana quinquefolia* (Gentianaceae) in temperate eastern North America (Fig. 9.3e). Seeds of the two desert species come out of dormancy during summer (C. Baskin et al., 1993c), while those of *G. quinquefolia* come out of dormancy during cold stratification. Thus, depending on the species, high or low temperatures are required for dormancy break, and seeds germinate in autumn/winter and spring, respectively.

Dormancy state(s) throughout the year

A characteristic of many seeds with non-deep PD, but not those with intermediate or deep

PD, is the ability to cycle between dormancy and non-dormancy or between conditional dormancy and non-dormancy on an annual basis. Much has been learned about the dynamics of seed dormancy by burying seeds in soil and exposing them to natural seasonal temperature cycles. In this procedure, samples (destructive sampling) of seeds are exhumed at monthly intervals and tested for germination over a range of alternating temperature regimes in both light and darkness. Buried seeds of a few species quickly gain the ability to germinate over a range of temperatures in light, and they maintain this capacity even though exposed to seasonal temperature changes (J. Baskin *et al.*, 1989). Thus, the buried seeds do not exhibit annual cyclic changes in state of dormancy. However, a light requirement for germination prevents seeds from germinating during burial. Consequently, seeds of these species can germinate at any time during the growing season if they are exposed to light and adequate soil moisture at the same time.

Buried seeds of many species undergo changes in their ability to germinate, even in light, and two broad patterns of changes have been identified. In one pattern, seeds exhibit an annual dormancy/non-dormancy cycle (Fig. 9.4). As seeds cycle between dormancy and non-dormancy, they pass through a continuum of conditional dormancy states. If seeds are coming out of dormancy, the continuum of conditional dormancy states is characterized by a period of several months during which there is a gradual increase in the temperature range for germination. However, compared with dormancy break, re-entrance into dormancy through conditional dormancy is characterized by a more abrupt decrease (curves have steeper slopes) in the temperature range for germination (J. Baskin and Baskin, 1983c). For example, seeds of the winter annual *Arabidopsis thaliana* are dormant at maturity in May, conditionally dormant from June to September, non-dormant in October, conditionally dormant in November and December and dormant from January to June or July, depending on the year (Fig. 9.4). Thus, even if seeds of *A. thaliana* are exposed to light, they cannot germinate at all times of the year. If seeds of obligate winter

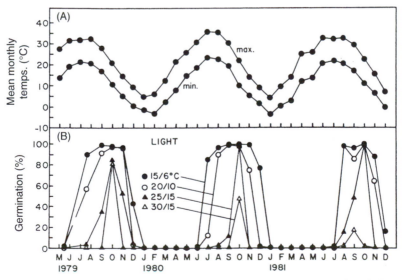

Fig. 9.4. Annual dormancy/non-dormancy cycle in buried seeds of the winter annual *Arabidopsis thaliana* (with Type 1 non-deep physiological dormancy) from Fayette County, Kentucky, USA. (A) Mean daily maximum and minimum monthly temperatures that seeds were exposed to during the time they were buried, (B) germination percentages (mean ± SE, if ≥5%) of seeds incubated in light at 15/6°C, 20/10°C, 25/15°C, 30/15°C and 35/20°C following various periods of burial. No seeds germinated in darkness. From C. Baskin and Baskin, 1998, with permission.

annuals fail to germinate in autumn, low temperatures of late autumn and/or winter induce the non-dormant seeds into dormancy again (J. Baskin and Baskin, 1973, 1984c). Consequently, even if seeds are exposed to light in spring, they cannot germinate.

Annual dormancy/non-dormancy cycles also occur in species whose seeds require cold stratification for dormancy break. As dormancy break occurs during winter, seeds exhibit a continuum of conditional dormancy states, during which there is a gradual decrease in the minimum temperature at which seeds will germinate (Fig. 9.3b). By spring, seeds are non-dormant and can germinate over the full range of temperatures possible for the species, variety or ecotype. If seeds fail to germinate in spring and/or early summer (actual germination season varies with the species), high temperatures induce the non-dormant seeds into dormancy again (e.g. C. Baskin et al., 1995a). Consequently, even if seeds are exposed to light in late summer or autumn, they cannot germinate.

In the second pattern of responses of seeds to seasonal temperature changes, seeds exhibit an annual conditional dormancy/non-dormancy cycle. For example, if seeds of facultative winter annuals fail to germinate in autumn, low temperatures during late autumn and/or winter may cause seeds to lose the ability to germinate at high but not low temperatures; these seeds are conditionally dormant. Consequently, seeds could germinate the following early spring, when temperatures are low (if exposed to light), but they cannot germinate in late spring or summer when temperatures are high. During summer, the conditionally dormant seeds become non-dormant again (J. Baskin and Baskin 1983d, 1984d). Another example of seeds with an annual conditional dormancy/non-dormancy cycle is found in seeds of summer annuals that retain the ability to germinate at medium to high, but not at low, temperatures all summer; these seeds are conditionally dormant. Consequently, seeds could germinate from late spring to early autumn but not in mid to late autumn. During winter, the conditionally dormant seeds become non-dormant again (C. Baskin et al., 1994a).

Germination in darkness

A light requirement for germination plays an important role in the persistence of permeable seeds in the soil. Obviously, the light requirement would prevent germination during the time(s) of the year when the range of temperatures over which seeds are capable of germinating overlaps with temperatures in the habitat. However, seeds of some species can germinate in darkness if they are exhumed (with no exposure to light) at the peak of non-dormancy, but not at other times of the year, e.g. Bidens polylepis (C. Baskin et al., 1995a) and Verbascum blattaria (J. Baskin and Baskin, 1981). Thus, like responses to temperature, ability to germinate in darkness is cyclic. For many species, we have observed that although exhumed seeds germinate in darkness, they do not germinate while buried in soil. Note that many precautions have been taken in our studies to ensure that there was no exposure to light (not even a green 'safe' light (C. Baskin and Baskin, 1998)) while seeds were being exhumed, placed in Petri dishes and wrapped with two layers of aluminium foil (see methods in C. Baskin et al., 1995a; J. Baskin and Baskin, 1981). We speculate that some factor(s) associated with the burial environment, other than darkness, prevents germination of buried seeds. The oxygen: carbon dioxide ratio in the soil (Bibbey, 1948) and/or volatile metabolites (Holm, 1972) are possible reasons for lack of germination while seeds are buried.

In some species, however, non-dormant seeds do germinate while they are buried in soil, e.g. Collinsia verna (J. Baskin and Baskin, 1983a) and Nemophila aphylla (C. Baskin et al., 1993a). If non-dormant seeds germinate while they are buried, this means that dormancy cycling is impossible. Any seeds of species such as C. verna and N. aphylla that persist in the soil at a site until the second (or later) germination season are those that failed to come out of dormancy prior to the time of the first germination season (C. Baskin et al., 1993a). The few that do not germinate during the first germination season may survive and do so during the second or later germination season.

Breaking Dormancy in Permeable Seeds with Underdeveloped Embryos

Morphological dormancy

The breaking of morphological dormancy is basically a matter of sowing the seeds and waiting for embryo growth and radicle/cotyledon emergence to occur. However, depending on the species, temperatures can be an important factor in regulating embryo growth and consequently germination. Embryo growth and germination occurred within a few weeks when seeds of *Isopyrum biternatum* were incubated at daily alternating temperature regimes of 15/6°C and 20/10°C but not at 25/15°C and 30/15°C (J. Baskin and Baskin 1986b). However, seeds of the oil palm *Elaeis guineensis* germinated best at temperatures of 35°C–40°C (Hussey, 1958). Light vs. darkness can also be an important factor in promoting germination of seeds in some species. For example, seeds of *Apium graveolens* required light for germination (Jacobsen and Pressman, 1979), but those of the cultivated de Caen type of *Anemone coronaria* germinated to higher percentages and at a faster rate in darkness than in light (Bullowa *et al.*, 1975).

Morphophysiological dormancy

Non-deep simple

In one group of species, including *Anemone coronaria*, *Chaerophyllum procumbens*, *Chaerophyllum tainturieri*, *Corydalis flavula*, *Corydalis ledebouriana*, *Corydalis solida*, *Hyacinthoides nonscripta*, *Paeonia californa*, *Spermolepias echinata* (see references in Table 5.2 of C. Baskin and Baskin, 1998) and *Papaver rhoeas* (C. Baskin *et al.*, 2002a), the PD component of this level of MPD is broken during summer, and embryo growth and germination occur in autumn. Although the high temperatures of summer promote the breaking of PD, lower temperatures of autumn are required for embryo growth and germination (J. Baskin and Baskin, 1990). Light may (J. Baskin and Baskin, 1990; C. Baskin *et al.*,

2002a) or may not (J. Baskin and Baskin, 1994) be required for embryo growth. However, in seeds of *C. tainturieri*, the light requirement for germination could be fulfilled after the breaking of PD had started in summer, but before PD was completely broken, and thus embryo growth and germination would take place in darkness in autumn (J. Baskin and Baskin, 1990). In a second group of species, including *Chamaelirium luteum* (C. Baskin *et al.*, 2001a) and *Thalictrum mirabile* (Walck *et al.*, 1999), the PD component of this level of MPD is broken during winter, and embryo growth and seed germination occur in spring.

Intermediate simple

Although this level of MPD occurs in the woody species *Aralia mandshurica* and *Dendropanax japonicum* (C. Baskin and Baskin, 1998), it has not been reported in wild flowers or other herbaceous species. Thus, this level of MPD will not be discussed here.

Deep simple

Herbaceous species whose seeds have this level of MPD include *Jeffersonia diphylla*, *Panax ginseng*, *Panax quinquefolia* and *Panax pseudoginseng* (see references in Table 5.4 of C. Baskin and Baskin, 1998). In many respects, seeds with deep simple MPD are more difficult to germinate than those with the other levels of MPD. Perhaps the reason for the difficulty lies in the fact that seeds require three different temperature treatments, but the investigator is not able to observe any external changes, e.g. emergence of the radicle, until completion of the three treatments, at which time both the radicle and cotyledons emerge. For example, seeds of *J. diphylla* must be kept moist and exposed to summer, autumn and winter temperature regimes in sequence before they germinate in spring (Table 9.1). Embryo growth in *J. diphylla* seeds occurs during autumn, but this growth does not occur unless seeds have received a warm, wet summer period. In our studies, 12 weeks at 30/15°C were sufficient to promote embryo growth of seeds subsequently moved to autumn (20/10°C) temperatures. After embryo growth occurred, seeds responded to cold stratification and germinated when moved to spring (20/10°C) temperatures (J. Baskin and Baskin, 1989).

Deep epicotyl

This level of MPD has been documented in many herbaceous species, including *Actaea pachypoda, Actaea spicata, Allium burdickii, Allium tricoccum, Allium ursinum, Asarum canadense, Asarum heterotropoides, Cimicifuga racemosa, Cimicifuga rubifolia, Disporum lanuginosum, Fritillaria ussuriensis, Hepatica acutiloba, Hydrophyllum appendiculatum, Hydrophyllum macrophyllum, Hydrophyllum virginianum, Lilium auratum, Lilium canadense, Lillium japonicum, Paeonia suffruticosa, Polygonatum commutatum, Sanguinaria canadensis, Trillium flexipes* and *Trillium sessile* (see references in Table 5.5 of C. Baskin and Baskin, 1998). Also, deep simple epicotyl MPD occurs in seeds of *Hexastylis heterophylla* (Adams *et al.*, 2003). The term 'epicotyl dormancy' is not an adequate description *sensu stricto* for dormancy in seeds of these species because both the radicle and the epicotyl are dormant. Further, the dormancy-breaking requirements of the radicle differ from those of the epicotyl.

Radicle dormancy is broken by warm stratification, and epicotyl dormancy is broken by cold stratification, but only after the radicle has emerged (e.g. J. Baskin and Baskin, 1985b, 1986c). Depending on the species, seeds require 4–6 weeks of moist high (30/15°C) temperature conditions to break radicle dormancy, after which radicles emerge if seeds are moved to autumn (20/10°C or sometimes 15/6°C) temperatures. Root growth and development occur during autumn, but the epicotyl remains inside the seed. The cold stratification treatment received during winter breaks epicotyl dormancy, and the epicotyl emerges in spring. Depending on the species, 5–8 weeks of cold stratification are required to break epicotyl dormancy (J. Baskin amd Baskin, 1983b, 1986c). In some ways, deep simple epicotyl MPD is like deep simple MPD; however, in the former the radicle emerges from the seed in

autumn, but in the latter it grows but remains inside the seed until spring.

Deep simple double

This level of MPD has been documented in seeds of *Arisaema dracontium, Arisaema triphyllum, Caulophyllum thalictroides, Clematis albicoma, Clematis viticaulis, Convallaria majalis, Polygonatum biflorum, Polygonatum commutatum, Smilacina racemosa, Trillium erectum, Trillium grandiflorum, Uvularia grandiflora* and *Uvularia perfoliata* (see references in Table 5.7 of C. Baskin and Baskin, 1998). Also, about half of the seeds of both *Sanguinaria canadensis* and *P. commutatum* studied by Barton (1944) had epicotyl dormancy and the other half had double dormancy.

In seeds with deep simple double MPD, both the radicle and epicotyl are dormant, and they both require cold stratification for dormancy break. However, the radicle and epicotyl do not come out of dormancy at the same time, and the first cold stratification treatment breaks radicle dormancy only. In *T. grandiflorum*, no embryo growth was observed while seeds were being cold-stratified, but roots emerged about 60 days after the seeds were moved to 25°C (Gyer, 1997). During summer, the root system develops slowly, and food reserves are translocated from the endosperm to a small corm (Pickett, 1913) or swollen hypocotyl (Takagi, 2001) via the cotyledon, which remains inside the seed. By the end of summer, a perennating bud forms at the apex of the corm or swollen hypocotyl. Dormancy of this bud is broken during winter, and the bud produces a leaf the following (second) spring.

Non-deep complex

This level of MPD has been documented in seeds of *Eranthis hiemalis, Erythronium albidum, Erythronium americanum, Erythronium rostratum, Osmorhiza claytonii* and *Osmorhiza longistylis* (see references in Table 5.8 of C. Baskin and Baskin, 1998). Dormancy break requires both warm and cold stratification and warm stratification must precede cold stratification. If seeds are cold-stratified, without first being warm-stratified, embryos do not grow. However, if seeds are given a warm stratification followed by a cold stratification treatment, embryo growth occurs during cold stratification. The optimum length of the warm stratification treatment is 4–6 weeks, and the optimum length of the cold stratification treatment is 6–8 weeks (J. Baskin and Baskin, 1984a, 1985c). When exposed to the natural summer, autumn and winter sequence of temperatures, embryo growth begins in early to mid autumn as temperatures, especially at night, decline enough (to ≤10°C) to be within the range of those effective for cold stratification (J. Baskin and Baskin, 1991, 1985c).

Seeds must be imbibed for the high temperatures of summer to be effective in dormancy break; thus, seeds remaining undispersed on plants during summer do not undergo the first phase of dormancy break. Consequently, if seeds are dispersed in late autumn, after temperatures are too low for warm stratification, the embryo does not grow during winter (J. Baskin and Baskin, 1984a, 1991). Seeds dispersed in late autumn do not germinate the following spring (first spring following seed maturation), and their germination is delayed until the second spring following seed maturation, after they have received warm followed by cold stratification.

Intermediate complex

This level of MPD has been documented in seeds of *Aralia continentalis*, *Stylophorum diphyllum* and 12 species of *Trollius* (C. Baskin and Baskin, 1998; Hitchmough *et al.*, 2000). The only requirement for dormancy break is a 10–12 week period of cold stratification, i.e. seeds can be cold-stratified as soon as they are mature. Further, GA will promote germination (Nikolaeva, 1977; C. Baskin and J. Baskin, unpublished).

Since seeds of *Aconitum heterophyllum* come out of dormancy under snow and since GA promotes germination (Singh *et al.*, 2000), we suspect they have intermediate complex MPD. However, no studies have been done on the requirements for embryo growth to confirm this. Seeds of *A. heterophyllum* are dispersed at the end of September in the habitat of this species at 3000–4000 m elevation in

the Himalayas (Singh *et al.*, 2000). Thus, it seems reasonable that seeds would receive little or no warm stratification prior to the onset of cold stratification and that cold stratification would be the only requirement for dormancy break. If seeds required warm followed by cold, they would have non-deep complex MPD.

Deep complex

This level of MPD is found in seeds of *Cryptotaenia canadensis*, *Delphinium tricorne*, *Erythronium grandiflorum*, *Frasera albacaulis*, *Frasera umpquaensis*, *Fritillaria eduardii*, *Fritillaria raddeana*, *Heracleum sphondylium*, *Ornithogalum arcuatum*, *Osmorhiza chilensis*, *Osmorhiza occidentalis*, *Thaspium pinnatifidum*, *Tulipa greigii* and *Tulipa tarda* (see references in Table 5.10 of C. Baskin and Baskin, 1998). Further, seeds of *Caltha leptosepala* (Forbis and Diggle, 2001), *Aconitum sinomontanum* (Dosmann, 2002) and *Erythronium japonicum* (Kondo *et al.*, 2002) have either intermediate or deep complex MPD, in which case seeds require only cold stratification for embryo growth and germination; however, the responses of seeds to GA have not been determined. If GA promoted germination, seeds of these species would have intermediate complex MPD, but if GA did not promote germination, seeds would have deep complex MPD.

Since the only requirement for dormancy break is cold stratification, this treatment can be successfully given as soon as seeds are mature. Both loss of PD and embryo growth occur during cold stratification. Depending on the species, 8 weeks (Walck *et al.*, 2002) to 24 weeks (C. Baskin *et al.*, 1995b) of cold stratification are required to break dormancy, and in many species seeds will germinate at the stratification temperature regime (e.g. Walck *et al.*, 2002; C. Baskin *et al.*, 1995b).

'Move-along Experiment' for Permeable Seeds

This experiment consists of two phenology studies that are run concurrently. One starts at summer conditions, and the other starts at winter conditions (Table 9.4). From each starting point, seeds are moved through a simulated yearly sequence of temperatures. Control seeds are kept continuously at the simulated early and late spring, early and late autumn, winter, and summer temperatures. Early spring and late autumn are the same temperature regime, which is 15/6°C in the south-central region of eastern USA; late spring and early autumn are the same, which is 20/10°C in the south-central region of eastern USA. We have used 25/15°C or 30/15°C to simulate summer, and 5°C, 5/1°C or 1°C, to simulate winter. Seeds are placed on a moist substrate in 18 Petri dishes, and three dishes of 50 seeds each are placed at each control. Also, three additional dishes each are placed at summer and at winter temperatures; these are the 'move-along' dishes (Table 9.4).

Seeds of *Erythornium americanum* subjected to a 'move-along experiment' were

shown to require warm followed by cold stratification to break dormancy (Table 9.4). Some of the control seeds kept at 15/6°C eventually germinated, but this is probably because they were receiving 12 h of warm (15°C) and 12 h of cold (6°C) stratification each day.

Breaking Dormancy in Seeds with Water-impermeable Coats

Physical dormancy

If the only impediment to germination is impermeability of the seed (or fruit) coat to water, the quickest way to elicit germination is to mechanically scarify the coat, i.e. cut a small hole in it. However, if large numbers of seeds are involved, cutting/filing a hole in each seed is time-consuming and not practical. Thus, various techniques have been developed in an attempt to find a quick way

Table 9.4. Outline of the 'move-along experiment' to determine whether permeable seeds require warm and/or cold stratification for dormancy break and germination (modified from C. Baskin and Baskin, 2004). Numbers in parentheses show cumulative germination percentages (mean ± SE) for seeds of *Erythronium americanum* subjected to a 'move-along experiment' (C. Baskin and J. Baskin, unpublished).

Cumulative time at each control (weeks)	Control temperature regimes				Time (weeks) at each phase of 'move along'	'Move alongs'	
	Early spring[a]	Late spring[b]	Summer	Winter		Start at summer	Start at winter
12	(0)	(0)	(0)	(0)	12	(0) ↓	(0) ↓
16	(0)	(0)	(0)	(0)	4	Early autumn (0) ↓	Early spring (0) ↓
20	(0)	(0)	(0)	(0)	4	Late autumn (0) ↓	Late spring (0) ↓
32	(1 ± 1)	(0)	(0)	(0)	12	Winter (93 ± 2) ↓	Summer (0) ↓
34	(7 ± 1)	(0)	(0)	(0)	4	Early spring (93 ± 2) ↓	Early autumn (0) ↓
38	(19 ± 3)	(0)	(0)	(1 ± 1)	4	Late spring (93 ± 2) ↓	Late autumn (0) ↓
52	(33 ± 3)	(1 ± 1)	(0)	(1 ± 1)	12	Summer (93 ± 2)	Winter (83 ± 4)

[a]Same as late autumn; [b]same as early autumn.

of making seeds permeable to water, including acid scarification, dipping in boiling water for a few seconds and heating in a laboratory drying oven for various times. Acid scarification makes the seed permeable by breaking down various portions of the seed coat, while heat treatments dislodge or break the water plug (C. Baskin and Baskin, 1998; Li et al., 1999b). Use of acid scarification or heat treatments requires that 'trial runs' be made to determine the appropriate period to make seeds permeable, but not kill them. For example, 60 min of dry heat at 80°C resulted in 85% germination of Iliamna corei seeds, but 60 min at 90°C, 100°C and 110°C caused a decrease in germination percentages due to seed death. Even 1 min at 120°C killed most of the seeds (J. Baskin and Baskin, 1997). Also, seeds of various species differ in the duration and intensity of a treatment required to break physical dormancy. Although seeds of I. corei germinated to 85% after 1 h of dry heat at 80°C, those of Dalea foliosa germinated to only 50% (J. Baskin and Baskin, 1998) after this treatment. Further, although wet or dry heat may break physical dormancy in seeds of some species (J. Baskin and Baskin, 1997), these treatments may have little or no effect on seeds of other species (J. Baskin et al., 1998a). In some species, one type of heat treatment will promote germination, but the other will not (Martin et al., 1975; Van Staden et al., 1994).

Seeds of some species with physical dormancy gradually become permeable with time, especially if stored dry at room temperatures. For example, most seeds of Sida spinosa (Malvaceae) became permeable during 4 months of dry storage at 25°C (Egley, 1976), and those of Aeschynomene virginica (Fabaceae) became permeable during 1 year of dry storage at room temperatures (J. Baskin et al., 1998b). Seeds of Geranium carolinianum (Geraniaceae) stored dry at simulated summer temperature regimes, i.e. May = 20/10°C, June = 30/15°C and July and August = 35/20°C, germinated to 86% in early September, while those alternately wet (5 days) and dry (10 days) from May to September germinated to 97%. Seeds kept continuously wet germinated to only 12% (J. Baskin and Baskin, 1974).

Some of the information obtained from studies of the environmental factors controlling germination of seeds with physical dormancy should be of interest/benefit to people establishing populations of wild flowers. Seeds of some wild flowers, as well as those of some of the 'weedy' species that could germinate in wildflower plantings, have seeds with physical dormancy.

Temperature

In nature, exposure of seeds with physical dormancy to high or to high-fluctuating temperatures plays a major role in causing the water plugs to move or be dislodged, thereby making the seeds permeable to water. For example, seeds of annual legumes (Stylosanthes spp.) lying on the soil surface in Australia where they are exposed to drying and mean monthly maximum and minimum temperatures of 67°C and 28°C, respectively, became permeable. Consequently, seeds germinated when it rained (McKeon and Mott, 1982). Thus, seeds of some wild flowers may become permeable during the hot, dry summer months. To ensure that physical dormancy will be broken in the first year after sowing, impermeable seeds that mature in spring and germinate in autumn should be sown as soon as they mature; therefore, they will be exposed to maximum heat and high temperature fluctuations during summer.

In some species whose seeds have physical dormancy, increases in the amplitude of the daily temperature fluctuations cause the water plugs to 'open.' Thus, seeds of some species become permeable and germinate when they are in openings (gaps), but not when they are under the shade of a forest canopy (e.g. Vázquez-Yanes and Orozco-Segovia, 1982). In non-forested areas, seeds would receive maximum daily temperature fluctuations, but under a forest canopy the difference between the daily maximum and minimum temperatures is greatly diminished, as is the maximum temperature.

Sensitivity to high temperatures and/or amplitude of daily temperature fluctuations can also serve as a means of detecting depth in the soil. That is, if seeds are deep in the soil, the amplitude of daily temperature fluctuations is quite small and water plugs are not dislodged. Seeds of *Abutilon theophrasti* planted at depths of 0, 2 and 6 cm germinated to highest percentages at 0 cm (Webster *et al.*, 1998). Seeds of *Chamaecrista chamaecristoides* subjected to a daily temperature fluctuation of 10°C germinated to 40%, while those subjected to fluctuations of 20°C to 35°C germinated to 80–90% (Martínez and Moreno-Casasola, 1998). Thus, if soil disturbance results in seeds being brought to the surface, the difference between day and night temperatures may trigger opening of the water plug and thus germination. In seeds of *Sida spinosa*, long periods of exposure to temperatures too low to promote germination (15/6°C, 20/10°C) increased sensitivity of seeds after they were shifted to high (30/15°C, 35/20°C) temperatures (J. Baskin and Baskin, 1984e).

Fire is also effective in promoting germination of seeds with physical dormancy in natural situations. However, seeds on the soil surface may be destroyed by fire (Hodgkinson and Oxley, 1990). If seeds with physical dormancy are 1–10 cm below the soil surface, heat from the fire may not kill them, and it may actually cause the water plugs to open. Thus, many seeds with physical dormancy germinate following fire (see references in Table 6.7 in C. Baskin and Baskin, 1998). However, using fire to promote germination, especially in the field, is difficult. Depending on fuel loads, intensity and duration of the fire and soil moisture, temperatures at the soil surface and in the first few centimetres of soil can vary greatly (Auld and Bradstock, 1996). Also, resistance and response of seeds to high temperatures can vary depending on seed size (Gashaw and Michelsen, 2002).

Thus, removal of vegetation, soil disturbance and fire could contribute to increased germination percentages of seeds with physical dormancy at a particular site that is being prepared for wild flowers. These activities could promote germination of desired species such as *Lupinus*, *Baptisia*, *Iliamna* and *Helianthemum*, but they might also stimulate germination of undesirable species such as *Melilotus*, *Leucaena* and *Rhus*.

It seems reasonable that low winter temperatures would play a role in making some seeds in temperate regions permeable, especially those of species like *Melilotus* that germinate in winter or early spring (C. Baskin and Baskin, 1998). However, only limited information is available on effects of naturally occurring low winter temperatures on the breaking of physical dormancy. After alternating freezing (−5°C or −15°C) and thawing (23°C) treatments, 23% of *Medicago sativa* seeds germinated, while 8% of the control seeds germinated (Midgley, 1926). Alternating freezing (−10°C or −20°C) and thawing (5°C) were not effective in breaking physical dormancy in seeds of *Dalea foliosa* (J. Baskin and Baskin, 1998), *I. corei* (J. Baskin and Baskin, 1997), *Senna* spp. (J. Baskin *et al.*, 1998a) or *Sida spinosa* (J. Baskin and Baskin, 1984e). Moist storage at 5°C and at −10°C broke physical dormancy in seeds of *Vicia villosa*, and seeds of *A. theophrasti* stored dry (30% RH) at 4°C became permeable at a rate of 0.8% per day (Cardina and Sparrow, 1997). However, drying as well as low temperatures may have played a role in dormancy break of the *A. theophrasti* seeds.

Microbes, soil abrasion and animals

It is often stated in various types of publications that in nature physical dormancy is broken by soil microbes, abrasion by soil particles and/or passage through the digestive tracts of various animals, especially birds and mammals. However, little or no evidence is available in the scientific literature that physical dormancy is broken either by microbes or by soil abrasion (J. Baskin and Baskin, 2000), and the role of animals in breaking physical dormancy is inconclusive (C. Baskin and Baskin, 1998).

Combinational dormancy

In seeds of the few herbaceous species known to have combinational dormancy, PD is of the non-deep type and broken prior to the time the seed coat becomes permeable to water (e.g. J. Baskin and Baskin, 1974; McKeon and Mott, 1984). Cold stratification is required to break PD in seeds of all temperate-zone woody species known to have combinational dormancy. Thus, the seed (or fruit) coat becomes permeable during summer and the seed imbibes water before the onset of winter, but not necessarily the first winter following seed maturation. PD is broken during winter and seeds germinate in spring (e.g. Afanasiev, 1944; Heit, 1967; Geneve, 1991). To ensure that seeds are permeable to water, they can be scarified (Keogh and Bannister, 1994; Li *et al.*, 1999a) prior to cold stratification. Also, GA treatment of scarified seeds, e.g. *R. aromatica* (Li *et al.*, 1999a) and *Cercis canadensis* (Geneve, 1991), may substitute for cold stratification in breaking PD of the seeds.

Conclusions

Seed dormancy is a problem in propagation of some wild flowers, as well as cultivated flowers, from seeds. Thus, identifying the kind of dormancy and obtaining information on how to break it can be important tools for greenhouse and/or nursery production of plants. A dichotomous key for the five classes of dormancy is presented, each class is described, and examples are provided of plant families in which each class occurs. Conditions required for embryo growth (dormancy break) in seeds with morphological dormancy are reviewed. Seeds or fruits with water-impermeable coats (physical dormancy) are discussed, and information on their dormancy-breaking requirements is summarized. Dormancy-breaking requirements in seeds with both physical and physiological (combinational) dormancy is also discussed. Physiological dormancy (PD) is the most frequently encountered class of dormancy on earth, and it is common in seeds of wild flowers. There are three levels of PD: non-deep, intermediate and deep, and non-deep is divided into five types. Information on dormancy break for seeds with each level of PD and for the five types of non-deep PD is provided. Although morphophysiological dormancy (MPD) accounts for seed dormancy in only about 12% of the species on earth, it is frequently encountered in seeds of herbaceous, mesic deciduous woodland

perennials. Eight levels of MPD have been distinguished, and the dormancy-breaking requirements of each are described. When dormant seeds are water-permeable and little or no information is available to serve as a guide to germinating them, we recommend using the 'move-along experiment' to determine whether high and/or low temperatures treatments are required for dormancy-break.

References

Adams, C.A., J.M. Baskin and C.C. Baskin. 2003. Epicotyl dormancy in the mesic woodland herb *Hexastylis heterophylla* (Aristolochiaceae). *Jour. Torrey Bot. Soc.* 130:11–15.

Afanasiev, M. 1944. A study of dormancy and germination of seeds of *Cercis canadensis*. *Jour. Agric. Res.* 69:405–419.

APG (Angiosperm Phylogeny Group) 1998. An ordinal classification for families of flowering plants. *Ann. Missouri Bot. Gard.* 85:531–553.

Auld, T.D. and R.A. Bradstock. 1996. Soil temperatures after the passage of a fire: do they influence the germination of buried seeds? *Aust. Jour. Ecol.* 21:106–109.

Ballard, W.W. 1987. Sterile propagation of *Cypripedium reginae* from seeds. *Amer. Orchid Soc. Bull.* 56:935–946.

Barton, L.V. 1944. Some seeds showing special dormancy. *Contrib. Boyce Thompson Inst.* 13:259–271.

Baskin, C.C. and J.M. Baskin. 1988. Studies on the germination ecophysiology of herbaceous plants in a temperate region. *Amer. Jour. Bot.* 75:286–305.

Baskin, C.C. and J.M. Baskin. 1994. Deep complex morphophysiological dormancy in seeds of the mesic woodland herb *Delphinium tricorne* (Ranunculaceae). *Intl. Jour. Plant Sci.* 15:738–743.

Baskin, C.C. and J.M. Baskin. 1995. Warm plus cold stratification requirement for dormancy break in seeds of the woodland herb *Cardamine concatenata* (Brassicaceae), and evolutionary implications. *Can. Jour. Bot.* 73:608–612.

Baskin, C.C. and J.M. Baskin. 1998. *Seeds: Ecology, Biogeography, and Evolution of Dormancy and Germination*. Academic Press, San Diego. California.

Baskin, C.C. and J.M. Baskin. 2004. Determining dormancy-breaking and germination requirements from the fewest seeds. *In*: E. Guerrant, K. Havens and M. Maunder (eds). Ex Situ *Plant Conservation: Supporting Species Survival in the Wild*. Island Press, Covelo, California. pp. 162–179.

Baskin, C.C., J.M. Baskin and E.W. Chester. 1991. Temperature response pattern during after-ripening of achenes of the winter annual *Krigia oppositifolia* (Asteraceae). *Plant Species Biol.* 6: 111–115.

Baskin, C.C., J.M. Baskin and G.R. Hoffman. 1992. Seed dormancy in the prairie forb *Echinacea angustifolia* var. *angustifolia* (Asteraceae): after-ripening pattern during cold stratification. *Intl. Jour. Plant Sci.* 153:239–243.

Baskin, C.C., J.M. Baskin and E.W. Chester. 1993a. Seed germination ecology of two mesic wood-land winter annuals, *Nemophila aphylla* and *Phacelia ranunculacea* (Hydrophyllacae). *Bull. Torrey Bot. Club* 120:29–37.

Baskin, C.C., J.M. Baskin and S.E. Meyer. 1993b. Seed dormancy in the Colorado Plateau shrub *Mahonia fremontii* (Berberidaceae) and its ecological and evolutionary implications. *Southw. Nat.* 38:91–99.

Baskin, C.C., P.L. Chesson and J.M. Baskin. 1993c. Annual seed dormancy cycles in two desert winter annuals. *Jour. Ecol.* 81:551–556.

Baskin, C.C., J.M. Baskin and E.W. Chester. 1994a. Annual dormancy cycle and influence of flooding in buried seeds of mudflat populations of the summer annual *Leucospora multifida*. *Ecosci.* 1:47–53.

Baskin, C.C., J.M. Baskin and O.W. Van Auken. 1994b. Germination response patterns during dormancy loss in achenes of six perennial Asteraceae from Texas, USA. *Plant Species Biol.* 9:113–117.

Baskin, C.C., J.M. Baskin and E.W. Chester. 1995a. Role of temperature in the germination ecology of the summer annual *Bidens polylepis* Blake (Asteraceae). *Bull. Torrey Bot. Club* 122: 275–281.

Baskin, C.C., S.E. Meyer and J.M. Baskin. 1995b. Two types of morphophysiological dormancy in seeds of two genera (*Osmorhiza* and *Erythronium*) with an Arcto-Tertiary distribu-tion pattern. *Amer. Jour. Bot.* 82:293–298.

Baskin, C.C., J.M. Baskin and E.W. Chester. 1996. Seed germination ecology of the aquatic winter annual *Hottonia inflata*. *Aquat. Bot.* 54:51–57.

Baskin, C.C., J.M. Baskin and E.W. Chester. 1998. Effect of seasonal temperature changes on ger-mination responses of buried seeds of *Agalinis fasciculata* (Scrophulariaceae), and a compari-son with 12 other summer annuals native to eastern North America. *Plant Species Biol.* 13:77–84.

Baskin, C.C., P. Milberg, L. Andersson and J.M. Baskin. 2000. Germination studies of three

dwarf shrubs (*Vaccinium*, Ericaceae) of North-ern Hemisphere coniferous forests. *Can. Jour. Bot.* 78:1552–1560.

Baskin, C.C., J.M. Baskin and E.W. Chester. 2001a. Morphophysiological dormancy in seeds of *Chamaelirium luteum*, a long-lived dioecious lily. *Jour. Torrey Bot. Soc.* 128:7–15.

Baskin, C.C., P. Milberg, L. Andersson and J.M. Baskin. 2001b. Seed dormancy-breaking and germination requirements of *Drosera anglica*, an insectivorous species of the Northern Hemisphere. *Acta Oecol.* 22: 1–8.

Baskin, C.C., P. Milberg, L. Andersson and J.M. Baskin. 2002a. Non-deep simple morpho-physiological dormancy in seeds of the weedy facultative winter annual *Papaver rhoeas*. *Weed Res.* 42:194–202.

Baskin, C.C., O. Zackrisson and J.M. Baskin. 2002b. Role of warm stratification in promoting germination of seeds of *Empetrum hermaphro-ditum* (Empetraceae), a circumboreal species with a stony endocarp. *Amer. Jour. Bot.* 89: 486–493.

Baskin, J.M. and C.C. Baskin. 1973. Delay of germination in seeds of *Phacelia dubia* var. *dubia. Can. Jour. Bot.* 51:2481–2486.

Baskin, J.M. and C.C. Baskin. 1974. Some eco-physiological aspects of seed dormancy in *Geranium carolinianum* L. from central Tennessee. *Oecologia* 16:209–219.

Baskin, J.M. and C.C. Baskin. 1981. Seasonal changes in germination responses of buried seeds of *Verbascum thapsus* and *V. blattaria* and ecological implications. *Can. Jour. Bot.* 59: 1769–1775.

Baskin, J.M. and C.C. Baskin. 1983a. Germination ecology of *Collinsia verna*, a winter annual of rich deciduous woodlands. *Bull. Torrey Bot. Club* 110:311–315.

Baskin, J.M. and C.C. Baskin. 1983b. Germination ecophysiology of eastern deciduous forest herbs: *Hydrophyllum macrophyllum. Amer. Midl. Nat.* 109:63–71.

Baskin, J.M. and C.C. Baskin. 1983c. Seasonal changes in the germination responses of buried seeds of *Arabidopsis thaliana* and ecological interpretation. *Bot. Gaz.* 144:540–543.

Baskin, J.M. and C.C. Baskin. 1983d. The germination ecology of *Veronica arvensis. Jour. Ecol.* 71:57–68.

Baskin, J. M. and C. C. Baskin. 1984a. Germination ecophysiology of the woodland herb *Osmorhiza longistylis* (Umbelliferae). *Amer. Jour. Bot.* 71: 687–692.

Baskin, J.M. and C.C. Baskin. 1984b. Germination ecophysiology of an eastern deciduous forest herb *Stylophorum diphyllum*. *Amer. Midl. Nat.* 111:390–399.

Baskin, J.M. and C.C. Baskin. 1984c. Role of temperature in regulating timing of germination in soil seeds reserves of *Lamium purpureum*. *Weed Res.* 24:341–349.

Baskin, J.M. and C.C. Baskin. 1984d. Effect of temperature during burial on dormant and nondormant seeds of *Lamium amplexicaule* and ecological implications. *Weed Res.* 24: 333–339.

Baskin, J.M. and C.C. Baskin. 1984e. Environmental conditions required for germination of prickly sida (*Sida spinosa*). *Weed Sci.* 32: 786–791.

Baskin, J.M. and C.C. Baskin. 1985a. The annual dormancy cycle in buried weed seeds: a continuum. *BioSci.* 35:492–498.

Baskin, J.M. and C.C. Baskin. 1985b. Germination ecophysiology of *Hydrophyllum appendiculatum*, a mesic forest biennial. *Amer. Jour. Bot.* 72: 185–190.

Baskin, J.M. and C.C. Baskin. 1985c. Seed germination ecophysiology of the woodland spring geophyte *Erythronium albidum*. *Bot. Gaz.* 146:130–136.

Baskin, J.M. and C.C. Baskin. 1986a. Temperature requirements for after-ripening in seeds of nine winter annuals. *Weed Res.* 26:375–380.

Baskin, J.M. and C.C. Baskin. 1986b. Germination ecophysiology of the mesic deciduous forest herb *Isopyrum biternatum*. *Bot. Gaz.* 147: 152–155.

Baskin, J.M. and C.C. Baskin. 1986c. Seed germination ecophysiology of the woodland herb *Asarum canadense*. *Amer. Midl. Nat.* 116: 132–139.

Baskin, J.M. and C.C. Baskin. 1987. Temperature requirements for after-ripening in buried seeds of four summer annual weeds. *Weed Res.* 27: 385–389.

Baskin, J.M. and C.C. Baskin. 1989. Seed germination ecophysiology of *Jeffersonia diphylla*, a perennial herb of mesic deciduous forests. *Amer. Jour. Bot.* 76:1073–1080.

Baskin, J.M. and C.C. Baskin. 1990. Germination ecophysiology of seeds of the winter annual *Chaerophyllum tainturieri*: a new type of morphophysiological dormancy. *Jour. Ecol.* 78:993–1004.

Baskin, J.M. and C.C. Baskin. 1991. Nondeep complex morphophysiological dormancy in seeds of *Osmorhiza claytonii* (Apiaceae). *Amer. Jour. Bot.* 78:588–593.

Baskin, J.M. and C.C. Baskin. 1994. Nondeep simple morphophysiological dormancy in seeds of the mesic woodland winter annual *Corydalis flavula* (Fumariaceae). *Bull. Torrey Bot. Club* 121:40–46.

Baskin, J.M. and C.C. Baskin. 1997. Methods of breaking seed dormancy in the endangered species *Iliamna corei* (Sherff) Sherff (Malvaceae), with special attention to heating. *Nat. Areas Jour.* 17:313–323.

Baskin, J.M. and C.C. Baskin. 1998. Greenhouse and laboratory studies on the ecological life cycle of *Dalea foliosa* (Fabaceae), a federal endangered species. *Nat. Areas Jour.* 18:54–62.

Baskin, J.M. and C.C. Baskin. 2000. Evolutionary considerations of claims for physical dormancy-break by microbial action and abrasion by soil particles. *Seed Sci. Res.* 10: 409–413.

Baskin, J.M. and C.C. Baskin. 2003a. New approaches to the study of the evolution of physical and physiological dormancy, the two most common classes of seed dormancy on earth. *In*: G. Nicolás, K.J. Bradford, D.Côme and H.W. Pritchard (eds). *The Biology of Seeds: Recent Research Advances*. CAB International, Wallingford, UK, pp. 371–380.

Baskin, J.M. and C.C. Baskin. (2003b) Classification, biogeography, and phylogenetic relationships of seed dormancy. *In*: H. Pritchard (ed.). *Seed Conservation: Turning Science into Practice*. Kew Botanic Gardens, London, UK, pp. 517–544.

Baskin, J.M., C.C. Baskin and M.T. McCann. 1988. A contribution to the germination ecology of *Floerkea proserpinacoides* (Limnanthaceae). *Bot. Gaz.* 149:427–431.

Baskin, J.M., C.C. Baskin and D.M. Spooner. 1989. Role of temperature, light and date seeds were exhumed from soil on germination of four wetland perennials. *Aquat. Bot.* 35:387–394.

Baskin, J.M., X. Nan and C.C. Baskin. 1998a. A comparative study of seed dormancy and germination in an annual and a perennial species of *Senna* (Fabaceae). *Seed Sci. Res.* 8:501–512.

Baskin, J.M., R.W. Tyndall, M. Chaffins and C.C. Baskin. 1998b. Effect of salinity on germination and viability of nondormant seeds of the federal-threatened species *Aeschynomene virginica* (Fabaceae). *Jour. Torrey Bot. Soc.* 125: 246–248.

Baskin, J.M., C.C. Baskin and X. Li. 2000. Taxonomy, anatomy and evolution of physical dormancy in seeds. *Plant Species Biol.* 15:139–152.

Bibbey, R.O. 1948. Physiological studies on weed seed germination. *Plant Physiol.* 23:467–484.

Bullowa, S., M. Negbi and Y. Ozeri. 1975. Role of temperature, light and growth regulators in germination in *Anemone coronaria* L. *Aust. Jour. Plant Physiol.* 2:91–100.

Cardina, J. and D.H. Sparrow. 1997. Temporal changes in velvetleaf (*Abutilon theophrasti*) seed dormancy. *Weed Sci.* 45:61–66.

Chaharsoghi, A.T. and B. Jacobs. 1998. Manipulating dormancy of capeweed (*Arctotheca calendula* L.) seed. *Seed Sci. Res.* 8:139–146.

Densmore, R.V. 1997. Effect of day length on germination of seeds collected in Alaska. *Amer. Jour. Bot.* 84:274–278.

Dosmann, M.S. 2002. Stratification improves and is likely required for germination of *Aconitum sinomontanum*. *HortTech.* 12:423–425.

Egley, G.H. 1976. Germination of developing prickly sida seeds. *Weed Sci.* 24:239–243.

Forbis, T.A. and P.K. Diggle. 2001. Subnivean embryo development in the alpine herb *Caltha leptosepala* (Ranunculaceae). *Can. Jour. Bot.* 79:635–642.

Forbis, T.A., S.K. Floyd and A. de Queiroz. 2002. The evolution of embryo size in angiosperms and other seed plants: implications for the evolution of seed dormancy. *Evolution* 56: 2112–2125.

Gashaw, M. and A. Michelsen. 2002. Influence of heat shock on seed germination of plants from regularly burnt savanna woodlands and grasslands in Ethiopia. *Plant Ecol.* 159:83–93.

Geneve, R.L. 1991. Seed dormancy in eastern redbud (*Cercis canadensis*). *Jour. Amer. Soc. Hort. Sci.* 116;85–88.

Gosling, P.G. 1988. The effect of moist chilling on the subsequent germination of some temperate conifer seeds over a range of temperatures. *Jour. Seed Technol.* 12:90–98.

Graham, A. 1999. *Late Cretaceous and Cenozoic History of North American Vegetation*. Oxford University Press, New York.

Gyer, J.F. 1997. Seed propagation of *Trillium grandiflorum*. *Comb. Proc. Intl. Plant Prop. Soc.* 47: 499–506.

Grushvitzky, I.V. 1967. After-ripening of seeds of primitive tribes of angiosperms, conditions and peculiarities. *In*: H. Borris (ed.). *Physiologie, Okologie and Biochemie der Keimung*. Volume 1. Ernst-Moritz-Arndt Universitat, Greifswald, Germany. pp. 329–345.

Heit, C. E. 1967. Propagation from seed. *Amer. Nurseryman* 125(12):10–11, 37–41, 44–45.

Hitchmough, J.D., J. Gouch and B. Corr. 2000. Germination and dormancy in a wild collected genotype of *Trollius europaeus*. *Seed Sci. Technol.* 28:549–558.

Hodgkinson, K.C. and R.E. Oxley. 1990. Influence of fire and edaphic factors on germination of the arid zone shrubs *Acacia aneura*, *Cassia nemophila* and *Dodonaea viscosa*. *Aust. Jour. Bot.* 38:269–279.

Holm, R.E. 1972. Volatile metabolites controlling germination in buried weed seeds. *Plant Physiol.* 50:293–297.

Hussey, G. 1958. An analysis of the factors controlling the germination of the seed of the oil palm, *Elaeis guineensis* (Jacq.). *Ann. Bot.* 22:259–284 + 2 plates.

Ichihashi, S. 1989. Seed germination of *Ponerorchis graminifolia*. *Lindleyana* 4:161–163.

Jacobsen, J.V. and E. Pressman. 1979. A structural study of germination in celery (*Apium graveolens* L.) seed with emphasis on endosperm breakdown. *Planta* 144:241–248.

Keogh, J.A. and P. Bannister. 1994. Seed structure and germination in *Discaria toumatou* (Rhamnaceae). *Weed Res.* 34:481–490.

Kondo, T., N. Okubo, T. Miura, K. Honda and Y. Ishikawa. 2002. Ecophysiology of seed germination in *Erythronium japonicum* (Liliaceae) with underdeveloped embryos. *Amer. Jour. Bot.* 89:1779–1784.

Li, X., J.M. Baskin and C.C. Baskin. 1999a. Anatomy of two mechanisms of breaking physical dormancy by experimental treatments in seeds of two North America *Rhus* species (Anacardiaceae). *Amer. Jour. Bot.* 86:1505–1511.

Li, X., J.M. Baskin and C.C. Baskin. 1999b. Physiological dormancy and germination requirements of seeds of several North American *Rhus* species (Anacardiaceae). *Seed Sci. Res.* 9:237–245.

Lodge, G.M. and R.D.B. Whalley. 1981. Establishment of warm- and cool-season perennial grasses on the north-west slopes of New South Wales. I. Dormancy and germination. *Aust. Jour. Bot.* 29:111–119.

Martin, A.C. 1946. The comparative internal morphology of seeds. *Amer. Midl. Nat.* 36: 513–660.

Martin, R.E., R.L. Miller and C.T. Cushwa. 1975. Germination response of legume seeds subjected to moist and dry heat. *Ecology* 56:1441–1445.

Martínez, M.L. and P. Moreno-Casasola. 1998. The biological flora of coastal dunes and wetlands: *Chamaecrista chamaecristoides* (Colladon) I. & B. *Jour. Coastal Res.* 14:162–174.

McKeon, G.M. and J.J. Mott. 1982. The effect of temperature on the field softening of hard seed of *Stylosanthes humilis* and *S. hamata* in a dry monsoonal climate. *Aust. Jour. Agric. Res.* 33:75–85.

McKeon, G.M. and J.J. Mott. 1984. Seed biology of *Stylosanthes*. *In*: H.M. Stace and L.A. Edye (eds). *The Biology and Agronomy of Stylosanthes*. Academic Press, Sydney, Australia. pp. 311–332.

McVean, D.N. 1955. Ecology of *Alnus glutinosa* (L.) Gaertn. II. Seed distribution and germination. *Jour. Ecol.* 43:61–71.

Midgley, A.R. 1926. Effect of alternate freezing and thawing on the impermeability of alfalfa and dodder seeds. *Jour. Amer. Soc. Agron.* 18: 1087–1098.

Miller, A.J., D.A. Young and J. Wen. 2001. Phylogeny and biogeography of *Rhus* (Anacardiaceae) based on its ITS sequence data. *Intl. Jour. Plant Sci.* 162:1401–1407.

Nickrent, D.L., A. Blarer, Y.-L. Qiu, D.E. Soltis, P.S. Soltis and M. Zanis. 2002. Molecular data place Hydnoraceae with Aristolochiaceae. *Amer. Jour. Bot.* 89:1809–1817.

Nikolaeva, M.G. 1969. Physiology of deep dormancy in seeds. [Translated from Russian by Z. Shapiro, National Science Foundation, Washington, DC]. Izdatel'stvo 'Nauka', Leningrad.

Nikolaeva, M.G. 1977. Factors controlling the seed dormancy pattern. *In:* A.A. Khan (ed.). *The Physiology and Biochemistry of Seed Dormancy and Germination.* North-Holland, Amsterdam, The Netherlands. pp. 51–74.

Pemadasa, M.A. and Lovell, P.H. 1975. Factors controlling germination of some dune annuals. *Jour. Ecol.* 63:41–59.

Pickett, F.L. 1913. The germination of seeds of *Arisaema. Indiana Acad. Sci. Proc.* 1913:125–128.

Pritchard, H.W., A.J. Haye, W.J. Wright and K.J. Steadman. 1995. A comparative study of seed viability in *Inga* species: desiccation tolerance in relation to the physical characteristics and chemical composition of the embryo. *Seed Sci. Technol.* 23:85–100.

Prothero, D.R. 1994. *The Eocene-Oligocene Transition: Paradise Lost.* Columbia Univiversity Press, New York.

Quick, C.R. 1935. Notes on the germination of *Ceanothus* seeds. *Madrono* 16:23–30.

Ransom, E.R. 1935. The inter-relations of catalase, respiration, after-ripening, and germination in some dormant seeds of the Polygonaceae. *Amer. Jour. Bot.* 22:815–825.

Roberts, H.A. and Boddrell, J.E. 1983. Field emergence and temperature requirements for germination in *Solanum sarrachoides* Sendt. *Weed Res.* 23:247–252.

Roberts, H.A. and Neilson, J.E. 1982. Seasonal changes in the temperature requirements for germination of buried seeds of *Aphanes arvensis* L. *New Phytol.* 92:159–166

Schütz, W. 2000. Ecology of seed dormancy and germination in sedges (*Carex*). *Perspect. Plant Ecol. Evol. Syst.* 3:67–89.

Singh, V., H. Nayyar, R. Uppal and J.J. Sharma. 2000. Effect of gibberellic acid on germination of *Aconitum heterophyllum* L. *Seed Res.* 28:85–86.

Takagi, H. 2001. Breaking of two types of dormancy in seeds of *Polygonatum odoratum* used as vegetables. *Jour. Jap. Soc. Hort. Sci.* 70:416–423.

Van Staden, J., K.M. Kelly, and W.E. Bell. 1994. The role of natural agents in the removal of coat-imposed dormancy in *Dichrostachys cinerea* (L.) Wight et Arn. Seeds. *Plant Growth Reg.* 14: 51–59.

Vázquez-Yanes, C. and A. Orozco-Segovia. 1982. Seed germination of a tropical rain forest pioneer tree (*Heliocarpus donnell-smithii*) in response to diurnal fluctuation of temperatures. *Physiol. Plant.* 56:295–298.

Vegis, A. 1964. Dormancy in higher plants. *Annu. Rev. Plant Physiol.* 15:185–224.

Walck, J.L., C.C. Baskin and J.M. Baskin. 1999. Seeds of *Thalictrum mirabile* (Ranunculaceae) require cold stratification for loss of nondeep simple morphophysiological dormancy. *Can. Jour. Bot.* 77:1769–1776.

Walck, J.L., S.N. Hidayati and N. Okagami. 2002. Seed germination ecophysiology of the Asian species *Osmorhiza aristata* (Apiaceae): comparison with its North American congeners and implications for evolution of types of dormancy. *Amer. Jour. Bot.* 89:829–835.

Walter, H. 1979. *Vegetation of the Earth and Ecological Systems of the Geo-biosphere.* 2nd edn. Translated from the third, revised German edition by Joy Wieser. Springer-Verlag, Berlin, Germany.

Webster, T.M., J. Cardina and H.M. Norquay. 1998. Tillage and seed depth effects on velvetleaf (*Abutilon theophrasti*) emergence. *Weed Sci.* 46: 76–82.

10 Flower Seed Longevity and Deterioration

Miller B. McDonald

Seed Biology Program, Department of Horticulture and Crop Science, Ohio State University, 2021 Coffey Road, Columbus, OH 43210-1086, USA

Introduction

Flower seed deterioration can be defined as 'deteriorative changes occurring with time that increase the seed's vulnerability to external challenges and decrease the ability of the seed to survive'. Three general observations can be made about seed deterioration. First, seed deterioration is an undesirable attribute of flower production. McDonald and Nelson (1986) estimated that 25% of the seed product is lost on an annual basis to deterioration. When one considers that the international value of seed is US$4.1 billion for 2002 (Schmidt, 2002), then the loss in overall seed revenue can approach US$1 billion worldwide. An understanding of seed deterioration, therefore, provides a template for improved flower seed production as well as increased floricultural profits. Second, the physiology of flower seed deterioration is a separate event from seed development and/or germination. Thus, the knowledge gained from understanding these events is unlikely to apply to the events that occur during deterioration. Third, seed deterioration is cumulative. As flower seed ageing increases, seed performance is increasingly compromised.

Predisposition for Seed Deterioration

Many factors contribute to making some flower seed crops more predisposed to seed deterioration than others. Among these are genetics, seed structure, seed chemistry, physical/physiological quality, seed treatments, and relative humidity and temperature.

Genetics

Certain flower seeds do not store well. All flower seeds can be classified into short (less than 1 year), medium (up to 2–3 years) and long (more than 3 years) storage categories (Table 10.1). If a seed has a short storage life, then it is prudent to purchase new seed on an annual basis. The relative storage life of medium storage seeds can be extended when appropriate production and storage practices are followed. In all cases, an understanding of the genetics of seed deterioration is an important consideration in flower seed companies, where extensive inventories of seed are maintained, and the quality should be constantly monitored to ensure that the value of the seed is not lost during storage.

Table 10.1. Relative storage life of flower seeds if maintained under satisfactory storage conditions. Short = less than 1 year, Medium = less than 3 years. Long = more than 3 years.

Short	Medium	Long
Anemone	*Achillea*	*Brassica*
Aquilegia	*Ageratum*	*Calendula*
Arabis	*Alyssum*	*Celosia*
Asclepias	*Antirrhinum*	*Centaurea*
Asparagus	*Brachycome*	*Chrysanthemum*
Aster	*Campanula*	*Cucurbita*
Begonia	*Capsicum*	*Gypsophila*
Bellis	*Cineraria*	*Lycopersicon*
Browallia	*Clarkia*	*Mimulus*
Calceolaria	*Coleus*	Morning glory
Callistephus	*Cyclamen*	Shasta daisy
Catharanthus	*Dahlia*	Sweet pea
Cleome	*Delphinium*	*Zinnia*
Coneflower	*Dianthus*	
Consolida	Dusty miller	
Coreopsis	*Euphorbia*	
Echinops	*Gaillardia*	
Fuchsia	*Gomphrena*	
Gaillardia	*Helianthus*	
Gerbera	*Heuchera*	
Geum	*Hibiscus*	
Helichrysum	*Lathyrus*	
Hippeastrum	*Lavandula*	
Iberis	*Lisianthus*	
Impatiens	*Lobelia*	
Iris	*Lobularia*	
Lantana	*Lotus*	
Liatris	*Lupinus*	
Lilium	Marigold	
Limonium	*Matthiola*	
Nemesia	*Nicotiana*	
Pansy	*Paeonia*	
Penstemon	*Papaver*	
Phlox	*Pelargonium*	
Primula	*Petunia*	
Salvia	*Portulaca*	
Sinningia	*Rudbeckia*	
Thunbergia	*Saintpaulia*	
Veronica	*Scabiosa*	
Vinca	*Schizanthus*	
Viola	*Sedum*	
	Snapdragon	
	Tagetes	
	Torenia	
	Verbena	

Seed structure

Flower seed structure is an important determinant of susceptibility to storage deterioration. In particular, the size/surface area ratio of the seed as well as seed coat permeability influence the rate at which water enters the seed. The more potential for water uptake, the greater the rate of seed deterioration.

Size/surface area ratio

Smaller seeds have a greater surface area to volume ratio compared with larger seeds. As a result, water moves into a smaller seed more rapidly than into a larger seed, making smaller seeds more prone to deterioration. In addition, seeds with a greater surface area (due to seed shape or the presence of appendages) increase the area exposed to relative humidity, thus increasing the potential for moisture uptake of the seed. Figure 10.1 illustrates these size/surface area differences in three flower seed crops (zinnia, verbena and petunia).

Seed chemistry

Generally, seeds high in oil content are more susceptible to seed deterioration than seeds high in starch and protein content. This association is even stronger when the high oil content is present in the embryo rather than the storage reserves. Thus, flower seeds such as impatiens that are high in oil content exhibit shorter seed storage life (Table 10.1). In addition, some flower seeds produce mucilage that surrounds their seed coat during imbibition. Mucilage is a polysaccharide with a high affinity for water. As a result, when relative humidities are high, the mucilage attracts the water, which can then be readily absorbed by the seed, thereby shortening seed storage life. An example of a flower seed high in mucilage is salvia (Fig. 10.2).

Fig. 10.1. Differences in size/surface area of three flower seed crops (zinnia – left, verbena – middle, petunia – right).

Physical/physiological seed quality

The history of the seed during maturation and harvest has an important bearing on how the seed survives during storage. It is often believed that a seed lot is uniform in its physical/physiological quality, but this is seldom the case. In fact, a flower seed lot represents a heterogeneous population of individuals, each differing in its ability to store well and produce a seedling. One of the principal reasons for this begins at the time the seed matures on the parent plant. Many flower crops produce seeds on indeterminate inflorescences where the most mature seeds are produced on the bottom of the inflorescence and still emerging flowers and young seeds exist at the top of the inflorescence. As a result, each maturing seed experiences a differing maturation environment as it establishes its maximum seed quality level. If the environment is stressful, the developing seeds will be of poor quality and will not store as well as those produced when the environment is less stressful. At some point, harvest occurs and all seeds are gathered at one time, producing the diverse population of individuals – each possessing its own unique storage potential. While this principle is most obvious for flower crops with indeterminate inflorescences, it still applies to those with determinate inflorescences where the sequence of seed maturation still varies with time (Fig. 10.3). Thus, a seed lot will have some seeds that store well while others will not. This population-based predisposition for seed deterioration is illustrated in Fig. 10.4. When a seed lot is fresh, seed performance is uniformly high because even poor quality seeds have not initiated the deterioration process. However, with increasing time in storage, poor quality seeds deteriorate more rapidly, high quality seeds less rapidly, and the majority of seeds display an intermediate deterioration rate, resulting in a normal distribution curve. Figure 10.4 emphasizes that seeds in a seed lot do not deteriorate uniformly over time in storage.

Seed treatments

Flower seeds are exposed to a variety of seed treatments to improve performance. These include treatments to break dormancy such as scarification, pelleting and priming. While each of these has a beneficial response, each also has the ability to compromise seed longevity.

Scarification

Scarification is a treatment that abrades the seed coat to facilitate the entry of water into the seed. It can be accomplished mechanically with the use of sandpaper or chemically with acids such as sulphuric acid. The result of scarification is the alleviation of seed dormancy and the initiation of germination. However, scarification time is optimized for the majority of seeds in a seed lot. Often, seeds that are small are scarified too much and large seeds not scarified enough. The damage caused during the scarification process both reduces seed storage potential and encourages the growth of storage fungi.

Fig. 10.2. Salvia seed exuding mucilage around the seed coat during imbibition.

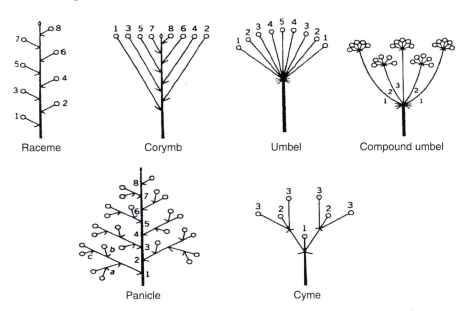

Fig. 10.3. Six differing inflorescence patterns characteristic of flower crops, demonstrating the sequence of flowering with the lowest numbers flowering first and the highest numbers flowering last.

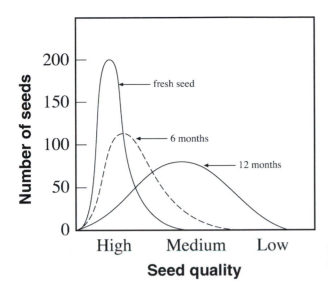

Fig. 10.4. Hypothetical distribution of loss in flower seed quality over a 12-month period.

Pelleting

Seed pellets may be one of the most important recent seed enhancement innovations in flower plug production because they improve seed plantability and performance. A seed pellet is a substance applied to the seed which obscures its shape, thereby making flat or irregularly shaped seeds more round, and making small and light seeds larger and heavier – thus enhancing precision planting and accurate placement by seeders. Most seeds are pelleted in a rotating drum to which the pelleting material and water are periodically added. Pellets are typically composed of fillers such as clays, diatomaceous earth, graphite, powdered perlite, or a combination of these and other materials. A binding or cementing agent is also applied at specific concentrations, which facilitates adhesion of the filler to the seed, thereby adding durability. The filler materials, as well as the binder, can be modified to regulate the water-holding capacity of the pellet. Since the pelleting process employs water as the solvent/binder for the pellet material and cementing agents, it can also be absorbed by the seed during the pelleting process. As a result, seed storage life can be reduced for pelleted seed if the pelleting process is not carefully controlled during its application.

Priming

Research has shown that hydrated flower seeds which were subsequently redried for normal handling have increased germination rate, expanded temperature ranges over which germination occurs, and greater uniformity of stand establishment. The key is to hydrate the seed so the moisture content is sufficient to initiate the early physiological events of germination, but not to the point of radicle protrusion. What many flower growers fail to realize, however, is that seeds subjected to a hydration treatment are very different physiologically from their non-hydrated counterparts. Because the metabolic events leading to germination have been initiated, these seeds are fragile and should be handled more carefully than conventional seed. As with all seeds exposed to hydration treatments prior to planting, growers need to understand that primed seeds have a limited shelf life and should be planted as soon as possible for optimum results.

Relative humidity and temperature

Relative humidity and temperature are the two most important factors determining flower seed deterioration. Relative humidity is important because it directly influences the moisture content of seeds in storage as they come to equilibrium with the amount of gaseous water surrounding them. Temperature is important because it: (i) determines the amount of moisture the air can hold – higher temperatures holding more water than lower temperatures; and (ii) enhances the rate of deteriorative reactions occurring in seeds as temperature increases. These relationships are so important that Harrington (1972) identified the following two 'rules of thumb' describing seed deterioration:

Rule 1. Each 1% reduction in seed moisture content doubles the life of the seed.
Rule 2. Each 5°C reduction in seed temperature doubles the life of the seed.

Harrington (1972) recognized that there were too many qualifications to the above 'rules of thumb' for them to be applied successfully. First, the rule regarding seed moisture content does not apply above 14% or below 6% seed moisture content. Seeds stored at moisture contents above 14% begin to exhibit increased respiration, heating and fungal invasion that destroy seed viability more rapidly than indicated by the moisture content 'rule'. Below 6% seed moisture, a breakdown of membrane structure hastens seed deterioration (this is probably a consequence of reorientation of hydrophilic membranes due to the loss of the water molecules necessary to retain their structural configuration). For the second rule, the qualifications are that at temperatures below 0°C the rule does not apply because many biochemical reactions associated with seed deterioration do not occur and further reductions in temperature have only a moderate effect in extending seed longevity. Finally, it should not be forgotten that these two factors, seed moisture content and temperature, interact. This was captured by another equation suggested by Harrington (1972) where the sum of the temperature in degrees Fahrenheit and the percentage relative humidity should not exceed 100. From this equation, one can see that as the temperature of the storage environment increases, the relative humidity must decrease.

Causes of Flower Seed Deterioration

Once the seed has been purchased and obtained from a seed company and is ready for planting, flower growers must be aware that other factors contribute to loss in seed quality during planting. Planting in a greenhouse is a particularly challenging situation because the two most important factors that influence seed deterioration are seed moisture content and temperature, as described earlier by the two 'rules of thumb.'

As stated, 'Rule 1' applies only at seed moisture contents between 6% and 14%. Based on this 'rule', the ideal seed moisture content for maximizing seed longevity would be between 5% and 6%. How can these moisture contents be achieved? Seed moisture content is a consequence of the relative humidity surrounding the seed. The best analogy is to imagine that the seed is a dry sponge thrown into a bucket of water. The sponge rapidly expands and absorbs the excess moisture. Seeds (because of their chemistry and membrane structure) do the same thing, just as they do when they imbibe free water prior to germination. For example, vinca seeds stored at 5°C and constant 11, 33, 52, 75 or 95% relative humidity will have a moisture content of 4, 6, 8, 11 and 15%, respectively (Carpenter and Cornell, 1990). This example demonstrates that growers can modify seed moisture content by varying the relative humidity surrounding the seed. This can be achieved by using differing salts as shown in Table 10.2. Similarly, during planting, flower growers need to minimize moisture uptake in a humid, warm greenhouse, for example, to ensure that much of the physiology associated with seed germination and deterioration does not occur. This problem is further compounded by the small seed size of many flower seed crops, which contributes to the rapid (in as little as 3 h) absorption of water compared with large seeds.

'Rule 2' applies at temperatures above 0°F. Temperature has both a direct and indirect effect on seed deterioration. The direct effect occurs because increasing temperature causes increased chemical reactions in the seed, thereby initiating both the earliest stages

Table 10.2. Percentage relative humidity of salts at 20°C, 25°C and 30°C.

Salt	20°C	25°C	30°C
Lithium chloride	11.2	11.2	11.2
Potassium acetate	23.2	22.7	22.0
Magnesium chloride	31.5	32.9	32.4
Potassium nitrate	49.0	48.2	47.2
Calcium nitrate	53.6	50.4	46.6
Sodium nitrate	65.3	64.3	63.3
Sodium chloride	75.5	75.5	75.6
Ammonium sulphate	80.6	80.3	80.0
Potassium nitrate	93.2	91.9	90.7

of germination and seed deterioration. The indirect effect occurs because air at high temperatures holds increasing quantities of water. As a result, warmer air provides more available water to the seed compared with cold air, even when the relative humidities are the same. This causes the seed moisture content to increase rapidly in a humid greenhouse and hastens the seed deterioration process described in 'Rule 1'. Never store seeds in the greenhouse or headhouse for periods longer than 1 month without first placing them in a refrigerator or freezer at low relative humidity and low seed moisture content.

Seed storage

After planting, flower growers are tempted to save unused seed. In many cases, they return seeds to the opened foil packet, which is placed in a refrigerator. This can culminate in rapid seed deterioration because the air in the refrigerator is cold, reducing the amount of water suspended in the air and thus increasing the relative humidity. As a result, the seed moisture content increases and, even though the temperature is low, seed deterioration occurs. Instead, if a flower grower is going to retain unused seed, the opened packets should be put into a closed container containing salts that modify the relative humidity (Table 10.2) prior to placing them in a refrigerator. Selecting the appropriate salt to achieve a seed moisture content of 5%–6% is the recommended method for achieving optimum seed storage

and minimizing seed deterioration. Whenever seeds are stored for a period of more than 1 year, they should be retested for germination and vigour to more accurately determine the planting value of the seed lot.

Precautions

There are, of course, various precautions that can easily be implemented to ensure that only the best quality flower seeds are provided to the grower. These can be adopted both within seed companies and by flower growers.

Seed companies

Flower seed companies have made important strides in reducing seed deterioration and improving seed quality. But there continues to be much variation in these production practices and fundamental seed quality issues still must be resolved. For example, one of the principal challenges confronting flower seed companies is the diversity of the flower seed species they must market, to say nothing of the differing varieties they handle. Each species/variety has its own specific production requirement needed to obtain the optimum quality seed. Sometimes these seed crops are contracted to outside sources, often in different parts of the world, where direct oversight of quality control is difficult. Even when the highest quality seed is obtained, however, seed companies could further help flower growers by providing meaningful germination and vigour test information on the seed lot. For example, the Association of Official Seed Analysts (AOSA) Rules for Testing Seeds and the criteria specified for germination testing of many flower seeds were developed well before the onset of the bedding plant industry. These germination criteria sometimes do not reflect the exacting germination requirements found in today's modern bedding plant operations, where exact temperature and relative humidity conditions are maintained in germination chambers and greenhouses. Germination information would be more valuable to these growers if the AOSA rules employed these same environmental conditions encountered in flower production. In addition, flower seed companies could incorporate vigour tests into their quality control programmes. The best example of this is the Ball Vigor Index (BVI). Flower growers have accepted this information as another important and desired component that determines the quality of a flower seed lot. Other vigour tests should also be developed to better assess the level of seed deterioration in order to provide higher quality seed to the grower (see chapter 16).

Because of the value of flower seeds, many seed companies have outstanding storage facilities. But relatively little research has been conducted on the best temperature/relative humidity conditions for flower seeds. Carpenter and Cornell (1990) reported recommendations (Table 10.3), but this information was developed using very few seed lots and seed moisture content information was not provided. More exhaustive studies are still required. As these criteria become better known, seed companies could then provide expiration dates on seed lots so that growers would know that after a specified date, the seed no longer has value. This could be accomplished by initiating cooperative seed storage trials with growers.

Growers

Growers can also take various precautionary steps to minimize seed deterioration. These include the following.

Determine seed needs 1 year in advance

Preparation in identifying the quantity of flower seeds needed in advance of planting can be of great assistance to seed companies in providing the highest quality seed of the desired variety. As soon as seed needs are known, this should be communicated to the seed company. They then can order the

Table 10.3. Best maximum time seed storage recommendations at specific relative humidity (%) and temperature (°C) for 12 flower crops (Carpenter and Cornell, 1990).

Ageratum houstonianum (Ageratum)
  5°C at 11–75% RH
  15°C at 11–32% RH
  25°C at 11–32% RH
Coreopsis
  5°C at 10–30% RH for 9 months
  15°C at 20–35% RH for 9 months
  25°C not recommended
Delphinium × *cultorum* (Delphinium)
  5°C at 30–50% RH for 6 months
  15°C at 20–40% RH for 6 months
  25°C not recommended
Pelargonium × *hortorum* (Geranium)
  5°C at 11–75% RH for 9 months
  15°C at 11–52% RH for 9 months
  25°C at 11–32% RH for 6 months
Gerbera jamesonii (Gerbera)
  5°C at 11– 2% RH for 12 months
  15°C at 11–32% RH for 12 months
  25°C at 11–32% RH for 9 months
Impatiens walleriana (Impatiens)
  5°C at 32–52% RH for 12 months
  15°C at 32% RH for 12 months
  25°C at 32% RH for 9 months
Tagetes erecta (Marigold)
  5°C at 11–52% RH for 12 months
  15°C at 11–52% RH for 9 months
  25°C at 11–32% RH for 8 months
Viola × *wittrockiana* (Pansy)
  5°C at 11–32% RH for 12 months
  15°C at 32–52% RH for 4.5 months
  25°C at 32–52% RH for 4.5 months
Petunia hybrida (Petunia)
  5°C at 25–35% RH for 12 months
  15°C at 32–45% RH for 12 months
  25°C at 32% RH for 12 months
Phlox drummondii (Phlox)
  5°C at 11–40% RH for 12 months
  15°C below 20% RH for 12 months
  25°C below 20% RH for 12 months
Salvia splendens (Salvia)
  5°C at 32% RH for 12 months
  15°C at 32% RH for 12 months
  25°C at 32% RH for 12 months
Catharanthus roseus (Vinca)
  5°C at 11–52% RH for 12 months
  15°C at 11–52% RH for 12 months
  25°C at 11–52% RH for 12 months

correct amount of seed at the best time following harvest thereby maintaining a fresh inventory. This advanced lead time also allows the seed company sufficient time to test the seed and monitor its quality prior to shipment to the grower.

Order seed quantities to fit each sowing

Large companies sequence planting times dependent on personnel, number of seeds to plant, variety and species. When planting schedules are established, it is best to then order or receive the exact number of seeds required for planting. Such periodic seed shipments allow growers to handle smaller seed package sizes, which reduces the risk of inadvertent mixing of varieties. This also minimizes the temptation to return unused seeds to storage, thus improving seed inventory management. Finally, by ordering exact seed quantities, the seed packages remain sealed longer, ensuring that only the freshest seeds are planted.

Follow the 'rules of thumb'

Flower seeds, on receipt, should be refrigerated as long as the storage container is not opened. Before opening the storage container, the grower should allow the seeds to come to room temperature. This process minimizes condensation of water on cold seeds, which causes deterioration. Once the seed package is opened, the seeds should be planted as soon as possible, particularly in greenhouse environments. If opened seeds are to be returned to storage, they should be stored in air-tight containers with a desiccant salt in a refrigerator. As a general rule, unopened seeds can be safely stored for up to 12 months and opened seeds should be planted within 6 months. Enhanced seeds (primed, pregerminated) should not be stored.

Physiology of Seed Deterioration

This complex milieu of interacting environmental factors makes the study of seed deterioration and its underlying physiology difficult. It is beyond the purview of this chapter to consider each of these factors critically and the reader is encouraged to examine a book (Priestley, 1986) as well as a comprehensive chapter (Copeland and McDonald, 2001) and reviews (Halmer and Bewley, 1984; McDonald, 1985; McDonald and Nelson, 1986; Smith and Berjak, 1995; McDonald, 1999) on the subject. In this chapter we will consider flower seed deterioration from a physiological perspective using examples from a variety of crops for which information is known. Starting with a high quality seed under optimum storage conditions, what happens to the seed as its quality is reduced?

Seed deterioration is not uniform

A general assumption is that seed deterioration occurs uniformly throughout a seed. But a seed is a composite of tissues that differ in their chemistry and proximity to the external environment. Thus, it should not be assumed that seed deterioration occurs uniformly throughout the seed. Perhaps the best examples that this does not occur come from the use of the tetrazolium chloride (TZ) test where living tissues in a seed turn red (AOSA, 2000). The challenge to the seed researcher/analyst is to decipher how important the living (or dead) tissues are to successful seedling establishment. When studies have been conducted on seeds using controlled natural and artificial ageing conditions, differences in the deterioration of seed tissues have been observed. For example, in wheat seeds, deterioration begins with the root tip and progressively moves upward through the radicle, scutellum and ultimately the leaves and coleoptile (Das and Sen-Mandi, 1988, 1992). Similar findings have been reported in maize, where root tip cells are the first to be damaged (Berjak et al., 1986), which causes the rate of radicle extension to be lower than coleoptile extension following ageing (Bingham et al., 1994). Similarly, in dicot seeds, root growth is more sensitive to accelerated ageing than is shoot growth (Hahalis and Smith, 1997) and the embryonic axis is more sensitive to deterioration than are the cotyle-dons (Chauhan, 1985; Seneratna *et al.*, 1988; Tarquis and Bradford, 1992). Thus, these findings demonstrate that the embryonic axis is more prone to ageing in monocot and dicot seeds and, of the axis structures, the radicle axis is more sensitive to deterioration than is the shoot axis. Further studies in flower seed deterioration are necessary in order to dem-onstrate whether similar ageing patterns exist.

Mechanisms of flower seed deterioration

Our quest to better understand flower seed deterioration has led to a variety of physiological proposals. Excellent and detailed considerations of these have been provided elsewhere (Smith and Berjak, 1995; McDonald, 1999). Among the changes are:

- Enzyme activities. Most of these studies search for markers of germination such as increases in amylase activity or changes in free radical scavenging enzymes such as superoxide dismutase, catalase, peroxidase and others.
- Protein or amino acid content. The con-sensus is that overall protein content declines while amino acid content increases with seed ageing.
- Nucleic acids. A trend of decreased DNA synthesis and increased DNA degradation has been reported. It is widely believed that degradation of DNA would lead to faulty translation and transcription of enzymes necessary for germination.
- Membrane permeability. Increased membrane permeability associated with increasing seed deterioration has been consistently observed and is the founda-tion for the success of the conductivity test as a measure of seed quality.

Free radical production

Each of these general findings represents the result – not the cause – of seed deteriora-tion. As evidence mounts, the leading candidate causing seed deterioration increas-ingly appears to be free radical production.

Free radical production, primarily initiated by oxygen, has been related to the peroxida-tion of lipids and other essential compounds found in cells. This causes a host of undesir-able events including decreased lipid con-tent, reduced respiratory competence, and increased evolution of volatile compounds ranging from hexanal to aldehydes (Wilson and McDonald, 1986b).

Free radicals – what are they and why are they important?

All atoms that make up molecules contain orbitals that occupy zero, one or two electrons. An unpaired electron in an orbital carries more energy than each electron of a pair in an orbital. A molecule that possesses any unpaired electrons is called a free radical. Some free radicals are composed of only two atoms ($O_2^{\bullet-}$) while others can be as large as protein or DNA molecules. Why is the free radical important in biological systems? It is because of the energetic 'lonely electron', which can: (i) detach from its host atom or molecule and move to another atom or mole-cule; or (ii) pull another electron (which may not have been lonely) from another atom or molecule. The most common free radical reaction is when one free radical and one non-free radical transfer one electron between them, leaving the free radical as a non-free radical, but the non-free radical is now a free radical. This initiates a chain of similar reac-tions that cause substantial damage during the interval that the reactions are occurring. Thus, free radicals can react with each other and with non-free radicals to change the structure and function of other atoms and molecules. If these are proteins (enzymes), lipids (membranes) or nucleic acids (DNA), normal biological function is compromised and deterioration increased. The positive association of free radicals with animal ageing has recently been reviewed (Beck-man and Ames, 1998). What still remains uncertain is their role in flower seed ageing.

How do free radicals cause lipid peroxidation?

The mechanism of lipid peroxidation is often initiated by oxygen around unsaturated or polyunsaturated fatty acids such as oleic and linoleic acids found commonly in seed membranes and storage oils. The result is the release of a free radical, often hydrogen (H•) from a methylene group of the fatty acid adjacent to a double bond. In other cases, the free radical hydrogen may combine with other free radicals from carboxyl groups (ROOH) leaving a peroxy-free radical (ROO•–). Once these free radicals are initiated, they continue to propagate other free radicals that ultimately combine, terminating the destructive reactions. In their wake, they create profound damage to membranes and changes in oil quality. As a result, long-chain fatty acids are broken into smaller and smaller compounds, some of these being released as volatile hydrocarbons (Wilson and McDonald, 1986a; Esashi et al., 1997). The final consequence is loss of membrane structure, leakiness and an inability to complete normal metabolism.

What is the influence of seed moisture content on free radical assault?

Lipid peroxidation occurs in all cells, but in fully imbibed cells water acts as a buffer between the autoxidatively generated free radicals and the target macromolecules, thereby reducing damage. Thus, as seed moisture content is lowered, autoxidation is more common and is accelerated by high temperatures and increased oxygen concentrations. Lipid autoxidation may be the primary cause of seed deterioration at moisture contents below 6%. Above 14% moisture content, lipid peroxidation may again be stimulated by the activity of hydrolytic oxidative enzymes, such as lipoxygenase, becoming more active with increasing water content. Between 6% and 14% moisture content, lipid peroxidation is likely to be at a minimum because sufficient water is available to serve as a buffer against autoxidatively generated free radical attack, but not enough water is present to activate lipoxygenase-mediated free radical production.

Lipoxygenases may contribute to cell degradation by modifying cell membrane composition. In higher plants, two major pathways involving lipoxygenase activity have been described for the metabolism of fatty acid hydroperoxides (Loiseau et al., 2001). One pathway produces traumatic acid, a compound that may be involved in plant cell wound response (Zimmerman and Coudron, 1979) and volatile C_6– aldehydes and C_6– alcohols shown to be correlated with seed deterioration (Wilson and McDonald, 1986a). The other pathway produces jasmonic acid, a molecule which may play a regulatory role in plant cells (Staswick, 1992; Sembdner and Parthier, 1993). Lipoxygenases have been identified and associated with almost every subcellular body in plants (Loiseau et al., 2001) so it is likely that they have important regulatory roles in development. This may include the deterioration of hydrated seeds through free radical production. For example, Zacheo et al. (1998) found increased lipoxygenase activity at high relative humidity (80%) and temperature (20°C) during natural ageing of almond seeds. Other correlative studies implicating lipoxygenases have been identified (McDonald, 1999). A direct study of the importance of lipoxygenases during orthodox seed deterioration employing mutants was reported by Suzuki et al. (1996, 1999). They found that a rice mutant deficient in lipoxygenase-3 had fewer peoxidative products and fewer volatile compounds during seed ageing compared with the wild type.

Thus, the mechanism of lipid peroxidation may be different under long-term ageing (autoxidation) compared with accelerated ageing (e.g. lipoxygenase) conditions. This supports the proposals by Wilson and McDonald (1986b) and Smith and Berjak (1995) that seeds are exposed to separate lipid peroxidative events during storage and during imbibition. It should be noted that oxygen is deleterious to seed storage based on this proposal, which is consistent with the success of hermetic seed storage and lipid peroxidation as a cause of membrane integrity loss.

Do free radicals attack only lipids?

Free radicals attack compounds other than lipids. Changes in protein structure of seeds have been observed and attributed to free radicals (McDonald, 1999). Soluble proteins and membrane proteins may be attacked by different classes of oxidants (and protected by different classes of antioxidants). The most reactive amino acids susceptible to oxidative damage appear to be cysteine, histidine, tryptophan, methionine and phenylalanine, usually in that order (Larson, 1997).

Free radicals are also suspected of assault on chromosomal DNA. Potential targets for oxidative damage in the DNA chain include the purine and pyrimidine bases as well as the deoxyribose sugar moieties (Larson, 1997). Specific damage to the bases may leave the strand intact, but modification of sugar residues can also lead to strand breakage. This may explain the increased propensity for genetic mutations as seeds age. Many of these mutations are first detected as chromosomal aberrations that delay the onset of mitosis necessary for cell division and germination. Others may influence the replication of mitochondria.

Why suspect free radical attack in mitochondria?

Three reasons exist to believe that free radical attack in mitochondria may be a prime cause of seed deterioration. First, mitochondria are the site of aerobic respiration. Thus, they are the prime 'sink' for oxygen, some of which can leak from the membranes during respiration to create free radicals. Second, mitochondria are indispensable to normal cell function. They use oxygen and substrates to generate energy. Third, an important manifestation of seed deterioration is reduced seedling growth, perhaps a consequence of less efficient mitochondrial function.

Mitochondria contain an inner membrane encased in another, outer, membrane and both membranes differ in many important ways. The inner membrane is intricately folded (structures called cristae) and has a much greater surface area than the outer membrane. The cristae are also the site of electron transport where lonely electrons can leak and cause damage to the extensive membrane surface, thereby compromising essential energy production necessary for germination. The space enclosed by the inner membrane is called the mitochondrial matrix.

This matrix is high in protein concentration, containing many enzymes as well as their cofactors critical for oxidative phosphorylation. The matrix also contains a small amount of DNA (mtDNA) and ribosomes for decoding the DNA. The outer membrane is not folded and has large holes in it that permit the passage of many large proteins.

Of these compounds and structures, mtDNA is the most critical for maintaining normal cell function and a review of its structure and function in plants has been provided (Hanson and Otto, 1992). To better understand this important role, it should be noted that mtDNA differs from nuclear DNA in two important ways. First, when a cell divides, both nuclear and mtDNA are separately replicated. Mitochondria can also divide in an active cell requiring the creation of a new copy of mtDNA. Thus, mtDNA is important for the production of new mitochondria in rapidly dividing and physiologically active cells such as those that occur during germination. Second, the enzymes encoded by mtDNA are absolutely essential for oxidative phosphorylation. Thus, maintenance of mtDNA is vital for actively respiring cells, those responsible for seedling growth. As a result, any challenges to mtDNA during storage would surely disrupt normal cellular growth and division.

Since it is now clear that mtDNA and mitochondria are essential for maintenance of cells during dry storage and growth of cells during germination, a crucial question is whether mtDNA or nuclear DNA is more prone to free radical attack. Studies have now documented that mtDNA suffers more spontaneous changes in its DNA sequence compared with nuclear DNA in animal cells, which results in the production of incorrect or truncated proteins (DeGrey, 1999). This greater susceptibility is attributed to:

- MtDNA being more exposed to free radical attack than nuclear DNA. This is because mitochondria are the principal site of oxygen utilization, which results in a greater level of free radical production.

- MtDNA being 'naked'. Nuclear DNA is protected by special proteins called histones, which must be degraded by free radicals prior to nuclear DNA exposure. MtDNA is not surrounded by these protective structures.

- The repair of nuclear DNA is more successful than mtDNA. Fewer repair enzymes exist around mtDNA.

How are seeds protected against free radical attack?

Seeds contain a complex system of antioxidant defences to protect against the harmful consequences of activated oxygen species. There are at least three approaches in seeds to protect against free radical attack. The first is an array of enzymes to neutralize activated oxygen species and, although these are unlikely to operate in dry seeds, their activity would be vital during imbibition. Specific enzymes exist which detoxify $O_2^{\bullet-}$, H_2O_2 and organic peroxides (McDonald, 2002). No enzymes have yet been found that detoxify $\bullet-OH$ or $^1O_2$. The second protective approach includes non-enzymatic compounds that react with activated oxygen species and thereby block the propagation of free radical chain reactions. The third type of defence is enzymes that specifically fix damage created by free radicals. These include DNA repair enzymes that involve a combination of base excision, nucleotide excision or DNA mismatch repair activity.

Repair of seed damage

Considerable evidence exists that repair of DNA (Rao *et al.*, 1987; Dell'Aquila and Tritto, 1990; Sivritepe and Dourado, 1994), RNA (Kalpana and Madhava Rao, 1994), protein (Petruzzeli, 1986; Dell'Aquila and Tritto, 1991), membranes (Powell and Harman, 1985; Tilden and West, 1985; Petruzzeli, 1986) and enzymes (Jeng and Sung, 1994) occurs during imbibition. Increasing seed moisture content hastens the repair process (Ward and Powell, 1983). Oxygen also increases the repair of high moisture (27% to 44%) lettuce (Ibrahim *et al.*, 1983) and high moisture (24% to 31%) wheat (Petruzzeli,

1986) seeds, suggesting that respiratory activity is an essential component of repair. This knowledge that repair occurs during imbibition has been practically adapted by the flower seed industry for many crops through seed priming. As a result, studies examining the physiological advantages/ disadvantages in extending seed performance are appropriate. Generally, it is accepted that repair of seeds deteriorated by lipid peroxidation occurs during hydration (priming). The repaired seed is then dried for normal handling, and the benefits of repair retained as the primed seed completes germination. However, primed seeds should still be planted quickly to obtain these benefits since their overall storage life is shortened by the priming treatment.

Most studies conclude that 'repair' has occurred. But when (during priming or after?), where (what seed part is repaired, if any?) and how (what is the mechanism?) 'repair' occurs is still not known.

When does repair occur?

The time when the beneficial effects of priming are achieved is unknown. It is generally thought that the hydration phase causes activation of essential metabolism associated with germination and the production of repair enzymes. These remain potentially active following subsequent drying and are quickly reactivated on imbibition, culminating in more rapid and uniform completion of germination. Other studies, however, suggest that the maximum beneficial effects of priming are achieved during the drying phase when enzymes are afforded sufficient time to effect repair and physiologically stabilize the seed. For example, the optimum effects of wheat seed osmopriming are observed 2 weeks after drying (Dell'Aquila and Tritto, 1990). Dell'Aquila and Bewley (1989) showed that protein synthesis is reduced in the axes of pea seeds imbibed in polyethyleneglycol and then dried, and is then increased on their return to imbibition. Further research is necessary to clarify whether the benefits of priming are achieved during the hydration or drying phases, or both.

Where is the location of repair?

The location of the beneficial priming response still needs clarification. Reversal of seed deterioration by priming generally occurs in the meristematic axis or the radicle tip (Fu *et al.*, 1988). Sivritepe and Dourado (1994) found that controlled humidification of aged pea seeds to 16.3%–18.1% just prior to sowing decreased chromosomal aberrations, reduced imbibitional injury and improved seed viability. Rao *et al.* (1987) reported a reversal of chromosomal damage (induced during seed ageing) with partial hydration of lettuce seeds by osmopriming to 33%–44%. This treatment also increased the rate of root growth and decreased the frequency of abnormal seedlings. In tomato, artificial ageing increased the percentage of aberrant anaphases in seedling root tips (Van Pijlen *et al.*, 1995). However, while osmo-priming partially counteracts the detrimental effects of artificial ageing on germination rate, uniformity and normal seedlings, it does not influence the frequency of aberrant anaphases in seedling root tips.

Priming also appears to increase germination metabolism in aged axes more than those that are not aged. For example, Dell'Aquila and Taranto (1986) demonstrated that primed embryos of aged wheat seeds had a faster resumption of cell division and DNA synthesis on subsequent imbibition. Clarke and James (1991) showed that accelerated ageing had an adverse effect on endosperm cells of leek seeds that resulted in their degradation and overall loss in seed viability during osmopriming. During germination, however, those seeds that were aged and then osmoprimed showed an increase in RNA species in the whole seeds and their embryos.

What is the mechanism of repair?

Priming appears to reverse the detrimental effects of seed deterioration. In sweetcorn, osmo- and matripriming results in decreased conductivity, free sugars and DNA content while RNA content increased (Sung and Chang, 1993). Natural ageing of French bean seeds stored for up to 4 years induced membrane disruption and leakage of UV-absorbing substances, which was ameliorated by hydropriming (Pandey, 1988, 1989). Lower electrical conductivity readings following hydropriming indicated reduced membrane leakage for aubergine and radish (Rudrapal and Nakamura, 1988b) and onion (Choudhuri and Basu, 1988) seeds. These beneficial effects may be due to the flushing of solutes from the seed during the priming procedure and prior to determination of leaked substances. As a practical result, primed seeds often perform better in disease-infested soils because of decreased electrolyte leakage and faster germination rate, which reduce the window of opportunity for fungal attack (Osburn and Schroth, 1988).

Priming is also thought to increase enzyme activity and counteract the effects of lipid peroxidation. Saha *et al.* (1990) showed that matripriming caused increased amylase and dehydrogenase activity in aged soybean seeds compared with raw seeds. In wheat, osmopriming increased protein and DNA synthesis (Dell'Aquila and Tritto, 1990). L-isoaspartyl methyltransferase enzymes were reported to initiate the conversion of detrimental L-isoaspartyl residues to normal L-isoaspartyl residues that accumulate in naturally aged wheat seeds (Mudgett and Clarke, 1993; Mudgett *et al.*, 1997). This enzyme is present in seeds of 45 species from 23 families representing most of the divisions of the plant kingdom (Mudgett *et al.*, 1997). Osmoprimed tomato seeds subjected to accelerated ageing showed restored activity of L-isoaspartyl methyltransferase to levels similar to non-aged controls, leading Kester *et al.* (1997) to suggest that this enzyme is involved in early repair of deteriorated seeds. Osmopriming reverses the loss of lipid-peroxidation-detoxifying enzymes, such as superoxide dismutase, catalase and glutathione reductase, in aged sunflower seeds, and these enzymes are present at the same activities as in unaged seeds (Bailly *et al.*, 1997).

Priming also reduces lipid peroxidation during subsequent seed storage. In onion seeds, Choudhuri and Basu (1988) demonstrated that hydropriming treatments effectively slowed physiological deterioration under natural (15 months) and accelerated ageing conditions, with the effect being dependent on seed vigour. This improved storability was associated with greater dehydrogenase activity and appreciably lower peroxide formation in cells. Similar findings were reported for hydroprimed aubergine and radish seeds, with the conclusion that hydropriming reduces free radical damage to cellular components (Rudrapal and Nakamura, 1988a). Jeng and Sung (1994) found that free radical scavenging enzymes such as superoxide dismutase, catalase and peroxidase and glyoxysome enzymes such as isocitrate lyase and malate synthase were increased by increasing hydration of artificially aged groundnut seeds. Chang and Sung (1998) also showed that martripriming with vermiculite of sweetcorn seeds enhanced the activities of several lipid peroxide-scavenging enzymes. Chiu *et al.* (1995) found that increasing hydration enhanced membrane repair in watermelon seeds and attributed this to the stimulation of peroxide scavenging enzymes that produced reduced glutathione, which may control ageing by counteracting

lipid peroxidation. Another possible antioxidant is glutathione, whose content has been shown to decrease with watermelon seed ageing as seeds are hydrated (Hsu and Sung, 1997).

Model of Flower Seed Deterioration and Repair During Priming/Hydration

As flower seeds deteriorate, a cascade of disorganization ensues, ultimately leading to complete loss of cell function. The current model of seed deterioration accepts lipid peroxidation as a central cause of cellular degeneration through free radical assault on important cellular molecules and structures. Figure 10.5 demonstrates some proposed events associated with seed deterioration during storage and their repair, or lack of repair, during hydration which can occur during imbibition or priming. Seeds also contain a variety of antioxidants including

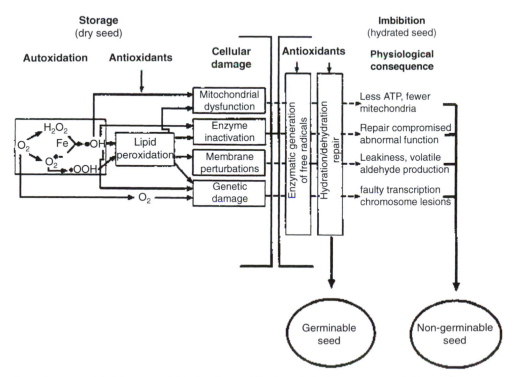

Fig. 10.5. A model of flower seed deterioration and its physiological consequences during seed storage and imbibition.

vitamins, polyphenols and flavonoids (Larson, 1997).

Storage

Low seed moisture content during storage favours free radical production by autoxidation. Through lipid peroxidation, these free radicals either directly or indirectly cause four types of cellular damage: mitochondrial dysfunction, enzyme inactivation, membrane perturbations and genetic damage. Thus, the amount of antioxidants in seeds might reduce the incidence of cellular damage due to free radical assault during seed storage.

Imbibition and priming

As time of seed storage increases, so does cellular damage. Imbibition and priming of the seed allow two events to occur. As imbibition proceeds, the cascade of cellular damage caused by autoxidation is exacerbated by free radical damage, induced less by autoxidation and more by free-radical-generating hydrolytic enzymes such as lipoxygenase. The presence of antioxidants may ameliorate this damage. In addition, upon hydration, anabolic enzymes associated with repair of cellular constituents counter these degenerative events. Their success determines whether a seed is capable of germinating and performing optimally. If unsuccessful, the cellular damage established during storage leads to unalterable detrimental physiological consequences, resulting in a non-germinable seed.

These findings demonstrate that many factors (external and internal) contribute to flower seed deterioration. Of these, seed moisture content and temperature have important roles that directly influence the biochemistry of deterioration. It is also apparent that seed deterioration is not uniform either among seeds or among seed parts (membranes being more prone to deteriorative events). At the cellular level, the mitochondria may be a central organelle susceptible to deteriorative events and their further study is warranted.

As oxygen 'sinks', they contain an extensive membrane structure for respiratory events which is particularly prone to free radical assault and lipid peroxidation. If these events occur, seed germination as measured by speed and uniformity of emergence would certainly be compromised. Fortunately, evidence exists that free radical attack can be reduced by free radical scavenger and antioxidant compounds found in seeds. In addition, specific 'repair' enzymes have been identified that potentially function during hydration, perhaps providing a mechanism for the success of priming as a flower seed enhancement technology. Clearly, all of this demonstrates that further studies are necessary to better understand the mechanism(s) of flower seed deterioration and its repair.

Conclusions

In conclusion, the cost of flower seeds represents only about 5% of the overall cost of flower production. When questions exist about their quality, growers should consider the overall risk (investment) involved in planting stored seed that fails to perform compared with the purchase of new seed. It should also be realized that while enhanced seeds are becoming more popular with (and expensive to) growers, they pose additional concerns relative to storage potential. Generally, seed storage concerns are best left to seed companies. Growers who fail to practice the 'rules of thumb' of successful seed storage or do not have appropriate storage facilities should be encouraged to buy new seeds for planting each production year.

References

AOSA. 2000. *Tetrazolium Testing Handbook.* Contribution No. 29. Association of Official Seed Analysts, Lincoln, Nebraska.

Bailly, C., A. Benamar, F. Corbineau and D. Come. 1997. Changes in superoxide dismutase, catalase and glutathione reductase activities in sunflower seeds during accelerated ageing and subsequent priming. *In*: R.H. Ellis, M. Black, A.J. Murdoch and T.D. Hong (eds). *Basic and Applied Aspects of Seed Biology*. Kluwer Academic Publishers. Dordrecht, The Netherlands. pp. 665–672.

Beckman, K.B. and B.N. Ames. 1998. The free radical theory of ageing matures. *Physiol. Rev.* 78:547–581.

Berjak, P., M. Dini and H.O. Gevers. 1986. Deteriorative changes in embryos of long-stored, uninfected maize caryopses. *South Afr. Jour. Bot.* 52:109–116.

Bingham, I.J., A. Harris and L. MacDonald. 1994. A comparative study of radicle and coleoptile extension in maize seedlings from aged and unaged seeds. *Seed Sci. and Technol.* 22:127–139.

Carpenter, W.J. and J.A. Cornell. 1990. Temperature and relative humidity recommendations for storing bedding plant seed. *Univ. Florida Bull.* 893.

Chauhan, K.P.S. 1985. The incidence of deterioration and its localization in aged seeds of soybean and barley. *Seed Sci. and Technol.* 13:769–773.

Chang, S.M. and J.M. Sung. 1998. Deteriorative changes in primed sweetcorn seeds during storage. *Seed Sci. and Technol.* 26:613–626.

Chiu, K.Y., C.S. Wang and J.M. Sung. 1995. Lipid peroxidation and peroxide-scavenging enzymes associated with accelerated ageing and hydration of watermelon seeds differing in ploidy. *Physiol. Plant.* 94:441–446.

Choudhuri, N. and R.N. Basu. 1988. Maintenance of seed vigour and viability of onion (*Allium cepa* L.). *Seed Sci. and Technol.* 16:51–61.

Clarke, N.A. and P.E. James. 1991. The effects of priming and acclerated ageing upon the nucleic acid content of leek seeds and their embryos. *Jour. Exp. Bot.* 42:261–268.

Copeland, L.O. and M.B. McDonald. 2001. *Principles of Seed Sci. and Technol.* Kluwer Press, New York.

Das, G. and S. Sen-Mandi. 1988. Root formation in deteriorated (aged) wheat embryos. *Plant Physiol.* 88:983–986.

Das, G. and S. Sen-Mandi. 1992. Triphenyl tetrazolium chloride staing pattern of differentially aged wheat embryos. *Seed Sci. and Technol.* 20:367–373.

DeGrey, A.D.W.J. 1999. *The Mitochondrial Free Radical Theory of Ageing*. R.G. Landes Company, Austin, Texas, 212 pp.

Dell'Aquila, A. and J.D. Bewley. 1989. Protein synthesis in the axes of polyethylene glycol treated pea seed and during subsequent germination. *Jour. Exp. Bot.* 40:1001–1007.

Dell'Aquila, A. and G. Taranto. 1986. Cell division and DNA synthesis during osmoconditioning

treatment and following germination in aged wheat embryos. *Seed Sci. and Technol.* 14:333–341.

Dell'Aquila, A. and V. Tritto. 1990. Ageing and osmotic priming in wheat seeds: effects upon certain components of seed quality. *Ann. Bot.* 65:21–26.

Dell'Aquila, A. and V. Tritto. 1991. Germination and biochemical activities in wheat seeds following delayed harvesting, ageing and osmotic priming. *Seed Sci. and Technol.* 19:73–82.

Esashi, Y., A. Kamataki and M. Zhang. 1997. The molecular mechanism of seed deterioration in relation to the accumulation of protein-acetaldehyde adducts. *In*: R.H. Ellis, M. Black, A.J. Murdoch and T.D. Hong (eds). *Basic and Applied Aspects of Seed Biology.* Kluwer Academy Publishers, Dordrecht, The Netherlands. pp. 489–498.

Fu, J.R., X.H. Lu, R.Z. Chen, B.Z. Zhang, Z.S. Liu, Z.S. Ki and C.Y. Cai. 1988. Osmoconditioning of peanut (*Arachis hypogaea* L.) seeds with PEG to improve vigour and some biochemical activities. *Seed Sci. and Technol.* 16:197–212.

Hahalis, D.A. and M.L. Smith. 1997. Comparison of the storage potential of soyabean (*Glycine max*) cultivars with different rates of water uptake. *In*: R.H. Ellis, M. Black, A.J. Murdoch and T.D. Hong (eds). *Basic and Applied Aspects of Seed Biology.* Kluwer Academy Publishers, Dordrecht, The Netherlands. pp. 507–514.

Halmer, P. and J.D. Bewley. 1984. A physiological perspective on seed vigour testing. *Seed Sci. and Technol.* 12:561–575.

Hanson, M.R. and F. Otto. 1992. Structure and function of the higher plant mitochondrial genome. *Int. Rev. Cytol.* 141:129–172.

Harrington, J.F. 1972. Seed storage and longevity. *In*: T.T. Kozlowski (ed). *Seed Biology, Vol. 3.* Academic Press, New York. pp. 145–240.

Hsu, J.L. and J.M. Sung. 1997. Antioxidant role of glutathione associated with accelerated ageing and hydration of triploid watermelon seeds. *Physiol. Plant* 100:967–974.

Ibrahim, A.E., E.H. Roberts and A.J. Murdoch. 1983. Viability of lettuce seeds. II. Survival and oxygen uptake in somatically controlled storage. *Jour. Exp. Bot.* 34:631–640.

Jeng, T.L. and J.M. Sung. 1994. Hydration effect on lipid peroxidation and peroxide-scavenging enzymes activity of artificially-aged peanut seed. *Seed Sci. and Technol.* 22:531–539.

Kalpana, R. and K.V. Madhava Rao. 1994. Absence of the role of lipid peroxidation during accelerated ageing of seeds of pigeonpea (*Cajanus cajan* (L.) Millsp.) cultivars. *Seed Sci. and Technol.* 22:253–260.

Kester, S.T., R.L. Geneve and R.L. Houtz. 1997. Priming and accelerated ageing affect L-isoaspartyl methyltransferase activity in tomato (*Lycopersicon esculentum* Mill) seed. *Jour Exp. Bot.* 48:943–949.

Larson, R.A. 1997. *Naturally Occurring Antioxidants.* Lewis Publishers, Boca Raton, Florida.

Loiseau, J., L.V. Benoit, M.H. Macherel and Y.L. Deunff. 2001. Seed lipoxygenases: occurrence and functions. *Seed Sci. Res.* 11:199–211.

McDonald, M.B. 1985. Physical seed quality of soybean. *Seed Sci. and Technol.* 13:601–628.

McDonald, M.B. 1999. Seed deterioration: physiology, repair and assessment. *Seed Sci. and Technol.* 27:177–237.

McDonald, M.B. 2004. Orthodox seed ageing and its repair. *In*: R. Sanchez and R. Benech-Arnold, (eds). *Seed Physiology: Applications to Agriculture* (in press).

McDonald, M.B. and L.O. Copeland. 1997. *Seed Production: Principles and Practices.* Chapman and Hall, New York.

McDonald, M.B. and C.J. Nelson (eds). 1986. *Physiology of Seed Deterioration.* CSSA Species Publication No. 11, Crop Science Society of America, Madison, Wisconsin.

Mudgett, M.B. and S. Clarke. 1993. Characterization of plant L-isoaspartyl methyltransferases that may be involved in seed survival: purification, cloning, and sequence analysis of the wheat germ enzyme. *Biochem.* 32: 111000–111111.

Mudgett, M.B., J.D. Lowenson and S. Clarke. 1997. Protein repair L-isoaspartyl methyltransferase in plants: phylogenetic distribution and the accumulation of substrate proteins in aged barley seeds. *Plant Physiol.* 114:1481–1489.

Osburn, R.M. and M.N. Schroth. 1988. Effect of osmopriming sugar beet seed on exudation and subsequent damping-off caused by *Pythium ultimum*. *Phytopath.* 78:1246–1250.

Pandey, D.K. 1988. Priming induced repair in French bean seeds. *Seed Sci. and Technol.* 16:527–532.

Pandey, D.K. 1989. Priming induced alleviation of the effects of natural ageing derived selective leakage of constituents in French bean seeds. *Seed Sci. and Technol.* 17:391–397.

Petruzzeli, L. 1986. Wheat viability at high moisture content under hermetic and aerobic storage conditions. *Ann. Bot.* 58:259–265.

Powell, A.A. and G.E. Harman. 1985. Absence of a consistent association of changes in membranal lipids with the ageing of pea seeds. *Seed Sci. and Technol.* 13:659–667.

Priestley, D.A. 1986. *Seed Ageing: Implications for Seed Storage and Persistence in the Soil.* Cornell University Press, Ithaca, New York.

Rao, N.K., E.H. Roberts and R.H. Ellis. 1987. Loss of viability in lettuce seeds and the accumulation of chromosome damage under different storage conditions. *Ann. Bot.* 60:85–96.

Rudrapal, D. and S. Nakamura. 1998a. Use of halogens in controlling eggplant and radish seed deterioration. *Seed Sci. and Technol.* 16: 115–122.

Rudrapal, D. and S. Nakamura. 1988b. The effect of hydration-dehydration pretreatments on eggplant and radish seed viability and vigour. *Seed Sci. and Technol.* 16:123–130.

Saha, R., A.K. Mandal and R.N. Basu. 1990. Physiology of seed invigoration treatments in soybean (*Glycine max* L.). *Seed Sci. and Technol.* 18:269–276.

Schmidt, D. 2002. Presentation at International Seed Federation Congress. Chicago, Illinois.

Sembdner, G. and B. Parthier. 1993. The biochemistry and the physiological and molecular actions of jasmonates. *Ann. Rev. Plant Physiol. Plant Mol. Biol.* 44:569–589.

Seneratna, T., J.F. Gusse and B.D. McKersie. 1988. Age-induced changes in cellular membranes of imbibed soybean axes. *Physiol. Plant* 73: 85–91.

Sivritepe, H.O. and A.M. Dourado. 1994. The effects of humidification treatments on viability and the accumulation of chromosomal aberrations in pea seeds. *Seed Sci. and Technol.* 22:337–348.

Smith, M.T. and P. Berjak. 1995. Deteriorative changes associated with the loss of viability of stored desiccation-tolerant and desiccation-sensitive seeds. *In*: J. Kigel and G. Galili, (eds). *Seed Development and Germination.* Marcel Dekker, New York. pp. 701–746.

Staswick, P.E. 1992. Jasmonate, genes and fragrant signals. *Plant Physiol.* 99:804–807.

Sung, F.J.M. and Y.H. Chang. 1993. Biochemical activities associated with priming of sweetcorn seeds to improve vigour. *Seed Sci. and Technol.* 21:97–105.

Suzuki, Y., T. Yasui, U. Matsukura and J. Terao. 1996. Oxidative stability of bran lipids from rice variety [*Oryza sativa* (L.)] lacking lipoxygenase-3 in seeds. *Jour. Agric. Food Chem.* 44:3479–3483.

Suzuki, Y., K. Ise, C.Y. Li, I. Honda, Y. Iwai and U. Matsukura. 1999. Volatile components in stored rice [*Oryza sativa* (L.)] of volatiles with and without lipoxygenase-3 in seeds. *Jour. Agric. Food Chem.* 47:1119–1124.

Tarquis, A.M. and K.J. Bradford. 1992. Prehydration and priming treatments that advance germination also increase the rate of deterioration of lettuce seeds. *Jour. Exp. Bot.* 43:307–317.

Tilden, R.L. and S.H. West. 1985. Reversal of the effects of ageing in soybean seeds. *Plant Physiol.* 77:584–586.

Van Pijlen, J.G., H.L. Kraak, R.J. Bino and C.H.R. De Vos. 1995. Effects of ageing and osmo-conditioning on germination characteristics and chromosome aberrations of tomato (*Lycopersicon esculentum* Mill.) seeds. *Seed Sci. and Technol.* 23:823–830.

Ward, F.H. and A.A. Powell. 1983. Evidence for repair processes in onion seeds during storage at high seed moisture contents. *Jour. Exp. Bot.* 34:277–282.

Wilson, D.O. and M.B. McDonald. 1986a. A convenient volatile aldehyde assay for measuring seed vigour. *Seed Sci. and Technol.* 14:259–268.

Wilson, D.O. and M.B. McDonald. 1986b. The lipid peroxidation model of seed deterioration. *Seed Sci. and Technol.* 14: 269–300.

Zacheo, G., A.R. Cappello, L.M. Perrone and G.V. Gnoni. 1998. Analysis of factors influencing lipid oxidation of almond seeds during accelerated ageing. *Lebensmittel-Wissenschaft und Technologie* 31:6–9.

Zimmerman, D.C. and C.A. Coudron. 1979. Identification of traumatin, a wound hormone, as 12-oxo-*trans*-10-dodecenoic acid. *Plant Physiol.* 63:536–541.

11   Flower Seed Production

Francis Y. Kwong

PanAmerican Seed Co., 622 Town Road, West Chicago, IL 60185-2698, USA

Commercial flower seed production is an international business involving highly specialized growers. Compared with the production of agronomic and vegetable seeds, it is done on a relatively small scale. However, a large number of species are being produced under very different cultural conditions. It is beyond the scope of this book to survey the technical aspects of production method for each species. Readers interested in pollination and seed harvest techniques for some common horticultural crop families can refer to literature on vegetable and row crops (Fehr, 1980; Erickson, 1983; Ashworth, 1991). This chapter will review how commercial flower seed production is carried out in general practice. Considerations on how to successfully produce high-quality seeds will also be discussed. Since there is a general lack of published reports on this subject matter, much of the information provided here is based on the author's direct experience.

Seed Producers

Most growers and home gardeners have little knowledge of the origins of their seeds. They buy seeds from seed distribution companies, which in turn purchase their seed supplies from breeding and production seed companies. In general, there are three categories of seed producers:

1. Breeding companies. Because of the relatively small scale of production and the high level of skills involved, many leading seed breeders set up their own production facilities where they can exercise full control over the production details. Most of these are greenhouse facilities specialized in the production of high-value bedding and cut flower seeds.

2. Contract seed production companies. These companies are located in areas with favourable climates and long traditions of seed production activities. Some of them are engaged in greenhouse production, while others specialize in open field crops. The greenhouse producers are usually seed growers themselves. Those that specialize in open field production often subcontract the production to other farmers in the areas in which they operate. They provide onsite supervision of the production activities for the original contracting parties.

3. Seed growers taking production contracts from either the breeding or contract production companies. These growers are usually not full-time seed producers. Their main business is crop production. The greenhouse contract growers are normally in the ornamental business, and some of them are experts in one to a few speciality crops. Many of the open field flower seed growers are vegetable or grain producers. Some of them also have experience in vegetable seed production.

Taking these three categories of seed producer together, there are a large number of growers participating in flower seed production. From a management standpoint, flower seed companies have the most direct control in their own proprietary production locations, and they can influence the quality of contract production through grower selection and the terms of the production agreements with them. Due to the vast diversity of crops required in different quantities, contract production is a necessary part of the seed business.

On an individual crop level, the production quantities for most flower seeds are small. A typical seed crop ranges from a few benches in the greenhouse for begonia to a few hectares in the field for marigold. The highly valued seeds of hybrid bedding plants and cut flowers are produced in units of grams or seed count, whereas the less expensive, open-pollinated crops are produced in pounds and kilograms. A seed producer usually grows a range of seed crops in a single facility. Some of the larger greenhouse production facilities operated by breeding companies are like plant factories. Each production station may cover over 20 ha (50 acres) of greenhouse production space, along with specialized facilities for seed drying and processing, and employ over 1000 workers. Figures 11.1 and 11.2 illustrate such a facility owned by PanAmerican Seed in Costa Rica. On the other end of the scale, a contract grower may produce a small number of crops in a family plot, utilizing part-time labour from family members.

Production Areas

There is a saying among the old-time seed producers that the ideal location for greenhouse seed production is in the tropical highlands, and the best place to do an open field seed crop is in an irrigated desert. While the technical validity of this statement is debatable, it describes the current major centres of flower seed production very well. The major areas for commercial production of flower seeds are listed in Table 11.1.

Most greenhouse operations set up by major breeding companies are indeed located in the tropical highlands, where the weather conditions are mild and even throughout the year. Claude Hope founded Linda Vista, a seed production company now managed by PanAmerican Seed, in Cartago, Costa Rica in 1953. Subsequently, Goldsmith Seed established similar production stations in Guatemala and Kenya. Sluis en Groot, a division of Syngenta, produces flower seeds in the highlands of Indonesia and Sri Lanka. These production areas supply the majority of the high-value bedding plant seeds (e.g. impatiens, petunia and geranium) in the

Fig. 11.1.   Some of the greenhouses of Linda Vista, a major hybrid seed producer in Cartago, Costa Rica.

Fig. 11.2. Petunia seed production at Linda Vista, Costa Rica.

Table 11.1. Major flower seed production areas.

Crop species	Seed production areas
Alyssum maritimum	USA
Antirrhinum majus	Costa Rica, France
Begonia × *hybrida*	Denmark, France, Germany, Holland
Cyclamen persicum	France, Germany, Holland
Delphinium elatum	USA
Dianthus chinensis	France, Japan, USA
Impatiens walleriana	Costa Rica, Guatemala, Indonesia
Lathyrus odoratus	USA
Lobelia erinus	France, USA
Matthiola incana	Holland, Mexico, USA
Pelargonium × *hortorum*	Indonesia, Kenya, Mexico, Sri Lanka
Primula acaulis	France, Germany, Holland, Japan
Salvia splendens	Chile, Costa Rica, France, Italy
Tagetes spp.	Chile, China, Guatemala, Mexico, USA
Viola tricolor	Chile, China, France, Holland
Zinnia spp.	France, USA

market today. There is also a long tradition of greenhouse seed production by contracts, particularly for cool-season crops like cyclamen, pansy and primula in Holland, Denmark, Germany and France. In the last decade, significant amounts of seeds have been produced in greenhouses located in Chile, China and India.

For open field production, the most important production countries are France, Holland, the USA and Mexico. Figure 11.3 illustrates some production fields located in the Lompoc Valley, California, one of the major flower seed production areas in the world. Historically, large quantities of flower seeds were also produced in African countries such as Kenya, Tanzania and Zimbabwe. Though still a substantial source of supply, the production in these areas has diminished greatly due to political and social changes over the last three decades. Increasing amounts of flower seeds

Fig. 11.3. Flower seed production fields in Lompoc, California.

are now being produced outdoors in China and some Eastern European countries.

Developing a stable seed production location is a long-term, capital- and labour-intensive undertaking. It requires investments in land improvement, building constructions and specialized machinery. It also requires extensive training of the workforce to acquire the necessary skills, especially when hybrid seed production is considered. In evaluating potential new production areas, seed companies examine not only the climatic data and labour costs, but the social and economic stability of the particular region, as well as facilities for transportation and communication.

Seed Production Procedures

Detailed descriptions of flower seed production procedures are not widely available. Most seed companies and seed growers have to develop their own cultural routines, and the experiences gathered are usually kept as proprietary information. Due to the very diverse locations where seed production is done, commercial seed production techniques are often the results of a fusion of the local cultural habits and agronomic experience.

Seed companies producing the same crops in multiple locations do not necessarily use identical procedures across locations. There is generally no one best method of managing a crop, but many good ways of getting the seed produced successfully. An effective production system depends on how well the different steps in the overall production process are linked up, rather than on applying advanced technologies in a few places.

A generalized production routine has the following components:

1. Parental plant culture.
2. Genetic quality control.
3. Pollination management.
4. Seed harvest and seed extraction.
5. Seed cleaning and conditioning.

Seed producers develop procedures that result in a reliable supply of good quality seeds. Historically, good seed quality is defined as seed that is genetically uniform, highly viable, and free from seed-borne pathogens. The recent advances in plug production technology have increased the demand for high-vigour seed. Professional bedding plant and cut flower growers demand not only seeds with high viability, but ones that will yield a very high percentage of uniform seedlings. This high seed vigour requirement puts an additional challenge before the seed producer. To meet this new challenge, the leading seed companies are investing research efforts in the production function. Continuous improvement of production techniques is now seen as a necessary part of the production work.

Substantial differences exist between greenhouse and open field seed production, both in the scale of production and the amount of control over the production environment. For these reasons, it is convenient to discuss methods used in these two types of production separately. The following sections will outline general methods used in the greenhouse production of hybrid seeds and the field production of open-pollinated flower seeds. Some considerations on how the different steps in the production process contribute to seed quality will also be discussed.

Greenhouse Production of Hybrid Seeds

Most hybrid flower seed crops are produced in greenhouses. The capital requirement for building suitable facilities is high, and the work is labour-intensive. In return for these investments, the producer gains protection from the elements and some flexibility in crop scheduling. Given the right training programmes, the labour pool can provide valuable human resources to ensure reliable high-quality seed production. Since the scale of production for each crop is typically small, close monitoring of plant development on a daily basis is possible. This can provide valuable data for technical improvement.

Parental plant culture

In principle, cultural handling of greenhouse-produced seed crops follows established guidelines for commercial bedding plants or cut flowers. However, most of the available published references are based on crop production in temperate climates, using largely peat-based growing media. When the production is done in areas where peat moss is not readily available, especially in the tropics,

these cultural guidelines are not directly applicable. A key initial study in developing seed production programmes in these areas is identifying a workable growing medium utilizing inexpensive components available locally. Coconut fibres and rice hulls are useful medium ingredients. Other organic materials and volcanic rocks are also included in these soil mixes. The use of a different growing medium greatly affects the irrigation and fertilization regimes required because of differences in water- and nutrient-holding capacities. The growers in each production location have to develop unique irrigation and plant nutritional programmes according to the soil mix being used and the climatic conditions of the area.

Parent plants of most hybrid flower seeds are raised from seed. The seeds are sown in seedling flats or plug trays in a specialized section of the greenhouse serving as the nursery. If a peat-based medium is used, cultural methods generally follow standard greenhouse bedding plant production (Ball, 1991; Styer and Koranski, 1997). Parents of male sterile lines, for example some impatiens and petunia varieties, are propagated by vegetative cuttings. Tissue culture propagated plants are sometimes used for special parent lines of primula and dianthus. These plants are also raised in the nursery. When a hybrid is produced from a cross between a seed- and a vegetative-raised parent, the timing in starting the plant materials has to be adjusted to ensure synchronization of flowering.

Well-developed young plants are transplanted into pots and put on benches in the production greenhouses. The types of pots commonly used include heavy-gauge plastic bags, thin-walled plastic pots and thick-walled Styrofoam pots. The size of the containers chosen is crop-dependent, with 5–10 litre capacities being the most widely used. Traditional greenhouse benches can be made from wood or metal frames. Some benches are specially designed to facilitate easy access to the plants during pollination and seed harvest. Figure 11.4 shows a bench of *Matthiola* seed production in Costa Rica.

Plant nutrition, disease control and pest management are the most important components in parental plant culture. Most seed

Fig. 11.4. *Matthiola* production benches at Linda Vista, Costa Rica.

producers have professionally trained horticulturists on the staff to manage these functions. Regular soil and foliar analyses are used
to provide data to guide nutritional management. Each location tends to develop its own
unique fertilizer programmes based on the
environmental conditions and the experience
of the grower. There is a general lack of detailed
information correlating plant nutritional status with seed yield or seed quality for most
flower seed crops. A higher level of phospho-
rus application has been reported to increase
seed yield in tomatoes, and increased mineral
nutrition (N and K) improves seed quality
(George *et al.*, 1980). Other studies show that
under prolonged nutrient or moisture stress,
seed yield is reduced before negative effects
on seed quality are seen. Calcium and minor
elements are important for proper seed development (Delouche, 1980). Nutritional imbal-
ance can also affect susceptibility to disease
infections (Huber and Graham, 1999).

Integrated pest management routines are
commonly used for effective insect and disease control. The soil medium, growing spaces
and containers are sterilized between crops.
Crop scouting is a daily routine. Yellow sticky
boards are widely used in the greenhouses to
detect a potential build-up of insect pests.
Both preventive and curative applications of

pesticides are used, based on the information
collected. Similar procedures are followed for
disease management. Fungicide applications
are used to control fungal disease. Virus- and
bacteria-infected plants are taken out once
symptoms are found on the plants. Workers at
the PanAmerican Seed production locations
are required to wash their hands frequently
with antiseptic soap, especially when they
move between greenhouses. Many seed
producers also disinfect stock seeds by
surface sterilization or hot-water treatment if
they suspect that they contain seed-borne
pathogens.

Genetic quality control

Genetic purity tests are conducted on stock
seed lots and only seeds of high genetic
purity are used in production. When the
parent plants begin to flower, they are
further checked for the presence of off-types.
Roguing is based on plant habit, foliage
colour, earliness to flower, flower colour and
flower form as some of the key features.
Breeding companies are responsible for the
purity of the stock plants in production contracts. All greenhouses used for hybrid seed

production are equipped with insect-proof screens to prevent accidental pollination by insects from the fields.

Pollination management

Pollination is the most essential part of seed production. It defines both the seed yield and genetic purity aspects of the commercial seeds. For hybrid seed production, pollination work is very labour-intensive and requires the most training. Consequently, it is also the most expensive part of the production process. The pollination process consists of three separate steps: (i) pollen collection; (ii) emasculation; and (iii) pollination. Detailed procedures vary greatly between crops depending on their unique flower morphology and flowering behaviour. A good description of the natural pollination mechanisms of different flowers, including aster, pansy, salvia and snapdragon can be found in Holma (1979). An account of commercial petunia pollination and seed production is given by Ewart (1984).

Pollen collection can be a simple procedure for plants that shed large quantities of loose pollen grains. For cyclamen, the flowers can be shaken by a mechanical device and the pollen collected on a flat dish or in a glass vial. For up-facing flowers like marigold, pollen can be harvested by a suction device. Some production facilities have installed centralized vacuum pumps to allow pollen collection by suction in the entire greenhouse (Fig. 11.5). The anthers of some flowers can be collected prior to anthesis. These anthers are then dried, ground, and pollen grains extracted from the remaining tissue by sieving. This method has been successfully applied in petunia and snapdragons. The stage of flower development at the point of anther collection and the anther drying conditions are major factors influencing pollen yield and quality. Extracted pollen can be cold-stored and used for weeks and months. The pollen should be held in air-tight containers such as glass vials during cold storage to maintain proper pollen moisture content. A frostless freezer is recommended for long-term storage (Hanna, 1994). When pollen quantities are limited, inert materials can be added to allow pollination of a large number of flowers. Some of the pollen diluents found to be useful include lycopodium spores, sieved soil, apple pollen, cornflour and talcum powder. For crops with short-lived pollen like pelargonium, fresh pollen is collected and used directly in making the

Fig. 11.5. Vacuum pump for pollen collection.

hybrid crosses. There are indications that pollen vigour may have direct influence on the vigour of the petunia (Mulcahy *et al.*, 1975) and dianthus (Mulcahy and Mulcahy, 1975) seeds produced. Large-scale verification of these findings has not been reported.

Emasculation is the process by which the anthers of each flower in the female parent line are manually removed. Next to genetic purity of the parental plants, this is the most important factor in obtaining genetically pure commercial hybrid seed. In order to completely prevent selfing, emasculation has to be carried out prior to anthesis, usually in a young bud stage. The petals are gently peeled back to expose the immature anthers, which are removed by hand. This process requires a great deal of precision and care, as the flower organs are very small and any damage to the stigmas will result in poor seed set. The need for emasculation is not as critical for protandrous flowers like impatiens, or heterostylous flowers like primula. Some female lines of petunia and snapdragon are male-sterile. Since no pollen is produced by the male sterile flowers, emasculation is not required.

The flowers are ready for pollination a day or two after emasculation. Flower-to-flower pollination is practised when fresh pollen is applied. Either the entire flower (e.g. impatiens) or the filament with the dehisced anthers attached from the male parent is held by hand and the pollen contents smeared on to the exposed stigma of the female flower. For crops pollinated with extracted pollen, the pollen is placed on the stigmatic surfaces by means of a brush (Fig. 11.6). Identifying the correct location of the stigma requires training. The receptive surface is located at the tip of the gynoecium for most flowers. In a pansy flower, however, it is found on the lower side of the globular head of the stigma. In pansy seed production, pollination using a needle was more efficient than using a brush, glass or petal (Pardo *et al.*, 1989). The timing of pollination is important for seed yield. Optimal seed set is obtained when the flowers are pollinated at the peak of stigma receptivity. This stage is easily identified in petunia when the stigmatic surface appears wet with exudates. Visually judging when the stigmas are ready for pollination is more difficult in other crops

Fig. 11.6. Petunia pollination. A small brush is used to transfer collected pollen to the stigma of the emasculated flower of the female parent.

and trial and error are required to arrive at a successful production routine. It has been observed that pollinating old flowers also negatively affects the quality of seed produced in impatiens (Kwong, 1991).

Seed harvest

Subsequent to successful pollination, the seed develops and matures on the mother plant. The seed harvest and seed drying steps are paramount in obtaining high-quality seed. Because of the small scale, most greenhouse-produced seeds are hand-picked seed pod by seed pod. Determining the correct stage of seed development for seed harvest is the first consideration. Immature seeds are usually low in seed vigour. Harvesting the seeds too late will allow the inclusion of deteriorated seeds. The timing of harvest may also influence the proportion of seeds with primary or secondary dormancy in the population.

Traditionally, seeds are harvested close to the time of seed dispersal, i.e. when the seed pods are dry and cracked open, or when the fruits are soft in the case of fleshy berries. Seeds are considered mature at the point of maximal dry weight accumulation (physiological maturity) and there may be a strong link between chlorophyll degradation and seed vigour (Kwong, 1991; Jalink et al., 1998). Physiologically mature seeds are desiccation-tolerant as they progress into a quiescent state. Conceptually, this should happen much earlier than the senescence of the fruit. Researchers at PanAmerican Seed attempted to identify the earliest seed harvest time by determining when seeds could withstand fast drying. They used sugars and heat-stable proteins (Blackman et al., 1992) as potential molecular markers in these studies. The results of the sugar studies on impatiens are given here as an example. To look for an appropriate marker, changes in the soluble sugar profile were assessed in germinating impatiens seed. Dry impatiens seeds contained a relatively high level of a raffinose-like

C-18 sugar, which has recently been identified as penteose (Gurusinghe and Bradford, 2001). Upon imbibition, the penteose level decreased quickly, with a corresponding increase first in sucrose followed by the C-6 sugars glucose and fructose (Fig. 11.7). When the time course of penteose accumulation was traced during seed development, seeds started producing penteose 14 days after pollination. The penteose level reached a plateau 19 days after pollination (Fig. 11.8). At this point, the impatiens seed pods were still fleshy and green. Seeds harvested after this stage of development, when properly dried, are of very high quality. In cucumber, Jing et al. (2000) reported that while the onset of germinability and desiccation tolerance occurred before maximum dry weight accumulation, improvement of seed vigour was seen after the completion of seed development.

All sizeable greenhouse seed production facilities are equipped with proper seed drying devices. Heated drying chambers are most commonly used (Brandenburg et al., 1961). After the seed pods are dried, the loose seeds are separated from other plant parts by using sieves. Seed pods that are hard and do not break open naturally are crushed mechanically before the sieving process.

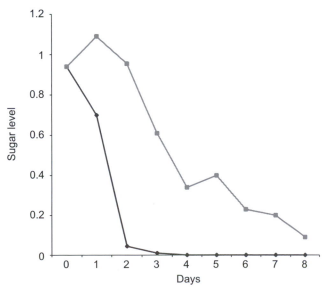

Fig. 11.7. Impatiens sugar profile during seed germination. Penteose (♦) degraded rapidly upon imbibition, closely followed by an increase in sucrose (■).

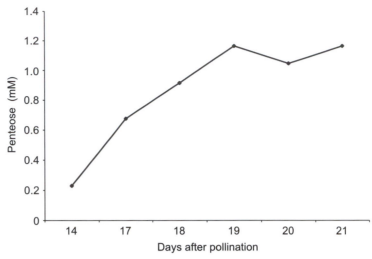

Fig. 11.8. Impatiens penteose content during seed maturation. The level of penteose started to increase at 14 days after pollination and peaked at 19 days after pollination.

Seed cleaning

There is little field debris in greenhouse-produced seed lots. Seed cleaning is generally a simple process and mainly involves removing the very small or light seeds in order to improve the overall quality of the seed lot. Hand screens and small air columns are commonly used to remove the small amounts of plant parts and small seeds.

Field Production of Open-pollinated Varieties

There are substantial differences between the management of open field and greenhouse flower seed production. In the open field, seed crops are produced in blocks of 1 or more hectares. It is less labour-intensive, but more equipment inputs are required. The general seed production techniques are similar to vegetable seed production. Vis (1980) outlined some general cultural methods for field production of flower seeds. Most of the crops produced in the field are open-pollinated varieties, though hybrids with self-incompatible parents can also be done. Unlike greenhouse production, which can be

a year-round activity, outdoor production is seasonal. Identifying locations with suitable climates and producing the crops in the appropriate season are keys to reliable seed supply. There are generally less precise controls in crop management. Some of the unique features and challenges of open-field flower seed production are discussed in the following paragraphs.

Site selection

Site selection is probably the most important factor in outdoor seed production. The production area should provide the required period of appropriate temperature and moisture conditions for the parent plants to develop and the seeds to ripen fully. There should be a dry period at harvest time to allow field drying. Areas in the mid-section of coastal California (around Lompoc) have a long tradition of excellent flower seed production. One can find different micro-climates within a short driving distance due to its mountainous terrain. Here, many different flower seed crops are produced in a relatively small region. The soil type is also an important factor in site selection. Some crops, e.g. pansy, tolerate heavy soils, while others, e.g. nasturtium, only do well in well-drained fields (Vis, 1980). It is impractical to change the basic characters of the soil on a large scale. Disease and pest pressures within the general production area are significant factors. One should consider the crop history of the plot as well as crops being grown in the neighbouring fields to make sure that insect pests and diseases (especially seed-borne diseases) can be avoided. Since most seed production in open fields is done by contracts, an area with growers experienced in seed production will increase success.

Isolation distance

An open-pollinated variety is one that is genetically stable and generally reproduced by self- or mass-pollination. In the open field, pollination is done by wind or insects, depending on the specific floral morphology and properties of the crop species. These natural means of pollen transfer are random in nature. Varieties of the same species will inter-cross with each other. To ensure variety purity, care has to be taken that different seed crops of the same species are not grown closely together. The isolation distance requirements generally range from 360 to 720 m or more, depending on whether the crop is mostly insect- or wind-pollinated (FAO, 1961; Reheul, 1987a,b). The topography of the production site, as well as the direction of prevalent winds, should be considered when determining by how far different varieties of the same species should be separated from each other. Some crops, like sweet pea, have flower structures that allow self-pollination as the flowers develop and mature. Different varieties of these crops can be produced as close as 100 m apart. Seed producers have to keep the required isolation distances in mind when they plan the placement of production fields. Different seed companies operating in the same production area normally exchange information on their production plans prior to planting so that unintended cross-pollination does not occur.

Crop culture

Agronomic practices used for flower seed production are generally similar between crops within a production area, but they vary greatly between production areas. Most of the growing techniques are based on direct farm experience and work habits of specific production locations. After field preparation, stock seeds are sown directly in the production plots. Some growers raise the young plants in a propagation area before transplanting them into the production fields. The parent plants are usually grown in beds to facilitate irrigation, fertilizer application, and fungicide and insecticide sprays.

The plants are checked for genetic uniformity when they begin to flower. Removal of off-types in the population is an intensive activity. This is usually accomplished by a seed company representative possessing detailed knowledge of variety characteristics. A few rounds of rogueing are often necessary to ensure high genetic purity because not all plants begin to flower at the same time.

Weeding is another labour-intensive aspect of field production. The need for it depends on the cleanliness of the production plots as well as adjacent fields. If left uncontrolled, weeds lower seed yield due to competition for nutrients. The presence of weed seeds complicates the seed-cleaning process, and noxious weeds, if not removed, can make the seed lot unmarketable. There is an increasing use of plastic mulches in seed production fields for weed control and moisture conservation.

Unpredictable weather conditions, as well as pest and disease pressures, require day-to-day judgements on irrigation and pest management needs. Good seed yield and quality occur when the environmental conditions are favourable. In years with more challenging natural conditions, only the skilled seed company field men and experienced growers deliver good crops. Low-vigour plants resulting from erratic irrigation and fertilization management are more susceptible to diseases and pests. Seed companies compete on their ability to consistently deliver high-quality seeds year after year.

Pollination

Pollination management for open-pollinated crops begins with selecting production locations naturally conducive to good seed set. Optimum climactic conditions must fit the crop's temperature and light requirements for flowering, pollen production and stigma receptivity. Insect-pollinated crops are best placed in locations where natural pollinator populations are high. In marginal cases, beehives can be placed in production fields to increase pollination activity (Fig. 11.9). Honey bees are the most common pollinators. Other commercially available insect pollinators include bumble bees, leaf-cutter bees and flies.

Application of fungicides and insecticides during the flowering period can negatively affect seed set. Some pesticides and fungicides cause damage to the stigma and interfere with pollen-tube development (Wetzstein, 1990; He *et al.*, 1995, 1996). Insecticides commonly used for insect pest control also kill pollinating insects and potentially lower seed yield.

Harvest and drying

Many popular bedding plant varieties continue to flower and set seed over a long period if the environmental conditions are favourable. Since the seed is harvested only once in the field, determining the proper time to harvest is a critical decision and is based on a compromise between optimum yield and potential seed quality. The onset of unfavourable climatic conditions, such as a rainy season or cold weather, also defines when the crop must be harvested. With the exception of very small production plots and production in countries with low labour costs, harvesting is done by machines. When the crop is judged ready for harvesting, the plants are cut and placed on canvas tarpaulin to dry in the field. The dried plant materials are threshed by a commercial combine (Fig. 11.10). Adverse field conditions, especially rain during the drying period, can cause seed deterioration. Covering the harvested seed materials before the rain, or moving them to dry indoors, are extra efforts required in these situations. There are also custom-made harvesting machines for special crops.

Fig. 11.9. Honey bee hives to increase pollination activities in a gazania production field in Lompoc, California.

Figure 11.11 shows a vacuum harvester used by PanAmerican Seed for crops like marigold and gazania. The field-harvested seed is partially cleaned by scalpers in the open air before being transported to the seed company mill. A crop that is harvested too early may germinate well initially, but the seed does not store well (Chin, 1981).

Seed cleaning

Field-grown seed contains substantial amounts of debris, from less than 20% to over 80% by volume, depending on the crop and the harvest method. The seed of low-growing plants cut at the soil line, e.g. alyssum, contains more field dirt. This seed is

Fig. 11.10.   Field dried alyssum being threshed by a combine in the open field.

Fig. 11.11.   Calendula seed being harvested by a custom-made vacuum harvester.

first put through an air-screen cleaner, which is the most widely used equipment for removing both plant parts and soil particles. Additional size separation by screening machines and density separation by gravity deck or air column may be needed before the seed is cleaned to a commercially acceptable standard. Since each batch of seed coming into the mill may be in a different state of cleanliness, the milling operation is largely experience-driven. The mill supervisor normally inspects the incoming seed lot visually before deciding on the sequence of equipment to use. Frequent machine adjustments may be required in order to obtain optimal results. There are general guidelines, but no fixed procedures, on how a seed crop is best handled. After the initial removal of plant and field debris, additional rounds of seed cleaning are required if the seed contains outer coat structures that impede singulation in packaging and sowing, or water uptake during germination. It is customary to remove hairy layers on the seed coats of gazania and anemone. The tail-like structure of marigolds is removed to facilitate sowing by automatic seeders. Seeds with hard seed coats (e.g. pelargonium) are routinely scarified.

Proprietary procedures are developed by different seed companies for these treatments.

Since the field-produced seed population is inherently more heterogeneous in maturity, the seed drying and conditioning processes have a great influence on seed quality. Some seed producers use portable seed dryers in the field. Others set up permanent drying facilities close to the major production areas (Figs 11.12 and 11.13). Still others rely entirely on natural drying in the fields. Not only are the environmental conditions during field drying important, but excessive heat in the containers during transit from the production area to the seed company warehouse can also cause seed deterioration. Additional drying may be required once the seed arrive in the seed company. Different kinds of dryer are being used, warm-air drying tunnels being the most common (Bradenburg *et al.*, 1961). Additional seed size or density grading may be done to improve the quality of the seed lots. Mechanical damage can occur during the seed conditioning process. Seeds damaged by improper drying or seed conditioning may not result in an immediate decrease in germination, but their shelf-life will be negatively affected.

Fig. 11.12.   Tunnel dryer commonly used for seed drying.

Fig. 11.13. Rotary dryer used for drying plant materials.

Key Challenges for the Seed Producer

The horticultural industry is becoming increasingly competitive. It demands an ever broader range of new products. In order to succeed, the speed of product development has to be accelerated as the average product life cycle decreases. Some breeding companies are now working closely with grower and retailer groups so they can identify consumer needs and respond quickly (Krinkels, 2002). This trend raises some new issues for seed producers. Some of the challenges facing them include:

1. High seed quality requirement. In response to commercial pressures, significant improvements in greenhouse seed production have been made. A reliable supply of high-vigour seed is now available in a range of high-value bedding plants, most notably impatiens and pansy. However, a great deal of additional work still remains in other crops, especially field-produced annuals and perennials.

2. New crops. Learning to produce a new crop is tedious and time-consuming. Experience with growing existing crops may not carry over to the new products. Details of crop culture, pollination methods, harvest and postharvest handling procedures must be identified. This is difficult because there is usually only one chance of learning per year. As the average market life of a product decreases, mistakes made in early production trials can be costly.

3. Small-scale production. When the total market is divided up by increasing numbers of unique products, the size of production for each product will necessarily decrease. This causes different problems for greenhouse and field production. The costs of training and general labour both increase for greenhouse production and specific crops may require different cultural regimes. When grown in the same greenhouse, more hand labour will be required to provide appropriate treatments. There are also more training and supervisory costs associated with pollinating and harvesting multiple crops compared with a single crop in the same area of production space. For open-field production, finding enough isolation to prevent cross-pollination can be a major problem. It is also difficult to find contract growers who are willing to produce crops in small plots.

4. Organic seed. There is an emerging market for certified organic crops, which mandates the use of organically produced seeds. Currently, most commercial flower seeds are produced using conventional agricultural practices. Organic certification requires production in fields with limited chemical usage. Most of the synthetic pest and disease control agents are disallowed in organic production. In order to produce organic seed, not only does the seed producer need to go through a 3- to 5-year transition period so that seed fields are rid of chemical residues, the producer must also learn an entire new set of crop management techniques, utilizing primarily natural products.

Very different sets of skills are required to address these new flower seed production challenges. Some general background knowledge is available, but much more research is needed to bring about solutions to these issues. It is likely, at least in the short term, that these challenges will be met by different specialists rather than one large seed-producing company. These issues offer the opportunity for new players to come into the flower seed industry and challenge existing seed producers to evaluate where they can deliver the best value.

References

Ashworth, S. 1991. *Seed to Seed*. Seed Savers Publication, Decorah, Iowa.

Ball, V. 1991. *The Ball Red Book*. 15th edn. Ball Publishing, Batavia, Illinois.

Blackman, S.A., R.L. Obendorf and A.C. Leopold. 1992. Maturation proteins and sugars in desiccation tolerance of developing soybean seeds. *Plant Physiol.* 100: 225–230.

Brandenburg, N.R., J.W. Simon and L.L. Smith. 1961. Why and how seeds are dried. *Year Book of Agriculture*, pp. 295–306.

Chin, H.F. 1981. The effect of time of harvest on seed storability and subsequent performance. *Acta Horticulturae* 111:249–253.

Delouche, J.C. 1980. Environmental effects on seed development and seed quality. *HortSci.* 15(6):775–780.

Erickson, E.H. 1983. Pollination of entomophilous hybrid seed parents. *In:* C.E. Jones and R.J. Little (eds). *Handbook of Experimental Pollination Biology*. Van Nostrand Reinhold Co. Inc., New York. pp. 493–535.

Ewart, L. 1984. Plant breeding. *In:* K.C. Sink (ed.). *Petunia*. Springer-Verlag, Berlin, Germany. pp. 180–212.

FAO. 1961. Semences agricoles et horticoles. Production, controle et distribution. *Etudes Agricoles de la FAO*, No. 55. Rome.

Fehr, W.R. 1980. Artificial hybridization and self-pollination. *In:* W.R. Fehr and H.H. Hadley (eds). *Hybridization of Crop Plants*. American Society of Agronomy–Crop Science Society of America, Madison, Wisconsin.

George, R.A.T., R.J. Stephens and S. Varis. 1980. The effect of mineral nutrients on the yield and quality of seeds in tomato. *In:* P.D. Hebblethwaite (ed.). *Seed Production*. Butterworths, London.

Gurusinghe, S. and K.J. Bradford. 2001. Galactosyl-sucrose oligosaccharides and potential longevity of primed seeds. *Seed Sci. Res.* 11: 121–133.

Hanna, W.W. 1994. Pollen storage in frostless and conventional frost-forming freezers. *Crop Sci.* 34:1681–1682.

He, Y., H.Y. Wetzstein and B.A. Palevitz. 1995. The effects of a triazole fungicide, propiconazole, on pollen germination, tube growth and cytoskeletal distribution in *Tradescantia virginiana*. *Sex. Plant Reprod.* 8: 210–216.

He,Y., A. Palevitz and H.Y. Wetzstein. 1996. Pollen germination, tube growth and morphology, and microtubule organization after exposure to benomyl. *Physiol. Plant.* 96:152–157.

Holm, E. 1979. *The Biology of Flowers*. Penguin Books, Harmondsworth, UK.

Huber, D.M. and R.D. Graham. 1999. The role of nutrition in crop resistance and tolerance to diseases. *In:* Z. Rengel (ed.). *Mineral Nutrition of Crops*. Haworth Press, New York.

Jalink, H., R. van der Schoor, A. Frandas, J.G. van Pijlen and R.J. Bino. 1998. Chlorophyll fluorescence of *Brassica oleracea* seeds as a non-destructive marker for seed maturity and seed performance. *Seed Sci. Res.* 8: 437–443.

Jing, H., J.H.W. Bergervoet, H. Jalink, M. Klooster, S. Du, R.J. Bino, H.W.M. Hilhorst and S.P.C. Groot. 2000. Cucumber (*Cucumis sativus* L.) seed performance as influenced by ovary and ovule position. *Seed Sci. Res.* 10: 435–445.

Krinkels, M. 2002. Cooperation will dominate the horticulture scene. *Prophyta Annual 2002*: 26–27.

Kwong, F.Y. 1991. Research needs in the production of high quality seeds. *In:* J. Prakask and R.L.M. Pierik (eds). Horticulture –

New Technologies and Applications, Kluwer Academic Publisher, Dordrecht, The Netherlands, pp. 13–20.

Mulcahy, D.L. and G.B. Mulcahy. 1975. The influence of gametophytic competition on sporophytic quality of *Dianthus chinensis*. *Theor. Appl. Genet.* 46:277–280.

Mulcahy, D.L., G.B. Mulcahy and E. Ottaviano. 1975. Sporophytic expression of gametophytic competition in *Petunia hybrida*. *In*: D.L. Mulcahy (ed.). *Gamete Competition in Plants and Animals*. Elsevier, Amsterdam.

Pardo, M.C., M.C. Cid and M. Caballero. 1989. Studies on F_1 seed production of pansies (*Viola × witrockiana* Gams) under shadehouses in Tenerife (Canary Islands). *Acta Horticulturae* 246:359–362.

Reheul, D. 1987a. L'isolation spatiale dans l'amelioration des plantes. 1. L'isolation spatiale des plantes pollinisees par le vent. *Revue de l'Agriculture* 40:5–14

Reheul, D. 1987b. L'isolation spatiale dans l'amelioration des plantes. 2. L'isolation spatiale des plantes pollinisees par les insects. *Revue de l'Agriculture* 40:15–23.

Styer, R.C. and D.S. Koranski. 1997. *Plug and Transplant Production*. Ball Publishing, Batavia, Illinois.

Vis, C. 1980. Flower seed production. *Seed Sci. Technol.* 8:495–503.

Wetzstein, H.Y. 1990. Stigmatic surface degeneration and inhibition of pollen germination with selected pesticidal sprays during receptivity in pecan. *Jour. Amer. Soc. Hort. Sci.* 115: 656–661.

12 Flower Seed Cleaning and Grading

Francis Y. Kwong,[1] Ruth L. Sellman,[1] Henk Jalink[2] and
Rob van der Schoor[2]

[1]PanAmerican Seed Company, 622 Town Road, West Chicago, IL 60185-2698, USA;
[2]Plant Research International, Bornsesteeg 65, 6708 PD Wageningen, The Netherlands

When a seed crop is delivered to the seed company from the production field or greenhouse, the harvested seed represents a heterogeneous population, in physical as well as physiological characteristics. Some of these variations can be accounted for by differences in seed maturity that exist in a once over harvest of crops with indeterminate flowering habit. However, even when each seed capsule is harvested by hand at the optimum stage of development (e.g. *Viola tricolor*, Fig. 12.1), differences still exist in seed size, colour and weight indicating that individual seed development within the capsule is not synchronized. Wulff (1986a) found direct influences of temperature, light and available nutrients on seed size of *Desmodium paniculatum*. A commercial seed lot is by necessity a composite of seeds harvested over a period of time. Since the environmental conditions are constantly changing during the production period, they add to the heterogeneous nature of the commercial seed batch. Depending on the crop, production location and harvest

Fig. 12.1. Mature pansy seed pod close to the time of seed harvest. Seeds within the pod vary both in size and colour.

method, the seed batch also contains varying amounts of soil particles and plant debris. The purpose of seed cleaning is to remove all field and plant debris, as well as empty and immature seeds, to obtain a high degree of physically pure and viable seeds.

The conversion to plug production in the bedding plant industry has placed increased emphasis on highly vigorous flower seeds (see Chapters 8 and 17). Direct correlations were found between the size of the *Desmodium* seed and the resultant leaf size, root weight and overall dry weight of seedlings. Larger seeds also tended to out-compete smaller seeds in seedling establishment when cultural conditions were suboptimal (Wulff, 1986b). In *Lupinus albus*, seedling dry weight was positively correlated with the seed weight (Huyghe, 1993). Extensive seed grading is now used by seed companies to improve the overall vigour of commercial flower seeds.

Mechanical cleaning of flower seeds has been used in Europe and the USA since the 1960s. Most of the widely used machines are laboratory and smaller models of machines developed for cleaning row crops. In general, seed separation is based on the size, density and surface textural and/or frictional differences among individual seeds. The use of colour sorters is less common for flower seeds. Descriptions of many of these seed cleaning machines can be found in reports by Harmond *et al.* (1968), Jensen (1987) and Vaughan *et al.* (1968). Van der Burg and Hendricks (1980) also provided an extensive listing of the methods used in cleaning flower seeds. Variations in these methods are still in use by seed companies. The sequences of equipment commonly used to process the major flower crops are listed in Table 12.1. A complete seed cleaning and grading process generally involves several machines utilizing principles

Table 12.1. Sequence of use of machinery for cleaning the major flower seeds after threshing. Adapted from van der Burg and Hendricks (1980). Additional screens may be used in the air-screen cleaner for size grading. For small seed lots, sizing can be done by stacked screens, and air columns can replace gravity tables for density separation.

Species	Air-screen cleaner	Indented cylinder	Gravity table	Belt grader
Alyssum maritimum	1		2	
Antirrhinum majus	1	2	3	4
Aquilegia hybrida	1		2	
Begonia semperflorens	1		2	
Calendula officinalis	1		2	3
Callistephus chinensis	1	2		
Celosia argentea	1		2	
Centaurea dealbata	1	2	3	
Coleus blumei	1		2	
Coreopsis lanceolada	1	2	3	
Cosmos bipinnatus	1			
Delphinium belladonna	1	2	3	
Dianthus chinensis	1		2	
Gazania × splendens	1	2	3	
Impatiens walleriana	1	2		
Lobelia erinus	1		2	
Matthiola incana	1	2		
Nemesia strumosa	1	2		
Petunia × hybrida	1		2	3
Portulaca grandiflora	1		2	
Primula vulgaris	1		2	
Salvia coccinea	1		2	
Tagetes patula	1	2		
Viola corneta	1		2	
Zinnia elegans	1	2		

based on different seed characters. There has been relatively little innovation in equipment design in seed separation based on the physical characteristics described by Jensen (1987). A density separator using mixtures of organic solvents has been introduced (Franken B.V., Goes, The Netherlands), but it is not widely used. A more recent approach to seed grading is to separate seeds based on chlorophyll fluorescence, which reflects the physiological status of the individual seed (Jalink *et al.*, 1998). When used in combination with the traditional physical separation techniques, these new technologies offer effective ways to deliver highly vigorous seed lots to the consumer.

This chapter will outline traditional seed cleaning and grading techniques commonly used by flower seed companies. It will also introduce the chlorophyll fluorescence (CF) and spectral sorting techniques. Data will be provided to illustrate the suitability of these methods in providing floricultural growers with the highest quality seeds.

Initial Seed Cleaning

Air screen cleaner

The air screen cleaner is of primary importance in most seed cleaning and separating processes. It is used extensively in field-produced flower seed crops, which often contain field debris. Sometimes it is the only piece of equipment used for the basic cleaning and grading of some flower seeds. More often, the air screen cleaner is used to pre-clean the product, allowing further seed grading to be more effective.

The air screen cleaner utilizes two basic concepts in physical separation – width and thickness – through the use of screens and density separation using air. The working principles of these two seed separation methods will be discussed in later sections. A simple air screen cleaner has three screens, although machines with four or five screens are also made (Fig. 12.2). A wide variety of

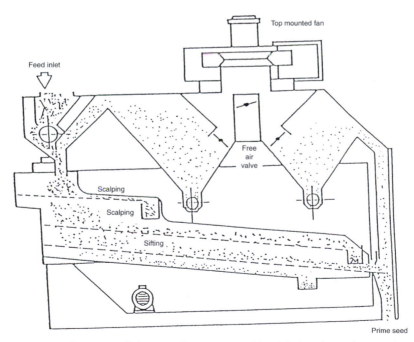

Fig. 12.2. Air screen cleaner. Small dust particles are removed by air before the seed passes through a series of three screens. The two scalping screens remove field materials that are larger than the seed, and the bottom screen removes particles that are smaller than the seed. The seed is then directed through a stream of air, which blows off light seed and debris similar in size to the seed. The prime seed collected may be subjected to further grading.

screens are available, ranging from wire-mesh of different dimensions to perforated metal screens with precision-drilled round or oblong holes. Typically, the first screen is a scalping screen designed to remove large unwanted materials before they pass through the body of the machine where actual sizing and density separation occur. These large particles include dirt clods and broken plant parts collected during field harvest. Greenhouse-produced seeds are usually hand harvested and do not contain much large debris. The second screen is the grading screen – it divides the material into two fractions, one larger and the other smaller than the screen's aperture. The large fraction typically contains the desired whole seed. This fraction is then passed through a column of air to remove all light particles. The small fraction consists of small inert particles and broken or small seed. Seed cleaning is finished in most instances after the screening and air separation. There are cases where the small-sized fraction may contain good seed. This material can be put through the machine again, using a different set of screens (with smaller apertures).

All air screen cleaners should have adjustable feed rate and screen movement (horizontal shaking), along with control of air velocity for the density separation. The inclination of the screens can also be adjusted in some air screen machines. The different controls allow the operator to fine-tune the functions of the machine on a lot-to-lot basis, since no two seed lots are identical. Key factors of success in using the air screen cleaner are: (i) matching the appropriate screen type and aperture to seed size; (ii) proper feed rate of materials into the machine; (iii) speed of screen movement; and (iv) proper air flow.

The air screen cleaner should be connected to an aspiration system that traps very small particles and dust present in the seed lot. The smaller laboratory type models are often equipped with a self-contained aspiration system which is designed to first trap the dust particles and then re-circulate the filtered air.

In order to prevent any unwanted seed mixing, all screens must be removed and cleaned thoroughly in between seed cleaning runs. Pressurized air is normally used to dislodge any seeds from screens and unseen crevices inside the machine. Extra time for cleaning is necessary for equipment with more than the basic three screens.

Dimensional Seed Sizing

Width and thickness

Seed sizing is a simple concept in which a heterogeneous population is physically separated by size to create a more homogeneous group or groups of seed. It is widely used by itself or in combination with other separation techniques. The traditional method for sizing the seed is to pass it through a set of stacked screens declining sequentially in aperture size. Typically, the screens are stacked vertically (Fig. 12.3), but horizontal placement of screens in a stair-step fashion is used as well. All of these techniques rely on mechanical movement of the screens to force the seed to travel across them in order to create a sieving action for separation.

A key factor of seed sizing is to optimally match the seed to the proper screen type and size. The screen types are generally made of a wire or nylon woven mesh or a perforated metal sheet. The mesh screens offer square holes while the perforated metal screens offer round or slotted holes (Fig. 12.4). The shape of the aperture in the screen can greatly influence the effect of the sizing process on the seed. A round-hole perforated screen and a wire-mesh screen of the same aperture size will not produce identical separation results (Lampeter, 1987). A perforated slotted-hole screen allows seed to be sized by the thickness of the seed, and not by its width. Available today are a wide range of precision-made screening equipment, screen types and aperture sizes.

A challenging factor in flower seed conditioning is extremely small seeds. For example, there are approximately 10,000 petunia seeds and 100,000 begonia seeds in 1 gram. The efficacy of sizing using mechanically shaken or vibrated screens is greatly diminished for particle size less than 1.0 mm. The mechanical movement of the screens requires the seed to travel across them for separation to occur. The

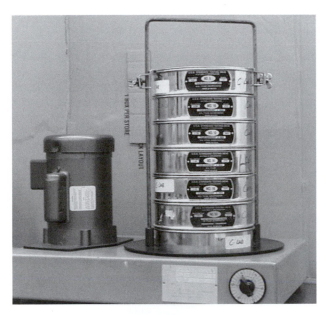

Fig. 12.3. Stacked sizing machine. Screens are stacked vertically from largest to smallest aperture size. As the machine shakes in a circular motion the seed falls through the screen until it reaches a screen with holes too small for it to go through. The seed collected above each screen represents a size fraction approximating to the screen aperture.

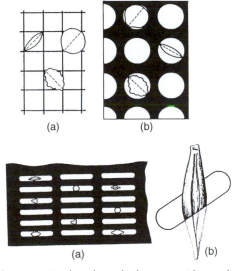

Fig. 12.4. Mesh and punched screens with round and oblong holes. The parameter of sizing in a sieving machine is dependent on the type of screen used. Seeds of the same width that do not pass through a wire mesh screen (top diagram) (a) will pass through a round-holed screen (b) of the same aperture size. A slotted screen (bottom diagram) (a) separates seeds according to thickness rather than width. (b) A long seed passing through a slotted hole.

screens interfere with small seed sizing in two ways: (i) when a mesh screen is used, the seed must travel across, not a smooth flat surface, but rather an expanse of 'hills and valleys' (Fig. 12.5); and (ii) when using a perforated metal screen, the seed must fall through a tunnel-like hole due to the thickness of the metal material. In both cases, the effectiveness and efficiency are compromised, and the probability of physical damage to the seed is increased.

For very small-seeded crops, a sonic siever can be used for size separation. This machine is used primarily in the pharmaceutical and food industries where grading of powders is necessary. The Sonic Siever incorporates no mechanical or vibratory movement in any of its parts. The mesh screens used in this machine are made of very fine wire, with exceptionally precise aperture sizes ranging from 0.02 mm to 6.30 mm. Selected screens are vertically stacked and placed into the machine's separating chamber. They are then compressed in order to seal them from any air leakage and the chamber is closed. When activated, sound waves are forced through the stacked screens. After passing through the bottom screen, they strike a flexible diaphragm which bounces them back up through the screen layers again. When they reach the top of the column, they collide with another diaphragm and are forced back down through the screens. There is a continual motion as the machine is operating. The seeds on the screens are physically affected by the constant up-and-down action of the sound waves, which causes them to bounce on the screens, allowing them to be sieved through the screen layers (Fig. 12.6). The factors that influence the separation process are the screen sizes used, the amplitude of the sound waves, which is measurable and controllable, and the amount of seed placed on the screens. The machine operates only as a batch treatment because a sealed system is needed. If there is too heavy a layer of seed on one or more screens, it can influence the movement of the sound waves thereby hindering the sieving motion. Thorough cleaning of both the machine and the screening materials, and in some cases control of static electricity, are important elements in the successful operation of the machine.

Length

Another approach to seed sizing is separation by length. The indent cylinder (Fig. 12.7) is a

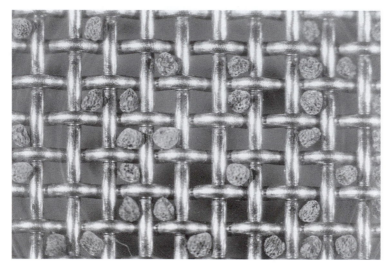

Fig. 12.5. Lisianthus seed on mesh screen with an aperture of 0.42 mm. A mechanical shaker is not an effective way to sieve these extremely small seeds: note the 'hills and valleys' that the seed must traverse.

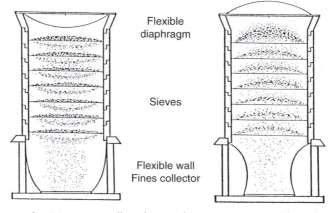

Flexible diaphragm

Sieves

Flexible wall
Fines collector

Fig. 12.6. Sonic siever for sizing very small seeds. Sound waves are generated from the top of the machine and pulsate through the stacked screens to an elastic diaphragm at the bottom. The diaphragm first inflates and then contracts in response, creating pulses of air that cause the seed to bounce on the screens, effecting size separation.

machine designed to do this quickly and accurately. In addition to length separation on clean seed, it is often used to separate damaged seed (broken, cracked, etc.) or weed seeds from field-produced crops. The indent cylinder is also used to separate de-hulled and in-hull seed as one function in a multi-step process.

The principle of separation for the indent cylinder is straightforward. The main part of the machine is the cylinder, which is drum-shaped with both sides open. It is usually made of metal and consists of many precision-made indents or pockets lining the inside wall (Fig. 12.8). They are generally available in millimetres or fractions of an inch depending on where they are manufactured. A general rule for the indent cylinder is that the indents are half the height of their width, e.g. a 12 mm indent is 12 mm in width and 6 mm deep.

During operation, the seeds are fed by vibratory feeder into the horizontally positioned cylinder that is rotating around a stationary central shaft. It is important that the rate of seeds entering the cylinder is high enough to maintain a good bed of seed along the bottom of the cylinder. If there is too little volume of seed in the cylinder, there will be a capacity issue, and if the volume is too big, the accuracy of separation will be compromised. The optimum volume of seed in the cylinder is dependent on seed size and the capacity of the machine. As seeds travel through the cylinder by gravity, any seeds that are shorter than the width of the indents will slide into them and be carried upwards with the rotation of the cylinder. This action is achieved through the centrifugal forces exerted on the seeds by the drum rotation. As the seeds are carried upwards they will at some point be overcome by gravity and fall from the pocket into a collection trough. The trough is attached to the central shaft and can be lowered or raised following the curve of the cylinder. Adjusting the trough to the proper position is a key element in gaining an optimum separation. The seeds that have fallen into the trough are carried to one collection bin. All seeds greater in length than the width of the indents will travel along the base of the cylinder as it is rotating and exit into a second collection bin.

Large-scale indent cylinders have the capacity to operate several cylinders simultaneously, allowing the seed lot to be separated into several lengths at one time. Smaller machines and laboratory models use a single cylinder at one time so running multiple separations requires changing the cylinder for various sizes.

A successful separation using the indent cylinder is dependent on several key factors.

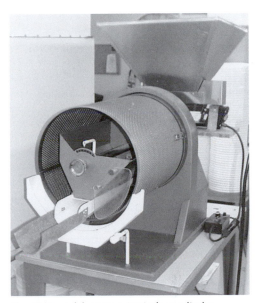

Fig. 12.7. A laboratory size indent cylinder separator commonly used for length separation.

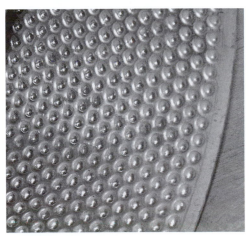

Fig. 12.8. A close-up view of an indent cylinder. The pockets are generally half as deep as they are wide.

Most importantly, there must be a difference in the length of the seeds to be separated – a physical difference in the width or the thickness cannot be adequately separated by this machine. Also, properly matching the correct cylinder to seed size and positioning the trough accurately are both fundamental to achieving good results. The rotation speed of the cylinder and the rate of seed feeding into the cylinder must be optimally set through visual determination.

Seed Grading by Density Separation

Separation by density is commonly incorporated as part of the routine cleaning and grading process in flower seed production procedures. It can be done as a single process or as part of a series of steps in the seed cleaning routine. For all practical purposes, it is a weight separation. However, the seed's ability 'to fly' in air should also be taken into consideration. Seeds of diverse physical forms will behave differently when introduced to density separation. A broad, flat seed will not act in the same manner as a thin, narrow seed. Density separation is used to remove not only broken seed, dirt and inert materials, but also partially filled, empty and immature seeds – innately present in seed lots but invisible to our eyes.

Many types of density separation equipment are available. Generally, the two most commonly used are the air column and gravity table. There are many different types of air column machine, most of which feature air being blown up through a column, although others use a vacuum to pull seed through a column as well. The gravity table incorporates a combination of air (or fluid in some models) and physical movement, which enables it to stratify the seed in order to divide it into different density fractions.

Air column

Air column separation works on the principle of terminal velocity, i.e. the maximum velocity attained by a free-falling body under a

given condition. Conventional air column separation (sometimes referred to as air stream separation) is achieved by allowing seed to fall, or fly, through a continual flow of air. A simple design is usually a freestanding machine consisting of a motor to create air velocity, a baffle to control it, a vertical tube or column to provide space for the physical separation and two collection bins. The top of the machine is equipped with a fine screen to allow for even air flow through the column and to contain the seed within the system (Fig. 12.9).

Typically, seed is fed into the column by a vibratory feeder. Air is flowing freely through the column and, as the seed falls, some of the lighter seeds are carried by the air flow to the top of the column where they are deposited

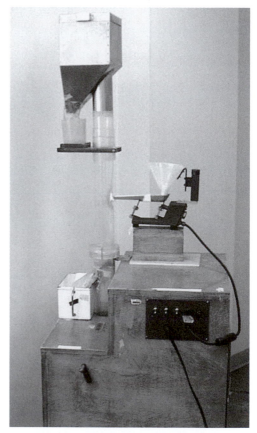

Fig. 12.9. A simple air column used for density separation of small seed lots. Two fractions (light and heavy) can be obtained in each run.

into a collection bin. The heavier seeds fall to the bottom of the column and are collected there. The column is constructed of a rigid, clear material (glass, plastic, acrylic, etc.), which allows for visually adjusting the air flow to obtain the desired separation. Control of the separation is accomplished by increasing or decreasing the air velocity in the column. Speed of seed feeding into the column can also affect the separation. If too much seed enters the column too fast, it begins to ricochet, bouncing aimlessly off the inner wall of the column, which greatly decreases the accuracy of the separation. Some air columns are designed to collect more than two fractions (light and heavy). This is accomplished by strategically placing additional air outlets along the length of the column. Adding extra outlets to a column, however, reduces the ability to keep the air flow constant. Traditional multi-column systems offer little control of the separations, typically allowing the operator to focus on just one column, while seed collected in the remaining columns is difficult to control.

Nowadays, there is a density separator that incorporates multiple air columns with individual air flow control for *each* column (Fig. 12.10). These models are offered in two, three and four column designs addressing different seed grading needs. Another improvement in the design of this machine is the change in column form. Square-shaped columns are used instead of round ones to achieve a more even air flow.

A successful density separation using an air column requires an experienced operator with a watchful eye. Factors that influence the separation are: (i) evenness of air flow; (ii) speed and volume of seed entering the column; and (iii) normal fluctuating environmental conditions. Conducting the separation in an environmentally controlled room increases ease and accuracy of separation.

Fig. 12.10. A multi-cut air column with individual air control for each column. Four density fractions can be obtained in each run.

is the volume and speed at which the seed can be separated. Gravity tables are available in an array of sizes and models designed to accommodate small to very large seed lots. Depending on their capacity, they generally have three to six collection locations with adjustable dividers. Gravity tables are used extensively for cleaning row crops, vegetable seeds and field-produced flower seeds because of their ability to separate large quantities with speed and accuracy.

According to Vaughan *et al.* (1968), seed separation on a gravity separator is based on the following principles of stratification:

1. Seed of the same size will be stratified and separated by differences in their specific gravity.
2. Seed of the same specific gravity will be stratified and separated by differences in their size.

Gravity table

The gravity table offers another technique for density separation. The advantage of using a gravity table as compared with an air column

3. A mixture of seeds differing in both size and specific gravity cannot be stratified and separated effectively.

The operation of the gravity table begins with the seed being fed by a vibratory feeder on to a flat, porous deck surface usually made of a woven wire mesh, nylon screen and canvas or cloth material. The speed at which the seed is fed on to the deck must match the speed at which it exits the deck in order to maintain an even distribution and suitable load of seed on the deck. Improper operation at this initial stage will result in poor separation.

A fan situated below the deck blows air up through the deck's surface. The velocity of air is controllable and the operator must determine the appropriate flow while the process is beginning. As the air blows up through the seed, the deck is continually moving in an oscillating motion. The effect of the air, in conjunction with the motion of the deck, causes the seed to stratify into three density layers – light, medium and heavy. The top layer is the light-density seed and the bottom layer is the heavy-density seed. The top layer is not in contact with the deck's surface and is therefore floating in the air. Since it is not affected by the deck's oscillating motion, it

will be moved forward by the additional seed volume being fed into the deck. The heavy-density seed, which is in full contact with the deck, is affected by the deck's motion and minimally affected by the air, if at all. It will be moved across the deck in the direction of the oscillating motion and be collected in the end bin. The medium-density seed is affected by *both* the air and the motion of the deck because it is slightly lifted off the surface of the deck. This combined influence of air and motion causes the seed to bounce. It, too, travels in the direction of the oscillating motion, but it moves at a slower rate due to the seed's limited contact with the deck's surface. Hence, it collects in the bin(s) between the light and heavy-density seed (Fig. 12.11).

The pitch and the slant of the deck also influence separation efficiency. These two angles can be adjusted according to differences in density of the seeds to be separated. For example, a crop that has been harvested in the field and passed through the scalping screens of a combine will contain not only whole seed and some broken seed, but also broken stems and other plant pieces, dirt clods and possibly small stones. This results in a very high variance in particle densities. In this case, it is advantageous to increase the pitch (or longitudinal slope) of the deck. The degree of

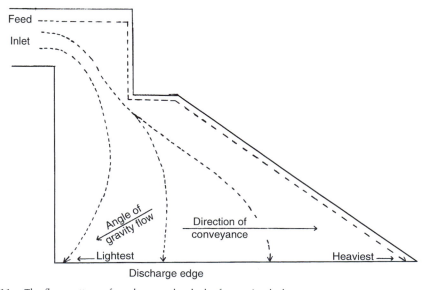

Fig. 12.11. The flow pattern of seed across the deck of a gravity deck.

tilt is made by visual observation of the mixed materials as they travel across the deck. The slant (or lateral slope) can also be increased to facilitate the removal of unwanted heavy particles. Conversely, if a seed lot is very clean, a level deck allows for a close density separation. Different models of gravity table may direct seed flow from front to back (rectangular decks) or from side to side (triangular decks), but the principles of operation are the same. Figure 12.12 shows a small gravity separator used for small flower seed lots.

Achieving an accurate density separation using a gravity table requires not only a familiarity of how the machine functions, but also a working knowledge of the principle of the separation technique. There are many who consider the precise operation of the gravity table an art. The factors which influence the separation are: rate of feed, air velocity, speed of oscillation, pitch and slant of the deck. Improper setting of any of these controls will negatively impact the final result. Seeds of the same species harvested by different methods may require different machine settings to achieve optimal grading. Finally, proper installation of the machine is imperative. A gravity table that is not level or stationary will not do its job properly.

Seed grading by surface texture

Separation of seeds by surface texture is a long-standing technique. It is a viable option in seed lots where there is no size difference and minimum variation in seed density. For example, a weed seed similar in size and density to a crop seed could be removed from the seed lot by this separation technique provided there is a difference in surface texture. Even though surface texture grading is generally a slower method than sizing or density separation, it may be the only mechanical process that produces the required results. Several kinds of equipment are available today that apply the principle of surface texture separation. The more widely used machines are the vibratory deck and the belt grader.

Vibratory deck

During seed separation on a vibratory deck (Fig. 12.13), the seed moves across a textured surface of the deck. The friction created is directly related to the surface texture of the seed and the texture of the deck. A rough-coated, irregular-shaped or broken seed will encounter greater friction than a

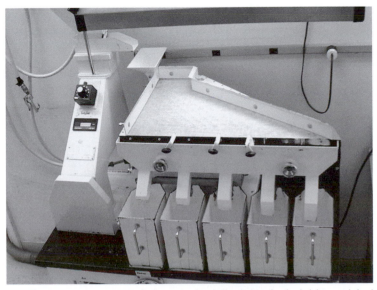

Fig. 12.12. A small gravity separator. The lightest seed is collected in the far left bin and the heaviest seed is collected at the far right. Up to five density fractions can be obtained.

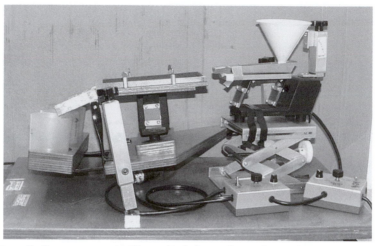

Fig. 12.13. A small vibratory separator. The seed is fed on to the sandpaper deck by a magnetic feeder. Seed separation is based on surface texture. Up to three fractions can be collected.

smooth-coated, oval or round shaped seed. A typical seed separation begins with the seed being fed on to the deck by a feeder. The seed should be positioned to fall on to the deck in a manner that allows full use of the deck's surface area. The deck is mechanically vibrated so the seed is in continual motion on the deck. The surface of the deck is made of a rough-textured material, usually sandpaper, cloth or canvas. Utilizing different deck materials controls the degree of roughness in order to successfully separate the seeds. The speed at which the seed crosses the deck is determined by the intensity of the vibration and the friction present between each seed and the deck's surface. A rough-coated, irregular-shaped or flat seed will move more slowly as its edges 'catch' on the rough surface. A smooth-coated or round seed will experience much less friction and therefore travel more freely across the deck's surface.

The slant and pitch of the deck plays an important role in the separation process. The slant of the deck will cause the rough-coated, irregularly shaped or flat seed to slowly 'climb upwards' to the high end of the slope. Conversely, the smooth-coated or round seed will roll down to the low end due to gravity and the low degree of friction between the seed and the deck. The pitch of the deck offers additional control in the speed at which the seeds traverse the deck. A typical separation run

commonly yields two or more groups of seeds, depending on the type and size of equipment used. Adjustable dividers allow the operator to make the most precise separation possible. Figure 12.14 illustrates the result of *Trachelium* seed grading done on a vibratory deck.

A successful separation using this technique requires: (i) noticeable difference in surface texture; (ii) deck covering matching the surface texture of the seed; (iii) optimized vibration speed; (iv) proper pitch and slant positions; and (v) proper speed of seed loading. The feeder should be placed close to the deck to prevent, or at least reduce, seed bouncing when it is loaded, thereby permitting more surface area of the deck to be used for seed separation. The surface area of the deck is usually small because it is difficult to obtain even distribution of vibration in large decks. An uneven vibration creates 'dead zones', which in turn reduces the efficacy of the separation.

Belt grader

A belt grader consists of four basic elements: a vibratory feeder, a rotating belt made of rubber, canvas, sandpaper or other materials, a mechanism to control the slant or pitch of the belt and a variable speed control to regulate the rate of rotation of the belt. Typically, the machine is manufactured with the belt

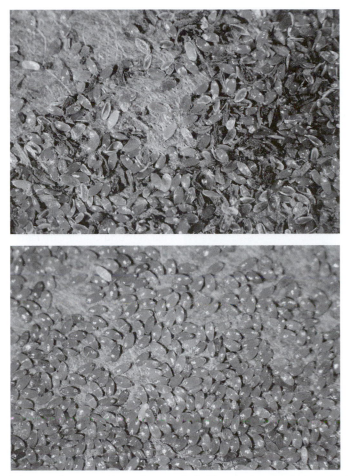

Fig. 12.14. *Trachelium* seed separated by a vibratory deck. The top photograph shows broken and irregular-shaped seed that has rough surface texture; the bottom photograph shows the smooth, good seed.

vertically inclined and rotating in an upward motion. There are also machines with horizontal belts that move from left to right. Belt graders have an advantage over the vibratory deck in that there is a significant increase in the volume of seed being processed.

Seed is fed on to the deck in a single layer by a vibratory feeder. As the belt rotates, smooth or round seeds roll down the incline at a speed faster than the set rotation of the belt and thereby exit at its base. The rough-coated, broken or irregular-shaped seed move more slowly because of the friction between them and the surface of the belt. Those moving slower than the rotation speed of the belt are brought towards the top of the inclined belt and collected there (Figs 12.15 and 12.16).

A successful separation using this machine requires: (i) matching the appropriate belt material to the seed; (ii) proper belt inclination; (iii) proper belt speed; and (iv) proper feed rate. Moreover, positioning where the seed is deposited on the belt and the distance of the fall from the feeder to the belt will also influence the result of the separation.

New Seed Conditioning Techniques

Optical sorting techniques

The idea behind optical sorting techniques is to illuminate the seed by light and to

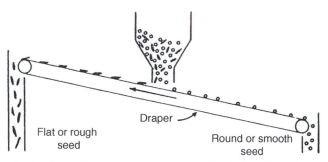

Fig. 12.15. Belt grader. Seed is loaded on to the mid-section of an inclined belt. The smooth seed rolls down the gradient into the bottom collection bin. Rough seed is carried by the textured belt to the upper collector.

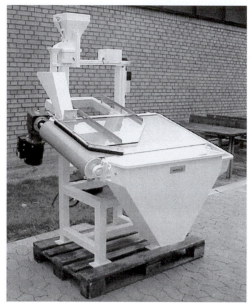

Fig. 12.16. A belt grader manufactured by Westrup (Slagelse, Denmark).

measure the spectral features of the reflected light. Light with a predefined spectrum is directed on to the seed. The light is partly absorbed in different wavelength intervals depending on different pigments present in the seed. The light that comes back from the seed is captured by a detector and analysed on its spectral information. The information can be embedded in a reflectance, transmittance or fluorescence spectrum (Pasikatan and Dowell, 2001). A reliable and simple measurement of a reflectance measurement is the colour measurement using a light source and colour card to define the desirable

colour of the seed. Light reflected from a background colour card causes a constant signal of a light detector. Seeds with a different colour or brightness from that of the background colour card will cause a change in the signal of the light detector. These colour sorters using a colour card were developed half a century ago. Early colour sorters became more advanced by replacing the colour card with an optical filter in front of the light detector. Using one well-defined colour, these sorters became known as monochromatic sorters. A refinement of the monochromatic sorter was the use of two or three different colours: bichromatic or trichromatic sorters, respectively. Using a trichromatic sorter based on a red (R), green (G) and blue (B) colour measurement, almost any colour can be described. The accuracy of a colour measurement can be increased by: (i) the use of more colour filters, for instance adding a near-infrared filter; (ii) using a CCD (charge coupled device) detector instead of single light detectors; or (iii) by ultimately measuring the reflectance spectrum (see paragraph: *New developments: hyper-spectral sorting*, below). The CCD detector consists of multiple light detectors called pixels and makes images of the passing seeds. This has the advantage that small local colour deviations on the seed show up as changes in signal of individual pixels of the CCD (Satake, USA; Sortex, UK). These changes per pixel increase the sensitivity and accuracy of the sorting procedure.

A fluorescence measurement uses only one wavelength or narrow wavelength interval to excite the pigment in the seed and the emission, called fluorescence, is measured at a larger wavelength shifted to the red (Taylor *et al.*, 1993; Jalink *et al.*, 1998). The advantage of a fluorescence measurement over a reflectance or transmittance measurement is that a fluorescence measurement is much more sensitive (typically by about a factor of 1000). Furthermore, a fluorescence measurement usually measures the amount of a known pigment, rather than a mixture of pigment, which is measured by a reflectance or transmittance method. The direct measurement of a single pigment at high sensitivity makes the fluorescence measurement suitable as a sorting technology for seeds when this pigment is correlated with the quality of seeds. A universal fluorescence method was developed based on the presence of chlorophyll in the seed or seed coat (Jalink, 1996). In general, the amount of chlorophyll is correlated with the maturity of the seed. Sorting seeds on the level of chlorophyll fluorescence (CF) results in fractions having different levels of maturity and thus quality. Almost all seeds contain chlorophyll and, by using the fluorescence property of chlorophyll, small amounts of chlorophyll present in single seeds can be measured. This opens up the possibility of applying and testing this new seed sorting technology on numerous seed species.

All the optical sorting technologies sort on a single seed basis. Each seed is measured and its spectral features compared with predetermined values. Based on this comparison, the seed can be classified and sorted into different fractions with each fraction having a different seed quality.

CF correlates with maturity

Generally, seeds show a decrease in total chlorophyll content during the maturation process. Coincidental with the decrease in chlorophyll content, the germination performance of seeds increases (Steckel *et al.*, 1989). It is known that chlorophyll shows prompt fluorescence when the molecule is excited at the proper wavelength (Schreiber, 1986). The CF sorting method makes use of these two findings. The method is based on a spectroscopic measurement designed for the assessment of maturity and quality of seeds. The CF method has been applied to tomato seeds to establish the relationship between the maturation status of the seeds and their CF intensity (Fig. 12.17) (Jalink *et al.*, 1999). An exponential decrease in CF signal was found with increasing days after anthesis. This is an important finding. It shows that the CF method is sensitive since a linear

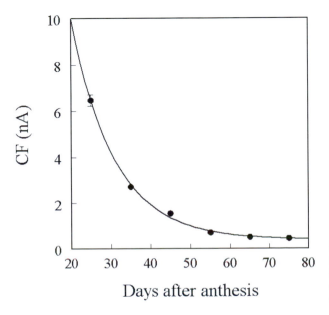

Days after anthesis

Fig. 12.17. Chlorophyll breakdown measured by CF of tomato seeds cultivar Moneymaker at different days after anthesis. Each point is an average of 100 seeds (with SE).

decrease in maturity results in an exponential decrease in CF signal. For practical applications, it means that, within a seed lot, differences in maturity status are exponentially amplified in the CF signal and can thus be very large. Using CF sorting, one measures the intensity of CF signal of each individual intact seed at high speed. Based on the magnitude of the chlorophyll fluorescence signal, the seeds can be sorted into various classes of maturity and linked to seed performance. Seeds with the lowest amount of CF will, in general, result in the highest percentage of germination, more uniform germination, higher speed of germination, fewer dead seeds and higher percentage of normal seedlings (Jalink *et al.*, 1998).

CF correlates with the presence of diseases

Maximum germination and normal seedlings are not the only important seed quality factors. The presence of certain diseases should also be considered. For barley (*Hordeum vulgare* L.) seeds, a relationship was established between the CF signal and the presence of pathogens (Konstantinova *et al.*, 2002). Seeds from the fraction with the highest CF signals were always the most heavily infected. It was not clear whether this was due to infected seeds being unable to break down chlorophyll or whether it was metabolically connected with the chlorophyll content. CF sorting of barley seeds improved their germination quality not only by removing less mature seeds, but also by removing seeds with the largest fungal infection levels. Applying CF sorting on flower seed lots could be used to lower the amount of pathogens in a seed lot and therefore increase the germination and quality of the seedlings.

CF correlates with controlled deterioration

A new application of CF sorting was found recently. White cabbage (*Brassica oleracea* L.) seeds showed a large increase in CF signal after a controlled deterioration (cd) treatment (Del'Aquilla *et al.*, 2002). The increase of CF of deteriorated, low-quality seeds could be a result of the increase in disorganization of the chloroplast membranes. Energy absorbed by

chlorophyll cannot be used for photosynthesis, since the water concentration is too low. As a result, the energy is mainly dissipated into heat and fluorescence, with the fluorescence channel dissipating only a little of the total absorbed energy. Thermal deactivation is most likely to be less efficient for disorganized chloroplast membranes, due to fewer interactions between neighbouring molecules. This implies that less energy is transferred into the thermal channel. A small decrease in the thermal deactivation will yield a large increase in fluorescence. Furthermore, after applying the cd treatment, the seeds with high CF signals did not germinate, but the quality of the seeds with low CF signals was high and equal to the seeds with low CF signals not subjected to a cd treatment. This result shows that CF analysis and sorting could be used on a routine basis to monitor the deterioration of flower seeds during storage and improve their quality after storage by sorting.

The CF measurement

A 3-D spectrum of a green *Lisianthus* seed showed two major fluorescence peaks (Fig. 12.18). One peak is centred around 690 nm and a second peak, lower in intensity, is centred around 730 nm. First, based on this measurement, a CF sorter could be designed that excites chlorophyll at 670 nm and measures the fluorescence around 690 nm, because this would result in the highest CF signal. However, using 690 nm for fluorescence measurement would mean that stray light from the laser would not be filtered out. At low fluorescence levels, a major part of the fluorescence signal is then due to reflected laser light. Second, at 690 nm, re-absorption of the emitted fluorescence light can occur by chlorophyll. At certain chlorophyll concentrations this could result in an increase in fluorescence signal with a decreasing amount of chlorophyll (Lichtenthaler, 1988). To avoid these two drawbacks, it was decided that the chlorophyll would be excited at 670 nm and the fluorescence would be measured at 730 nm. CF analysis and CF sorting is accomplished by using a linear vibrator to align the seeds. The linear vibrator feeds the seeds one by one to the optical CF unit, being made up of a laser diode and a lens system with an interference filter and photodiode (Fig. 12.19). The laser diode produces radiation with a wavelength of 670 nm that coincides with the peak absorption of chlorophyll *a*. The laser generates a laser spot of 1×8 mm^2 and 10 mW laser power at the end of the chute (Fig. 12.20). The chlorophyll fluorescence from the seeds is first captured by a lens and then filtered by an interference filter at 730 nm (half bandwidth of 10 nm) and then focused by a second lens on to a photodiode, which converts the fluorescence into an electrical signal. The used fluorescence measuring method is capable of measuring the low intensity of fluorescence in normal daylight conditions. This is accomplished by modulation of the laser light intensity at a fixed frequency and measuring the resulting fluorescence as an alternating photodiode signal at the same frequency. A lock-in amplifier is used as a very narrow electronic filter that suppresses background signals. The result is that the lock-in amplifier converts the alternating current of the

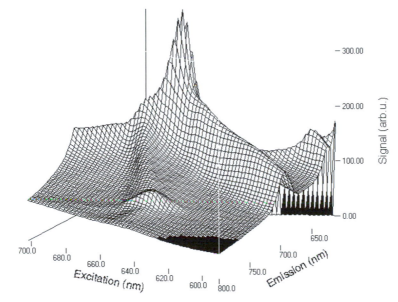

Fig. 12.18. 3-D fluorescence spectrum of green *Lisianthus* seed obtained with a spectrofluorophotometer (Shimadzu RF5000) that was automated with a PC and in-house developed software.

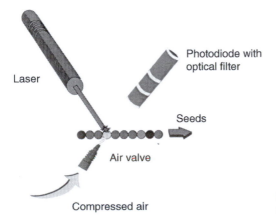

Fig. 12.19. Schematic arrangement of a CF measurement.

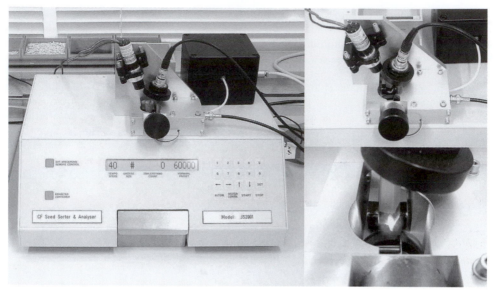

Fig. 12.20.   CF sorter equipment with detail of the optical CF unit and laser excitation of a seed at the end of the linear vibrator.

photodiode into a signal that is proportional to the fluorescence intensity (Stanford, 1996, 1997). This CF signal of the lock-in amplifier is fed into a computer either to obtain a frequency histogram, which describes the maturity distribution of the seed lot, or to control the ejector for seed sorting. For seed sorting, the ejector is controlled by the computer and ejects seeds by means of a pulse of air, depending on an adjustable threshold (cut-off) value of the CF signal called the sorting level. Different CF fractions are created by first sorting seeds with high CF values, with the cut-off value for ejecting seeds set at such a value that about 10–25% of the seeds are ejected. This results in two fractions. For the next fractions, the sorting cycle is repeated until the last two fractions are sorted with seeds with the lowest and nearest lowest CF signals.

Application and results of CF sorting on flower seeds

Flower seeds of verbena, pelargonium and pansy were sorted in different fractions using the CF-sorter. The result showed no clear influence of CF sorting on the maximum germination, G_{max}, and the normal seedling score, N (Table 12.2). However, a strong effect was found after a cd treatment (Powell and Matthews, 1981). Large differences were observed in G_{max} and N between the different CF fractions. The highest score for both the G_{max} and N was obtained for fraction 3 for verbena, fraction 2 and 3 for pelargonium and fraction 4 for pansy. The seeds of verbena and pelargonium of fraction 4 may have been over-mature and therefore of lower quality and seeds of fraction 1 of these three species were not fully mature. The general trend is that seeds gain their tolerance towards stress at a late phase of chlorophyll breakdown, therefore at the late phase of maturation. However, when the seeds are over-mature, the quality can also decline.

Many flower species contain chlorophyll: impatiens, primula, matthiola, verbena, pansy, pelargonium, etc. However, there is an exception: sunflower seeds show no measurable CF signal. It is not clear why some seed species show no CF signal, while other seed species do show CF signals.

In conclusion, seeds of high quality that were not subjected to a cd treatment could be improved slightly in G_{max} and N. After a cd treatment, the potential use of CF sorting became obvious and possible improvements

Table 12.2. Results on the germination and seedling evaluation of *Verbena*, *Pelargonium* and *Viola* seeds before (control) and after CF sorting (fraction 1–4). Germination and seedling quality of four replicates of 50 seeds of a total of 200 seeds were tested on moist filter paper at 20°C and a 16-h dark/8-h light cycle (ISTA, 1996). *Verbena* seeds were treated with KNO_3 (0.2%) at 10°C for 3 days for dormancy breaking. Seeds were visually inspected for root tip emergence. After 5 and 10 days the seedlings were evaluated and the number of normal seedlings, fresh seeds, dead seeds and abnormal seedlings were scored according to standard ISTA (1996) rules. The data were statistically analysed using Student's *t*-test (level of significance: $\alpha = 0.05$). Seed vigour was tested by a controlled deterioration (cd) test. The seeds were first equilibrated for 3 days at 20°C/85% RH in a cabinet with circulating air. Seeds were transferred to small aluminium foil packets, hermetically sealed and stored in a cabinet at 40°C for 14, 2 and 14 days for *Verbena*, *Viola* and *Pelargonium*, respectively. After the cd treatment the seeds were dried back for 3 days at 20°C/32% RH to their original moisture content.

	Control	Fraction 1 High CF	Fraction 2 Medium CF	Fraction 3 Low CF	Fraction 4 Very low CF
Verbena					
Fraction %	–	17.5	37.3	28.5	16.7
CF signal range (pA)	–	> 250	125–250	90–125	< 90
G_{max}	93.5 ± 1	95 ± 1	93.5 ± 1	93 ± 3	91 ± 4
N	91 ± 2	91.5 ± 2	89.5 ± 2	89.5 ± 2	88 ± 3
CD G_{max}	69.5 ± 6	47 ± 3[a]	76 ± 6	82 ± 2[a]	73 ± 4
CD N	52 ± 6	29.5 ± 4[a]	67 ± 8[a]	75 ± 2[a]	65 ± 2[a]
Pelargonium					
Fraction %	–	17.8	49.6	23.1	9.5
CF signal range [pA]	–	> 400	125–400	85–400	< 85
G_{max}	97.5 ± 1	97.5 ± 1	98 ± 2	99.5 ± 0.5[a]	100 ± 0[a]
N	82 ± 2	79 ± 1[a]	88.5 ± 2[a]	89 ± 3[a]	84 ± 1
CD G_{max}	76.5 ± 4	63 ± 3[a]	79 ± 7	83.5 ± 6	72.5 ± 3
CD N	56.5 ± 5	49 ± 3	61.5 ± 8	57 ± 7	43.5 ± 4[a]
Viola					
Fraction %	–	18.5	29.9	20.3	31.3
CF signal range [pA]	–	> 70	50–70	35–50	< 35
G_{max}	99.5 ± 0.5	98.0 ± 0.1[a]	98.5 ± 0.5[a]	99.5 ± 0.5	100 ± 0
N	98 ± 1	94.5 ± 0.5[a]	97.5 ± 1	98.0 ± 0.1	100 ± 0[a]
CD G_{max}	89 ± 4	65.5 ± 2[a]	86 ± 1	96 ± 2[a]	94.5 ± 1[a]
CD N	82 ± 4	54 ± 3[a]	74 ± 2[a]	85 ± 4	90 ± 2[a]

G_{max}, maximum germination; N, normal seedlings; CD, controlled deterioration treatment.
[a]Significantly different from the non-separated control at $\alpha = 0.05$.

for flower seeds can be high. This shows that CF sorting for vigour of a seed lot can be improved. It also indicates that seeds gain tolerance towards stress at the late phase of maturation. This is an important result of CF sorting: the seed lot can become more stress-tolerant. CF sorting can also be beneficial in removing seeds with the heaviest fungal infection. Furthermore, CF analysis and sorting can be used to monitor the quality of the seeds during storage and remove seeds which are deteriorated during storage. Advantages of the CF method for determining seed maturity and seed quality are the separation of

seeds based on their physiological status, the high sensitivity due to the high quantum yield of chlorophyll and its fluorescence nature, the method being non-destructive since it is an optical method, and the high speed at which the fluorescence is generated and measured (within 10^{-6} s).

New developments: hyper-spectral sorting

A spectral reflectance or transmittance measurement is analysed on its spectrum using several different wavelength intervals (see for review on seeds: Pasikatan and Dowell, 2001). Multi- and hyper-spectral seed sorting are differentiated according to their spectral domain. Multi refers to sorters having a small number of non-adjacent spectral wavelength intervals (typically four intervals) with low resolution (about $\Delta\lambda \approx 100$ nm), while hyper refers to sorters having a large number of adjacent wavelength intervals (typically 100–500 intervals with medium resolution – $\Delta\lambda \approx 1$–10 nm). An example of a multi-spectral sorter is the bichromatic or trichromatic sorter. With the availability of modern spectrophotometers, it is now possible to measure a large spectrum range (300–2100 nm) of the reflected or transmitted light of a seed within a few milliseconds. Using sophisticated software, the spectrum can be analysed and individual spectra of seeds compared. Such a hyper-spectral seed sorter would incorporate a fibre-optic spectrophotometer. This instrument can easily contain up to 2000 detectors in the form of a linear array detector. To measure a broad wavelength range, a broadband light source is needed, such as a halogen lamp. The light is directed on to the passing seeds. A part of the reflected light is captured and directed by a fibre to a spectrophotometer. In the spectrophotometer, the light is directed on to a grating. The angle at which the light leaves the grating is wavelength-dependent. By measuring the intensity of the light at different angles by multiple photodiodes the spectra can be simultaneously measured. This instrument replaces the different colour filters with a grating, which projects the different wavelength intervals on to the individual photodiodes of the linear array detector. For practical reasons, a linear array CCD is used instead of multiple photodiodes. This device couples a small detection area per photodiode with a high-speed readout of the signals. The advantages of these spectrophotometers compared with conventional spectrophotometers with rotating grating are the real-time capturing of the spectrum, its compact size and lower price. Optical fibre spectrophotometers can contain different wavelength ranges from UV, visible to near-infrared (300–2100 nm).

The hyper-spectral measurement

A prototype of a hyper-spectral seed sorter in the Vis/NIR region (400–1700 nm) was reported by Baker *et al.* (1999). It separated wheat kernels that were infested with parasitized rice weevils from kernels that were uninfested at a speed of 2 seeds/s. Another hyper-spectral seed sorter was developed in Wageningen, The Netherlands, at Plant Research International. Spectral analysis and spectral sorting are accomplished by using the same linear vibrator as the CF measurement to align the seeds. The linear vibrator feeds the seeds one by one to the optical spectral unit, being made up of a central single fibre to illuminate the seed and six fibres centred around the central fibre to capture the reflected light. A halogen lamp light source is connected to the central fibre and delivers white light. The reflected light from the seed is guided by the six fibres to the spectrophotometer. The measured spectrum ranges from 400 nm to 900 nm. The computer program algorithm detects whether the signal at a certain predetermined wavelength is changing and uses this to detect a single seed. At the signal maximum, the whole spectrum is captured. This measured spectrum is corrected for the spectrum of the halogen lamp, the spectral response of the spectrophotometer and the distance between the fibre and the seed. Initial results were obtained by measuring the average spectrum of normal light brown and discoloured *Cyclamen persicum* Mill. seeds (Fig. 12.21). These discoloured seeds must be removed from the seed lot, because they have a higher probability of having pathogens. In a sorting procedure, the normalized spectrum of a measured seed is compared with the average normalized spectrum of high-quality test seeds. Based on the difference between the two spectra, seeds can be sorted. This approach results in a sorted seed lot that resembles the spectrum of the test set within certain pre-set limits. The capacity of this equipment is about 5 seeds/s, depending on the vibratory feeder and the mechanism of seed separation. At higher speed than 5 seeds/s, the seeds are not fed to the optical measuring unit in such a way that they can be measured on a single seed basis. The electronics allow a speed of about 50 seeds/s, mainly limited by the used time constant of 20 ms for the measurement of a spectrum.

The Future of Seed Conditioning

The development of seed cleaning and grading methods began with mechanical separation of seed particles based on differences in their physical properties. It has evolved in recent years to include the application of optical techniques that enable detection of minute differences in seed composition. Advances in computer technology have greatly aided the speed and precision of new seed-cleaning equipment.

Mechanical seed separation techniques are very effective in cleaning out field debris from the crop seed because large differences in size, shape, weight or surface texture between the particles generally exist. These methods are also suitable for removal of broken and immature seeds that are usually smaller and lighter. The use of size and density separation as general tools to upgrade the germination performance of flower seed, however, is not always effective (Kwong, 1991). The major advantage of mechanical seed cleaning equipment discussed above is the speed of separation, allowing large quantities of seeds to be graded in a short time. Chlorophyll fluorescence seed separation is closely linked to seed maturity level. It can be a valuable tool in improving the overall seed vigour in flower seed lots, especially in cases when the entire crop is harvested at one time. The main drawback for this method at the moment is the low throughput of the operation. The speed of separation is limited not by the detection and separation devices *per se*, but by the slow feeding mechanism which conveys seeds in single file to the detector.

Effective seed cleaning and grading in the future will probably continue to be carefully designed routines that combine components of traditional mechanical techniques and newly developed methods such as optical separation. As commercial flower growers continue to demand high seed vigour, the search for better seed grading technologies is likely to continue. In order to deliver commercial flower seed lots with high physical and physiological uniformity, more research will be needed in two critical areas:

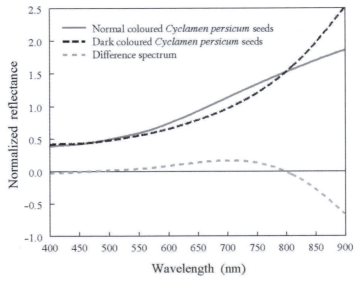

Fig. 12.21. Normalized spectra of normal light-brown coloured and dark-brown coloured *Cyclamen persicum* seeds. The spectra were obtained by averaging 100 spectra of individual seeds at a speed of 5 seeds/s.

1. Elimination of deteriorated seed. Current seed grading techniques mainly screen out immature seeds which are smaller, lighter or greener. A mature seed that deteriorates during postharvest handling and storage is not likely to be removed by the techniques outlined in this chapter. Studies in seed physiology are needed to identify key markers of seed deterioration. Application of the emerging multi-spectral sorting techniques could provide the working tool for seed separation based on these markers.

2. Efficient conveying system for single seed sorting. As mentioned before, computer technologies have made it possible to measure minute differences in seed content in less than 1 millisecond. However, current optical separations based on single seeds are still very slow (usually less than 10 seeds per second). This is due to the lack of an effective mechanism to convey the seeds to the equipment separator in single file. This is a particular problem for high-value flower seeds that are very small, e.g. petunia with 10,000 seeds per gram. Practical solutions for this problem can come from either building machines with multiple separators, or developing superfast, precision conveying devices. The former approach necessitates the use of very inexpensive optical detectors and sorters to make it commercially viable.

More than any other steps in bringing high quality, high value flower seeds to the commercial grower and home gardener, seed conditioning requires a multidisciplinary approach, involving both seed physiologists and experts in various engineering fields.

References

Baker, J.E., F.E. Dowell and J.E. Throne. 1999. Detection of parasitized rice weevils in wheat kernels with near-infrared spectroscopy. *Biological Control* 16:88–90.

Del'Aquilla, A., R. van der Schoor and H. Jalink. 2002. Application of chlorophyll fluorescence in sorting controlled deteriorated white cabbage (*Brassica oleracea* L.) seeds. *Seed Sci. Technol.* 30:689–695.

Harmond, J.E., R. Brandenburg and L.M. Klein. 1968. Mechanical seed cleaning and handling. *Agriculture Handbook No. 354.* USDA-ARS, Washington, DC.

Huyghe, C. 1993. Growth of white lupin seedlings during the rosette stage as affected by seed size. *Agronomie* 13:145–153

ISTA. 1996. International Seed Testing Association, International Rules for Seed Testing, Rules 1996, *Seed Sci. and Techol.* 24.

Konstantinova, P., R. van der Schoor, R. van den Bulk and H. Jalink. 2002. Chlorophyll fluorescence sorting as a method for improvement of barley (*Hordeum vulgare* L.) seed health and germination. *Seed Sci. Technol.* 30:411–421.

Jalink, H. 1996. US patent no.: 6,080,950 (priority date: 2 May, 1996, date of patent: 27 June, 2000). Method for determining the maturity and quality of seeds and an apparatus for sorting seeds.

Jalink, H., R. van der Schoor, A. Frandas, J.G. van Pijlen and R.J. Bino. 1998. Chlorophyll fluorescence of *Brassica oleracea* as a non-destructive marker for seed maturity and seed performance. *Seed Sci. Res.* 8:437–443.

Jalink, H., R. van der Schoor, Y.E. Birnbaum and R.J. Bino. 1999. Seed chlorophyll content as an indicator for seed maturity and seed quality. *Acta Hort. ISHS 1999*, 504, 219–227.

Jensen, K.M. 1987. Small scale cleaning machines. *In*: *ISTA Handbook for Cleaning of Agricultural and Horticultural Seeds on Small-scale Machines.* ISTA, Zurich. pp. 57–78

Kwong, F.Y. 1991 Research needs in the production of high quality seeds. *In*: J. Prakash and R.L.M. Pierik (eds). *Horticulture – New Technologies and Applications.* Kluwer Academic Publishers, Dordrecht, The Netherlands. pp. 13–20.

Lampeter, W. 1987. Grading properties of seeds. *In*: *ISTA Handbook for Cleaning Agricultural and Horticultural Seeds on Small-scale Machines.* ISTA, Zurich. pp. 8–40.

Lichtenthaler, H.K. 1988. *In vivo* chlorophyll fluorescence as a tool for stress detection in plants. *In*: H.K. Lichtenthaler (ed.). *Application of Chlorophyll Fluorescence in Photosynthesis*, Kluwer Academic Press, Dordrecht, The Netherlands. pp. 129–142.

Pasikatan, M.C. and F.E. Dowell. 2001. Sorting systems based on optical methods for detecting and removing seeds infested internally by insects or fungi: a review. *App. Spectro. Rev.* 36(4):399–416.

Powell, A.A. and S. Matthews. 1981. Evaluation of controlled deterioration, a new vigour test for small seeded vegetables. *Seed Sci. Technol.* 9:633–640.

Rideal, G. 1996. Absolute precision in particle size analysis. *American Laboratory* November:46–50

Schreiber, U. 1986. Detection of rapid induction kinetics with a new type of high frequency modulated chlorophyll fluorometer. *Photosynthesis Res.* 9:261–272.

Stanford Research Systems. 1996–1997. *Scientific and Engineering Instruments*. pp. 169–179.

Steckel, J.R.A., D. Gray and H.R. Rowse. 1989. Relationships between indices of seed maturity and carrot seed quality. *Ann. Appl. Biol.* 114:177–183.

Taylor, A.G., D.B. Churchill, S.S. Lee, D.M. Bilsland and T.M. Cooper. 1993. Color sorting of coated *Brassica* seeds by fluorescent sinapine leakage to improve germination. *Jour. Amer. Soc. Hort. Sci.* 118(4):551–556.

van der Burg, W.J. and R. Hendricks. 1980. Cleaning flower seeds. *Seed Sci. Technol.* 8: 505–522

Vaughan, C.E., B.R. Gregg and J.C. Delouche. 1968. Seed processing and handling. *Seed Technology Laboratory Handbook No. 1.* Mississippi State University, State College, Mississippi.

Wulff, R.D. 1986a. Seed size variation in *Desmodium paniculatum*. I. Factors affecting seed size. *Jour. Ecol.* 74:87–97

Wulff, R.D. 1986b. Seed size variation in *Desmodium paniculatum*. II. Effects on seedling growth and physiological performance. *Jour. Ecol.* 74: 99–114.

13 Flower Seed Priming, Pregermination, Pelleting and Coating

G. Tonko Bruggink
Research Leader Seed Technology Vegetables, Syngenta Seeds B.V.,
Westeinde 62, 1601 BK Enkhuizen, The Netherlands

Introduction

With the increased industrialization of young plant production, the importance of easy sowing and predictable, synchronized seedling emergence has grown. Unfortunately, seeds do not always perform in ways that enable successful young plant production under industrial conditions. The shape of seeds can be such that sowing is not easy, fungi can attack the germinating seed or emerging seedling, or germination can be too slow, too irregular or the final percentage of seeds germinating is too low.

Universities, institutes and seed companies have developed numerous treatments to improve seed characteristics. This chapter describes the backgrounds and benefits of such treatments. Issues considered will be:

- Priming – to improve speed and uniformity of germination, especially under adverse conditions.
- Pregermination – to obtain close to 100% usable plants, irrespective of the initial seed quality.
- Pelleting – to improve the sowability of seeds.
- Coating – to fight fungi during and shortly after germination.

Priming

Background

Although priming is known by many names, such as osmoconditioning, matriconditioning, liquid priming, the term 'priming' will be used here to describe the different germination-enhancing presowing treatments which do not result in radicle emergence. The term 'pregermination', which is sometimes used to describe priming treatments, is reserved for treatments that lead to radicle protrusion. The rationale behind priming is to have seeds complete the first steps in the germination process before they are actually sown. To do so, seeds are partially hydrated to a point where germination processes begin, but radicle emergence does not occur (Heydecker and Coolbear, 1977).

The idea is not new; it was probably practised in more primitive ways for thousands of years by immersing seeds in water for a period of time before they were sown (Evenari, 1980/81). This allowed easy imbibition, washing off inhibiting substances, and the initiation of the first steps in the germination process. Seeds treated in such a way germinated faster and more reliably and also had a higher percentage of seeds reaching the stage of a usable plant. It is generally agreed that priming also reduces the natural spread in germination from seed to seed, resulting in an increase in uniformity of the final product (Finch-Savage, 1995).

Commercial priming techniques

Commercial development of germination-enhancing treatments is of relatively recent date. The first work published in this field was from the early 1970s. Heydecker *et al.* (1973) described a method to incubate seeds on blotting paper moistened with polyethylene glycol (PEG). Later, methods to incubate seeds in aerated PEG solutions were described (Darby and Salter, 1976; Bujalski *et al.*, 1989), which enabled commercial application of the technique. In a similar way, solutions of inorganic salts have been used to create the desired osmotic potential (Cantliffe, 1981), or solutions of other substances like mannitol (Georghiou *et al.*, 1987). When incubated in a PEG or salt solution of sufficient concentration, seeds do not imbibe water to the same degree as they would in pure water (Fig. 13.1). When water is freely available, seeds generally imbibe until they reach a moisture level of around 50% on a fresh weight basis. Compare this

with dry seeds in storage, which have moisture contents generally between 5% and 10%. Under conditions of full imbibition, sooner or later germination will occur providing that temperature and aeration are sufficient and the seeds are not dormant. In order to enable elongation of the radicle, the seed takes up more water, leading to a higher moisture content.

In contrast, seeds incubated in PEG solution are restricted in their water uptake (Fig. 13.1). Seed moisture content rises to around 40%, though this value is species and PEG concentration dependent. Water uptake in the seeds is restricted because of the negative water potential of PEG. Consider this as a suction force exerted by the solution. In general, water potentials between −1.0 and −1.5 MPa are used. The water potential of the seeds will equilibrate with that of the solution surrounding it, resulting in a relatively low seed moisture content. Because of the lower moisture content, metabolic processes in the seed proceed at a lower rate than in water and emergence of the radicle is prevented. In this way, a priming treatment can proceed for a long period of time, e.g. several weeks, without the occurrence of radicle emergence.

Although PEG is often considered an osmoticum, it is really the matric potential exerted by PEG which prevents full imbibition of the seeds. Other osmotic treatments have been described in which salt solutions prevent full imbibition of the seeds. Priming in a solution offers the possibility of adding growth regulators to the solution. Finch-Savage (1991) described bulk priming for impatiens, primula, petunia and verbena. For petunia, priming alone without further additions appeared to be the best treatment. For primula and verbena, the addition of gibberellic acid gave the best results. For impatiens, the combination of gibberellic acid and benzyladenine gave the best performance. Plant growth regulators were more effective when added to the priming solution than when applied as a pre-soak.

Priming in PEG has several disadvantages. First, the PEG solution must be disposed of after treatment in an environmentally acceptable fashion. The quantity of PEG

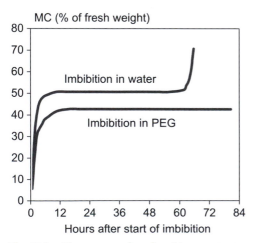

Fig. 13.1. Time course of seed moisture content during imbibition in water and imbibition in PEG solution.

used can be considerable: usually around 300 grams PEG in 1 litre of water. The amount of solution can be up to ten times the amount of seed being primed. Aeration of the seeds during the process can be a problem because of the low solubility of oxygen in PEG solution. Therefore, other ways of priming to obtain similar results have been explored.

A refinement of PEG priming is membrane priming, as described by Rowse et al. (2001). Seeds are separated by a membrane from the PEG solution which is contained within the walls of a double-walled cylinder. Advantages of this system are the good aeration of seeds, the reduction in use of PEG, and the possibility of priming small amounts of seeds. The improved aeration was especially beneficial for verbena, which has a high oxygen requirement and for mucilaginous seeds such as pansy (Rowse, 2001).

Another approach to ensuring the correct moisture content of seeds during the priming process is solid matrix priming. In this treatment, seeds are mixed with a moisture-containing 'matrix', e.g. peat (Taylor et al., 1988; Eastin, 1990; Khan, 1992). Generally, the seeds and the 'matrix' are contained in a drum which rotates around a horizontal axis. In this way an equal distribution of moisture is ensured. It is essential to keep the moisture content of the seed–matrix mixture at such a level that no germination can occur. At the end of the priming treatment, usually lasting several days, the contents are dried and the seeds separated from the solid matrix. During this process the addition of fungicides and growth regulators is possible. Instead of peat, many other moisture-retaining compounds have been identified. Examples are vermiculite, diatomaceous silica and expanded clay (Gray, 1994). An important practical consideration is the ease of separating seeds and matrix at the end of the process.

A comparable method is drum priming in which only seeds and water are used. Seeds are mixed with the right amount of water to raise their moisture content to the desired level for priming. This can be done in an essentially non-stirred situation (Wiebe and Tiessen, 1979) or seeds can be kept in a stirring motion in a drum to ensure a uniform distribution of moisture (Rowse, 1996).

The question of which type of treatment to use under specific conditions is not easily answered. As mentioned, the disposal of PEG solution is a limitation for PEG-based systems. Drum priming is not practical if only small quantities of seed are to be treated, as is often the case with flower seeds. The reason for this is that the control of seed moisture is generally carried out by weighing the container with the seeds. For small quantities, this is often inaccurate. In general, the method that is most practical under industrial conditions is chosen and different seed companies have different preferences.

Probably more important than the priming system chosen is the choice of variables such as seed moisture content (or water potential of the solution or matrix), temperature during treatment and duration of the treatment. In all of these cases it is essential that germination does not occur during treatment because this would reduce the survival of the seeds after drying. Choosing either the correct water potential/moisture content or duration of the priming treatment may prevent germination. An attempt at predicting the average time to germination (t_{50}) after a priming treatment, depending on the variables chosen during the treatment, was made by Bradford and Haigh (1994). Even with this work, determining optimal priming conditions remains largely an empirical process. Most methods described in the literature are carried out at temperatures which are more or less 'optimal' for germination of the species, usually between 15°C and 25°C. The water potential of the solution (or of the seeds) is usually in the range –1.0 to –2.0 MPa, and duration from several days to weeks, though usually not more than 2 weeks.

To obtain the maximum benefit of priming, seeds should be sown directly after treatment. However, this is not usually possible from a logistic point of view. Therefore, seeds are dried upon completion of the priming treatment. After a priming treatment in PEG solution, rinsing to remove adhering PEG should precede drying. Upon drying, most – but usually not all – of the advantages gained during priming are retained. Seeds generally germinate faster than before treatment, but not as fast as non-dried seeds directly after treatment.

Drying of seeds can be carried out in many different ways, usually by exposure to moving air of controlled temperature and relative humidity until seeds are back to the moisture content of untreated seeds. There are contradicting reports on the relevance of drying for retaining the maximum priming effect. Parera and Cantliffe (1994b) reported better performance for sweetcorn seeds dried at 30°C or 40°C after priming compared with those dried at 15°C or 20°C. Nascimento and West (2000) found no differences between muskmelon seeds dried at 18, 28 or 38°C after priming. In contrast, Brocklehurst and Dearman (1983) found for carrot, celery and onion seeds a faster germination after priming when seeds were dried at 15°C compared with 30°C. It seems, therefore, that the best conditions for drying are species-specific.

A potential disadvantage of primed seeds is their reduced shelf-life. Although the literature shows both positive and negative effects of priming treatments on seed storability (Parera and Cantliffe, 1994a), there is a general consensus that storability is reduced after priming, as was described by Bradford et al. (1993), which shows the relationship between germination rate and deterioration rate of seeds.

The reduction in shelf-life complicates the logistics of treating seeds. One solution is to require more accurate control of temperature and moisture than for untreated seeds. In general, storage of primed seeds for more than 1 year is not advisable, though this conclusion is species- and treatment-dependent. The reasons for the reduction in shelf-life are not precisely known. It may be that the protective mechanisms in a seed, which enable dry storage for many years, are partly degraded during the priming treatment. This makes sense since these mechanisms are no longer needed after germination.

In recent years, we have developed methods at Syngenta Seeds to improve the storability of primed seeds. After the priming treatment, but before final drying, seeds are exposed to additional temperature and moisture stress (Bruggink and van der Toorn, 1995; Gurusinghe et al., 2002). Although the mechanism leading to the improvements in shelf-life is unclear, it is likely that similar protective mechanisms are induced that were present before the priming treatment.

Physiology of priming

During a priming treatment, the first steps of the germination process are started or completed. These may include leakage of inhibiting substances, breakdown of reserves, build-up of enzymes necessary for endosperm breakdown, etc. For some species, morphological changes have been observed. In carrot, growth of the embryo occurs during priming (Gray et al., 1990), but this seems to be more the exception than the rule.

Surprisingly little is known about the processes occurring during priming or the physiological differences between primed and non-primed seeds. It remains largely unknown whether different processes take place during the first stages of germination as occur during priming. Recently it was shown (Gallardo et al., 2001) that priming of Arabidopsis seeds leads to synthesis and degradation of different proteins than occur during germination. This study compared osmotic priming with imbibition in water for such durations that caused the resultant speed of germination to be the same. The abundance of certain heat shock proteins (HSPs) increased during the priming treatment in PEG, whereas during incubation in water for 1 day, followed by drying, HSPs declined rapidly. The presence of these HSPs might ensure proper folding of other proteins because of their purported chaperone activity, and thus act in protecting the seeds. Catalase activity increased, especially during incubation in water, presumably to alleviate oxidative stress occurring during germination. With the advent of new techniques such as proteomics and cDNA arrays, research in the field of seed physiology will be greatly facilitated (van der Geest, 2002), and progress will be made in clarifying what exactly happens during a priming treatment.

Different explanations can be found for the changes in potential longevity of seeds due to a priming treatment. Two possible mechanisms are changes in the amount or composition of sugars and protective proteins in the seeds. The presence of specific sugars in seeds has been one of the purported mechanisms which enable seeds to survive in the dry state. Oligosaccharides in particular contribute to good storability of seeds (Obendorf, 1997). During germination and priming, these oligosaccharides are degraded and sugars like sucrose, fructose and glucose are formed. The fate of these sugars during priming has recently been studied in more detail for impatiens seeds (Buitink *et al.*, 2000; Gurusinghe and Bradford, 2001). In both studies, the content of the protective sugars declined during priming, whereas that of sucrose increased. Although the correlation between sugar composition and shelf-life was striking, no mechanistic relationship could be established. Buitink *et al.* (2000) used spin probe techniques to measure molecular mobility and glass transition temperature in the seeds. They established a clear correlation between these parameters and seed moisture content, explaining the negative effects of elevated seed moisture on shelf-life, but not between these parameters and sugar composition of the seeds. They concluded that the altered sugar composition of seeds was not responsible for the reduced longevity after priming.

Similarly, Gurusinghe and Bradford (2001) found that improvements in longevity as a result of post-priming treatments did not correlate with changes in sugar composition. They concluded that it was unlikely that changes in oligosaccharide content alone were responsible for the reduction in longevity due to priming or its restoration by post-priming treatments. The correlation between specific proteins and longevity of seeds before and after priming and post-priming treatments has also been studied by Gurusinghe *et al.* (2002). They found that changes in heat shock proteins did not consistently correlate with longevity after priming or post-priming treatments. However, the contents of BiP (binding protein) did correlate with longevity. This makes BiP an interesting candidate for further research on seed longevity in relation to priming treatments.

Benefits of primed seed

The perception of primed seeds is often that priming leads to higher percentages of usable plants. Although this may be true in certain cases, the biggest benefit is in the increased speed and uniformity of germination. There are also indications that the effect of a priming treatment is better when seed batches of good quality are used. This was shown for carrot and tomato seeds (Parera and Cantliffe, 1994a). In addition, seed companies would prefer not to spend the costs of priming on seedlots of poor or mediocre quality. So, in general, the use of primed seed lots results in a high percentage of usable plants.

Despite the differences in technologies for priming seeds, one thing is clear: primed seeds show faster germination than unprimed seeds. The advantages of this increase in speed of germination are also obvious to the bedding plant producer requiring a shorter period in the germination chamber and less space requirement. A shorter period until germination also means less risk of failure, and the success or failure of sowing is known earlier. However, priming is reported to have other advantages, such as an increase in the uniformity of the seedlings. Figure 13.2 shows two possible situations. In the first, germination is faster after treatment, but the slowest seeds are advanced more than the faster ones. In this way, germination becomes more uniform. In the second, all seeds are advanced by the same number of days, resulting in faster but not more uniform germination.

Rowse *et al.* (2001) showed that with membrane priming both situations occurred depending on the species. For primula, they obtained an increase in speed and uniformity after priming. For impatiens and verbena, a clear increase in speed of germination, but only a small increase in uniformity was observed. Petunia showed an increase in speed, no change in uniformity, and a decrease in total germination. For salvia, there was a decrease in speed, uniformity and final total germination. This clearly shows that optimal conditions must be defined for each

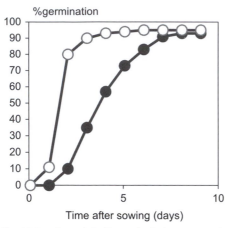 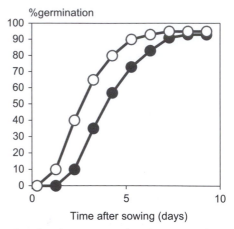

Fig. 13.2. Potential effects of priming on germination rate. Closed circles = untreated seeds; open circles = primed seeds. Left: germination after priming is faster and more uniform, the slowest seeds are advanced more than the faster ones. Right: germination is just faster after a priming treatment; all seeds are advanced in the same way.

species, though it remains unclear whether an increase in uniformity can be found for each species.

Another effect of priming is the better performance of seeds under adverse conditions. Seeds that show thermo-inhibition often benefit from a priming treatment. The best-known case is that of lettuce (Cantliffe *et al.*, 1984), where the upper temperature limit for germination was increased by a priming treatment. A similar, though less obvious, phenomenon is found in pansy seeds. Carpenter and Boucher (1991) showed that PEG-primed pansy seeds germinated 51% at 35°C vs. only 10% for untreated seeds. Similar results were found by Yoon *et al.* (1997), who primed pansy seeds in a range of salt solutions. Priming in $CaCl_2$ solution gave good germination at high temperature compared with priming in other salt solutions or in PEG.

In addition, under low temperature conditions, the effects of priming may be very pronounced, though this situation is more likely to occur in field crops than in flower crops. Because of the slow progress of germination under low temperatures, the absolute reduction in time to germination becomes more obvious, and threats to emergence, such as attack by fungi or dehydration, are reduced.

Pregerminated Seed

Background

Pregermination is a process that, in contrast to priming, allows seeds to develop until the emergence of a radicle, at which time the process is stopped. Figure 13.3 illustrates the time course of seed moisture content in this situation. The theoretical benefits of

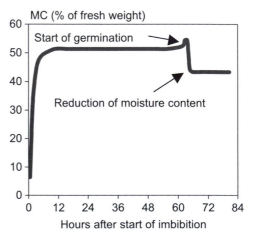

Fig. 13.3. Time course of seed moisture content during pregermination. At the start of germination the process is stopped and seeds are dried back slightly to prevent further radicle growth.

Fig. 13.4. Seedlings produced from untreated impatiens seeds (left) and from pregerminated and separated seeds (right).

pregermination are clear. Sowing seeds which have already germinated saves time because the germination process is completed. Ideally, the process allows part of the seed lot to germinate, and these seeds are collected while allowing the other seeds to develop until radicle protrusion. In this way, only the (most) viable seeds are selected, which are all at the same stage of development. In theory, close to 100% emergence and very high uniformity are possible. Figure 13.4 illustrates this for impatiens seeds.

Generally, the most viable seeds show the earliest germination. Finch-Savage (1986) showed for cauliflower, leek and onion that seeds which needed less time between imbibition and germination showed higher seedling length, less variation in length and fewer abnormal seedlings compared with slower-germinating seeds. Similar results with several flower species have been obtained (Fig. 13.5). Of course, non-germinating seeds are most easily eliminated in such a process. If it is possible to have all seeds at the same stage of development at the end of the process, the natural spread in germination times of seeds is eliminated, resulting in extremely high uniformity.

Pregermination was first developed for commercial use in the UK, mainly for application on field-sown vegetables such as cabbages, leek and carrots (Finch Savage, 1989; Finch-Savage and McQuistan, 1988; Finch-Savage and McKee, 1989). Commercialization of the process has been advanced by Syngenta Seeds, first for impatiens, followed by pansy, under the tradenames of PreNova and PreMagic.

Pregermination techniques and physiology

The easiest way to obtain pregerminated seeds is by placing seeds in aerated water for several days at a temperature conducive to germination. Once seeds have started to germinate, and supposing that not all seeds start germinating at the same time, there are two difficult problems to resolve. First, the germinated seeds need to be separated from the ungerminated seeds. Second, the pregerminated seeds need to be treated in such a way that they can be handled, stored and sown. Figure 13.6 shows examples of pregerminated seeds of impatiens and pansy.

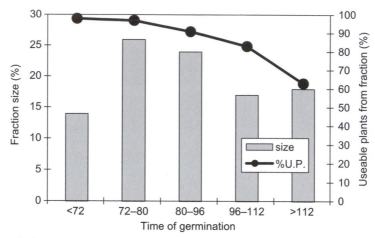

Fig. 13.5. Speed of germination and %UP for impatiens seeds. For seeds in subsequent germination speed classes, the size of the class and the resulting percentage of usable plants is shown (after Bruggink and van der Toorn, 1997).

Fig. 13.6. Pregerminated seeds of impatiens (left) and pansy (right).

Techniques for separation of seeds fall largely into the category of 'company secrets', though a method of separation based on density differences has been described. Taylor and Kenny (1985) separated a mixture of germinated and ungerminated *Brassica* seeds by placing them in a maltrin solution with density >1. Because of water uptake during germination, the specific density of seeds decreases, allowing them to float in densities where ungerminated seeds sink. In this way, a higher concentration of germinated seeds is obtained in the selected fractions. Other ways to separate pregerminated seeds can be envisaged. These can be based on the visible presence of a radicle, on differences in length, width, weight and other physical properties.

Another obstacle is storage of pregerminated seeds and finding a way of drilling them. Pregerminated seeds are extremely vulnerable, not only because they are easily mechanically damaged, but also because after germination seeds have lost their tolerance to desiccation. Attempts to store pregerminated seeds in water at temperatures near zero have been described (Brocklehurst and Dearman, 1980; Gray *et al.*, 1981) but did not enable storage for more than several days or weeks, complicating the logistics of such a product considerably. Initially, techniques for sowing such seeds relied on introducing the seeds into a gel, a process called 'fluid-drilling' technology. For seeds sown in rows in the field, this could be an acceptable solution. But for precision sowing of seeds in a tray, it is not suitable.

Approaches to store pregerminated seeds by lowering their moisture content have been described (Finch-Savage and McKee, 1989; Finch-Savage, 1989). When *Brassica* seeds were dried to a moisture content of 20–30%, their root growth stopped and they did not suffer desiccation damage and could be stored at low temperature for a considerable period of time (Finch-Savage and McQuistan, 1988). Our experience is that seeds treated in this way can still easily suffer desiccation damage during later handling and sowing (Bruggink and van der Toorn, 1997).

A breakthrough in this respect was the development of a method to reinduce desiccation tolerance in seeds after they germinated (Bruggink and van der Toorn, 1995). Exposing germinated seeds to specific temperature and moisture conditions led to an increase in desiccation tolerance. This increase was accompanied by a rise in sucrose content and the formation of certain protective proteins. More work in this field was done by Leprince *et al.* (2000), who focused on the downregulation of respiration during drying. This downregulation was different between germinated seeds with and without desiccation tolerance. Recently, it was shown by Buitink *et al.* (2003) that re-establishment of dessication tolerance in radicles of *Medicago* was correlated with the inclusion of MtDHN, a dehydrin. Nevertheless, a full understanding of the mechanisms leading to desiccation tolerance, both in germinated and ungerminated seeds, is still lacking.

Germinated seeds in which desiccation tolerance has been re-induced can either be stored at relatively high moisture contents or be desiccated to a moisture content similar to that of untreated seeds. In the first situation, seeds have a restricted shelf-life of only a few weeks. They need oxygen to respire and have to be stored cool, but their emergence is extremely rapid because they have to imbibe only a small amount of water before they can continue to grow. The presence of desiccation tolerance is, in this case, only relevant during sowing, where damage might occur if the seeds are exposed to air for a prolonged time. In contrast, if germinated seeds are completely desiccated, they can be treated in the same way as ordinary seeds in terms of storage and handling, provided that the radicles are sufficiently short to prevent mechanical damage. The presence or absence of desiccation tolerance can easily be shown by a tetrazolium test. Figure 13.7 shows tetrazolium tests of germinated and desiccated impatiens seeds, with and without induction of desiccation tolerance.

Benefits of pregerminated seed

The benefits of using pregerminated seeds are based on their rapid emergence, often making

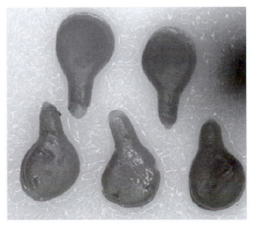

Fig. 13.7. Tetrazolium staining of impatiens seeds that have been either directly dried after germination (left), or that have been exposed to desiccation-tolerance-inducing conditions before desiccation (right).

the use of a germination chamber unnecessary. Because of the selection of germinated seeds in the same stage of development, their emergence is very uniform and a high percentage of seeds will develop into a usable seedling. Because the development of methods for obtaining pregerminated seeds is time-consuming, their availability is still restricted to only a few species.

Pelleting and Film Coating

Background

Pelleting and film coating are treatments which physiologically have little effect on seeds, but are mainly concerned with improving their sowability and with protection against pests. Butler (1993) defined pelleting as the application of a layer of inert materials that may obscure the original shape and size of the seed. This results in significant weight increase and improved plantability. For many seeds, automated planting would not be possible without pelleting. In contrast, in film coating, only a very thin layer is applied, usually containing crop protection agents. For flower seeds, this concerns only

fungicides. For vegetable seeds, insecticides may also be applied.

Pelleting

A pellet increases the seed weight considerably. As an example, begonia and petunia seeds have average seed weights of around 0.01 and 0.1 mg, respectively. In Syngenta Seeds, the pelleting process increases this weight to 0.9 and 1.2 mg, an almost 100-fold and more than tenfold increase (Fig. 13.8).

The build up of a pellet is achieved by placing a batch of seeds in a rotating drum or pan to which pelleting material and water are added at a determined rate (Fig. 13.9). At the end of the process the pelleted seeds are dried to enable storage.

The material used for pelleting can consist of any material that leads to a good build-up, results in a pellet with sufficient firmness for sowing, and afterwards does not negatively influence germination. A range of materials has been described for this purpose; pellets often consist of different layers. Among these materials are clay, diatomaceous earth and wood flour. Depending on the type of material, it may be necessary to add binders

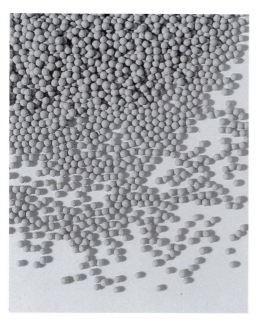

Fig. 13.8. The difference in size between petunia seeds that were pelleted (left) or film-coated (right).

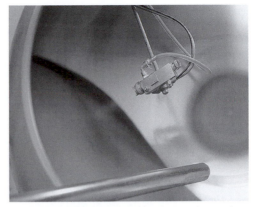

Fig. 13.9. Pelleting of flower seeds in a pelleting pan (left) and a detail of the addition of water during the process (right).

like methylcellulose. Apart from the type of material, the particle size also influences the final pellet characteristics because particle size influences the size of the capillaries necessary for water entry to the seed.

After sowing, it is important that the pellet does not negatively influence germination of the seeds. Germination could be retarded by reduced water uptake as a consequence of the presence of a pellet, or by the reduced entry of oxygen to the seed (Sachs *et al.*, 1981). A pellet that splits upon contact with water prevents such negative influences, but pellets that incorporate oxygen-liberating substances are also produced (Langan *et al.*, 1986).

A special type of pellet is the multipellet, which incorporates more than one seed in a pellet. This is especially useful for lobelia because more than one seed is sown per cell. Despite these developments, the application of new pelleting treatments remains empirical, and the successful production of pellets is an art.

Film coating

Coating of seeds with pesticides initially started as a means of protecting seeds from soil-borne fungi which cause pre-emergence and damping-off diseases, such as *Pythium*, *Phytophthora* and *Rhizoctonia*. This practice was first applied to seeds of cereals and later adopted for vegetables (Callan, 1975).

Maude (1978) made a distinction between application for short-term protection and for long-term protection. In the latter case, considerably higher dosages are applied, leading to protection during a whole growth season, e.g. in the case of white rot in onions. A next step in vegetables was the protection against insects during the growth period. Examples of this are chlorpyrifos coating on *Brassica* seeds, a protection against the cabbage root fly, and the imidacloprid coating of lettuce against aphids. It seems that application in the case of flower seeds generally focuses on short-term protection against damping-off diseases.

The requirements for a good seed treatment process with fungicides or insecticides were summarized by Elsworth and Harris (1973) and comprise:

1. The correct ratio of chemical to seed should be present.
2. Chemicals should be uniformly divided between the seeds.
3. Chemicals should adhere strongly enough to the seeds to avoid losses during handling.
4. Treatment and treated seeds should be safe for the operators working with them.
5. There should be no environmental pollution.

Before the advent of film coating of seeds, these requirements were not fulfilled. Seeds were often just mixed with fungicides to prevent emergence problems like 'damping-off'.

Thiram was, and in many countries still is, the main crop protection product used. Seeds treated this way were dusty, with associated health risks for those working with the seeds. The use of polymers to stick the active ingredients to the seed was a logical step, using equipment developed for the pharmaceutical industry. Today, a film coating is generally applied by spraying a liquid mixture of a formulated active ingredient, a colourant, a polymer and often also a filler. Powder application is also a possibility.

In most cases the application of the active ingredient is the main purpose of the film-coating treatment. The application of a colourant helps in improving the visibility of seeds after sowing. Flow characteristics of the seeds may also be improved due to a coating, allowing easier handling of the seeds during sowing.

Application of the solution on the seeds can be carried out in several ways. For high-volume crops, like cereals, generally continuous throughput equipment is used. For low-volume crops, like vegetables and flowers, batch treaters are used (Halmer, 1994), which can handle volumes of from 100 grams to several 100 kilograms.

The use of a fluidized bed coating machine is shown in Fig. 13.10, but the solution can also be sprayed on the seeds in a rotating drum. In both cases, dry air is forced through the system to allow continuous drying of the seeds. If the rate of application is in the same range as the rate of drying, the moisture content of the seeds will remain low during the entire process, seeds will not stick together, and no additional drying is needed afterwards. The amount of active ingredient applied is generally in the range of several grams per kilogram of seeds, though for insecticides considerably higher dosages can be used. Because of the proximity of the active ingredient to the seed, care should be taken that no phytotoxic effects result from the application.

Major obstacles in applying crop protection products to seeds are legal limitations. Each country has its own regulations concerning which active ingredients are allowed on seeds of different species.

The future of new developments in the application of pesticides on seeds will largely

Fig. 13.10. Equipment for film coating according to the fluidized bed method.

depend on the ability to obtain registrations. For flower seeds, the use of higher dosage fungicides for long-term protection could prove beneficial. In certain cases, the use of insecticides might be feasible, allowing long-term protection against aphids.

Conclusion

Priming, pelleting and film coating of seeds are technologies which have reached a level that allows large-scale commercial application for vegetable and flower seeds. The use of these treatments has led to increased efficiency in the production of young plants.

Academic research in the field of seed treatments is not as widespread as it was 10 or 20 years ago. Therefore, major technological breakthroughs affecting the use of seed treatments may not occur. For pregermination, the situation is different. This technology is relatively new and restricted to only a few species. Development of pregermination technology for more species might lead to further improvements in the number of usable plants that can be obtained from a batch of seeds.

Nevertheless, the increasing understanding of processes occurring during germination and treatment of seeds may in future lead to new ways to improve the performance of seeds by focusing on seed quality during breeding, defining better conditions during seed production, and developing more efficient seed treatments.

References

Bradford, K.J. and A.M. Haigh. 1994. Relationship between accumulated hydrothermal time during seed priming and subsequent seed germination rates. *Seed Sci. Res.* 4:63–69.

Bradford, K.J., A.M. Tarquis and J.M. Duran. 1993. A population based threshold model describing the relationship between germination rates and seed deterioration. *Jour. Exper. Bot.* 44: 1225–1234.

Brocklehurst, P.A. and J. Dearman. 1980. Effects of aeration during cold storage of germinated vegetable seeds prior to fluid drilling on seed viability. *Ann. Appl. Bot.* 95:261–266.

Brocklehurst, P.A. and J. Dearman. 1983. Interactions between seed priming treatments and nine seed lots of carrot, celery and onion. I. Laboratory germination. *Ann. Appl. Bot.* 102(3):577–584.

Bruggink, G.T. and P. van der Toorn. 1995. Induction of desiccation tolerance in germinated seeds. *Seed Sci. Res.* 5(1):1–4.

Bruggink, G.T. and P. van der Toorn. 1997. Induction of desiccation tolerance in germinated *Impatiens* seeds enables their practical use. Basic and applied aspects of seed biology. *Proc. 5th Int. Workshop on Seeds* 1997:461–467

Bruggink, G.T., J.J.J. Ooms and P. van der Toorn. 1999. Induction of longevity in primed seeds. *Seed Sci. Res.* 9(1):49–53

Buitink, J., M.A. Hemminga and F.A. Hoekstra. 2000. Is there a role for oligosaccharides in seed longevity? An assessment of intracellular glass stability. *Plant Physiol.* 122:1217–1224.

Buitink, J., B. Ly Vu, P. Satour and O. Leprince. 2003. The re-establishment of desiccation tolerance in germinated radicles of *Medicago truncata* Gaertn. seeds. *Seed Sci. Res.* 13:273–86.

Bujalski, W., A.W. Nienow and D. Gray. 1989. Establishing the large scale osmotic priming of onion seeds using enriched air. *Ann. Appl. Bot.* 115:171–176.

Butler, R. 1993. Coatings, films and treatments. *Seed World* October:19–24.

Callan, I.W. 1975. Achievements and limitations of seed treatments. *Outlook on Agriculture* 8: 271–274.

Cantliffe, D.J. 1981. Seed priming of lettuce for early and uniform emergence under conditions of environmental stress. *Acta Horticulturae* 122: 29–38.

Cantliffe, D.J., J.M. Fischer and T.A. Nell. 1984. Mechanism of seed priming in circumventing thermodormancy in lettuce. *Plant Physiol.* 75: 290–294.

Carpenter, W.J. and J.F. Boucher. 1991. Priming improves high-temperature germination of pansy seed. *HortSci.* 26:541–544.

Darby, R.J. and P.J. Salter. 1976. A technique for osmotically pre-treating and germinating quantities of small seeds. *Ann. Appl. Bot.* 83:313–315.

Eastin, J.A. 1990. Solid matrix priming of seeds. US Patent no. 4,912,874.

Elsworth, J.E. and D.A. Harris. 1973. The 'Rotostat' Seed Treater: a new application system. Proceedings of the 7th British Insecticide and Fungicide Conference. pp. 349–355.

Evenari, M. 1980/81. The history of germination research and the lesson it contains for today. *Israel Jour. Bot.* 29:4–21.

Finch-Savage, W.E. 1986. A study on the relationship between seedling characcters and rate of germination within a seed lot. *Ann. Appl. Bot.* 108: 441–444.

Finch-Savage, W.E. and C.I. McQuistan. 1988. The potential for newly germinated cabbage seed survival and storage at sub zero temperatures. *Ann. Bot.* 62:509–512.

Finch-Savage, W.E. and J.M.T. McKee. 1989. A study on the optimum drying conditions for cabbage seed following selection on the basis of a newly emerged radicle. *Ann. Appl. Bot.* 113: 415–424.

Finch-Savage, W.E. 1989. A comparison of Brussels sprout seedling establishment from ungerminated and low moisture content germinated seeds. *Ann. Appl. Bot.* 113:425–429.

Finch-Savage, W.E. 1991. Development of bulk priming plant growth regulator seed treatments and their effect on the seedling establishment of four bedding plant species. *Seed Sci. Technol.* 19(2):477–486.

Finch-Savage, W.E. 1995. Influence of seed quality on crop establishment, growth and yield. *In*: A.S. Basra (ed.). *Seed Quality: Basic Mechanisms and Agricultural Implications*. Food Products Press, New York.

Gallardo, K., C. Job, S.P.C. Groot, M. Puype, H. Demol, J. Vandekerckhove and D. Job. 2001. Proteomic analysis of *Arabidopsis* seed germination and priming. *Plant Physiol.* 126: 835–848.

Georghiou, K., C.A. Thanos and H.C. Passam. 1987. Osmoconditioning as a means of counteracting the aging of pepper seeds during high-temperature storage. *Ann. Bot.* 60:279–285.

Gray, D. 1994. Large scale seed priming techniques and their integration with crop protection treatments. *In:* T. Martin (ed.). *British Crop Protection Council Monograph. Seed Treatment: Progress and Prospects.* 57:353–362.

Gray, D., F. Tognoni and D. Bartlett. 1981. Fluid sowing of tomatoes: the effects of exposure of pregerminated tomato seeds to low temperatures on emergence and growth. *Jour. Hort. Sci.* 56:207–210.

Gray, D., J.R. Steckel and L.J. Hands. 1990. Responses of vegetable seeds to controlled hydration. *Ann. Bot.* 66:227–235.

Gurusinghe, S. and K.J. Bradford. 2001. Galactosyl-sucrose oligosaccharides and potential longevity of primed seeds. *Seed Sci. Res.* 11:121–133.

Gurusinghe, S., A.L.T. Powell and K.J. Bradford. 2002. Enhanced expression of BiP is associated with treatments that extend storage longevity of primed tomato seeds. *Jour. Amer. Soc. Hort. Sci.* 127:528:534.

Halmer, P. 1994. The development of quality seed treatments in commercial practice: objectives and achievements. *In:* T. Martin (ed.). *British Crop Protection Council Monograph. Seed Treatment: Progress and Prospects.* 57:363–374.

Heydecker, W. and Coolbear, P. 1977. Seed treatments for improving performance: survey and attempted prognosis. *Seed Sci. and Technol.* 5:353–425.

Heydecker, W., J. Higgins and R.J. Gulliver. 1973. Accelerated germination by osmotic seed treatment. *Nature* 246:42–44.

Khan, A.A. 1992. Preplant physiological seed conditioning. *Hort. Rev.* 13:131–181.

Langan, T.D., J.W. Pendleton and W.S. Oplinger. 1986. Peroxide coated seed emergence in water-saturated soil. *Ag. Jour.* 78:769–772.

Leprince, O., F.J.M. Harren, J. Buitink, M. Alberda and F.A. Hoekstra. 2000. Metabolic dysfunction and unabated respiration precede the loss of membrane integrity during dehydration of germinating radicles. *Plant Physiol.* 122(2): 597–608.

Maude, R.B. 1978. Vegetable seed treatments. *In:* K.A. Jeffs (ed.). *Seed Treatment. CIPAC Monograph 2.* Collaborative International Pesticides Analytical Council. pp. 91–101.

Nascimento, W.M. and S.H. West. 2000. Drying during muskmelon (*Cucumis melo* L.) seed priming and its effects on seed germination and deterioration. *Seed Sci. Technol.* 28(1): 211–215.

Obendorf, R.L. 1997. Oligosaccharides and galactosyl cyclitols in seed desiccation tolerance. *Seed Sci. Res.* 7(2):63–74.

Parera, C.A. and D.J. Cantliffe. 1994a. Presowing seed priming. *Hort. Rev.* 16:109–141.

Parera, C.A. and D.J. Cantliffe. 1994b. Dehydration rate after solid matrix priming alters seed performance of shrunken-2 corn. *Jour. Amer. Soc. Hort. Sci.* 119(3):629–635.

Rowse, H.R. 1996. Drum priming: a non-osmotic method of priming seeds. *Seed Sci. Technol.* 24:281–294

Rowse, H.R., J.M.T. McKee and W.E. Finch-Savage. 2001. Membrane-priming: a method for small samples of high value seeds. *Seed Sci. Technol.* 29:587–597.

Sachs, M., D.J. Cantliffe and T.A. Nell. 1981. Germination of clay-coated sweet pepper seeds. *Jour. Amer. Soc. Hort. Sci.* 106:385–389.

Taylor, A.G. and T.J. Kenny. 1985. Improvement of germinated seed quality by density separation. *Jour. Amer. Soc. Hort. Sci.* 110:347–349.

Taylor, A.G., D.E. Klien and T.H. Whitlow. 1988. SMP: Solid matrix priming of seeds. *Scientia Horticulturae* 37:1–11.

van der Geest, A.H.M. 2002. Seed genomics: germinating opportunities. *Seed Sci. Res.* 12: 145–153.

Wiebe, H.J. and H. Tiessen. 1979. Effects of different seed treatments on embryo growth and emergence of carrot seeds. *Gartenbauwissenschaft* 44:280–284.

Yoon B.H., H.J. Lang and B.G. Cobb. 1997. Priming with salt solutions improves germination of pansy seed at high temperatures. *HortSci.* 32(2):248–250.

14 Laboratory Germination Testing of Flower Seed

Marian Stephenson[1] and Jolan Mari[2]

[1]California Department of Food and Agriculture, 3294 Meadowview Road, Sacramento, CA 95832, USA; [2]PanAmerican Seed Company, 622 Town Road, West Chicago, IL 60185-2698, USA

Flowers have long been regarded as a speciality in the seed industry and the expertise in evaluation of flower seed quality has resided with relatively few individuals. Seed testing knowledge and techniques have been communicated within the speciality, but not widely published. The standard germination test is the most commonly used seed quality evaluation method. The basic principle in germination testing of seed is to provide optimum laboratory conditions for germination and subsequent seedling development. An estimate of germinative capacity is expressed as the percentage of seeds tested that develop seedlings capable of producing plants under favourable conditions.

Two major associations publish official procedures for laboratory testing of seed quality: the Association of Official Seed Analysts (AOSA) and the International Seed Testing Association (ISTA). These standardized testing procedures are the basis for documentation of seed quality in commerce. Seed companies, botanical gardens, and horticultural and weed societies provide additional germination methods for less common species. Compilations of germination testing procedures for many species not included in the AOSA *Rules for Testing Seeds* were published by Chirco and Turner (1986) and Chirco (1987); those procedures, with revisions, are now accessible on the AOSA website

(http://www.aosaseed.com/reference.html). Atwater's (1980) review of the relation of germination and dormancy patterns to seed morphology is useful in developing procedures for seeds of herbaceous ornamentals that do not germinate readily. Baskin and Baskin's (2001) synthesis of literature on the ecological basis of germination requirements and dormancy breaking includes guidelines for germinating 'seeds you know nothing about'.

Early efforts to incorporate methods for germination testing of flower seed into the laboratory protocols generally accepted in the USA (the AOSA *Rules for Testing Seeds* – hereafter, the Rules) met with inconclusive success. Procedures for 79 kinds were compiled by the AOSA Research and Methods Committee and presented for adoption in 1924 (Toole, 1925); methods for only 24 kinds were included in the 1937 Rules. The 1944 Rules omitted all mention of flower seed testing (Heit, 1958).

After the New York State Seed Law – which included tentative germination testing methods for many flower seed species – was presented at the AOSA national meeting in 1954, those methods were published in the Rules in an independent table, 'Tentative methods of testing of flower seed for laboratory germination and hard seed'. Some 110 kinds were included (AOSA, 1954). Commercial seed analysts involved in testing flower seed collaborated in comparison testing of the tentative methods and, in 1957, the Society of Commercial Seed Technologists (SCST) established a formal flower seed committee (Meyr, 1998).

In 1958, the amended methods became a permanent part of the Rules and AOSA formed a subcommittee for flower and herb seed (Meyr, 1998). By 1998 laboratory germination conditions for 272 kinds of flowers were incorporated into the AOSA Rules (Meyer, 1998).

The International Seed Testing Association (ISTA) adopted new rules for testing flower seeds for both germination and purity in 1977 (Meyr, 1998) and added germination test conditions for some 350 flower, spice, herb and medicinal species in 1985 (ISTA, 1985).

Detailed descriptions of the seedling structures essential to produce a normal plant under favourable conditions *and* of the defects to those structures which render a seedling incapable of continued development into a normal plant are necessary for the standardization of laboratory germination test results. Such descriptions and illustrations have been established for many agricultural and horticultural crops (AOSA, 2003b; Don, 2003). Guidelines for evaluating flower seedlings were not developed simultaneously with germination methods (Meyer, 1998). Lubbock (1892a,b) described and illustrated seedling structures of genera in 164 families, among them seedlings of some genera cultivated for their flowers. The third edition of the *ISTA Handbook for Seedling Evaluation* (Don, 2003) includes a number of flower genera, most of which fall into the generalized description for an epigeal dicot where the primary root is essential. Cyclamen, dianthus, freesia and helianthus seedlings are illustrated; photographs include normally developing seedlings and seedlings 'that lack the potential to develop into satisfactory plants' (Don, 2003). A collaborative ISTA–AOSA project in progress (Ripka, 2003) will produce a handbook on flower seed testing.

The 1992 AOSA *Seedling Evaluation Handbook* had no seedling descriptions or evaluation guidelines for 38 families of flower seeds included in the AOSA Rules (Meyer, 1998). Since that time, general descriptions

for the evaluation of flower seedlings in six families have been adopted by AOSA. Illustrated descriptions have been developed for three families (AOSA, 2003b).

The desirability of common international rules for analysis of seeds has long been recognized. In 1926, former AOSA President M.T. Munn attended the ISTA Executive Committee meeting in Copenhagen, Denmark, reportedly 'in the hope of obtaining agreement as to a proposal for common international rules for seed analyses' (Dorph-Petersen, 1926). Flower seed testing methods of the two rule-making bodies became the focus of comparative studies in 1979 (Meyr, 1998). At present, innumerable differences between the two sets of testing methods continue to confront seed analysts, scientists and regulatory officials striving to meet the challenges of a global market. Together the methods and explanatory notes presented in this chapter represent the amalgamation of practical experience and scientific research of more than two generations of flower seed testing professionals.

Germination Test Conditions

An official laboratory germination procedure for a particular kind of seed specifies the temperature at which to hold the seeds during germination; the type of substrate on which to germinate the seeds; light conditions during the test; moisture level to be maintained, if wetter or drier than typical; germination promoting treatments to apply to dormant seeds; and estimated duration of the test. Four hundred seeds are tested in a standard germination test (AOSA, 2003a; ISTA, 2003). In the case of seed mixtures, AOSA specifies a test of 200 seeds of any kind constituting 15% or less of the mixture.

Temperature

The seeds of many cultivated species will germinate over a fairly broad range of temperatures. In principle, the temperature specified in an official procedure for a standard germination test is the optimum temperature. That is, germination of the species being tested occurs most rapidly at the specified temperature. In fact, more than one temperature regime is listed for many species. In part, the use of alternative temperatures is permitted to accommodate efficient operation of laboratories; the more seed kinds that can be tested at the same temperature, the fewer germination chambers are required (Ashton, 2001). For example, the AOSA Rules permit testing of *Centranthus ruber*, Jupiter's-beard, at either a constant temperature of 15°C or 20°C or at an alternating temperature regime of 15°C–25°C. The order in which alternative temperatures are listed does not indicating priority; any of those listed may be used. The temperatures most frequently used in flower seed germination testing and the Fahrenheit equivalents are listed in Table 14.1.

AOSA specifies that temperature variation due to the germination apparatus must not exceed ±1°C; ISTA permits variation of ±2°C. ISTA Rules note that the prescribed temperatures are those the seed is exposed to on or inside test media. It is important to continuously monitor the temperature inside germinators and to calibrate and regularly check the accuracy of temperature chart recorders.

Substrates

Laboratory germination tests for the flowers discussed here are conducted either on paper products specifically manufactured for seed testing or in sand. Ashton (2001) described the types of paper substrate commonly used – paper towelling, blotters, pleated paper, creped cellulose and filter paper – and discussed the physical qualities of sand suitable for use.

AOSA Rules do not specify qualities of the paper products used in seed testing, other than to prescribe check tests if a paper

substrate appears to be toxic to developing seedlings. Particular types to be used, e.g. blotter or towel, are specified for each kind of seed. Figures 14.1 and 14.2 illustrate two types of germination substrate.

ISTA Rules permit the use of paper substrates including filter paper, blotters and towels and do not specify a particular type of paper product for each species tested (see Fig. 14.3). ISTA also provides general specifications for the composition of the paper substrate, its texture, strength, moisture capacity, pH, storage and sterilization, and also warns of the possibility of the presence of phytotoxic substances in paper products and how to evaluate the paper using sensitive species.

Moisture

Water is used to moisten substrates except where otherwise noted. Tap water is generally used; distilled or deionized water should be used where water quality is not satisfactory (i.e. unacceptable contamination by organic or inorganic impurities). Some laboratories routinely use distilled or deionized water as a means of further reducing variability in test procedures. Solutions of potassium nitrate (KNO_3) or gibberellic acid (GA_3), where prescribed, are always made using distilled or deionized water.

Particular attention must be given to the level of moisture in the test substrate at the start of a germination test. Because the optimum amount of water to be applied to the substrate varies with the species and the substrate used, AOSA and ISTA Rules provide only general directions for judging the initial moisture level. The substrate must be moist enough to supply needed moisture to the seeds at all times. This requires taking the precaution that the substrate cannot dry out during the test (e.g. enclosing the test in a container and ensuring adequate humidity within the germination chamber). Monitoring the moisture level of the substrate between sowing time and first count is critical. ('First count' refers to the number of days after initiation of germination when the first fully developed seedlings are evaluated.) Since excessive moisture will restrict aeration

Table 14.1. Temperatures commonly used in laboratory germination testing.

°C	5	10	15	20	25	20–30
°F	41	50	59	68	77	68–86

Fig. 14.1. Blotters and paper towels manufactured specifically for seed germination testing are two substrates commonly used in testing flower seed. Placement of the seed on the substrate varies with seed size and the number of seeds in each replicate. A test of *Gomphrena* sp. might consist of four replicates of 100 seeds each, each set of 100 seeds placed on top of two blue blotters in a covered plastic box.

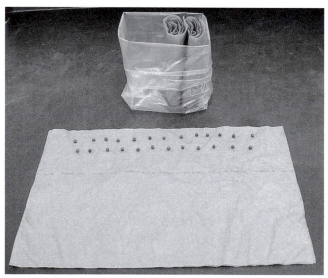

Fig. 14.2. Arrangement of seeds on the substrate is not prescribed. *Lathyrus odoratus*, sweet-pea, might be tested as 16 replicates of 25 seeds each. In this example, seeds have been placed in two rows along the long edge of two paper towels. The towels will be folded so that the lower edge meets the upper, then folded horizontally to form a roll. The manner of folding must be such that the seeds remain well spaced and do not fall to the bottom of the roll, and yet are not so tightly bound that seedling growth is restricted.

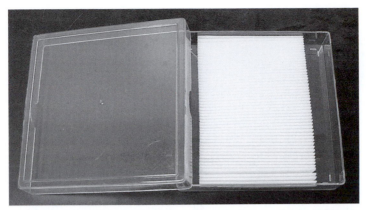

Fig. 14.3. Pleated paper, a strip of paper folded accordion-style, is recognized in ISTA Rules as an alternative substrate whenever BP (between paper) or TP (top of paper) is specified. The pleated paper has 50 folds; common procedure is to place two seeds in each trough, one at either end, to produce a replicate of 100 seeds.

of the seeds, blotters or other paper substrates should not be so wet that when pressed with a fingertip a film of water forms around the finger. When 'moisture on the dry side' is specified in a test procedure, the moistened substrate should be pressed against a dry absorbent surface, such as a dry paper towel or blotter, to remove excess moisture before placing seed on the substrate.

Light/dark

Where prescribed in AOSA procedures, light should be provided by a cool white fluorescent source and seeds should be germinated on top of the substrate. Illuminance for non-dormant seed and during seedling development may be as low as 25 foot-candles (ft-c). Seeds should be illuminated for 8 h in every 24. Where alternating temperatures are used, the period of illumination should be during the high temperature period. Illuminance for dormant seed should be 75–125 ft-c (750–1250 lux) (AOSA, 2003a). Two germination chambers equipped with fluorescent tubes are shown in Figs 14.4 and 14.5. When light is applied during a germination test, developing seedlings will orientate toward the source of illumination (phototropism). If not rotated during the test, the seedlings in the chamber with side lighting (Fig. 14.4) will 'lean' to one side (toward the light).

Cotyledons may not spread. With overhead lighting (Fig. 14.5) seedlings will grow in a more upright position. Distance of the seedlings from the light affects the degree of elongation of the seedling. Placing tests on a slant, rather than flat, during germination makes seedling evaluation easier, since roots will orientate downward on the surface of the substrate, but not *into* the substrate.

ISTA Rules note that seeds of most of the species for which germination methods are provided will germinate in either light or darkness (see Fig. 14.6). However, illumination from an artificial source or natural light produces better developed seedlings, which are more easily evaluated. Where either light or dark is specifically required for germination, it is indicated (ISTA, 2003).

First count, final count (days)

The 'first count' refers to the approximate number of days after initiation of a germination test when some seedlings may have reached a stage of development permitting accurate evaluation; seedlings having the essential structures necessary for further development may be removed from the test (and recorded). The number of days stated for the first count is approximate. AOSA permits a deviation of 1–3 days; ISTA Rules specify that the time to first count must be

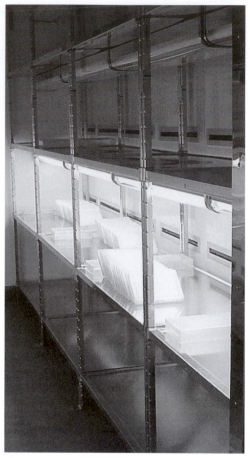

Fig. 14.5. In walk-in germination chambers, lights may be installed on the walls or overhead.

Fig. 14.4. In freestanding germination units, light panels are situated on the sides, back and/or on the inside of the door.

sufficient to permit the seedlings to reach a stage of development which permits accurate evaluation. Furthermore, preliminary counts of flower seedlings are suggested only when necessary or advisable for efficiency. Many flower seedlings can be judged more accurately and critically if left for the full duration of the test, especially in tests of 7–14 days (AOSA, 2003a).

Both AOSA and ISTA provide for the termination of a test prior to the number of days listed for final count if the analyst is positive that the maximum germination of the sample

has been attained. If, at the end of the prescribed test period, some seedlings are not sufficiently developed for positive evaluation, AOSA permits a 2-day extension of the test. ISTA Rules permit greater extensions of tests: if, at the time of the final count, some seeds have just begun to germinate, the prescribed test period may be extended by 7 days or up to half the prescribed period for longer tests.

Dormancy-breaking treatments

Seeds that remain ungerminated at the end of a standard germination test may be dead, 'hard' – unable to imbibe moisture due to an impermeable seed coat, or dormant.

Fig. 14.6. The appearance of seedlings is affected by the presence or absence of light, its intensity and duration, during germination and development. Emerging seedlings of *Centaurea cyanus* after 54 h in dark (left) and in light (right).

It is the convention to report the percentage hard seed remaining at the end of a germination test. Percentage germination and percentage hard seed are both listed, separately, on labels of seed in commerce. If determination of the viability of hard seed is necessary or desirable, seeds can be scarified (the seed coat nicked to permit entry of moisture – see Fig. 14.7) and re-tested at the normal temperature for the species being tested.

Flowers that may have hard seeds include those in the families Fabaceae (e.g. *Lathryrus odoratus*, sweet-pea; *Lupinus* spp., lupin), Geraniaceae (e.g. *Pelargonium* spp., geranium), and Malvaceae (e.g. *Alcea rosea*, hollyhock).

'Dormant' and 'fresh' are terms used to describe seeds that have imbibed moisture, are viable, but do not germinate under the moisture, temperature and light conditions prescribed for the particular kind of seed. The physical and physiological causes of dormancy are discussed elsewhere in this book, as is the use of tetrazolium chloride staining to determine the viability of firm, ungerminated seed at the end of a germination test.

Many garden flower seeds exhibit 'non-deep physiological dormancy' (Baskin and Baskin, 2001). Relatively short periods of cold stratification (prechill) and use of dilute potassium nitrate solution (KNO_3) are common dormancy-breaking techniques used in laboratory seed testing.

Prechill treatment consists of placing seeds in contact with the moist substrate and holding at a low temperature for a prescribed period before placing at the temperatures specified for germination of non-dormant seeds. The temperature to be used and the duration of prechill treatments are specified for each kind where prescribed in the AOSA Rules. ISTA Rules specify that the temperature for prechill be between 5°C and 10°C for up to 7 days and that the duration of the prechill period may be extended or the test may be rechilled after a period of incubation

Fig. 14.7. The impermeable seed coat of hard seeds may be slightly nicked to permit imbibition. An appropriate point to chip the seed coat of *Lupinus* sp., lupin, above, is the edge of the cotyledons (left). Slight damage to the cotyledons will not prevent normal seedling development. Nicking the seed coat overlying critical structures of the embryo – the radicle–hypocotyl juncture (centre) or the radicle (right) – must be avoided.

under the conditions prescribed for non-dormant seeds.

When the use of KNO_3 is specified, a two-tenths per cent (0.2%) solution is used in place of water to moisten the substrate.

Seedling evaluation

Evaluation of seedlings is the most difficult aspect of seed germination testing to standardize. Figures 14.8 and 14.9 illustrate two common types of seedling development.

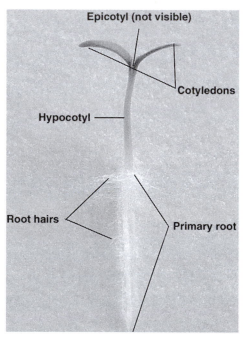

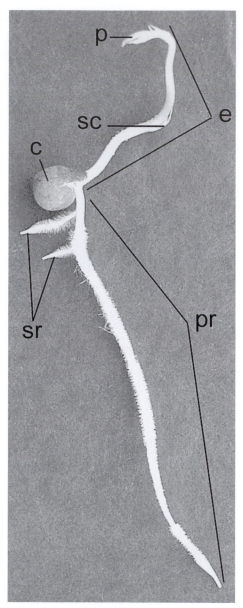

Fig. 14.8. Many annual garden flowers are epigeal dicots, e.g. *Amaranthus* sp. The embryos have two cotyledons, the cotyledons rise above the soil as the seed germinates and the hypocotyl elongates. The leaf-like cotyledons become photosynthetic upon exposure to light. The epicotyl, the part of the seedling above the cotyledons, is protected between the cotyledons as the seedling moves through the soil and is usually not visible at the stage of development at which seedlings are evaluated in laboratory germination tests. If the cotyledons are damage-free at the point of attachment to the hypocotyl, the epicotyl is considered to be undamaged. The seedling size at time of evaluation in a laboratory germination test may range from 2 mm length (e.g. begonia) to 10 cm (e.g. annual dahlia).

Fig. 14.9. In hypogeal dicots, such as *Lathyrus odoratus*, sweet-pea, the cotyledons remain below the surface during seedling development and it is the epicotyl that elongates to emerge above the soil surface. The epicotyl is hooked as it pushes through the soil, protecting the primary leaves and terminal bud, straightening only after emergence. The hypocotyl does not elongate significantly. **c** – cotyledons (enclosed in seed coat); **e** – epicotyl; **p** – developing primary leaf; **pr** – primary root; **sc** – scale leaf; **sr** – secondary roots.

Table 14.2.   Explanation of abbreviations and conventions.

Temperature	Single numbers indicate constant temperatures, e.g. 20°C.
	Two numbers separated by a dash indicate alternating temperatures. In a 24-h cycle, a test is held at the lower temperature for 16 h, at the higher one for 8 h, e.g. 20–30°C
Substrates	BP = between paper – ISTA
	P = covered Petri dishes with either:
	two layers of blotters,
	three thicknesses of filter paper, or,
	top of sand – AOSA
	T = paper towelling, used either as folded towel tests or as rolled towel tests in horizontal or vertical position – AOSA
	TB = top of blotters – AOSA
	TP = top of paper – ISTA
Dormancy-breaking treatments	ISTA Rules for Seed Testing (Table 5A, ISTA, 2003) do not distinguish between treatments recommended for breaking seed dormancy and 'additional directions' required for standard germination tests – that is, conditions other than specified temperatures, substrate types and test durations. In the summaries of procedures in this chapter, use of KNO_3 to moisten substrates is noted with 'Substrates' and light requirement is specifically noted ('Light'); both are included under the heading 'Dormancy-breaking treatments' as well. Prechill treatment is noted under 'Dormancy-breaking treatments'.

Descriptions of 'types' of seedling development are insufficient as guidelines for seed analysts, who must examine carefully each seedling in a test and determine whether it is capable of developing into a plant 'under favourable conditions'. The analyst first must be able to recognize malformations or injuries that are the result of test conditions, and not inherent to the seed. When defects do occur, the analyst must assess their impact on continuing development of the seedling. Plates 14.1–14.24 illustrate seedlings of 24 genera and examples of deviations from normal development. Descriptions of the illustrations are included in the text relevant to each genus. Ongoing efforts toward development of detailed guidelines for evaluation of flower seedlings have been described above.

On the following pages the germination procedures and seedling characteristics for 24 flower genera of horticultural importance are summarized. See Table 14.2 for explanation of abbreviations and conventions used to describe germination procedures. The scientific names of the genera and species are those appearing in the 2003 AOSA and ISTA Rules. Common names associated with the species are those recognized in AOSA Handbook 25: *Uniform Classification of Weed and Crop Seeds*, and by the United States Department of Agriculture Germplasm Resources Information Network (http://www.ars-grin.gov/npgs/tax/index.html).

Alstroemeriaceae

Germination conditions: **Alstroemeria** × *hybrida*, alstroemeria, Peruvian lily, lis de Incas, goldene Inkalilie[1]

Temperature:	15°C – JMari[2]
Substrates:	T – JMari[2]
Light/dark:	Not specified
First count, final count (days):	28, 42 – JMari[2]
Comments:	Pour 55°C water over seeds and let soak for 24 h – JMari[2]
Dormancy-breaking treatments:	None
Evaluation guidelines:	**AOSA** – The AOSA *Seedling Evaluation Handbook* (AOSA, 2003b) does not include descriptions of seedlings of Alstroemeriaceae; the family is not included in descriptions for 'miscellaneous agricultural and horticultural' kinds.
	ISTA – The *ISTA Handbook for Seedling Evaluation* (Don, 2003) does not include *Alstroemeria* spp.
	A hypogeal monocot, with a well-developed primary root, a compact cotyledon and a primary shoot with elongated internodes and scale leaves developing before foliar leaves (adapted from Tillich, 1995).
Illustrations:	**Plate 14.1 Alstroemeriaceae** *Alstroemeria*: **A1–A2**. A 28-day-old seedling germinated in rolled towels has a long primary root and an elongated primary shoot. Close examination is required to confirm that foliar leaves are developing. **B**. A 37-day-old seedling grown on top of blotters under light has shorter internodes and an expanded photosynthetic foliar leaf. **C**. A scale leaf is the first visible development of the shoot; seedlings must demonstrate full shoot development to be considered capable of producing a plant. **D**. Upon examination, the shoot of this seedling is found to lack developing foliar leaves. The seedling is not capable of continued normal development. *Note*: the rate of development among seedlings is not uniform. Seedling sizes are not comparable due to varying degrees of magnification.

[1]*Alstroemeria* spp. not included in AOSA (2003a) or ISTA (2003).
[2]J. Mari, personal observation.

Amaranthaceae
Amaranthus spp., *Celosia* spp., *Gomphrena* sp.

Germination conditions: **Celosia argentea**, feather cockscomb, red-spinach, Silber-Brandschopf; **C. argentea** 'Childsii', cockscomb, celosia; **C. argentea** var. **cristata**, cockscomb, amarante crête de coq, célosie crête de coq, Hahnenkamm, keito, celósia-branca, crista-de-galo, borlón, cresta de gallo; **C. argentea** 'Thompsonii', cockscomb, celosia

Temperature:	20°C	*C. argentea* – ISTA
	20–30°C	*C. argentea* – ISTA *C. argentea* 'Childsii', *C. argentea* 'Thompsonii', *C. argentea* var. *cristata* – AOSA

Substrates:	P – AOSA TP – ISTA
Light/dark:	Light – AOSA Not specified – ISTA
First count, final count (days):	None, 8 – AOSA 3–5, 14 – ISTA
Comments:	Sensitive to drying in test – AOSA
Dormancy-breaking treatments:	Prechill – ISTA
Evaluation guidelines:	**AOSA** – The AOSA *Seedling Evaluation Handbook* (AOSA, 2003b) does not include descriptions of seedlings of Amaranthaceae; the family is not included in descriptions for 'miscellaneous agricultural and horticultural' kinds.
	ISTA – A dicotyledon with epigeal germination. The shoot system consists of an elongated hypocotyl and two cotyledons with the terminal bud lying between. There is no epicotyl elongation within the test period and the epicotyl and terminal bud are usually not discernible. The root system consists of a primary root, usually with root hairs, which must be well developed, as secondary roots cannot be taken into account.
Illustrations:	**Plate 14.2 Amaranthaceae** *Celosia*: **A**. A fully developed seedling has a long primary root with root hairs, a long hypocotyl and lanceolate cotyledons. **B**. Two seedlings may develop from one seed. Examined closely, these two are found to be entirely separate, each capable of continued development. Only one seedling is counted in calculating the germination percentage. **C1**. It is not clear how much root development has occurred in this seedling. **C2–C3**. Removal of the seed coat and careful uncoiling reveal that the primary root is developing normally; its tip has emerged from the seed coat. Roots trapped in tough seed coats are less likely to occur when tests are placed on a slant rather than flat during germination. The friction of surrounding medium (as in a sand test) is also likely to reduce the incidence of trapped roots. *Note*: seedling sizes are not comparable due to varying degrees of magnification.

Apocynaceae
Catharanthus sp.

Germination conditions: ***Catharanthus roseus***, rose periwinkle, vinca, Cape periwinkle, Madagascar periwinkle, old-maid, rose periwinkle, rosy periwinkle, pervenche de Madagascar, chatas, chula, pervinca de Madagascar[1]

Temperature:	20–30°C – AOSA
Substrates:	B, TB – AOSA
Light/dark:	Light – AOSA
First count, final count (days):	6, 23 – AOSA
Comments:	Maintain good moisture supply – AOSA

Dormancy-breaking treatments:	None

Evaluation guidelines:	**AOSA** – The AOSA *Seedling Evaluation Handbook* (AOSA, 2003b) does not include descriptions of seedlings of Apocynaceae; the family is not included in descriptions for 'miscellaneous agricultural and horticultural' kinds.
	ISTA – The *ISTA Handbook for Seedling Evaluation* (Don, 2003) does not include *Catharanthus* spp.
Illustrations:	**Plate 14.3 Apocynaceae** *Catharanthus*: **A1–A2**. A fully developed seedling has a long primary root with root hairs, an elongated hypocotyl and two oblong cotyledons. **B**. A seedling with a stubby root is incapable of continued development to produce a plant whether or not the cotyledons are intact. *Note*: seedling sizes are not comparable due to varying degrees of magnification.

[1]*Catharanthus* sp. not included in ISTA (2003).

Asteraceae

Achillea spp., *Ageratum* spp., *Amberboa* sp., *Anaphalis* sp., *Anthemis* spp., *Arctotis* spp., *Aster* spp., *Baileya* sp., *Balsamorhiza* sp., *Bellis* sp., *Brachycome* sp., *Buphthalmum* sp., *Calendula* sp., *Callistephus* sp., *Carthamus* sp., *Centaurea* spp., *Coreopsis* spp., *Cosmos* spp., *Dahlia* spp., *Dimorphotheca* spp., *Doronicum* spp., *Echinacea* spp., *Echinops* sp., *Erigeron* sp., *Gaillardia* spp., *Gazania* sp., *Glebionis* spp., *Helenium* spp., *Helianthus* spp., *Heliomeris* sp., *Heliopsis* spp., *Inula* sp., *Layia* sp., *Leontopodium* sp., *Liatris* spp., *Machaeranthera* spp., *Matricaria* spp., *Pericallis* sp., *Ratibida* spp., *Rudbeckia* spp., *Sanvitalia* sp., *Tanacetum* spp., *Tithonia* sp., *Xerochrysum* sp., *Zinnia* spp.

Germination conditions: ***Dahlia*** spp., dahlia; ***Dahlia pinnata***

Temperature:	15°C	*D. pinnata* – ISTA
		Dahlia spp. – AOSA
	20°C	*D. pinnata* – ISTA
	20–30°C	*D. pinnata* – ISTA
Substrates:	B, T, TB, – AOSA	
	BP, TP – ISTA	
Light/dark:	Not specified	
First count, final count (days):	4, 14 – AOSA	
	4–7, 21 – ISTA	
Comments:	Sensitive to drying in test	*Dahlia* spp – AOSA
Dormancy-breaking treatments:	Prechill	*D. pinnata* – ISTA
Evaluation guidelines:	**AOSA** – (Generalized description for all Asteraceae other than lettuce; *Carthamus tinctorius* and *Helianthus annuus* are the only flower species in Asteraceae for which guidelines are specified.) Epigeal dicot with cotyledons which expand and become thin, leaf-like and photosynthetic; a hypocotyl that elongates and carries the cotyledons above the soil surface; and a long primary root with secondary roots usually developing within the test period. The epicotyl usually does not show any development within the test period. It is acceptable for the primary root to be weak or stubby if sufficient secondary or adventitious roots have developed.	

ISTA – A dicotyledon with epigeal germination. The shoot system consists of an elongated hypocotyl and two cotyledons with the terminal bud lying between. There is no epicotyl elongation within the test period and the epicotyl and terminal bud are usually not discernible. The root system consists of a primary root, usually with root hairs, which must be well developed, as secondary roots cannot be taken into account.

Illustrations:

Plate 14.4 Asteraceae *Dahlia*: **A.** A fully developed seedling has a long primary root with root hairs, a short hypocotyl and spatulate cotyledons. **B.** This seedling lacks a primary root. **C.** Close examination reveals the stunted primary root and the point of origin of roots emerging above it. *Note*: seedling sizes not comparable due to varying degrees of magnification.

Germination conditions: ***Senecio cineraria***, dusty-miller; ***Senecio cruentus*** [syn. ***Pericallis cruenta***], common cineraria; ***Senecio elegans***, purple ragwort

Temperature:	20°C	*S. cineraria, S. cruentus, S. elegans* – ISTA *P. cruenta* – AOSA
	20–30°C	*S. cineraria, S. cruentus, S. elegans* – ISTA
Substrates:	P – AOSA TP – ISTA	
Light/dark:	Light – AOSA Not specified – ISTA	
First count, final count (days):	None, 14 – AOSA 4–7, 21 – ISTA	
Comments:	None	
Dormancy-breaking treatments:	Prechill – ISTA	
Evaluation guidelines:	**AOSA** – (Generalized description for all Asteraceae other than lettuce; *C. tinctorius* and *H. annuus* are the only flower species in Asteraceae for which guidelines are specified.) Epigeal dicot with cotyledons which expand and become thin, leaf-like and photosynthetic; a hypocotyl that elongates and carries the cotyledons above the soil surface; and a long primary root with secondary roots usually developing within the test period. The epicotyl usually does not show any development within the test period. It is acceptable for the primary root to be weak or stubby if sufficient secondary or adventitious roots have developed.	

ISTA – A dicotyledon with epigeal germination. The shoot system consists of an elongated hypocotyl and two cotyledons with the terminal bud lying between. There is no epicotyl elongation within the test period and the epicotyl and terminal bud are usually not discernible. The root system consists of a primary root, usually with root hairs, which must be well developed, as secondary roots cannot be taken into account.

Illustrations:

Plate 14.5 Asteraceae *Senecio*: **A1–A2**. A fully developed seedling has a long primary root and hypocotyl and two expanded cotyledons. **B1–B2**. Unshed seed coats must be removed to examine the enclosed cotyledons. Here, removal exposes decayed cotyledons. At least 50% of the original cotyledon tissue must be functional for a seedling to continue normal development. **C**. Trapped in the seed coat, the primary root has not elongated; the hypocotyl is short and swollen. **D1–D2**. A granular hypocotyl requires further examination to determine if the tissue disruption affects central conducting tissues. *Note*: seedling sizes are not comparable due to varying degrees of magnification.

Balsaminaceae
Impatiens balsamina, Impatiens walleriana

Germination conditions: **Impatiens balsamina**, balsam, garden balsam, rose balsam, touch-me-not, balsamine des jardins, impatience, Balsamine, Gartenspringkraut, tsuri-fune-so, balsamina, chachupina, chico, madama; **Impatiens walleriana**, impatiens, busy-Lizzie, patience-plant, patient-Lucy, sultana, zanzibar balsam, balsamina, chino

Temperature:	20°C	*I. balsamina, I. walleriana* – AOSA – ISTA
	25°C	*I. walleriana* – AOSA
	20–30°C	*I. balsamina* – AOSA
		I. balsamina, I. walleriana – ISTA
Substrates:	BP, TP, KNO$_3$	– ISTA
	P	*I. walleriana* – AOSA
	TB	*I. balsamina* – AOSA
Light/dark:	Light	*I. balsamina, I. walleriana* – ISTA
	Light	*I. balsamina* (in addition to KNO$_3$ on sensitive stocks), *I. walleriana* – AOSA
First count, final count (days):	0, 8	*I. balsamina* – AOSA
	4–7, 21	– ISTA
	7, 18	*I. walleriana* – AOSA
Comments:	*I. walleriana*	New crop seed sensitive to higher temperatures – AOSA
	Impatiens spp.	Some colour forms dormant at temperatures above constant 27°C – AOSA

Comments (continued): Normal root growth patterns include both development of multiple secondary roots during initial growth period and predominance of a primary root with secondary roots developing later. The former pattern is usual for *I. balsamina* – AOSA

Dormancy-breaking treatments: Light, KNO$_3$, prechill – ISTA
KNO$_3$, prechill at 5°C – AOSA

Evaluation guidelines: **AOSA** – Epigeal dicot with cotyledons which expand and become thin, leaf-like and photosynthetic; a hypocotyl that elongates and carries the cotyledons above the soil surface; and a long primary root with one to many secondary roots usually developing within the test period. Primary and secondary roots are not always readily distinguishable. The epicotyl usually does not show any development within the test period. AOSA (2003b) includes descriptions specific to *Impatiens* spp., with illustrations.

ISTA – A dicotyledon with epigeal germination. The shoot system consists of an elongated hypocotyl and two cotyledons with the terminal bud lying between. There is no epicotyl elongation within the test period and the epicotyl and terminal bud are usually not discernible. The root system consists of a primary root, usually with root hairs, and often secondary roots, which are taken into account in seedling evaluation if the primary root is defective (Don, 2003; Ripka, 2003).

Illustrations:

Plate 14.6 Balsaminaceae *Impatiens*: **A1–A2**. In this fully developed seedling with four roots, the primary is indistinguishable from secondary roots. The cotyledons are rounded-to-kidney-shaped. **B**. A seedling with a markedly shortened hypocotyl and no strong root is considered incapable of normal development, despite visible root initiation. **C**. The hypocotyl has not yet straightened nor the cotyledons expanded in this seedling grown in a side-lit chamber. *Note*: seedling sizes are not comparable due to varying degrees of magnification.

Begoniaceae
Begonia Semperflorens-Cultorum group, *Begonia* × *tuberhybrida*, *Begonia* spp.

Germination conditions: ***Begonia*** Semperflorens-Cultorum group, bedding begonia, perpetual begonia, wax begonia, bégonia semperflorens, Semperflorens-Begonie, begonia perpetua; ***Begonia*** × ***tuberhybrida***, hybrid tuberous begonia, tuberous begonia, bégonia tubéreux, Knollenbegonie, begonia tuberosa; ***Begonia*** spp., begonia

Temperature:	20°C	*Begonia* Semperflorens-Cultorum group, *B*. × *tuberhybrida* – ISTA *Begonia* spp. – AOSA
	20–30°C	*Begonia* Semperflorens-Cultorum group, *B*. × *tuberhybrida* – ISTA
Substrates:	P, TB – AOSA TP – ISTA	
Light/dark:	Light – AOSA Not specified – ISTA	
First count, final count (days):	7–14, 21 – ISTA 14, 21 – AOSA	
Comments	None	
Dormancy-breaking treatments:	Prechill – ISTA	
Evaluation guidelines:	**AOSA** – The AOSA *Seedling Evaluation Handbook* (AOSA, 2003b) does not include descriptions of seedlings of Begoniaceae; the family is not included in descriptions for 'miscellaneous agricultural and horticultural' kinds.	
	ISTA – A dicotyledon with epigeal germination. The shoot system consists of an elongated hypocotyl and two cotyledons with the terminal bud lying between. There is no epicotyl elongation within the test period and the epicotyl and terminal bud are usually not discernible. The root system consists of a primary root, usually with root hairs, which must be well developed, as secondary roots cannot be taken into account.	

Illustrations: **Plate 14.7 Begoniaceae** *Begonia*: **A.** This tiny seedling has two
                                        rounded-to-kidney-shaped cotyledons, an elongated hypocotyl and a
                                        primary root of indeterminable length with an obvious swirl of root
                                        hairs. **B.** The growing point lies between the two cotyledons and is
                                        usually not discernible during the test period. **C.** The radicle has
                                        emerged but not elongated during early seedling development. **D.** Root
                                        hairs may be the only readily visible means of attachment to the
                                        substrate. **E1–E2.** The primary root may be difficult to see against the
                                        fibres of the paper substrates. Seedlings of tuberous begonia have
                                        hypocotyls that are succulent in appearance. Seedlings of fibrous
                                        begonia have white or pigmented hypocotyls and longer primary roots.
                                        – JMari[1]
                                        Note: seedling sizes are not comparable due to varying degrees of
                                        magnification.

[1]J. Mari, personal observation.

Campanulaceae
Campanula spp., *Lobelia* spp., *Platycodon* sp.

Germination conditions: ***Campanula carpatica***, Carpathian bellflower, tussock bellflower; ***Campanula fragilis***; ***Campanula garganica***, Adriatic bellflower; ***Campanula glomerata***, clustered bellflower, Dane's-blood, yatsushiroso, toppklocka; ***Campanula lactiflora***; ***Campanula medium***, Canterbury bells; ***Campanula medium*** 'Calycanthema', cup and saucer bellflower; ***Campanula persicifolia***, peach bellflower; ***Campanula portenschlagiana***, ***Campanula pyramidalis***, chimney bell-flower; ***Campanula rapunculus***, rampion; ***Campanula rotundifolia***, blue bells of Scotland, bellflower

Temperature:	20°C	*C. carpatica, C. fragilis, C. garganica, C. glomerata, C. lactiflora, C. medium, C. persicifolia, C. portenschlagiana, C. pyramidalis, C. rapunculus* – ISTA *C. medium, C. medium* 'Calycanthema', *C. rotundifolia* – AOSA
	20–30°C	*C. carpatica, C. fragilis, C. garganica, C. glomerata, C. lactiflora, C. medium, C. persicifolia, C. portenschlagiana, C. pyramidalis, C. rapunculus* – ISTA *C. carpatica, C. medium, C. medium* 'Calycanthema', *C. persicifolia* – AOSA
Substrates:	BP, TP	*C. carpatica, C. fragilis, C. garganica, C. glomerata, C. lactiflora, C. medium, C. persicifolia, C. portenschlagiana, C. pyramidalis, C. rapunculus* – ISTA
	P	*C. carpatica, C. persicifolia* – AOSA
	TB	*C. medium, C. medium* 'Calycanthema', *C. rotundifolia* – AOSA
Light/dark:	Light	– ISTA
	Light	*C. carpatica, C. persicifolia, C.rotundifolia* – AOSA
First count, final count (days):	4–7, 21	– ISTA
	6, 12	*C. medium, C. medium* 'Calycanthema' – AOSA
	6, 16	*C. carpatica, C. persicifolia* – AOSA
	7, 14	*C. rotundifolia* – AOSA
Comments:	None	
Dormancy-breaking treatments:	Light, prechill – ISTA	

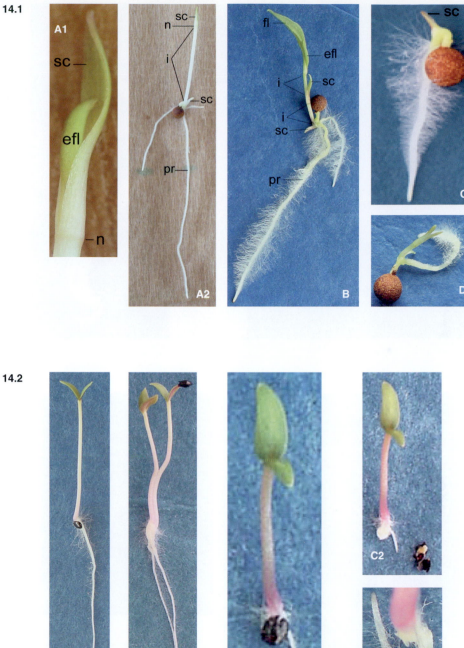

Plate 14.1. Alstroemeriaceae *Alstroemeria*: **A1–A2.** 28-day-old seedling germinated in paper towelling. **B.** 37-day-old seedling grown on top of blotters. **C.** A scale leaf is the first visible development of the shoot. **D.** Shoot lacking developing foliar leaves. efl = emerging foliar leaf, fl = foliar leaf, i = internode, n = node, pr = primary root, sc = scale leaf.

Plate 14.2. Amaranthaceae *Celosia*: **A.** Fully developed seedling. **B.** Two intertwined but completely separate seedlings developing from one seed. **C1.** Root trapped inside seed coat. **C2.** Normally developing primary root released from seed coat and (**C3**) uncoiled.

Plate 14.3. Apocynaceae *Catharanthus*: **A1–A2.** Fully developed seedling and expanded cotyledons. **B.** Stubby root, adhering seed coat.

Plate 14.4. Asteraceae *Dahlia*: **A.** Fully developed seedling. **B.** Primary root lacking. **C.** Closer view of stunted primary root and roots emerging above it.

14.5

14.6

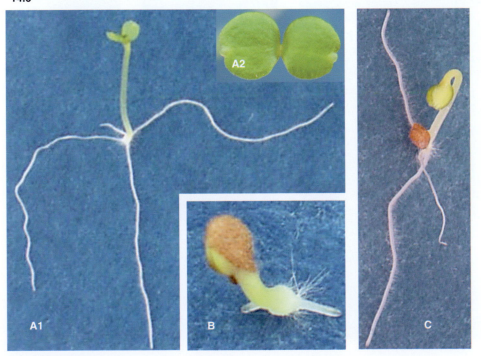

Plate 14.5. Asteraceae *Senecio*: **A1–A2.** Fully developed seedling and expanded cotyledons. **B1.** Unshed seed coat. **B2.** Seed coat removed to reveal decayed cotyledons. **C.** Short, stubby hypocotyl and stunted primary root. **D1–D2.** Granulated hypocotyl tissue.

Plate 14.6. Balsaminaceae *Impatiens*: **A1.** Fully developed seedling. **A2.** Rounded-to-kidney shaped cotyledons. **B.** Seedling incapable of normal development. **C.** Hypocotyl remains hooked, cotyledons unexpanded in seedling grown in side-lit chamber.

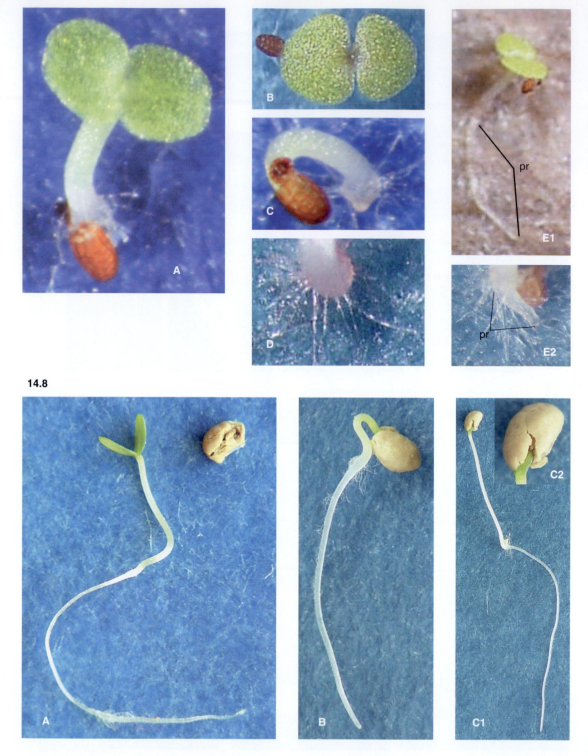

Plate 14.7. Begoniaceae *Begonia*: **A.** Fully developed seedling; root not visible. **B.** Terminal bud lies between the cotyledons. **C.** Radicle has emerged, not elongated. **D.** Root hairs readily visible. **E1–E2.** Primary root tiny, may not be readily visible against fibres of paper substrates.

Plate 14.8. Campanulaceae *Campanula*: **A.** Fully developed seedling with completely shed pelleting material. **B.** Primary root and elongating hypocotyl have emerged from pellet. **C1–C2.** Unshed pelleting material must be removed to determine condition of cotyledons.

14.9

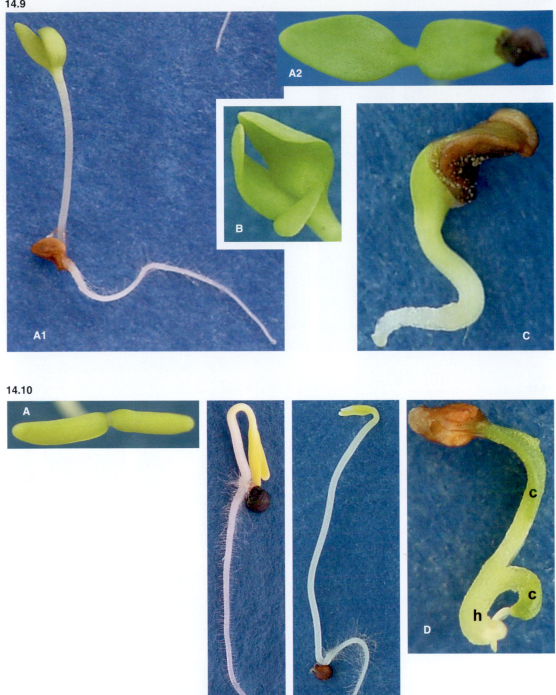

14.10

Plate 14.9. Caryophyllaceae *Dianthus*: **A1–A2.** Fully developed seedling and expanded cotyledons. **B.** Seedling with three cotyledons. **C.** Markedly shortened hypocotyl, stubby primary root.
Plate 14.10. Caryophyllaceae *Gypsophila*: **A.** Expanded cotyledons. **B.** Developing seedling has hooked hypocotyl and long primary root with root hairs. **C.** Short but sufficient primary root. **D.** Shortened hypocotyl; primary root lacking. c = cotyledon, h = hypocotyl.

14.11

14.12

Plate 14.11. Crassulaceae *Sedum*: **A1–A2.** Fully developed seedling with rounded cotyledons. **B.** Developing seedling with tightly adhering seed coat. **C.** Otherwise normal seedling appears to lack primary root.

Plate 14.12. Geraniaceae *Pelargonium*: **A1–A2.** Fully developed seedling and expanded cotyledons. **B.** Seedling with one attached cotyledon, lesion on hypocotyl, and trapped primary root. **C.** Primary root trapped in seed coat, no visible secondary root development. **D.** Longitudinal section of embryo. c = cotyledons, r = radicle.

14.13

14.14

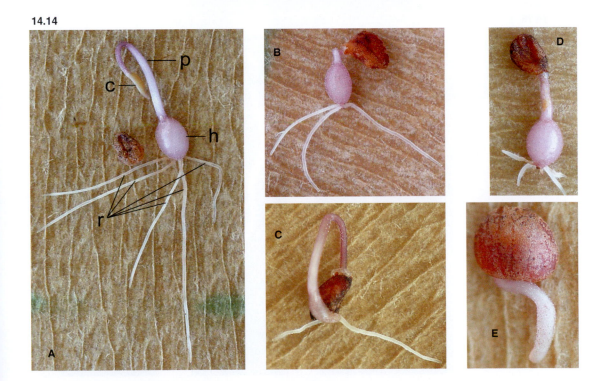

Plate 14.13. Papaveraceae *Papaver nudicaule*: **A1.** Linear cotyledons. **A2.** Adhering seed coat. **B.** Stubby primary root. **C.** Early seedling development.
Plate 14.14. Primulaceae *Cyclamen*: **A.** Fully developed 28-day-old seedling. **B.** Cotyledon blade remains enclosed in seed coat. **C.** No distinct tuber formed. **D.** Stubby seminal roots. **E.** No tuber formation, no roots.
c = cotyledon blade, h = tuberous hypocotyl, p = cotyledon petiole, r = seminal roots.

Plate 14.15. Primulaceae *Primula acaulis*: **A1.** Cotyledons with shed seed coat. **A2.** Tightly adhering seed coat. **A3.** Fully developed seedling. **B.** Early stage of seedling development.

Plate 14.16. Ranunculaceae *Anemone*: **A1–A2.** Fully developed seedling with rounded cotyledons. **B.** Unshed seed coat, no hypocotyl hook. **C.** Short root, not proportional to rest of seedling.

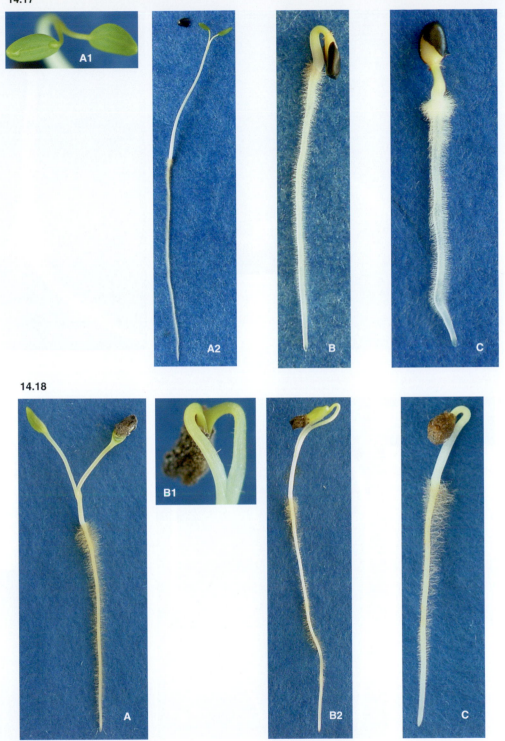

Plate 14.17. Ranunculaceae *Aquilegia*: **A1–A2.** Fully developed seedling with oval cotyledons. **B.** Early stage of seedling development. **C.** Thickened, unelongated hypocotyl.
Plate 14.18. Ranunculaceae *Delphinium*:* **A.** Fully developed seedling with long-petioled cotyledons. **B1–B2.** Early stages of development; cotyledons visible. **C.** Tightly adhering seed coat. *D. cardinale* has short petioles, long hypocotyl.

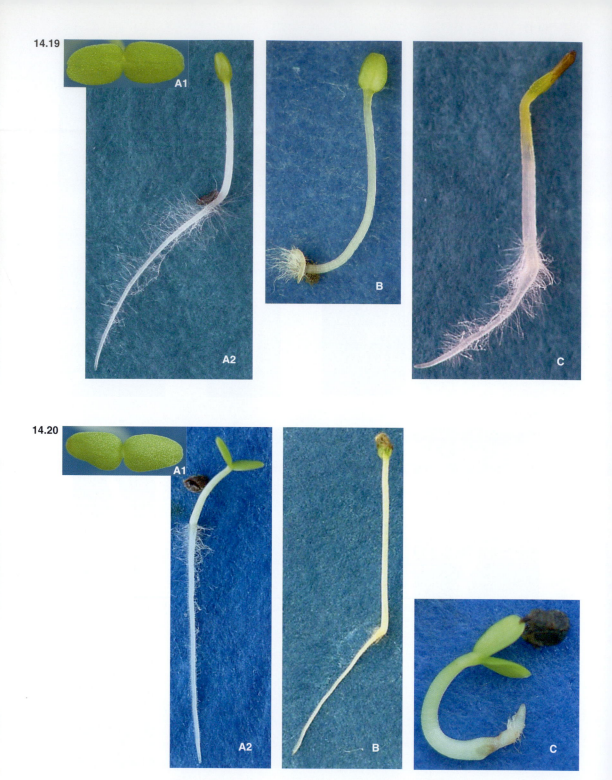

Plate 14.19. Scrophulariaceae *Nemesia*: **A1–A2.** Fully developed seedling with oblong cotyledons. **B.** Stubby primary root. **C.** Watery seedling.

Plate 14.20. Scrophulariaceae *Penstemon heterophyllus*: **A1–A2.** Fully developed seedling with broadly lanceolate, obtuse-tipped cotyledons. **B.** Adhering seed coat. **C.** Lesion at hypocotyl/root juncture and stubby primary root.

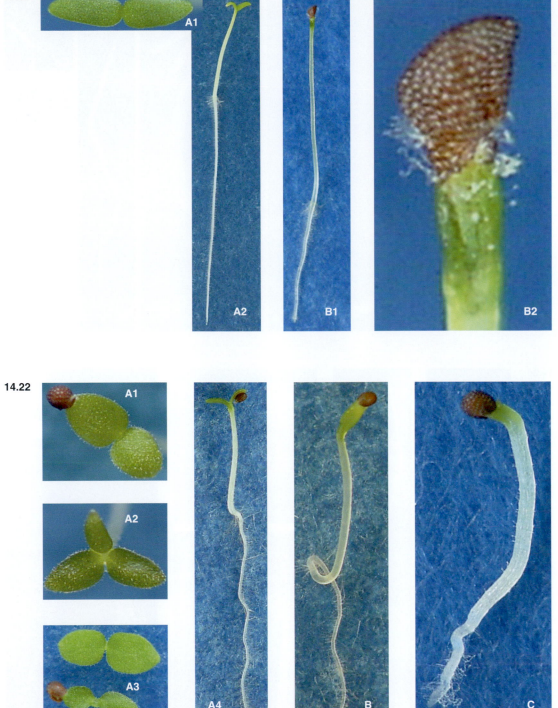

Plate 14.21. Solanaceae *Nierembergia*: **A.** Fully developed seedling and expanded cotyledons. **B1–B2.** Watery seedling with unshed seed coat.

Plate 14.22. Solanaceae *Petunia*: **A1.** Expanding cotyledons with shed seed coat. **A2.** Three cotyledons may develop. **A3.** Cotyledon sizes compared. **A4.** Fully developed seedling. **B.** Transparent or 'glassy' seedling. **C.** Watery seedling, not phototropic, short root and unshed seed coat.

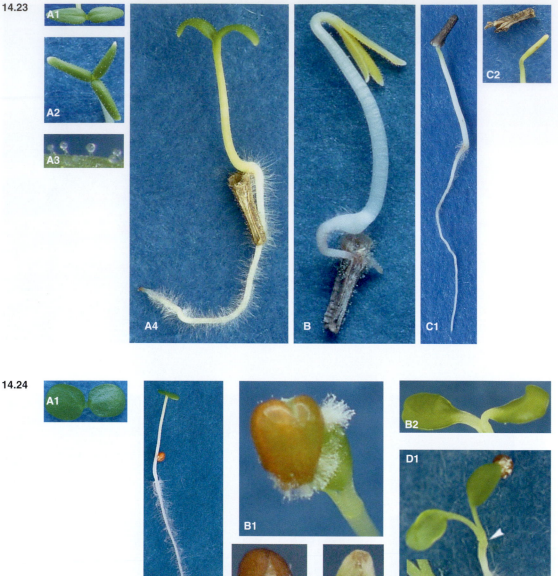

Plate 14.23. Verbenaceae *Glandularia*: **A1.** Narrowly lanceolate cotyledons. **A2.** Three cotyledons. **A3.** Glandular hairs. **A4.** Fully developed seedling. **B.** Short, stubby primary root, three cotyledons. **C1.** Adhering seed coat. **C2.** Seed coat removed to reveal only minor necrosis.

Plate 14.24. Violaceae *Viola*: **A1–A2.** Fully developed seedling with rounded cotyledons. **B1–B2.** Adhering seed coat removed to free functional cotyledons. **C1–C2.** Adhering seed coat removed to reveal watery, discoloured cotyledons. **D1–D2.** Shortened, granular hypocotyl.

Evaluation guidelines:	**AOSA** – The AOSA *Seedling Evaluation Handbook* (AOSA, 2003) does not include descriptions of seedlings of Campanulaceae; the family is not included in descriptions for 'miscellaneous agricultural and horticultural' kinds.	

ISTA – A dicotyledon with epigeal germination. The shoot system consists of an elongated hypocotyl and two cotyledons with the terminal bud lying between. There is no epicotyl elongation within the test period and the epicotyl and terminal bud are usually not discernible. The root system consists of a primary root, usually with root hairs, which must be well developed, as secondary roots cannot be taken into account.

Illustrations:	**Plate 14.8 Campanulaceae** *Campanula*: (pelleted)[1] **A**. A fully developed seedling has two cotyledons, an elongated hypocotyl and a long primary root with root hairs. It has completely shed the pelleting material that surrounded the seed. **B**. In a normally developing seedling, the primary root and the elongating hypocotyl have emerged from the pellet, the hypocotyl forming a 'hook' that pushes through the soil surface carrying the cotyledons to the surface. **C1–C2**. Unshed pelleting material must be removed for close examination of the condition of the cotyledon tissue. At least 50% of the original cotyledon tissue must be attached and free of necrosis or decay for the seedling to continue normal development. *Note*: seedling sizes are not comparable due to varying degrees of magnification.

[1]Pelleting refers to substances – commonly clay, diatomaceous earth, graphite, or powdered perlite – applied to seeds to create planting units that are more uniform in size and shape than the raw seed. The size, shape and weight of pelleted seed facilitates precise placement of seed by planters.

Caryophyllaceae

Cerastium sp., *Dianthus* spp., *Gypsophila* spp., *Lychnis* spp., *Sagina* sp., *Saponaria* spp., *Silene* sp., *Vaccaria* sp.

Germination conditions: ***Dianthus*** × ***allwoodii***, sweet wivelsfield; ***Dianthus barbatus***, sweet-william, amerika-nadeshiko, mani-saythu-pan; ***Dianthus caryophyllus***, border carnation, carnation, clove pink, oeillet, Nelke, oranda-nadeshiko, zaw-hmwa-gyi, clavel; ***Dianthus chinensis***, Chinese pink, Japanese pink, dianthus, Indian pink, shi zhu, kara-nadeshiko, zaw-hmwa-gale, rainbow pink, sekichiku; ***Dianthus chinensis*** 'Heddewigii' and ***Dianthus chinensis*** 'Heddensis', china pinks; ***Dianthus deltoides***, maiden pink, meadow pink; ***Dianthus plumarius***, clove pink, cottage pink, garden pink, pink, clavelina, laced pink, zaw-hmwa

Temperature:	20°C	*D.* × *allwoodii*, *D. barbatus*, *D. caryophyllus*, *D. chinensis*, *D. plumarius* – AOSA – ISTA *D. deltoides* – ISTA
	20–30°C	*D. barbatus*, *D. caryophyllus*, *D. chinensis*, *D. deltoides*, *D. plumarius* – ISTA *D. deltoides* – AOSA
Substrates:	BP, TP – ISTA P (specified for *D. deltoides*) – AOSA TB – AOSA	
Light/dark:	Light (specified for *D. deltoides*) – AOSA Not specified – ISTA	

First count, final count (days):	none, 7	*D. chinensis*, *D. plumarius* – AOSA
	none, 8	*D.* × *allwoodii*, *D. barbatus*, *D. caryophyllus* – AOSA
	none, 10	*D. deltoides* – AOSA
	4–7, 14	*D. barbatus*, *D. caryophyllus*, *D. chinensis*, *D. deltoides*, *D. plumarius* – ISTA

Comments:	*D. caryophyllus*, *D.chinensis*, *D. chinensis* 'Heddewigii', *D. chinensis* 'Heddensis'	Prone to seed injury and broken seedlings – AOSA

Dormancy-breaking treatments: Prechill – ISTA

Evaluation guidelines:

AOSA – Epigeal dicot with leaf-like cotyledons. The hypocotyl elongates and carries the cotyledons above the soil surface. The epicotyl usually does not show any development within the test period. A primary root; root hairs may develop within the test period. Secondary roots will not compensate for a missing, weak or stubby primary root. AOSA (2003b) includes descriptions specific to *Dianthus* spp. with illustrations.

Seedlings producing three cotyledons, if otherwise normal, will continue to develop normally. – AOSA

Common defects in seedlings due to mechanical damage include: hypocotyls exhibiting *minor* twisting, and broken cotyledons. The severity of the twisting and the extent of the cotyledon damage (50% or more of the original area must be functional) must be evaluated to determine if a seedling is capable of continued development. – AOSA

Declining vigour may manifest as shortened roots and/or hypocotyls; if growth is proportional and adequate to support the seedling, continuing normal development can be expected. – AOSA

Watery hypocotyls may occur when the test conditions are too wet. – AOSA

ISTA – A dicotyledon with epigeal germination. The shoot system consists of an elongated hypocotyl and two cotyledons with the terminal bud lying between. There is no epicotyl elongation within the test period and the epicotyl and terminal bud are usually not discernible. The root system consists of a primary root, usually with root hairs, which must be well developed, as secondary roots cannot be taken into account. The *ISTA Handbook for Seedling Evaluation* (Don, 2003) includes photographs.

Illustrations:

Plate 14.9 Caryophyllaceae *Dianthus*: **A1–A2**. A fully developed seedling has two cotyledons, an elongated hypocotyl and a primary root. **B**. A seedling with three cotyledons will continue normal development if the rest of the seedling structures are normal. **C**. A markedly shortened hypocotyl and stubby primary root indicate that this seedling will not develop further. *Note*: seedling sizes are not comparable due to varying degrees of magnification.

Germination conditions: ***Gypsophila elegans***, long-petalled baby's-breath, annual baby's-breath; ***Gypsophila paniculata***, perennial baby's-breath; ***Gypsophila pacifica***, Pacific baby's-breath; ***Gypsophila repens***, baby's breath

Temperature:	15°C	*G. elegans, G. paniculata, G. repens* – ISTA *G. elegans, G. pacifica, G. repens* – AOSA
	20°C	*G. elegans, G. paniculata* – AOSA – ISTA *G. repens* – ISTA
Substrates:	BP, TP – ISTA TB – AOSA KNO$_3$ (specified for *G. elegans* and *G. paniculata*) – AOSA	
Light/dark:	Light (specified for *G. elegans, G. paniculata*) – AOSA Light (specified for *G. elegans, G. paniculata, G. repens*) – ISTA	
First count, final count (days):	none, 7 none, 8 4–7, 14	*G. paniculata* – AOSA *G. elegans, G. pacifica, G. repens* – AOSA *G. elegans, G. paniculata, G. repens* – ISTA
Comments:	*G. elegans*	Some cultivars may be sensitive to temperatures above 18°C – AOSA
	G. pacifica, G. repens	Sensitive to temperatures above 18°C – AOSA
Dormancy-breaking treatments:	Light – ISTA	
Evaluation guidelines:	**AOSA** – Epigeal dicot with leaf-like cotyledons. The hypocotyl elongates and carries the cotyledons above the soil surface. The epicotyl usually does not show any development within the test period. A primary root; root hairs may develop within the test period. Secondary roots will not compensate for a missing, weak or stubby primary root. AOSA (2003b) includes descriptions specific to Caryophyllaceae with illustrations.	
	Seedlings with twisted hypocotyls may occur as the result of processing damage; if the twisting is not severe, such seedlings can be expected to continue development into plants. – AOSA	
	ISTA – A dicotyledon with epigeal germination. The shoot system consists of an elongated hypocotyl and two cotyledons with the terminal bud lying between. There is no epicotyl elongation within the test period and the epicotyl and terminal bud are usually not discernible. The root system consists of a primary root, usually with root hairs, which must be well developed, as secondary roots cannot be taken into account.	
Illustrations:	**Plate 14.10 Caryophyllaceae** *Gypsophila*: **A.** The expanded cotyledons of a mature seedling are linear in shape. **B.** A developing seedling has a hooked hypocotyl and a long primary root with root hairs. **C.** Despite a short primary root, this seedling has the essential structures needed to produce a plant. Shortened roots may be a sign of an older seed lot and declining vigour. **D.** Mechanical damage or physiological deterioration may cause shortened hypocotyls and lacking primary root. *Note*: seedling sizes are not comparable due to varying degrees of magnification.	

Crassulaceae
Kalanchoe sp., *Sedum* spp., *Sempervivum* spp.

Germination conditions: ***Sedum acre***, gold-moss sedum, golden-carpet[1]

Temperature:	15°C – AOSA
Substrates:	P – AOSA
Light/dark:	Light (8 h or more) – AOSA
First count, final count (days):	None, 14 – AOSA
Comments:	None
Dormancy-breaking treatments:	None
Evaluation guidelines:	**AOSA** – The AOSA *Seedling Evaluation Handbook* (AOSA, 2003b) does not include descriptions of seedlings of Crassulaceae; the family is not included in descriptions for 'miscellaneous agricultural and horticultural' kinds.
	ISTA – The *ISTA Handbook for Seedling Evaluation* (Don, 2003) does not include *Sedum* spp.
Illustrations:	**Plate 14.11 Crassulaceae** *Sedum*: **A1–A2**. A fully developed seedling has a long primary root with root hairs, an elongated hypocotyl and two rounded cotyledons. **B**. This developing seedling has a tightly adhering seed coat. At the termination of the test, if the seedling is otherwise normal, the seed coat must be removed to determine the condition of the cotyledons. At least 50% of the cotyledon tissue must be attached and free of necrosis or decay. **C**. An otherwise normal seedling appears to lack a primary root. Close examination is required to determine if the primary root has developed but is not visible because it has penetrated the substrate. *Note*: seedling sizes are not comparable due to varying degrees of magnification.

[1]*Sedum* sp. not included in ISTA (2003).

Geraniaceae
Pelargonium spp.

Germination conditions: ***Pelargonium*** spp, geranium; ***Pelargonium*** zonal hybrids

Temperature:	20°C – AOSA – ISTA 20–30°C – ISTA
Substrates:	B, T, TB – AOSA BP, TP – ISTA
Light/dark:	Not specified
First count, final count (days):	7, 14 (clipped or scarified seed) – AOSA 7, 28 – AOSA – ISTA
Comments:	Hard seeds Pierce seed or file off fragment of testa – ISTA
Dormancy-breaking treatments:	None

Evaluation guidelines:	**AOSA** – A seedling is considered normal if it possesses those essential structures indicative of its ability to produce a plant under favourable conditions. AOSA (2003b) includes no evaluation guidelines specific to flowers in the family Geraniaceae.
	ISTA – A dicotyledon with epigeal germination. The shoot system consists of an elongated hypocotyl and two cotyledons with the terminal bud lying between. There is no epicotyl elongation within the test period and the epicotyl and terminal bud are usually not discernible. The root system consists of a primary root, usually with root hairs, which must be well developed, as secondary roots cannot be taken into account.
Illustrations:	**Plate 14.12 Geraniaceae** *Pelargonium*: **A1–A2**. A fully developed seedling has two cotyledons, a long hypocotyl and a long primary root. **B**. A seedling with only one attached cotyledon can continue normal development if it is otherwise normal. Here, the primary root and a lesion on the hypocotyl must be closely examined. The depth of the lesion must be determined. If it does not extend into central conducting tissue, the lesion will not impede continued development. The seed coat and detached cotyledon are obscuring the root. Only if the primary root has developed within the adhering seed coat is the seedling considered capable of development into a plant. Long soft hairs cover both the pigmented hypocotyl and cotyledons. **C**. The primary root of this seedling is trapped in the seed coat and no secondary roots are developing. It is considered incapable of continued normal development. However, care must be taken that test conditions are sufficiently moist throughout the duration of the germination test to enable the developing seedling to shed the tough seed coat. **D**. The tip of the radicle of the *Pelargonium* embryo lies in a 'pouch' formed by the tough seed coat. Externally, the protuberance of the pouch renders the radicle especially vulnerable to damage during seed scarification. *Note*: seedling sizes are not comparable due to varying degrees of magnification.

Papaveraceae
Eschscholzia spp., *Hunnemannia* sp., *Papaver* spp.

Germination conditions: ***Papaver alpinum***, alpine poppy, Austrian poppy, pavot de Burser, pavot des Alpes, bursers Alpen-Mohn; ***Papaver glaucum***, tulip papaver, tulip poppy; ***Papaver nudicaule***, Icelandic poppy, Arctic poppy, pavot d'Islande; ***Papaver orientale***, Oriental poppy, pavot d'Orient; ***Papaver rhoeas***, corn poppy, Shirley poppy, Flanders poppy, field poppy, coquelicot, Klatsch-Mohn, amapola

Temperature:	10°C	*P. alpinum*, *P. glaucum*, *P. nudicaule* – ISTA
	15°C	*P. alpinum*, *P. glaucum*, *P. nudicaule*, *P. rhoeas* – ISTA
		P. glaucum, *P. nudicaule*, *P. rhoeas* – AOSA
	20°C	*P. orientale*, *P. rhoeas* – ISTA
	20–30°C	*P. orientale* – AOSA
		P. orientale, *P. rhoeas* – ISTA
Substrates:	P, TB	*P. rhoeas* – AOSA
	P, TB, KNO₃	*P. glaucum*, *P. nudicaule*, *P. orientale* – AOSA
	TP, KNO₃	*P. alpinum*, *P. glaucum*, *P. nudicaule*, *P. orientale*, *P. rhoeas* – ISTA

| Light/dark: | Light | *P. glaucum, P. nudicaule, P. rhoeas* – ISTA |
| | Light | *P. orientale* – AOSA |

First count, final count (days):	None, 8	*P. rhoeas* – AOSA
	4–7, 14	*P. alpinum, P. glaucum, P. nudicaule,*
		P. orientale, P. rhoeas – ISTA
	6, 12	*P. orientale* – AOSA
	6, 14	*P. glaucum, P. nudicaule* – AOSA

| Comments: | None | |

Dormancy-breaking treatments:	KNO_3	*P. alpinum* – ISTA
	Light, KNO_3	*P. glaucum, P. nudicaule* – ISTA
	Prechill, KNO_3	*P. orientale* – ISTA
	Light, prechill, KNO	*P. rhoeas* – ISTA

Evaluation guidelines:

AOSA – The AOSA *Seedling Evaluation Handbook* (AOSA, 2003b) does not include descriptions of seedlings of Papaveraceae; the family is not included in descriptions for 'miscellaneous agricultural and horticultural' kinds.

ISTA – A dicotyledon with epigeal germination. The shoot system consists of an elongated hypocotyl and two cotyledons with the terminal bud lying between. There is no epicotyl elongation within the test period and the epicotyl and terminal bud are usually not discernible. The root system consists of a primary root, usually with root hairs, which must be well developed, as secondary roots cannot be taken into account.

Illustrations:

Plate 14.13 Papaveraceae *Papaver nudicaule*: **A1**. Intact cotyledons are linear in shape. **A2**. This seedling has an adequately elongated hypocotyl and primary root. The adhering seed coat must be removed and the cotyledons examined. If at least 50% of the total cotyledon tissue is free of necrosis or decay, the seedling can be expected to continue normal development to produce a plant. **B**. A stubby primary root will prevent continued normal development of this seedling. **C**. In early seedling development, the hypocotyl forms a hook, which pushes through the soil, straightening after reaching the surface. *Note*: seedling sizes are not comparable due to varying degrees of magnification.

Primulaceae
Anagallis spp., *Cyclamen* sp., *Primula* spp.

Germination conditions: ***Cyclamen africanum***, cyclamen; ***Cyclamen persicum***, florist's cyclamen

Temperature	15°C	*C. persicum* – ISTA
	20°C	*C. africanum* – AOSA
		C. persicum – ISTA

| Substrates: | BP, TP, S, KNO_3 – ISTA | |
| | P,T – AOSA | |

| Light/dark: | Not specified – AOSA – ISTA | |

| First count, final count (days): | 14, 28 – AOSA | |
| | 14–21, 35 – ISTA | |

Comments: Soak in water 24 h – ISTA
Provide good moisture supply during test – AOSA
Pour 55°C water over seeds and let soak for 24 h – JMari[1]

Dormancy-breaking treatments: KNO$_3$ – ISTA

Evaluation guidelines: **AOSA** – Epigeal dicot. Swollen tuberous hypocotyl and a single cotyledon (normally there is no second cotyledon) borne on a petiole, the terminal bud lying at its base. Several seminal roots, developing more or less simultaneously at the distal end of the hypocotyl. AOSA (2003b) includes written descriptions of seedling defects specific to *Cyclamen*.
ISTA – Dicotyledon with epigeal germination. The shoot system consists of a swollen, tuberous hypocotyl and a single cotyledon borne on a petiole, the terminal bud lying at its base. The root system consists of several seminal roots, developing more or less simultaneously at the distal end of the hypocotyl. The *ISTA Handbook for Seedling Evaluation* (Don, 2003) includes descriptions of defects and photographs.

Illustrations: **Plate 14.14 Primulaceae** *Cyclamen*: **A.** This 28-day-old seedling has formed a tuber at the base of the hypocotyl, several seminal roots, a stout petiole and the cotyledon blade, which has emerged from the seed coat, but not unfolded. The blade is usually not visible within the laboratory test period. **B.** In a developing 28-day-old seedling, the cotyledon blade is usually still enclosed in the seed coat. The petiole at the point of entry into the seed coat must be free of decay. **C.** A seedling without a distinct tuber will not continue development to produce a plant. **D.** With a tuber, but stubby seminal roots, this seedling will not be expected to develop into a plant. **E.** The hypocotyl emerged from this seed, but did not form a tuber and has no roots. *Note*: seedling sizes are not comparable due to varying degrees of magnification.

Germination conditions: ***Primula auricula***, bear's ear; ***Primula denticulata***, drumstick primula; ***Primula elatior***, oxlip; ***Primula japonica***, kurin-so; ***Primula* × *kewensis***; ***Primula malacoides***, baby primrose, fairy primrose; ***Primula obconica***, German primrose, poison primrose, e bao chun; ***Primula praenitens***, Chinese primrose, zang bao chun; ***Primula veris***, cowslip; ***Primula vulgaris***, primrose; ***Primula* spp.**, primula

---

Temperature:	15°C	*P. auricula, P. denticulata, P. elatior, P. japonica, P.* × *kewensis, P. malacoides, P. obconica, P. praenitens, P. veris, P. vulgaris* – ISTA *Primula* spp. – AOSA
	20°C	*P. auricula, P. denticulata, P. elatior, P. japonica, P.* × *kewensis, P. malacoides, P. obconica, P. praenitens, P. veris, P. vulgaris* – ISTA
	20–30°C	*P. auricula, P. denticulata, P. elatior, P. japonica, P.* × *kewensis, P. malacoides, P. obconica, P. praenitens, P. veris, P. vulgaris* – ISTA

Substrates: P – AOSA
TP, KNO$_3$ – ISTA

Light/dark: Light – AOSA
Not specified – ISTA

| First count, final count (days): | 7–14, 28 | *P. auricula, P. denticulata, P. elatior, P. japonica, P.* × *kewensis, P. malacoides, P. obconica, P. praenitens, P. veris, P. vulgaris* – ISTA |
| | 12, 28 | *Primula* spp. – AOSA |

| Comments: | Provide good moisture supply during test – AOSA |
| | Species and lots vary in response – AOSA |

Dormancy-breaking treatments: Prechill, KNO_3 – ISTA

Evaluation guidelines:
AOSA – Epigeal dicot. The hypocotyl elongates carrying the cotyledons above the soil surface. The epicotyl usually does not show any development within the test period. A primary root; secondary roots will not compensate for a missing or weak primary root. AOSA (2003b) includes descriptions of defects that will prevent continued development.

ISTA – A dicotyledon with epigeal germination. The shoot system consists of an elongated hypocotyl and two cotyledons with the terminal bud lying between. There is no epicotyl elongation within the test period and the epicotyl and terminal bud are usually not discernible. The root system consists of a primary root, usually with root hairs, which must be well developed, as secondary roots cannot be taken into account.

Illustrations:
Plate 14.15 Primulaceae *Primula acaulis*: **A1–A3**. A fully developed seedling has a long primary root, an elongated hypocotyl and two cotyledons. Tightly adhering seed coats must be removed to examine cotyledons; at least 50% of the original cotyledon tissue must be attached and free of necrosis or decay. **B**. Radicle emergence has occurred and the hypocotyl is beginning to elongate in this early stage of development. *Note*: seedling sizes are not comparable due to varying degrees of magnification.

[1]J. Mari, personal observation.

Ranunculaceae
Anemone spp., *Aquilegia* spp., *Consolida* sp., *Delphinium* spp., *Nigella* spp., *Ranunculus* spp.

Germination conditions: **Anemone coronaria**; **Anemone pulsatilla**, anemone, European pasqueflower, pasqueflower, pulsatilla; **Anemone sylvestris**, snowdrop windflower

Temperature:	15°C	*A. coronaria, A. sylvestris* – ISTA
		A. pulsatilla – AOSA
	20°C	*A. coronaria, A. sylvestris* – ISTA

| Substrates: | TB – AOSA |
| | TP – ISTA |

Light/dark: Not specified

| First count, final count (days): | 7–14, 28 – ISTA |
| | 14, 28 – AOSA |

| Comments: | *A. pulsatilla* | Sensitive to temperatures above 15°C – AOSA |
| | Substratum must be monitored throughout test to ensure that it does not dry out – J Mari[1] |

Dormancy-breaking treatments:	*A. coronaria, A. sylvestris* Prechill – ISTA
Evaluation guidelines:	**AOSA** – The AOSA *Seedling Evaluation Handbook* (AOSA, 2003b) does not include descriptions of seedlings of Ranunculaceae; the family is not included in descriptions for 'miscellaneous agricultural and horticultural' kinds.
	ISTA – A dicotyledon with epigeal germination. The shoot system consists of an elongated hypocotyl and two cotyledons with the terminal bud lying between. There is no epicotyl elongation within the test period and the epicotyl and terminal bud are usually not discernible. The root system consists of a primary root, usually with root hairs, which must be well developed, as secondary roots cannot be taken into account.
Illustrations:	**Plate 14.16 Ranunculaceae** *Anemone*: **A1–A2**. A fully developed seedling has a long primary root with root hairs (both root and root hairs are tan to brown), a long hypocotyl and two rounded cotyledons. Otherwise normally developing seedlings with three cotyledons are capable of continuing development to produce plants. **B**. A seedling having all the essential structures, but with an unshed seed coat that cannot easily be removed with forceps is considered to be deficient. The lack of a hypocotyl 'hook' during early development is also indicative of a weak seedling. **C**. An adequate primary root is at least 50% the length of the hypocotyl and has root hairs. – J Mari[1] The root of this seedling is deficient because it is short and not proportional to the rest of the seedling. *Note*: seedling sizes are not comparable due to varying degrees of magnification.

Germination conditions: ***Aquilegia alpina***, alpine columbine; ***Aquilegia canadensis***, honeysuckle, meeting-houses, red-bell, wild columbine; ***Aquilegia chrysantha***, columbine; ***Aquilegia coerulea***, Rocky Mountain columbine; ***Aquilegia*** × ***cultorum, Aquilegia longissima***, long-spur columbine; ***Aquilegia vulgaris***, columbine, European columbine, European crowfoot, granny's-bonnet; and ***Aquilegia*** spp., columbine, long-spurred hybrids

Temperature:	15°C	*A. alpina, A. canadensis, A. chrysantha, A.* × *cultorum, A. vulgaris* – ISTA
	20–30°C	*A. alpina, A. canadensis, A. chrysantha, A. coerulea, A.* × *cultorum, A. longissima, A. vulgaris, Aquilegia* spp. – AOSA, ISTA
Substrates:	BP, TP	– ISTA
	P	– AOSA
Light/dark:	Light	
First count, final count (days):	6, 16	*A. alpina* – AOSA
	7, 21	*Aquilegia* spp. – AOSA
	7–14, 28	*A. alpina, A. canadensis, A. chrysantha, A.* × *cultorum, A. vulgaris* – ISTA
	14, 30	*A. chrysantha, A. coerulea, A. longissima* – AOSA
Comments:	*Aquilegia* spp.	Germination may be complete with some strains in 14 days – AOSA
	Substrate must be monitored throughout test to ensure that it does not dry out – J Mari[1]	
Dormancy-breaking treatments:	*A. alpina*	Prechill 14–21 days at 3–5°C on medium moistened with KNO₃ prior to testing – AOSA

A. alpina	
A. canadensis	
A. chrysantha	Light, prechill – ISTA
A. × *cultorum*	
A. vulgaris	

A. chrysantha	Prechill sensitive stocks at 5°C for 3–4 weeks –
A. coerulea	AOSA
A. longissima	

Evaluation guidelines:

AOSA – The AOSA *Seedling Evaluation Handbook* (AOSA, 2003b) does not include descriptions of seedlings of Ranunculaceae; the family is not included in descriptions for 'miscellaneous agricultural and horticultural' kinds.

ISTA – A dicotyledon with epigeal germination. The shoot system consists of an elongated hypocotyl and two cotyledons with the terminal bud lying between. There is no epicotyl elongation within the test period and the epicotyl and terminal bud are usually not discernible. The root system consists of a primary root, usually with root hairs, which must be well developed, as secondary roots cannot be taken into account.

Illustrations:

Plate 14.17 Ranunculaceae *Aquilegia*: **A**. A fully developed seedling has two oval cotyledons, a long hypocotyl and a long primary root with root hairs present. Root and root hairs are white. **B**. In early development, a normally developing seedling has a long white root, an elongating hypocotyl forming a 'hook', and cotyledons that are partly to completely visible. **C**. A thickened, unelongated hypocotyl not in proportion to the root length is abnormal. *Note*: seedling sizes are not comparable due to varying degrees of magnification.

Germination conditions: ***Delphinium*** × ***belladonna***, ***Delphinium cardinale***, cardinal larkspur, scarlet larkspur, perennial delphinium; ***Delphinium*** × ***cultorum***, ***Delphinium elatum***, Pacific Giant, gold medal, belladonna, bellamosum, chinensis and other types; ***Delphinium formosum***, ***Delphinium grandiflorum***, cui que

---

Temperature:	10°C	*D.* × *belladonna*, *D. cardinale*, *D.* × *cultorum*, *D. formosum*, *D. grandiflorum* – ISTA
	15°C	*D.* × *belladonna*, *D. cardinale*, *D.* × *cultorum*, *D. formosum*, *D. grandiflorum* – ISTA *D. cardinale* – AOSA
	20°C	*D.* × *belladonna*, *D. cardinale*, *D.* × *cultorum*, *D. formosum*, *D. grandiflorum* – ISTA *D. elatum* (belladonna, bellamosum and chinensis types) – AOSA
	20–30°C	*D. elatum* (Pacific giant, gold medal and other types) – AOSA
Substrates:	B, T, TB – AOSA BP, TP – ISTA	
Light/dark:	Light (specified for *D.* × *belladonna*, *D.* × *cultorum*, *D. formosum*, *D. grandiflorum*) – ISTA Not specified – AOSA	

First count, final count (days):	7, 16	*D. elatum* (belladonna, bellamosum types) – AOSA
	7, 18	*D. elatum* (chinensis types) – AOSA
	7–10, 21	*D.* × *belladonna, D. cardinale, D.* × *cultorum, D. formosum, D. grandiflorum* – ISTA
	8, 18	*D. elatum* (Pacific giant, gold medal and other types) – AOSA
	12, 28	*D. cardinale* – AOSA

| Comments: | *D. cardinale* | Very sensitive to warm temperature – AOSA |
| | *D. elatum* (all types) | Sensitive to drying in test; it is critical to monitor the moisture level of the substratum between planting time and first evaluation time and throughout the duration of the test – AOSA – JMari[1] |

Dormancy-breaking treatments: Fresh seed may germinate better at alternating temperatures than at a constant temperature – JMari[1]

D. cardinale	Prechill – ISTA
D. × *belladonna*	Light, prechill – ISTA
D. × *cultorum*	
D. formosum	
D. grandiflorum	

Evaluation guidelines:

AOSA – The AOSA *Seedling Evaluation Handbook* (AOSA, 2003b) does not include descriptions of seedlings of Ranunculaceae; the family is not included in descriptions for 'miscellaneous agricultural and horticultural' kinds.

ISTA – A dicotyledon with epigeal germination. The shoot system consists of an elongated hypocotyl and two cotyledons with the terminal bud lying between. There is no epicotyl elongation within the test period and the epicotyl and terminal bud are usually not discernible. The root system consists of a primary root, usually with root hairs, which must be well developed, as secondary roots cannot be taken into account.

Illustrations:

Plate 14.18 Ranunculaceae *Delphinium:*[2] **A.** A fully developed seedling has two long-petioled expanded cotyledons, a very short hypocotyl and a long primary root with root hairs. Root and root hairs are tan or brownish. **B1–B2.** In early development, the petioles form a 'hook', and the cotyledons may be partly to completely visible. **C.** A seedling with all the essential structures, but with a tightly adhering seed coat that cannot easily be removed with forceps is abnormal. Petioles are not visible in this photograph. *Note:* seedling sizes are not comparable due to varying degrees of magnification.

[1]J. Mari, personal observation.
[2]Seedlings of *D. cardinale* (larkspur) differ from those of other species in having a long hypocotyl and short petioles. – JMari, personal observation.

Scrophulariaceae
Antirrhinum spp., *Calceolaria* spp., *Collinsia* sp., *Digitalis* spp., *Linaria* spp., *Mimulus* sp., *Misopates* sp., *Nemesia* spp., *Penstemon* spp., *Pseudolysimachion* sp., *Torenia* sp., *Veronica* sp.

Germination conditions: ***Nemesia strumosa, Nemesia versicolor, Nemesia* spp.**, nemesia

| Temperature: | 15°C | *N. strumosa, N. versicolor* – ISTA |
| | | *Nemesia* spp. – AOSA |

| | 20°C | *N. strumosa, N. versicolor* – ISTA |

Substrates: BP, TP – ISTA
                                   P, TB – AOSA

Light/dark: Light – ISTA
                                   Not specified – AOSA

First count, final count (days):   5, 10 – AOSA
                                   5–7, 21 – ISTA

Comments: Sensitive to temperatures above 18°C – AOSA

Dormancy-breaking treatments: Light, prechill – ISTA

Evaluation guidelines: **AOSA** – Epigeal dicot. The hypocotyl elongates carrying the cotyledons above the soil surface. The epicotyl usually does not show any development within the test period. A primary root; secondary roots will not compensate for a weak, missing or stubby primary root. AOSA (2003b) includes descriptions of defects that will prevent continued development.

Glassy or watery seedlings may develop as a result of environmental conditions, e.g. lack of light during germination. – AOSA Such seedlings are considered to be capable of continued development if they have all the essential structures to produce a plant, *and* orientate toward light. – JMari[1]

ISTA – A dicotyledon with epigeal germination. The shoot system consists of an elongated hypocotyl and two cotyledons with the terminal bud lying between. There is no epicotyl elongation within the test period and the epicotyl and terminal bud are usually not discernible. The root system consists of a primary root, usually with root hairs, which must be well developed, as secondary roots cannot be taken into account.

Illustrations: **Plate 14.19 Scrophulariaceae** *Nemesia*: **A1–A2**. A fully developed seedling has a long primary root with root hairs, an elongated hypocotyl and two oblong cotyledons. **B**. A stubby primary root will not support continued development to produce a plant. **C**. Seedlings with a watery, or glassy, appearance may be the result of test conditions. If a watery seedling has the structures essential to produce a plant under favourable conditions and orientates toward the light source, it is considered capable of continued normal development. Seedlings that appear watery due to deficiencies inherent to the seed will not be upright and orientated toward the light source. *Note*: seedling sizes are not comparable due to varying degrees of magnification.

Germination conditions: ***Penstemon barbatus***, beardlip penstemon; ***Penstemon eatonii***, Eaton's penstemon, firecracker penstemon; ***Penstemon grandiflorus***, large beardtongue, penstemon; ***Penstemon hartwegii***, penstemon; ***Penstemon hirsutus***, hairy beardtongue, northeastern beardtongue; ***Penstemon*** hybrids, penstemon; ***Penstemon laevigatus***, eastern beardtongue; ***Penstemon palmeri***, Palmer penstemon; ***Penstemon penlandii***, Penland's beardtongue; ***Penstemon strictus***, Rocky Mountain penstemon; ***Penstemon*** spp., penstemon or beardtongue

---

Temperature: 10–20°C   *P. eatonii, Penstemon* spp. (all species not named in other AOSA germination procedures) – AOSA

	15°C	P. barbatus, P. eatonii, P. grandiflorus, P. hirsutus, P. laevigatus, P. palmeri, P. strictus, Penstemon spp. (all species not named in other AOSA germination procedures) – AOSA P. barbatus, P. hartwegii, Penstemon hybrids – ISTA
	15–25°C	P. penlandii – AOSA
	20–30°C	P. barbatus, P. hartwegii, Penstemon hybrids – ISTA
Substrates:	P – AOSA	
	P, 500 ppm GA₃	P. eatonii, Penstemon spp. not named in other AOSA procedures (method 1) – AOSA
	TP – ISTA	
Light/dark:	Light	P. barbatus, P. grandiflorus, P. hirsutus, P. laevigatus, P. palmeri, P. strictus – AOSA P. eatonii, Penstemon spp. not named in other AOSA procedures (method 2) – AOSA
	Not specified – ISTA	
First count, final count (days):	7, 14	P. eatonii (method 1), P. penlandii, Penstemon spp. not named in other AOSA procedures (method 1) – AOSA
	7, 21	P. barbatus, P. hartwegii, Penstemon hybrids – ISTA P. strictus – AOSA
	8, 18	P. barbatus, P. grandiflorus, P. hirsutus, P. laevigatus – AOSA
	14, 28	P. eatonii (method 2), P. palmerii, Penstemon spp. not named in other AOSA procedures (method 2) – AOSA
Dormancy-breaking treatments:	Prechill	P. barbatus, P. hartwegii, Penstemon hybrids – ISTA
	Prechill on substrate moistened with 500 ppm GA₃ for 60 days at 2–5°C; determination of post-test viability of ungerminated seeds required.	P. eatonii (method 1), Penstemon spp. not named in other AOSA procedures (method 1) – AOSA
	Submerge seeds in aerated GA₃ (350 ppm) for 24 h at room temperature, drain, blot off excess moisture and place on substrate for germination. Clip seed coats of ungerminated seeds after 14 days and continue test for 7 additional days.	P. penlandii (method 1) – AOSA

Comments:	*P. eatonii*, *P. penlandii* and *Penstemon* spp. not named in other procedures must be tested by two methods (method 1 and method 2 are the same for *P. eatonii* and *Penstemon* spp.; different methods apply to *P. penlandii*). – AOSA
	A TZ test on 400 seeds may be conducted as an alternative to method 1 for *P. eatonii* and *Penstemon* spp. (the TZ viability test is described elsewhere in this book). – AOSA
	Where two test methods are prescribed, the germination results using method 2 are reported as percentage germination. If the percentage germination is less than that obtained using method 1, the difference is reported as dormant seed. – AOSA
Evaluation guidelines:	**AOSA** – Epigeal dicot. The hypocotyl elongates carrying the cotyledons above the soil surface. The epicotyl usually does not show any development within the test period. A primary root; secondary roots will not compensate for a missing or weak primary root. AOSA (2003b) includes descriptions of defects that will prevent continued development.
	ISTA – A dicotyledon with epigeal germination. The shoot system consists of an elongated hypocotyl and two cotyledons with the terminal bud lying between. There is no epicotyl elongation within the test period and the epicotyl and terminal bud are usually not discernible. The root system consists of a primary root, usually with root hairs, which must be well developed, as secondary roots cannot be taken into account.
Illustrations:	**Plate 14.20 Scrophulariaceae** *Penstemon heterophyllus*: **A1–A2**. A fully developed seedling has a long primary root, an elongated hypocotyl and two broadly lanceolate, obtuse-tipped cotyledons. **B.** The exposed portion of the cotyledons appears to be undamaged. The adhering seed coat must be removed to determine whether at least 50% of the original cotyledon tissue is free from necrosis or decay. **C.** A lesion at the hypocotyl/root juncture and a stubby primary root render this seedling incapable of continued normal development. *Note*: seedling sizes are not comparable due to varying degrees of magnification.

[1]J. Mari, personal observation.

Solanaceae
Atropa sp., *Browallia* spp., *Brugmansia* sp., *Capsicum* spp., *Nicotiana* spp., *Nierembergia* spp., *Petunia* spp.

Germination conditions: ***Nierembergia hippomanica***, chuscho; ***Nierembergia*** spp., cupflower, nierembergia

Temperature:	20°C	*N. hippomanica* – ISTA
	20–30°C	*N. hippomanica* – ISTA
		Nierembergia spp. – AOSA
Substrates:	P – AOSA	
	TP – ISTA	
Light/dark:	Not specified	
First count, final count (days):	5–7, 21	*N. hippomanica* – ISTA
	6, 14	*Nierembergia* spp. – AOSA

Comments: None

Dormancy-breaking treatments: None

Evaluation guidelines: **AOSA** – Epigeal dicot. Cotyledons expand and become thin, leaf-like and photosynthetic. The hypocotyl elongates and carries the cotyledons above the soil surface. The epicotyl usually does not show any development within the test period. A primary root; secondary roots or root hairs may develop within the test period for some species, but will not compensate for a missing, weak or stubby primary root. AOSA (2003b) includes descriptions of defects that will prevent continued development.

Glassy-looking hypocotyls may be the result of unfavourable test conditions, such as too much moisture. – AOSA

Primary roots in some small-seeded species may appear to be stunted. If the root and root hairs are sufficient to hold the seedling upright and the seedling is otherwise normal, it is capable of continuing development to produce a plant. – AOSA

ISTA – A dicotyledon with epigeal germination. The shoot system consists of an elongated hypocotyl and two cotyledons with the terminal bud lying between. There is no epicotyl elongation within the test period and the epicotyl and terminal bud are usually not discernible. The root system consists of a primary root, usually with root hairs, which must be well developed, as secondary roots cannot be taken into account.

Illustrations: **Plate 14.21 Solanaceae** *Nierembergia*: **A.** A fully developed seedling has two narrowly lanceolate, obtuse-tipped cotyledons, a hypocotyl and a primary root with root hairs. Elongation of root and hypocotyl varies with test conditions. **B1–B2.** A watery seedling incapable of shedding the seed coat is considered to be incapable of continued development. *Note*: seedling sizes not comparable due to varying degrees of magnification.

Germination conditions: ***Petunia*** × ***hybrida*** and ***Petunia*** spp., petunia

Temperature	20°C	*Petunia* spp. – AOSA *Petunia* × *hybrida* – ISTA
	20–30°C	*Petunia* spp. – AOSA *Petunia* × *hybrida* – ISTA
Substrates	P, TB – AOSA TP, KNO$_3$ – ISTA	
Light/dark	Light – AOSA Not specified – ISTA	
First count, final count (days):	5–7, 14 – ISTA 7, 14 – AOSA	
Comments:	Some commercial seed testing laboratories test *Petunia* spp. at a constant 25°C – JMari[1]	
Dormancy-breaking treatments:	Prechill, KNO$_3$ – ISTA Certain cultivars require KNO$_3$ and prechill – AOSA	

Evaluation guidelines:

AOSA – Epigeal dicot. Cotyledons expand and become thin, leaf-like and photosynthetic. The hypocotyl elongates and carries the cotyledons above the soil surface. The epicotyl usually does not show any development within the test period. A primary root; secondary roots or root hairs may develop within the test period for some species, but will not compensate for a missing, stubby or weak primary root. AOSA (2003b) includes descriptions of defects that will prevent continued development.

The hypocotyls of some seedlings incapable of normal development may appear watery or glassy. However, glassy-looking hypocotyls may also develop under unfavourable test conditions, e.g. condensation inside the test container or insufficient light. Glassy seedlings not orientated toward the light source and which fail to shed the seed coat are considered to be incapable of producing a plant. Such seedlings may also have short roots. – AOSA – JMari[1]

Seedling characteristics vary among cultivars; dwarf compact, ruffled and double-flowered cultivars are short-rooted types; testing at 20°C may facilitate seedling evaluation. – AOSA

Primary roots in some small-seeded species (e.g. *Petunia*) may appear to be stunted. If the root and root hairs are sufficient to hold the seedling upright and the seedling is otherwise normal, it is capable of continuing development to produce a plant. – AOSA

ISTA – A dicotyledon with epigeal germination. The shoot system consists of an elongated hypocotyl and two cotyledons with the terminal bud lying between. There is no epicotyl elongation within the test period and the epicotyl and terminal bud are usually not discernible. The root system consists of a primary root, usually with root hairs, which must be well developed, as secondary roots cannot be taken into account.

Illustrations:

Plate 14.22 Solanaceae *Petunia*: **A1–A4**. A fully developed seedling has two rounded cotyledons, an elongated hypocotyl and a primary root with root hairs. Degree of elongation varies with test environment. Seedlings with three cotyledons will develop normally. Smaller cotyledons in combination with a transparent hypocotyl and shorter root indicate a deficient seedling. **B**. A transparent, or 'glassy', seedling with all the essential structures of a normal seedling may be the result of laboratory test conditions, e.g. excessive moisture or light intensity or duration. If this seedling is phototropic it is considered capable of continued normal development. **C**. This watery seedling is not phototropic, has a short root and is incapable of shedding the seed coat. It is considered incapable of developing into a plant. *Note*: seedling sizes are not comparable due to varying degrees of magnification.

[1]J. Mari, personal observation.

Verbenaceae
Glandularia sp., *Lantana* sp., *Verbena* spp.

Germination conditions: ***Glandularia*** × ***hybrida***, ***Verbena*** × ***hybrida*** (syn), ***Verbena*** × hybrids, verbena; ***Verbena bonariensis***, purple-top vervain, purpletop, South American vervain, tall vervain; ***Verbena canadensis***, ***Glandularia canadensis*** (syn), rose vervain; *Verbena rigida*, purple vervain

Temperature:	15°C	*V. bonariensis, V. canadensis, V.* × hybrids, *V. rigida* – ISTA
	20°C	*G.* × *hybrida, V.* × hybrids – AOSA – ISTA
	20–30°C	*G.* × *hybrida* – AOSA *V. bonariensis, V. canadensis, V.* × hybrids – ISTA

Substrates:	TB – AOSA TP, KNO$_3$ – ISTA

Light/dark:	Light – AOSA Not specified – ISTA

First count, final count (days):	7–10, 28 – ISTA 8, 18 – AOSA

Comments:	None

Dormancy-breaking treatments:	Prechill, KNO$_3$ – ISTA

Evaluation guidelines:	**AOSA** – The AOSA *Seedling Evaluation Handbook* (AOSA, 2003b) does not include descriptions of seedlings of Verbenaceae; the family is not included in descriptions for 'miscellaneous agricultural and horticultural' kinds.
	ISTA – A dicotyledon with epigeal germination. The shoot system consists of an elongated hypocotyl and two cotyledons with the terminal bud lying between. There is no epicotyl elongation within the test period and the epicotyl and terminal bud are usually not discernible. The root system consists of a primary root, usually with root hairs, which must be well developed, as secondary roots cannot be taken into account.

Illustrations:	**Plate 14.23 Verbenaceae** *Glandularia*: **A1–A4**. A fully developed seedling has narrowly lanceolate cotyledons and a long hypocotyl, all with glandular hairs (**A3**), and a long primary root with root hairs. If otherwise normal, seedlings with three cotyledons will continue development to produce a plant. **B**. The primary root of this seedling is short and stubby, insufficient for development of a plant. **C1–C2**. Adhering seed coats must be removed to examine cotyledons. These exhibit only minor necrosis, far less than 50% of the total cotyledon tissue, and are expected to function normally. Enclosed in the seed coat, the cotyledons have not yet developed chlorophyll. *Note*: seedling sizes are not comparable due to varying degrees of magnification.

Violaceae
Viola spp.

Germination conditions: ***Viola cornuta***, viola, bedding pansy, horned pansy, horned violet, tufted pansy, pensamiento; ***Viola odorata***, English violet, florist's violet, garden violet, sweet violet, banafsha, nioi-sumire; ***Viola tricolor***, pansy, European wild pansy, field pansy, heart's ease, johnny-jump-up, love-in-idleness, miniature pansy, wild pansy, pensée, sanshiki-sumire

Temperature:	10°C	*V. odorata* – ISTA
	20°C	*V. cornuta, V. tricolor* – AOSA – ISTA
	20–30°C	*V. cornuta, V. tricolor* – AOSA – ISTA

Substrates:	TB – AOSA TP, KNO$_3$ – ISTA
Light/dark:	Not specified
First count, final count (days):	4–7, 21 – ISTA 7, 21 – AOSA
Comments:	None
Dormancy-breaking treatments:	Prechill, KNO$_3$ – ISTA Prechill 7 days at 5°C with KNO$_3$ – AOSA

Evaluation guidelines:

AOSA – Epigeal dicot; cotyledons which expand and become thin, leaf-like and photosynthetic; endosperm. The hypocotyl elongates and carries the cotyledons above the soil surface. The epicotyl usually does not show any development within the test period. A primary root with root hairs usually developing within the test period. Secondary roots will not compensate for a defective primary root. AOSA (2003b) includes descriptions specific to *Viola* spp. with illustrations.

Seedlings producing three cotyledons, if otherwise normal, will continue to develop normally. – AOSA

Unshed seed coats must be removed and cotyledons evaluated; if seed coat cannot be removed, the seedling is considered incapable of continuing development. – AOSA

Certain *V. tricolor* cultivars have characteristically short primary roots. If root and hypocotyl are proportionally balanced, such seedlings will continue normal development. – AOSA

ISTA – A dicotyledon with epigeal germination. The shoot system consists of an elongated hypocotyl and two cotyledons with the terminal bud lying between. There is no epicotyl elongation within the test period and the epicotyl and terminal bud are usually not discernible. The root system consists of a primary root, usually with root hairs, which must be well developed, as secondary roots cannot be taken into account.

Illustrations:

Plate 14.24 Violaceae *Viola*: **A1–A2**. A fully developed seedling has an elongated hypocotyl and primary root, with root hairs, and rounded cotyledons **B1–B2**. Adhering seed coats must be removed to examine the cotyledons. With the seed coat removed, the freed cotyledons can be seen to be functional. **C1–C2**. An adhering seed coat with visible portion of cotyledons appearing watery. Upon removal of the seed coat, the cotyledons are found to be watery, discoloured, and not spreading. **D1–D2**. A seedling with a shortened, granular hypocotyl is considered to be unable to continue development into a plant. *Note*: seedling sizes are not comparable due to varying degrees of magnification.

References

[AOSA] Association of Official Seed Analysts. 1954. Rules for testing seeds adopted by Association of Official Seed Analysts. *In*: B.E. Clark (ed.) Proceedings of AOSA 44th annual meeting. 28 June–2 July 1954; Ames, Iowa. pp. 59–64.

[AOSA] Association of Official Seed Analysts. 2002. Uniform classification of weed and crop seed. Contribution no. 25 to the *Handbook on Seed Testing*. Association of Official Seed Analysts. 274 pp.

[AOSA] Association of Official Seed Analysts. 2003a. *Rules for Testing Seeds*. Association of Official Seed Analysts. 166 pp.

[AOSA] Association of Official Seed Analysts. 2003b. *Seedling Evaluation Handbook*. Contribution no. 35 to the handbook on seed testing. Rev. edn. Association of Official Seed Analysts. 128 pp.

Ashton, D. 2001. Germination testing. *In*: M. McDonald, T. Gutormson and B. Turnipseed, (eds). *Seed Technologist Training Manual: Proficiency and Professionalism through Training*. Society of Commercial Seed Technologists. 344 pp.

Atwater, B.R. 1980. Germination, dormancy and morphology of the seeds of herbaceous ornamental plants. *Seed Sci. Technol.* 8:523–533.

Baskin, C.C. and J.M. Baskin. 2001. *Seeds: Ecology, Biogeography, and Evolution of Dormancy and Germination*. Academic Press, San Diego, California. 666 pp.

Bekendam, J. and R. Grob. 1979. *ISTA Handbook for Seedling Evaluation*. The International Seed Testing Association (ISTA), Zurich. 130 pp.

Chirco, E. 1987. Request for testing information: preliminary methods for testing less common flower seed species. *Association of Official Seed Analysts Newsletter* 61:21–30.

Chirco, E. and T.L. Turner. 1986. Species without AOSA testing procedures. *Association of Official Seed Analysts Newsletter* 60:2–70.

Don, R. 2003. *ISTA Handbook on Seedling Evaluation*. 3rd edn. The International Seed Testing Association [ISTA], Bassersdorf, Switzerland. Unpaginated.

Dorph-Petersen, K. 1926. Report of the Work of the International Seed Testing Association from 6 September 1925 to 31 October 1926. Six mimeographed pages, 8¼″ × 13″, printed front and back. Located at: California State Seed Laboratory, Sacramento, California.

Heit, C.E. 1958. A new concept and new fields in germination testing. *In*: D. Isely (ed.). *Fifty Years of Seed Testing, 1908–1958*. Association of Official Seed Analysts. 111pp.

[ISTA] International Seed Testing Association. 1985. International rules for seed testing. *Seed Sci. Technol.* 13(2):455–462.

[ISTA] International Seed Testing Association. 2003. *International Rules for Seed Testing*, ed. 2003. ISTA, Bassersdorf, Switzerland.

Lubbock, J. 1892a. *A Contribution to our Knowledge of Seedlings*. Vol. I. D. Appleton & Co., New York. 608 pp.

Lubbock, J. 1892b. *A Contribution to our Knowledge of Seedlings*. Vol. II. D. Appleton & Co., New York. 646 pp.

Meyer, D.J.L. 1998. Seed quality problems commonly encountered in the laboratory for vegetable and flower seeds. *Seed Technol.* 20(2):136–161.

Meyr, A. 1998. Flower seed committee report. *In*: A.L. Meyr and C.J. Weidmaier (eds). *History of the Society of Commercial Seed Technologists, Inc., 1922–1998*. Society of Commercial Seed Technologists. 130 pp.

Ripka, Z. 2003. Flower seed testing: preparation work of the ISTA handbook on flower seed testing. *ISTA News Bulletin* 125:32–33.

Tillich, H.-J. 1995. Seedlings and systematics in monocotyledons. *In*: P.J. Rudall, P.J. Cribb, D.F. Cutler and C.J. Humphries (eds). *Monocotyledons: Systematics and Evolution*. Vol. 1. Royal Botanic Gardens, Kew, London. 825 pp.

Toole, E.H. 1925. Changes in 'Rules for Seed Testing.' *Proceedings of AOSA Annual Meeting* 17:35–37.

15 Tetrazolium Testing for Flower Seeds

Annette Miller

USDA/ARS National Center for Genetic Resources Preservation,
1111 South Mason Street, Fort Collins, CO 80521-4500, USA

Introduction

The tetrazolium test is a biochemical viability test. Biochemical viability tests indicate the presence or absence of living cells in seeds. This type of test detects metabolic activity and cell respiration in hydrated tissues. Living cells respire. During respiration, enzymes assist the conversion of sugars to energy which is used for the construction of proteins, membranes and new cells. Biochemical tests detect this activity by introducing special compounds into the system and allowing reduction of these compounds to new forms with altered physical properties. To be useful in seed testing, the altered physical property should be a visible colour change that experienced seed analysts can interpret.

The term 'viability' in seed testing refers to the capacity to produce a normal plant. Germination and biochemical tests measure different aspects of seed viability. Germination evaluates the process of cell elongation and division and tissue differentiation of the emerging plant. Biochemical tests are a snapshot of life activity. A germination test indicates growth speed and the presence or lack of dormancy. Biochemical tests deduce dormancy only when the biochemical test is done in conjunction with a germination test. A biochemical test alone cannot provide dormancy information. Germination tests verify that the emerging cotyledons or epicotyl are photosynthetic and functioning normally.

Since the embryonic tissues are not yet photosynthetic, the biochemical test does not evaluate this capacity.

Detection of life in a desiccation-tolerant dry seed is not practical today. Respiration is so low that detection would require long periods of time and particularly sensitive instruments to detect it. For practical purposes, dry seeds do not respire. Most growth and respiration occurs when tissues are hydrated. This is a limitation of biochemical viability tests; they must be done on hydrated tissue.

Cell walls and organelle membranes in dehydrated cells are contracted and slightly dislocated. Water imbibition initiates cell wall expansion and organelle membrane repair. Mitochondria are the organelles where most of the cell's respiration takes place. Mitochondrial membranes contain enzymes called dehydrogenases. In the biochemical test known as the tetrazolium test, the tetrazolium solution, which is colourless, is added to the seed tissue and the tetrazolium compound is chemically reduced as a result of dehydrogenase activity (Fig. 15.1). After reduction, the compound's light absorption properties are changed and the reduced compound, called formazan, appears red. In this test, the red-stained tissue is considered direct evidence of active respiration. As a consequence of preparing and examining seeds for the biochemical test, the analyst may also learn about nutritive tissue soundness, pathogen presence, embryo maturity and embryo turgidity.

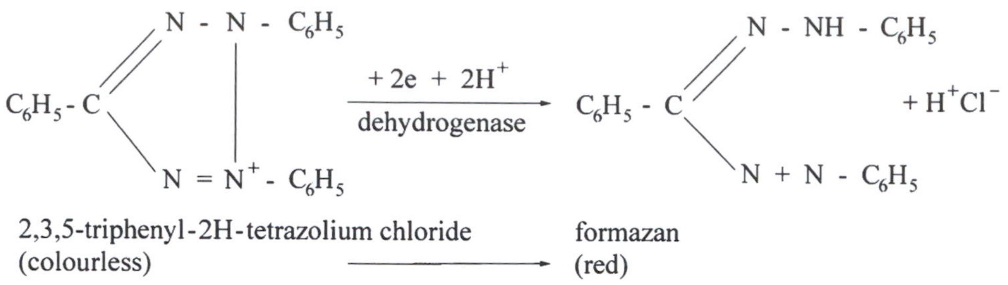

$$C_6H_5-C \diagup\diagdown \begin{matrix} N - N - C_6H_5 \\ | \\ N = N^+ - C_6H_5 \end{matrix} \xrightarrow[\text{dehydrogenase}]{+ 2e + 2H^+} C_6H_5 - C \diagup\diagdown \begin{matrix} N - NH - C_6H_5 \\ \\ N + N - C_6H_5 \end{matrix} + H^+Cl^-$$

2,3,5-triphenyl-2H-tetrazolium chloride   formazan
(colourless)   (red)
                    ──────────────────►

Fig. 15.1. Reduction of 2,3,5-triphenyl-2H-tetrazolium chloride to formazan red.

History of Tetrazolium Testing

Steiner (1997b) and Moore (1969) describe the historical development of biochemical viability tests. Many compounds have been used for biochemical viability tests of seeds (Grabe, 1959). A suitable reagent must have several attributes. The compound must undergo a visible chemical change when it is in the presence of actively respiring cells. The compound must be non-toxic, inexpensive, readily available, water soluble, and have a relatively low formula weight. Low formula weight and water solubility facilitate transfer across membranes and into tissue.

Kozo Hasegawa was the first to use chemicals such as tellurium and selenium for biochemical viability tests in seeds in the early part of the 20th century. He introduced this idea to Franz Erich Eidmann, W. Schmidt and Georg Lakon, who developed the concept further. They saw the importance of evaluating staining patterns in embryonic structures instead of simply noting the presence of stain. Lakon, with help from chemists Helene Bulat and Richard Kuhn, replaced the poisonous selenium salts with safer tetrazolium salts. Many tetrazolium derivatives change colour after reduction in living tissue. The salt that best fits the multiple criteria of low expense, low formula weight, low toxicity and wide availability is 2,3,5-triphenyl-2H-tetrazolium chloride (TTC). This has been the compound of choice in seed testing for over 50 years.

Handbooks by the International Seed Testing Association (Moore, 1985; Leist and Krämer, 2003) and the Association of Official Seed Analysts (Peters, 2000) have attempted to standardize the test procedures and seed viability evaluation using this compound. The need for standardization in seed testing methods and the development of tetrazolium testing handbooks have resulted in the almost exclusive use of TTC (TTC is the compound's abbreviation, while TZ refers to the tetrazolium test in general or any solution made with the TTC compound). Other biochemical reagents are used in research. In 1955, only 240 tetrazolium salts were known. Today, over 1000 tetrazolium salts are known (Steiner, 1997a). Scientists may yet find a better, less toxic, less expensive compound for biochemical testing. Opportunities for investigating new chemicals that detect seed viability are expanding as the science of biochemistry advances.

Tetrazolium Tests – the Current Status

The flower seed industry uses tetrazolium testing for many purposes. Tetrazolium (TZ) viability test information is used for year-to-year seed lot carry-over decisions, conditioning decisions, detection of dormancy (when combined with a germination test) and supplemental label information. Seed lots are sometimes sold using TZ information, but only on the condition that a germination test will be done by the time the seed is shipped. The US Federal Seed Act, most US state seed laws, Canadian Seed Law, and International Seed Testing Association certificates for international seed commerce all require that germination be listed on a seed lot label. These legal requirements are the main reasons why TZ information is rarely used as a stand-alone primary viability indicator.

There are practical and scientific reasons for using TZ information as a supplemental viability indicator. While TZ tests can be done in 1–2 days versus the 1–2 weeks needed for a germination test, they take more hands-on analyst time than does a germination test. Automation is standard for germination planting but is impractical for TZ test seed preparation. For most species, each seed must be skilfully cut before staining. During evaluation, further dissection may be needed. TZ test evaluation frequently requires 10–40× magnification using a dissection stereomicroscope. Germination evaluation, however, is usually done without further manipulation and with little or no magnification. Even when only 200 seeds are tested, as is commonly done in TZ tests, the analyst time required can be from two to ten times as long as the time spent performing a germination test on 400 seeds.

Conducting a Tetrazolium Test

The International Seed Testing Association (ISTA) and the Association of Official Seed Analysts (AOSA) handbooks (Moore, 1985; Peters, 2000) and working sheets (Leist and Krämer, 2003) describe methods of testing seeds for hundreds of taxa. This chapter will provide additional illustrations and notes for important families in ornamental plant commerce.

Sampling

Primary viability (germination or tetrazolium) tests should be performed using 400 seeds in replicates of 100 seeds or fewer (AOSA, 2003; ISTA, 2004) while supplementary tests are traditionally performed on 200 seeds. As is true for germination tests, 400 seeds is the number tested to reduce unavoidable sampling error to an acceptable level (Steiner, 1997c). Maximum lot size limits specified by ISTA Rules (ISTA, 2004) also help ensure that seed test results are

meaningful. AOSA and ISTA both provide sampling procedures and recommendations for obtaining a representative sample from a seed lot.

The analyst's first task in the laboratory is determination of pure seed as defined by AOSA (2003) or ISTA (2004). Differences in pure seed determination dramatically influence viability test results. For example, the AOSA *Rules for Testing Seed* (AOSA 2003) define pure seed for some species as including shrivelled and immature seeds. If the analyst avoids these seeds when counting 400 seeds for the test, the results will not reflect the true viability of the sample. Random sampling of finely conditioned and graded seed lots is straightforward. Ungraded lots from species with little domestication are more difficult to sample properly. Species with a wide range of seed size or chaffy appendages may cause differential or biased selection (as opposed to random selection) of seeds for a viability test. Adequate mixing prior to seed selection reduces this bias. There are three parts to the tetrazolium test: preconditioning, staining and evaluation.

Preconditioning

Preconditioning is the next step after seeds have been counted for the test. As noted above, the tetrazolium test must be conducted on hydrated tissue. During preconditioning, seeds are imbibed (hydrated), preferably slowly, on moist media (germination towels, blotters or filter paper). If seeds have hard seed coats that inhibit water uptake (as opposed to rigid seed coats that do not inhibit imbibition, e.g. Lamiaceae), the seed coats must be chipped or nicked (Fig. 15.2). Some families with seeds that have hardseededness are Fabaceae, Convolvulaceae and Malvaceae.

Staining

The staining portion of the tetrazolium test may or may not require cutting (Figs 15.3, 15.4, 15.5) or piercing (Fig. 15.6) the seed prior to immersion in the tetrazolium solution (TZ). Most seeds need to be cut in a way that exposes the embryo and nutritive tissue for adequate tetrazolium solution uptake. All cuts should be done with a slicing motion. A chopping motion with a scalpel or razor blade can cause artefacts. Artefacts are damaged areas that stain abnormally and are a direct result of cutting or embryo extraction efforts. When making lengthwise (termed 'longitudinal' in the handbooks) cuts, some analysts prefer to make an incomplete cut that leaves the two halves together for staining and prevents the embryo from floating out of the seed during staining. One can cut down deep, leaving part of the seed coat intact (Fig. 15.3), or cut three-quarters of the length of the seed, leaving the two halves connected at the cotyledon end. If the seeds are completely cut in half the analyst will either have to keep

both halves together by placing them in a shallow dish on a blotter soaked in the TZ solution, or discard half and test the remaining half. Accurate cutting is crucial for the latter method. Species with permeable seed coats are placed intact into the TZ solution.

The TZ solution consists of 2,3,5 triphenyl-2H-tetrazolium chloride (TTC) salt in water or in water with buffer additives to maintain a neutral pH. A 1.0% solution is generally used to stain intact seeds or seeds with peripheral cuts (Figs 15.4, 15.5) or a needle pierce (Fig. 15.6). A 0.1% solution is sufficient for seeds that have been bisected through the embryo (Fig. 15.3). Vials or shallow dishes containing the seeds and the TZ solution are typically placed in a 35°C incubator for as short a time as several hours to as long as 2 days for some species. Staining proceeds at room temperature. Higher temperatures allow shorter staining times. TZ concentration, seed preparation method, and seed physiology influence staining time and

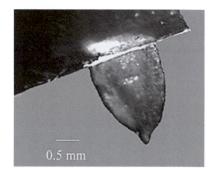

Fig. 15.4. Example of a lateral cut to allow TZ solution uptake in *Dianthus* seed.

Fig. 15.2. Example of a chipped or nicked hard seed coat of *Lupinus* (Fabaceae) to allow water imbibition (preconditioning).

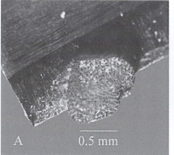
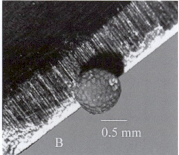

Fig. 15.3. Examples of a longitudinal cut of (A) *Salpiglossis* and a deep cut of (B) *Petunia* to allow TZ solution uptake.

intensity. Very large seeds may benefit from low TZ concentration (0.075%), moderate temperature and extended staining time. This evens out the stain in the tissue and ameliorates evaluation problems that occur from dense tissue that would otherwise overstain on the outside and remain unstained inside (Néto *et al.*, 1998).

Manipulation of the seeds after staining may be necessary. Some seeds with opaque seed coats or covering structures can be 'cleared' by immersion in glycerol or lactic

acid for at least a day. Clearing by this method eliminates the need for further dissection or cutting by making the opaque structures transparent.

Uncleared seeds that were previously uncut or were only peripherally cut may need fresh cuts to examine the embryo and surrounding tissues. Seeds with an incomplete cut that keeps the two halves together are spread apart with dissecting needles or forceps for evaluation (Fig. 15.7). Bisected seeds need no further dissection.

Fig. 15.5. Example of an edge cut for TZ solution uptake in *Iris* seed. Only nutritive tissue is exposed. The embryo is embedded within the nutritive tissue and is intact for even staining. After staining, the seeds will be cut again to expose the embryo for examination.

Fig. 15.6. Example of the piercing technique for TZ solution uptake in *Lobelia* seeds.

Evaluation

Evaluation is the last step. Criteria for evaluating stained seeds are given in several references (Leist and Krämer, 2003; Peters, 2000; Leadly and Hill, 1971). The embryo is examined for a uniform or a gradual graded stain that is most intense at the cut surfaces. Uneven staining or unstained areas are evaluated as they affect the seed's essential structures: the radicle, cotyledons, hypocotyl–cotyledon juncture (Fig. 15.8) and the epicotyl, if developed. Unstained or damaged areas are permitted on many kinds of seeds as long as the affected areas are away from critical junctures and are superficial. The extent of damaged tissue allowed is described in the TZ testing reference materials (Leist and Krämer, 2003; Peters, 2000). Nutritive tissue, if present, may or may not be living. If living, it must be normally stained. If non-living and unstained, it must be free of blemishes and texture differences that indicate pathogen presence or deterioration (Plate 15.1).

Fig. 15.7. Two halves of *Penstemon* seed spread apart for evaluation.

Critical areas
(*Dianthus*)

Cotyledon/
hypocotyl
juncture
unstained

Radicle
unstained

Fig. 15.8. Examples of critical areas to evaluate in *Dianthus* embryos.

Each seed is examined and scored as 'viable' or 'non-viable'. The result of a TZ test is stated as percentage viable and should never be stated as germination.

Embryo characteristics

Embryos have a wide range of shapes and states of differentiation. Learning the embryo orientation, embryo shape and the characteristics of any surrounding nutritive tissue is a fundamental part of learning how to conduct TZ tests. Taxa within families often show like seed characteristics. Learning the characteristics of a family can be important when testing an unfamiliar species. In the following section, seed anatomy illustrations are provided for 26 families.

Betty Ransom Atwater, founder of Ransom Seed Laboratory, placed seeds into two broad groups in her studies of dormancy and morphology of herbaceous ornamental plant seeds (Atwater, 1980). Endospermic seeds have significant nutritive tissue (endosperm or perisperm) present at maturity, while non-endospermic seeds have residual or no nutritive tissue and have well-developed embryos (Fig. 15.9). Within the endospermic group, rudimentary type embryos consist of a small mass of undifferentiated cells embedded in a large mass of nutritive tissue. Miniature embryos, however, fill the length of the seed, but the seeds are exceptionally small (less than 2 mm). Linear (embedded within nutritive tissue) and peripheral (curved around central nutritive tissue) embryos are larger. Atwater (1980) grouped non-endospermic types of seeds by the nature of their covering structures (e.g. mucilaginous, hard-seeded, woody).

Embryo size can vary within a sample. This is due to embryo or seed maturity. Lots from species that exhibit indeterminate flowering can exhibit a wide range of seed maturity. Seeds that may be the same size and shape may have drastically different embryo sizes (Fig. 15.10). Germination of seed lots with this wide range of embryo sizes can be less uniform and may require extended incubation to achieve maximum germination. The TZ test can identify this situation.

Morphological differences influence our confidence in TZ viability test results, although quantified evidence is not available. For example, compare a normal seedling with a TZ-stained rudimentary or miniature embryo. Our confidence about the ability of these two 'viable' seeds to produce normal plants is not the same. We are more confident that the fully formed seedling will produce a normal plant than we are about the plant-producing potential of the small undifferentiated embryo (Fig. 15.11).

Large embryo evaluation presents its own set of difficulties. While less magnification is needed, very large embryos may have defects that are missed in a TZ test. In many cases the analyst is examining the exterior of the embryo and only one cut plane. This limited view can hide deterioration in a critical place. Novice analysts often anticipate that larger seeds will be easier to test. However, all analysts spend more time evaluating larger embryos than smaller ones. Large embryo evaluation is more detailed. The analyst must answer many questions about each seed

Embryo types

Endospermic:

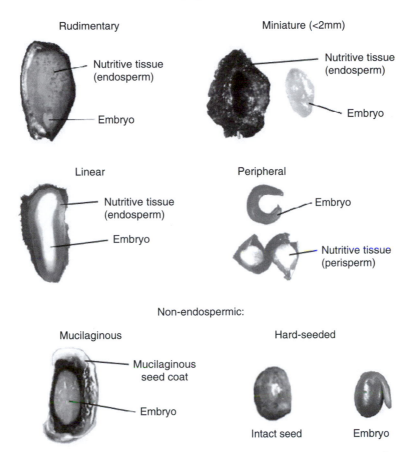

Rudimentary

Nutritive tissue
(endosperm)

Embryo

Miniature (<2mm)

Nutritive tissue
(endosperm)

Embryo

Linear

Nutritive tissue
(endosperm)

Embryo

Peripheral

Embryo

Nutritive tissue
(perisperm)

Non-endospermic:

Mucilaginous

Mucilaginous
seed coat

Embryo

Hard-seeded

Intact seed   Embryo

Fig. 15.9.   Examples of Atwater's embryo types (Atwater, 1980). Photographs by Annette Miller.

Fig. 15.10.   Differential embryo maturity in *Carum* seeds.

before determining whether a seed is viable or non-viable. Does the damage extend into the conducting tissue of the radicle? What proportion of the cotyledons is unstained? Is damage located near a critical juncture? In contrast, rudimentary embryos must stain entirely. There are no decisions to make about the proportion of allowable damage in a small mass of undifferentiated tissue.

Normally stained viable tissue is a bright cherry-red colour. Experience teaches the analyst which staining anomalies might be the result of artefacts from seed preparation (Fig. 15.12). Abrasions may stain darker, while seeds cut with a chopping motion may

Results of two kinds of viability tests (*Lobelia*):

Germination
- Normal seedling

Tetrazolium test
- Stained embryo

Photosynthetic cotyledons

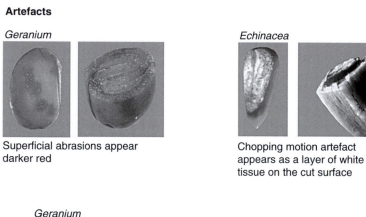

Normal growth

Healthy root and hypocotyl development

Miniature embryo with little tissue differentiation

0.2 mm

7 cm

Fig. 15.11. Results of two kinds of viability test. Our confidence that a seedling will produce a plant is much higher than it is for the small undifferentiated embryo.

Artefacts

Geranium

Superficial abrasions appear darker red

Echinacea

Chopping motion artefact appears as a layer of white tissue on the cut surface

Geranium

This example shows the importance of evaluating both halves. The radicle on the left is a thin slice of the outer shealth of the radicle and does not contain conducting tissue. The apparently unstained radicle is an artefact of an imperfect cut. This seed is viable.

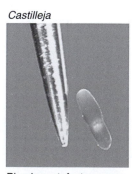

Castilleja

Piercing artefact appears white

Fig. 15.12. Types of artefacts commonly encountered as a result of seed preparation for a TZ test.

have a thin layer or flakes of white unstained tissue on the cut surface.

Fungi (Plate 15.1), bacterial pathogens and bruised tissue (Fig. 15.12) often stain darker, purplish red. Lesions and broken structures (Fig. 15.13) can stain darker or lighter than the surrounding tissue, depending on the cause and timing of the damage relative to when the TZ test was done. Severe heat damage appears as unstained or pale areas. Areas adjacent to severe heat damage are sometimes stained an orange-red colour. Freeze damage causes darker purple-red stain because ruptured cell walls permit more TZ solution penetration into damaged tissue. Abrupt changes in stain intensity are a cue for the analyst to examine the tissue more closely. Tissue turgidity is also examined. Healthy or turgid embryos maintain their shape as they are handled (Fig. 15.14). Flaccid and unhealthy tissue is limp and does not maintain its shape or spring back when manipulated with a needle or forceps.

Seed preparation and evaluation skills improve with experience. Training with other experienced analysts on the job, through workshops and in college courses, is essential. Comparisons with germination tests, plug growouts and field performance results are instructive and help hone the analyst's evaluation skills. These comparisons also provide the basis for ongoing improvement of TZ testing handbooks and references.

Limitations

TZ tests have limitations. External treatments like pelleting and coating on seeds of all sizes can cause germination abnormalities that are undetectable in the TZ test. The TZ test does not measure the capacity for normal photosynthesis and albinos stain normally. Pathogen presence can confound comparisons between TZ and germination results. Fungi and bacteria stain, but viruses do not. The TZ test will not detect how pathogens on the seed coat affect germination. These are some reasons why the germination test is still required by law for primary viability information.

Benefits

Managers and customers appreciate the timeliness of TZ results and the TZ test's usefulness as a diagnostic tool. When seeds have lengthy germination periods it is an advantage to have preliminary viability information for customers. The TZ test, in combination with the germination test, is for many species the only way to measure dormancy. Supplemental, timely TZ test information for buyers and seed lot managers is a benefit to both the seed industry and consumers.

Tetrazolium tests detect living cells by detecting respiration. Researchers in the 20th century developed the test and learned how to evaluate seed tissues. They wrote handbooks with the aim of standardizing the test. These standardization efforts continue today.

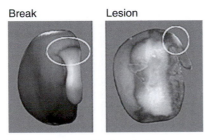

Fig. 15.13. Breaks and lesions in *Lupinus*.

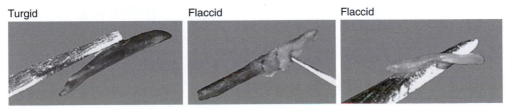

Fig. 15.14. Turgid tissues are firm when touched with a needle while flaccid tissues sink and are limp.

The flower seed industry uses the tetrazolium test to detect dormancy and provide supplementary viability assessments that aid managers and seed brokers. Analysts who perform TZ tests must understand the correct sampling, preconditioning, staining and evaluation procedures as described in this chapter and other seed testing references (Moore, 1985; Peters, 2000; AOSA, 2003; Leist and Krämer, 2003; ISTA, 2004). The tetrazolium test has become a staple of the seed analyst's toolkit. Its value grows each year as analysts gain experience testing more species and techniques are improved and standardized. Future collaborations among seed analysts, biochemists, researchers and customers will cultivate these refinements and broaden the TZ test's use as an important indicator of seed viability.

Tetrazolium Test Notes and Photographs for 26 Plant Families

Embryo types and staining characteristics for 26 families are shown in Plates 15.2 to 15.27. The notes and photographs for each family are intended as supplementary information for those who have the AOSA and ISTA tetrazolium testing handbooks and working sheets (Moore, 1985; Peters, 2000; Leist and Krämer, 2003). Leadley and Hill (1971) cover 47 flower seed genera in their TZ testing reference. These handbooks and references are recommended for anyone who performs tetrazolium tests. It is hoped that seed analysts who have conducted tetrazolium tests on cereals and vegetable seeds will find this a useful introduction to families not often encountered in other areas of seed commerce.

Families represented in the plates are:

Acknowledgements

All photographs were taken at the USDA/ARS National Center for Genetic Resources Preservation (NCGRP), Fort Collins, Colorado and are provided courtesy of the USDA. Many thanks for editing assistance are due to Jennifer Crane, Miller McDonald, Paul Miller, Nick Rotindo, Laurie Thoma, Diane Skogerboe and Loren Wiesner. John Waddell provided help with photography. Sincere appreciation is extended to the NCGRP Information Systems staff. Special thanks are due to my NCGRP Seed Quality Laboratory co-workers and supervisor Patricia Conine. Several laboratories and companies provided seeds for the photographs: Applewood Seed Company, Arkansas Valley Seed Solutions, PanAmerican Seed, STA Laboratories, Colorado Seed Laboratory, Gully Greenhouse and USDA/ARS NCGRP. Thank you to Miller McDonald for support and encouragement.

References

[AOSA] Association of Official Seed Analysts. 2003. *Rules for Testing Seeds*. AOSA, Las Cruces, New Mexico.

Atwater, B.R. 1980. Germination, dormancy and morphology of the seeds of herbaceous ornamental plants. *Seed Sci. Technol.* 8: 523–573.

Grabe, D. 1959. Preliminary evaluation of tetrazolium derivatives for testing seed viability. *Proc. Assoc. Official Seed Analysts* 49(1):131–133.

[ISTA] International Seed Testing Association. 2004. *International Rules for Seed Testing*, Edn. 2004. Zurich, Switzerland.

Leadley, P.R. and J.M. Hill. 1971. *Use of the Tetrazolium Chloride Method for Determining the Viability of a Range of Flower Species.* 16th ISTA Congress, Washington, DC. 7–12 June, 1971. Preprint No. 25. ISTA, Zurich, Switzerland.

Leist and Krämer. 2003. *Working Sheets on Tetrazolium Testing.* Vols 1 and 2. ISTA, Zurich, Switzerland.

Moore, R.P. 1969. History supporting tetrazolium seed testing. *Proc. Intl Seed Testing Assoc.* 34:233–243.

Moore, R.P. (ed.) 1985. *Handbook on Tetrazolium Testing.* ISTA, Zurich, Switzerland.

Néto, J.B.F., F.C. Krzyzanowski and N.P. da Costa. 1998. *The Tetrazolium Test for Soybean Seeds.* EMBRAPA, Brazil.

Peters, J. (ed.) 2000. Tetrazolium testing handbook. Contribution No. 29 to the *Handbook on Seed Testing*, revised 2000. AOSA, Las Cruces, New Mexico.

Steiner, A.M. 1997a. Chemistry of tetrazolium salts and biochemistry of tetrazolium reduction. *In*: R. Don *et al. Proceedings of the ISTA Tetrazolium Workshop held at the OSTS Edinburgh, June 1997.* ISTA, Zurich, Switzerland. pp. 55–68.

Steiner, A.M. 1997b. History of the development of biochemical viability determination in seeds. *In* R. Don, *et al. Proceedings of the ISTA Tetrazolium Workshop held at the OSTS Edinburgh, June 1997.* ISTA, Zurich, Switzerland. pp. 7–16.

Steiner, A.M. 1997c. Strategy for establishing tolerances for topographical tetrazolium test results. *In*: R. Don *et al. Proceedings of the ISTA Tetrazolium Workshop held at the OSTS Edinburgh, June 1997.* ISTA, Zurich, Switzerland. pp. 47–54.

15.1

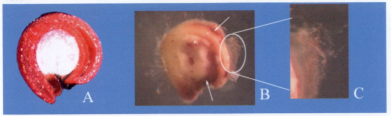

15.2
Celosia, Amaranthus, Gomphrena

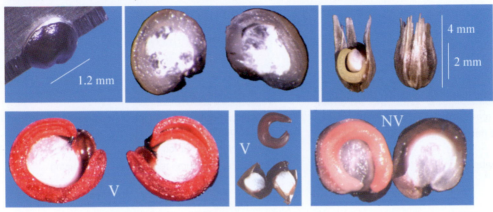

15.3
Vinca

Plate 15.1. Reduction of TTC by pathogens. **A.** Healthy stained *Amaranthus* seed. **B.** Pathogens visible on *Amaranthus* seed. Active dehydrogenases in live fungi convert TTC to formazan causing infected areas to appear red. Upper arrow points to infected cotyledon area. Lower arrow indicates infected perisperm. **C.** Stained fungal hyphae visible against the magnified background in image B.

Note: Plates 15.2 to 15.27 show cutting or piercing procedures, internal anatomy of typical genera and examples of viable and nonviable seeds. V = viable, NV = nonviable.
Plate 15.2. Amaranthaceae. Peripheral embryo with non-living perisperm.
Plate 15.3. Apocynaceae. Linear embryo with living endosperm.

15.4
Philodendron

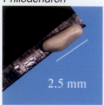     

2.5 mm

15.5
Echinacea

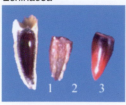 

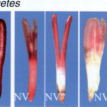

1. Fruit coat (pericarp)
2. Seed coat (testa) is membranous and tightly adhering
3. Stained embryo

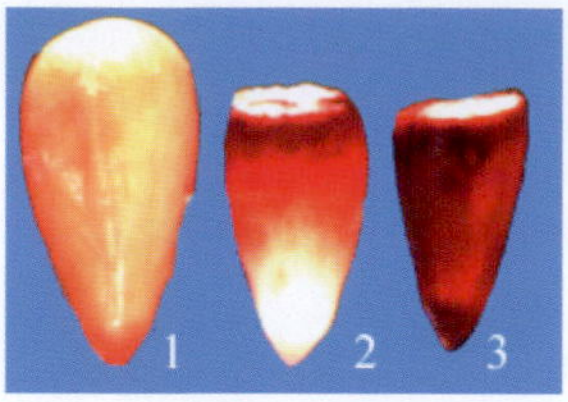

1. Membranous seed coat intact, insufficient TZ solution uptake
2. Membrane cut at top
3. All outer structures removed prior to staining

Echinacea

Tagetes

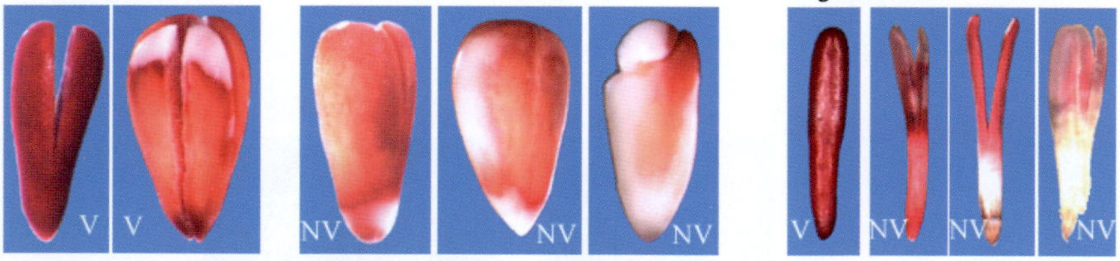

Aster laevis   *Aster novae-angliae*   *Solidago*

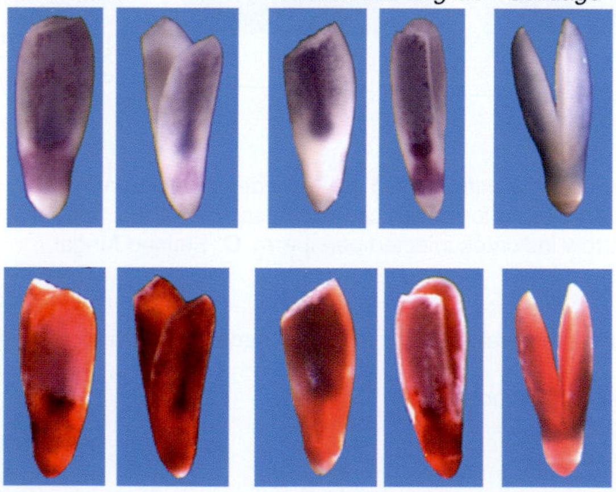

Natural purple pigmentation.
Pigmented areas stain darker.
All embryos are viable.

Before stain

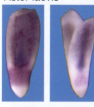

After stain

Plate 15.4. Araceae. Linear embryo with living endosperm.
Plate 15.5. Asteraceae. Non-endospermic type. Seed unit a fruit (achene).

15.6
Impatiens

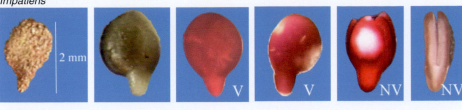

15.7
Begonia

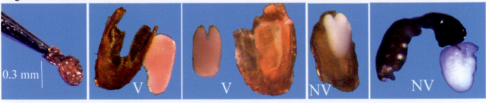

15.8
Platycodon

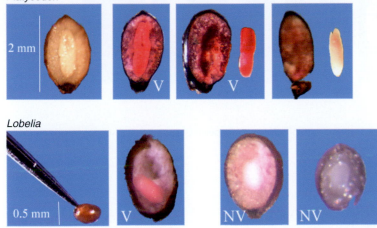

Lobelia

Plate 15.6. Balsaminiaceae. Non-endoplasmic type.
Plate 15.7. Begoniaceae. Miniature embryo with little or no nutritive tissue.
Plate 15.8. Campanulaceae. Miniature embryo with living endosperm.

15.9
Dianthus

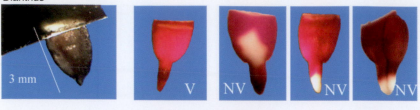

Gypsophila

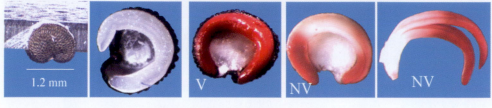

15.10
Sedum

Plate 15.9. Caryophyllaceae. Peripheral embryo with non-living endosperm or spatulate non-endospermic embryo.
Plate 15.10. Crassulaceae. Non-endospermic type.

15.11
Lupinus

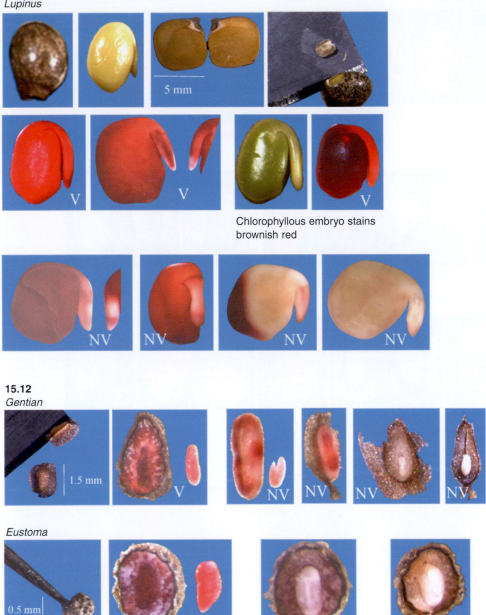

Chlorophyllous embryo stains
brownish red

15.12
Gentian

Eustoma

Plate 15.11. Fabaceae. Non-endospermic type (commonly). Endosperm non-living if present. Often hard seeded.
Plate 15.12. Gentianaceae. Miniature embryo with living endosperm.

15.13
Geranium, Pelargonium

radicle

3 mm

clip here

V   V   NV   V   V NV

Chlorophyllous embryo
stains normally

15.14
Sinningia

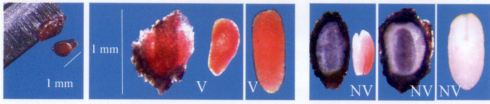

1 mm

1 mm

V   V   NV   NV NV

15.15
Salvia

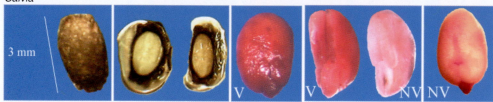

3 mm

V   V   NV NV

Plate 15.13. Geraniaceae. Non-endospermic type with folded cotyledons. Sometimes hard seeded.
Plate 15.14. Gesneriaceae. Miniature embryo with living endosperm.
Plate 15.15. Lamiaceae. Non-endospermic type. Seed coat mucilaginous.

15.16
Alstroemeria

Hemerocallis

3.5 mm

V

7 mm

V

Iris

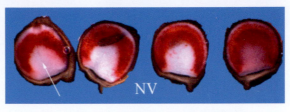

V

NV

8 mm

Arrows indicate central unstained or poorly stained area of concern.
<50% = V, >50% = NV

15.17
Godetia

2 mm

V

NV

NV

NV

15.18
Papaver

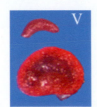

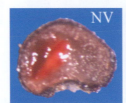

V

NV

NV

1 mm

15.19
Phlox

3 mm

V

V

NV

NV

NV

Plate 15.16. Order Liliales. Linear embryo with living endosperm. Edge cut technique used.
Plate 15.17. Onagraceae. Non-endospermic type.
Plate 15.18. Papaveraceae. Linear embryo with living endosperm.
Plate 15.19. Polemoniaceae. Linear embryo with living endosperm.

15.20
Cyclamen

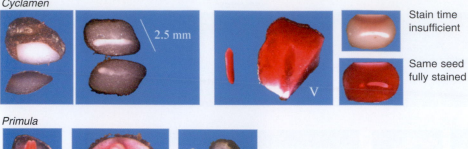

2.5 mm

Stain time
insufficient

Same seed
fully stained

V

Primula

V V NV

15.21
Aquilegia

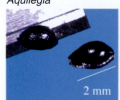   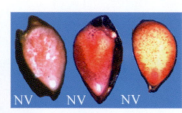

2 mm

2 mm

0.3 mm

V NV NV NV

Ranunculus

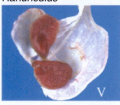  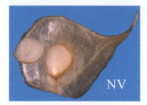

V V NV

15.22
Rosa

5 mm

V V NV

Plate 15.20. Primulaceae. Linear embryo with living endosperm. Edge slice technique used.
Plate 15.21. Ranunculaceae. Rudimentary embryo with a large amount of living endosperm.
Plate 15.22. Rosaceae. Usually non-endospermic type with rigid endocarp, or may have a small amount of living endosperm.

15.23
Heuchera

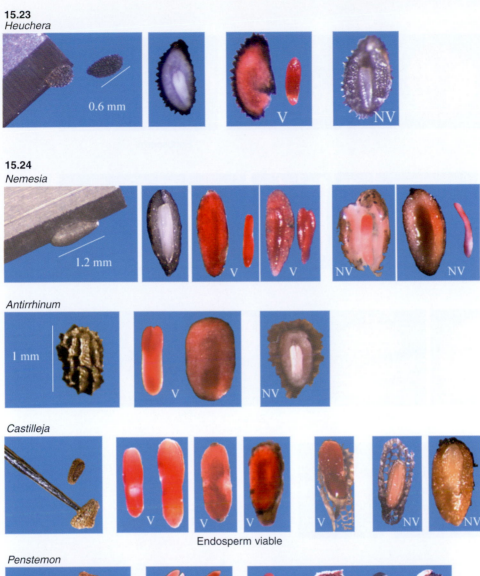

0.6 mm

V

NV

15.24
Nemesia

1.2 mm

V

V

NV

NV

Antirrhinum

1 mm

V

NV

Castilleja

V

V

V

V

NV

NV

Endosperm viable

Penstemon

V

NV

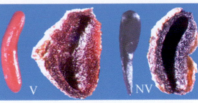

V

NV

Plate 15.23. Saxifragaceae. Miniature embryo with living endosperm.
Plate 15.24. Scrophulariaceae. Miniature or linear embryo with living endosperm.

15.25
Petunia

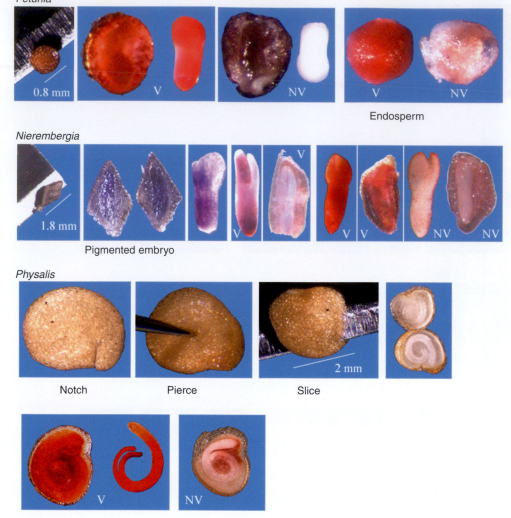

Plate 15.25. Solanaceae. Miniature, linear or coiled embryo with living endosperm.

15.26
Verbena

1. Seed coat
2. Endosperm
3. Embryo

15.27
Viola

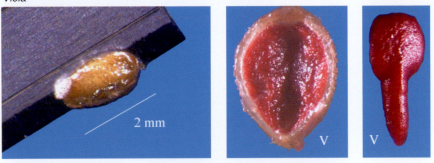

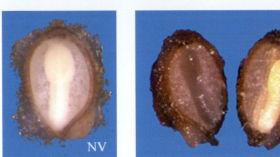

Plate 15.26. Verbenaceae. Linear embryo with living endosperm.
Plate 15.27. Violaceae. Linear embryo with living endosperm.

16   Vigour Testing in Flower Seeds

Robert L. Geneve

Department of Horticulture, N318 Agriculture Science North, University of Kentucky, Lexington, KY 40546-0091, USA

Introduction

Vigour is a well-established characteristic of seeds and can be defined as the inherent ability of seeds within a seed lot to establish normal (or usable) seedlings under diverse growing environments. It has been studied in detail for most agronomic crops and is routinely evaluated in commercial seed lots. Vigour testing in horticultural crops has not received the same level of attention as agronomic crops, although most flower seed producers incorporate some form of vigour assessment in their quality assurance programmes. The purpose of this chapter is to provide an overview of vigour testing, detail those tests applicable to flower seeds and provide a perspective on future directions for flower seed vigour assessment.

The basis for seed vigour testing

Evaluating seed lots for vigour differs from standard germination testing in several important aspects. Standard germination evaluates the seed's ability to produce a normal seedling under near-optimal germination conditions (ISTA, 1999). Seed laws require that standard germination testing results accompany the sale of commercial seeds. Acceptance of standards for purity and germination testing allows for orderly seed commerce within and between countries. Standard germination measures the potential of a seed to germinate to a minimum standard (i.e. normal seedlings). However, values obtained from standard germination testing do not always provide an adequate, subsequent determination of seedling emergence under field conditions. This is partly due to the near-optimal environment used for laboratory tests, and the fact that analysts evaluating seedlings for standard germination do not discriminate between strong and weak seedlings. This, combined with extended times allowed for second counts in germination testing, can overestimate the commercial value of a seed lot. Therefore, seed lots with comparable germination percentages can vary widely in their vigour.

Seed vigour is defined by the Association of Official Seed Analysts (AOSA, 1983) as 'those seed properties which determine the potential for rapid, uniform emergence, and development of normal seedlings under a wide range of field conditions'. Similarly, the International Seed Testing Association (ISTA) describes vigour as 'the sum total of those properties of the seed which determine the level of activity and performance of the seed or seed lot during germination and seedling emergence' (Perry, 1981).

Standard germination is an important first determinant of seed quality. Seed lots that do not germinate at high percentages under near-optimal conditions will obviously continue to perform poorly under commercial growing conditions. Vigour testing provides additional information about the potential for a seed lot to germinate under the less than optimal germination conditions found in most agricultural production systems. The ultimate goal of vigour testing is to provide a predictive measure of seedling emergence under commercial growing conditions. However, seed vigour is apparently a multifaceted trait that is often not emulated by current vigour tests. Also, field-growing conditions can range from near-optimal to severe, making correlation between vigour tests and field emergence variable. For a number of agronomic crops, vigour tests have been established that provide adequate measures of potential field emergence, but more importantly they are used to rank seed lots within or between growing seasons for quality (Hampton and Coolbear, 1990). Seed lots that do not reach a benchmark level assessed by single or multiple vigour tests are rejected even if the standard germination percentage is high.

Importance of vigour testing to flower seeds

Unlike their agronomic counterparts, most commercial flower crops are not germinated under field conditions. Rather, they are sown into soilless media in controlled environments (Styer and Koranski, 1997). These range from the relatively variable environment within greenhouses to controlled growth rooms that provide near-optimal germination conditions (Cantliffe, 1998). The variation in germination conditions for commercial flower seeds under greenhouse production is minimal compared with outdoor seed beds that have fluctuating soil moisture, variable temperature and potential disease pressure. Therefore, for many years it was assumed that germination percentage was an adequate evaluation of the potential for greenhouse seedling emergence. Historically, this was true in many cases. However, the commercial greenhouse industry made a significant change in transplant production systems by moving away from germinating seeds in community flats to producing unitized

transplants in small plug cells (Hartmann et al., 2002). In general, this has led to a two-tiered production system, with large commercial greenhouse firms specializing in plug production of small seedlings that are subsequently sold to transplant producers who finish the crop as larger-celled transplants for the consumer.

The flower industry has probably exceeded any other agricultural commodity group in its need for high-quality seeds. This emphasis started with the evolution of the bedding plant industry and its reliance on plug production. Twenty years ago, most flower growers planted seeds in rows in flats and selected those seedlings that emerged for subsequent handling. The performance of seed under this production regime was less critical because failure of seeds to germinate was superseded by the selection of those that did. However, technical problems were encountered because plants and plant roots became intertwined in the flats, leading to difficulties in separation and subsequent transplant shock. As a result, plug trays were developed where each seedling was self-contained in an individual cell that was subsequently transplanted into an individual pot. This shift from community flat to plug production created important requirements by growers for higher quality seeds that included:

- Each seed placed in a plug cell needed to germinate. To do otherwise resulted in unfilled plugs, inefficient use of greenhouse space and the costly, time consuming need to refill empty cells. As an example, each unfilled cell for impatiens results in a US$0.03–0.05 market loss to the grower.
- All seeds planted in a plug tray needed to germinate rapidly and uniformly. Rapid emergence is essential for faster greenhouse turnaround and frees up more greenhouse space for additional plants. Uniform emergence permits more accurate timing of shipments and creates a more desirable tray appearance to the buyer.

Today's marketplace demands 100% filled cells in a plug tray. As a result, plug growers pay particular attention to seed germination and are willing to pay the increased costs associated with high-quality seeds. The seed industry has responded by providing growers with a vast array of differing products, many requiring specific handling requirements to achieve optimum performance. The modern plug grower must understand the factors that comprise seed quality and compromise its performance.

Specialization has led to increased capital investment in modern greenhouses, automated seeders and sophisticated transplanting robots (Styer and Koranski, 1997). This has challenged the seed industry to provide seeds that perform under these demanding production systems (Karlovich, 1998). In some cases, this requires post-production processing of seeds to provide uniform sizes for seeding (i.e. sizing or pelleting) or techniques to accelerate germination (i.e. priming or pregermination) (Taylor et al., 1998).

For the plug producer, the plug flat is the production unit. Plug flats usually contain either 288 or 512 seedlings per sheet. A high-quality plug sheet has a usable seedling in each plug cell. A usable seedling is defined as a vigorous seedling at the proper stage of development for transplanting. Therefore, uniform seedling emergence is very important for producing a quality plug flat. Remarkably, even under the near-optimal germination conditions provided by sophisticated greenhouse environments, flower seed lots with greater than 95% standard germination can produce

seedlings that emerge at variable times (Fig. 16.1).

It is apparent that seed vigour assessment can have dramatic implications for the seed producer and plug grower. In addition to providing an estimate of seedling emergence uniformity, seed vigour can also be important in additional ways. Seed vigour can be used as a predictor of seed storage longevity and can be used to market seeds (Hampton and Coolbear, 1990).

Seed vigour is lost prior to germination in stored seeds (Delouche, 1965). Compared with most agronomic crops, flower seeds have a high dollar value per seed. Therefore, storage life can be extremely important for extracting the potential value from each seed lot. Seed deterioration models for agronomic crops have demonstrated the importance of initial seed vigour of a seed lot on its subsequent storage life (Tang et al., 1999). Therefore, vigour assessment prior to storage provides a better estimate of future storage life in flower seeds. In addition, periodic vigour testing of seed lots in storage can provide an early indicator of seed deterioration and prompt germplasm repositories to initiate a new seed production cycle to preserve germplasm. Additional discussion of seed deterioration, storage and germplasm preservation are found in Chapters 10 and 17, this volume.

Seed laws do not require vigour testing results to be reported for commercial seed

Fig. 16.1. Seedling emergence in high (left) and low (right) vigour seed lots of pansy.

lots. In most cases, commercial seed producers and brokers do not provide vigour test results to seed consumers. They do not want to expose themselves to grower expectations for field emergence based on seed vigour test values or alienate buyers who are not obtaining the highest vigour lots. However, there are examples of seed brokers using seed vigour test results to market 'high-quality' seeds to consumers (Hampton and Coolbear, 1990). For flower plug producers, seeds can be a significant part of plug production costs, but labour still overshadows this cost if usable seedlings are not produced and plugs have to be back-filled by hand (Styer and Koranski, 1997). Large plug producers have indicated a willingness to pay the extra cost of high-vigour seeds if they result in consistent high percentages of usable seedlings in a plug sheet (Aylsworth, 1996). Currently, there is only one seed broker providing a seed package vigour index on a limited number of bedding plant species (Ball Seed Vigor Index; see next section). Seed producers are also using in-house vigour assessment to identify their 'best' seed lots. These high-vigour seed lots are provided to their largest buyers or are reserved for post-production processing such as pelleting. There is currently no industry standard for flower seed vigour testing. This can lead to problems between seed producers and brokers when vigour is used as the benchmark for rejecting a seed lot and they fail to agree on the vigour of the seed lot because of differing testing methods.

Causes of Poor Vigour

Causes of poor vigour may be due to the inherent genetics of a cultivar, conditions during seed production and conditioning, or seed storage environment and duration. Recognizing a genetic component to vigour requires breeders to evaluate seed germination and vigour as a component of their screening process for new cultivars. Within a cultivar, differences between seed lots for vigour can be based on the production location or between production seasons.

Seed vigour associated with the production environment and seed conditioning

The major production events that negatively impact seed vigour include:

1. Environment during production.
2. Stage of seed development at harvest.
3. Abuse during harvest and seed conditioning.
4. Post-production processing enhancements such as pelleting or priming.

Problems during general seed production are not unique to flower seed production and are problems encountered during seed production for most crops (McDonald and Copeland, 1997). These include environmental extremes such as high or freezing temperature and excessive or limiting moisture during the production cycle. The bulk of the research addressing specific problems during production, harvesting and seed conditioning for flower crops is proprietary research conducted by individual seed companies (Watkins, 1998).

For many flower crops, seeds on the same plant or within the same inflorescence do not mature at the same time. In others, seeds may need to be harvested early before the fruit shatters and makes seed recovery difficult. Therefore, a seed lot is often a plurality of seeds with different germination potential. The challenge during seed conditioning is in eliminating those seeds with low germination potential. Density or size separation are common steps used during seed processing to create a more homogeneous seed lot. Depending on the flower crop, higher density seeds can have higher germination percentage and vigour than smaller, less dense seeds as shown with primrose (*Primula*) (Khademi et al., 1993).

Seed vigour associated with seed storage environment and duration

During seed storage, the vigour of a seed lot is reduced prior to seed viability as indicated by standard germination (Hampton and TeKrony, 1995). Therefore, the environmental conditions during seed storage and the length of time in storage are major factors affecting seed vigour in stored seed lots. Conditions used to store commercial seed lots should be determined with both seed vigour and viability as criteria (Carpenter *et al.*, 1995b; Chapter 10, this volume).

Advances in technology promise the development of additional tools to identify problem seeds within a seed lot for poor germination and vigour. For example, non-destructive, high throughput analysis of individual seeds has been developed for some agronomic seeds. Internal characteristics of seeds can be evaluated by machine vision using cameras that use wavelengths other than the visual spectrum (Shadow and Carrasco, 2000). For example, near-infrared spectroscopy can be used to evaluate a number of seed characteristics including seed moisture, oil composition and contaminating fungi. Similar equipment is being developed to identify the stage of maturity of individual seeds using chlorophyll fluorescence (Oluoch *et al.*, 1999). These and other technologies have the potential to improve the overall vigour of a seed lot by identifying and eliminating low-vigour seeds prior to sale.

Methods of Seed Vigour Testing

General aspects of vigour testing

Seed testing laws for horticultural crops were enacted in 1889 in Florida. The USDA (1897) recognized the need for seed testing rules to be standardized across states. These were formalized into standardized tests for germination sanctioned by the newly formed Association of Official Seed Analysts (AOSA, 1917). Even as standard rules for seed testing were being accepted, it was recognized that standard germination results did not evaluate all of the seed quality properties in a seed lot. Terms like 'germination energy' and 'shooting strength' were being used to describe properties of a seed important for field establishment (McDonald, 1993). Limitations of the standard germination test for predicting field emergence in horticultural crops were

first reported in the 1920s. For example, Munn (1926) observed field germination of garden peas to be significantly lower than percentages determined by standard germination tests. Subsequently, it was recognized that identification of abnormal seedlings in the standard germination test was a better predictor of field emergence in small-seeded crops, such as onion (Clark, 1943). The concept of germinating small-seeded horticultural crops under stressful conditions as a test for field emergence was demonstrated in early work with onion where laboratory tests were conducted in soil boxes (Munn, 1938).

Claude Heit, at the New York Experiment Station in Geneva, was a pioneer in testing flower seeds for standard germination and vigour (Heit, 1972). He conducted numerous studies on the specific requirements for testing flower seeds and, in 1946, summarized germination testing in over 100 species of flowers (Heit, 1946). His work with baby's breath (*Gypsophila*) species illustrates his keen observations for flower seed testing. He reported that there was a difference in seed germination requirements for different cultivars within a single baby's breath species (Heit, 1952), a problem still facing modern seed testing for numerous flower species. His studies were based on 8 years of observations for germination of several cultivars. He showed that white flowering cultivars of baby's breath were not sensitive to germination temperatures between 15°C and 30°C during germination tests, while cultivars with pink in the flowers germinated significantly better at 15°C. Heit (1946) also recognized seed vigour describing 'vitality' in sweet pea (*Lathyrus*) seed lots. Subsequently, he noted that evaluation of abnormal seedlings in the standard germination test was a better predictor of field emergence in cornflower (*Centaurea*) and baby's breath (Heit, 1952). Problems with conditions for conducting standard germination tests and identifying abnormal seedlings in flower seeds (see Chapter 14) can also be problematic for vigour testing. As will be discussed later, the accepted industry standard for vigour testing in agronomic crops mostly relies on germination percentage following imposition of stress (Hampton and TeKrony, 1995).

Since these early attempts at measuring 'seed vitality', numerous attempts have been made to establish a single test to evaluate seed vigour that could be used by seed testing laboratories in a similar way to standard germination tests. For a vigour test to become a standard test, McDonald (1980) suggested some basic requirements. A vigour test should: (i) provide a more sensitive index of seed quality than a standard germination test; (ii) provide a consistent ranking of seed lots in terms of potential performance; (iii) be objective, rapid, simple, and economically practical; and (iv) be reproducible and interpretable.

These requirements for vigour testing were poorly described for a number of agronomic and vegetable species, prompting ISTA in 1981 (Perry, 1981) and the AOSA in 1983 to publish guidelines for vigour testing in seeds. These methods have subsequently been revised by Hampton and TeKrony (1995) to include additional vegetable and tree seeds. They group vigour tests into three categories that include: (i) single tests based on germination behaviour; (ii) physiological or biochemical indices of vigour; and (iii) multiple testing procedures.

Single seed vigour tests include measures of seedling growth or normal germination percentage after a stress imposition. Seedling growth tests include measures of time to radicle protrusion, seedling growth rate after radicle protrusion, and sorting seedlings into strong or weak growing categories. Vigour tests that evaluate normal germination after stress include accelerated ageing, controlled deterioration, cool and cold tests.

Biochemical tests include various measures of metabolic activity in seeds. These include aspects related to respiration (i.e. ATP synthesis, tetrazolium staining), membrane repair (i.e. electrolyte leakage) and volatile production (i.e. ethylene). Of these tests, only electrolyte leakage measured by conductivity tests has been used extensively to evaluate commercial seed lots. Conductivity has been useful for evaluating field emergence in large-seeded legumes and certain sweetcorn cultivars (McDonald, 1980).

Multiple testing addresses the premise that seed vigour is a complex response to variable environments and that all seeds in a seed lot may not respond in the same way to those environments. Therefore, multiple testing uses several different vigour tests in assessing the overall vigour of a seed lot.

Specific vigour tests

Specific application of vigour tests to flower seeds has been limited compared with agronomic seeds. This section describes vigour tests applied to flower seeds or having potential for assessing vigour in flower seeds as indicated by their use in small-seeded vegetable species. Vigour tests will be organized according to the property measured by the test (Table 16.1). This includes vigour evaluation using biochemical indicators, standard germination percentage (usually following stress imposition), germination speed (measured as time to radicle or seedling emergence) and seeding growth (measured as length or area).

Vigour evaluation using biochemical indicators

TETRAZOLIUM.   Of the numerous biochemical vigour tests, only tetrazolium testing and electrolyte conductivity have been used to any extent (Hampton and Coolbear, 1990). Interpreting the topology of tetrazolium staining has generally restricted its use as a general vigour test. It has more utility for viability testing or other specialized uses such as determining localized cold temperature injury in corn. Tetrazolium evaluation of flower seeds has been reported for determining viability (Leadley and Hill, 1972). This has proven difficult because of the small size of most flower seeds and it is doubtful that tetrazolium testing could be routinely useful for testing flower seeds for vigour. However, it has been useful for determining viability (Khademi et al., 1993) and dormancy (Bratcher et al., 1993; Wartidiningsih and Geneve, 1994b) in herbaceous perennial seeds. Some herbaceous perennials display poor-quality seed lots or variable seed lots with a proportion of the seeds having some form of dormancy (Geneve, 1998). For example, Wartidiningsih and Geneve (1994b) obtained commercial

Table 16.1.   Categories of seed vigour testing arranged according to the germination parameters used to evaluate the seed lot.

Vigour test category	Vigour test	Unit of measure
Biochemical	1. Tetrazolium 2. Electrolyte leakage	Tetrazolium uses topology of red stain in embryo. Electrolyte leakage uses electrical conductivity (μmhos/g).

Germination percentage	1. Abnormal seedlings
	2. Cold test
	3. Thermal gradient germination
	4. Ageing tests
	  a. Controlled deterioration
	  b. Accelerated ageing
	  c. Saturated salt accelerated ageing
	  d. Natural ageing
		Percentage of normal seedlings under standard germination conditions. Some studies report only radicle protrusion percentage. Some tests impose a stress (temperature and/or moisture) prior to standard germination. Thermal gradient germination uses variable temperature during germination rather than standard germination conditions. Natural ageing uses K_i from models for seed deterioration in storage.

Germination speed	1. Germination rate 2. Seedling emergence	T_{50}; mean time to germination. Expressed as unit of time (days or hours) to reach 50% radicle or seedling emergence.
Seedling growth	1. Seedling size 2. Seedling growth rate 3. Vigour index	Linear (cm) or area (mm^2) after a specified time or rate calculated over time (cm or mm^2) per unit time (hour). Vigour index uses growth plus a measure of uniformity.

purple coneflower (*Echinacea purpurea*) seed lots that germinated between 22% and 90%. Germination percentage in each seed lot was improved by chilling stratification, suggesting that some seeds in the seed lot were dormant. In a subsequent study to evaluate seed production effects on seed quality in this species, tetrazolium testing proved useful for assessing seed viability vs. dormancy for seeds collected from different inflorescence positions and their implications on seed vigour expression (Wartidiningsih and Geneve, 1994a). Although using routine tetrazolium testing for flower seed vigour cannot be recommended, establishing viability and dormancy in herbaceous perennial seeds could be important for those vigour tests that use germination percentage following imposed stress (see cold and ageing tests in subsequent sections).

ELECTROLYTE LEAKAGE.   Conductivity measures electrolyte leakage from seeds as an indirect measure of initial membrane repair during imbibition (McDonald, 1980). It is an established vigour test, especially for large-seeded legumes like pea (*Pisum sativum* L.). Electrolyte conductivity has not received wide

use for vigour assessment in flower seeds. Fay *et al.* (1993b) evaluated bulk conductivity in three black-eyed Susan (*Rudbeckia fulgida*) seed lots that differed in vigour as indicated by greenhouse seedling emergence. They found that the lowest vigour seed lot showed the greatest electrolyte leakage. Based on these limited results and successful use of electrolyte conductivity as a vigour determinant for small-seeded vegetables like cabbage and onion, McDonald (1998), suggests that electrolyte conductivity deserves additional scrutiny for flower seed vigour testing.

Vigour evaluation using aspects of standard germination percentage

ABNORMAL SEEDLINGS.   From a historical standpoint, the majority of specific research results published on flower seeds were discovered at Cornell University in New York by Claude Heit between 1940 and 1970 (see references in Heit, 1972) and in California by Betty Atwater (Atwater, 1939; Bodger Seeds, 1935). Many of these methods for testing flower seeds were added to the AOSA official rules in 1960. These still constitute the

majority of methods for the approximately
270 different flower crops in the latest version
of the *Rules for Testing Seeds* (AOSA, 1998).

However, Heit (1957) recognized that
standard germination tests did not provide an
adequate prediction of seedling emergence in
flower crops. He subsequently compared laboratory germination tests of numerous flower
species and cultivars with field and greenhouse seedling stand establishment. His basic
conclusion over a range of flower types was
that the number of abnormal seedlings determined in laboratory tests under near-optimal
germination conditions could be used as a predictor of field emergence. He did, however,
provide the caveat that field conditions could
be severe enough to override the genetic factors responsible for seed vitality (i.e. vigour).

For example, in laboratory and field germination of over 100 annual and perennial carnation (*Dianthus*) seed lots, 'field stand of all
dianthus stocks was dependent on the speed
of germination and vigor of the seedlings.
Slow, weak germinating seed samples were a
partial or complete failure. These abnormal,
weak seedlings have no field value and should
not be counted as germinated' (Heit, 1957).
After evaluating over 200 seed lots of asters
(*Aster*), he concluded that those seed lots with
less than 5% abnormal seedlings and high
laboratory germination percentages provided
good field emergence, while those with 6%
to 15% abnormal seedlings showed irregular
field performance, and seed lots with more
than 15% abnormal seedlings were often
complete failures in the field (Table 16.2).

One striking observation in the results
with flower seeds reported in the 1950s was
the relatively low germination percentages

in commercial seed lots as compared with
modern standards. Many of the seed lots had
percentages below 90% or worse. In 1957,
Heit reported that aster, strawflower (*Calendula*), *Nasturtium* and spiderflower (*Cleome*)
were reported to have 60% (162/267), 22%
(10/46), 58% (48/83) and 78% (18/23) of
their seed lots with less than 80% germination, respectively. These tests did not exclude
abnormal seedlings from germination percentages and therefore would be even lower
under the current rules for seed testing.
Advances in production, conditioning and
seed storage have contributed to higher seed
quality in most modern flower seed stocks.
Therefore, these results correlating abnormal
seedlings with field emergence may not reflect
contemporary problems with seed vigour,
where comparative seed lots with high standard germination differ in seedling emergence
performance. However, standard germination
can still be substantially lower than agronomic standards (>90%) for minor flower
crop species, especially some of the herbaceous perennials. In these cases, abnormal
seedlings could be an important indicator of
subsequent seedling establishment.

Modern rules for seed testing exclude
abnormal seedlings from final germination
percentage results. There are general guidelines provided by ISTA and AOSA for determining normal seedlings, but these are subject
to interpretation by the seed analyst. To help
analysts determine abnormal seedlings, images
of abnormal seedling types were published for
major agronomic crops (AOSA, 1992). Corresponding reference images for horticultural
crops are generally lacking. Although
abnormal seedling counts no longer

Table 16.2. Use of abnormal seedlings to determine vigour classes in aster seed lots (data from Heit, 1957).

Vigour class	Number of samples	Laboratory germination		Field emergence	
		Germination range (%)	Mean germination (%)	Germination range (%)	Mean germination (%)
Strong, vigorous	80	61–97	80.2	34–81	58.8
Weak, slow germinating	54	21–84	54.0	4–47	29.0
0–5% abnormal	116	51–97	82.3	27–81	60.0
6–14% abnormal	115	24–86	68.0	10–74	44.8
> 15% abnormal	22	14–70	48.2	8–56	30.7

constitute a vigour assessment beyond the standard germination test, determining normal seedlings in flower crops between seed testing laboratories presents a problem for reproducible standard germination percentages and evaluation of normal seedlings following stress tests for vigour. McDonald (1997) conducted a refereed evaluation of two seed lots of impatiens (*Impatiens walleriana*) and four seed lots of pansy (*Viola* × *wittrockiana*) across eight seed laboratories. Generally, seed laboratories evaluated these seed lots similarly, but differences in impatiens standard germination between laboratories ranged from 21% to 29% for the same seed lot. For pansy, differences ranged between 6% and 11%. It was concluded that differences observed between laboratories were due to differences in standard germination conditions, as well as the analysts' differing views on evaluating abnormal seedlings. Germination percentages varied even more between these laboratories when stress vigour tests were used, where abnormal seedlings constituted a greater percentage of the germinating seedlings. Although Heit (1957) published images of abnormal seedlings for several flower crops, it seems imperative that illustrative plates for normal vs. abnormal seedlings from modern cultivars of major flower crops be made available to aid analysts in calculating germination in a standardized fashion.

COLD TEST.   The cold test has proven useful as a vigour test for several crops, but is the preferred test for maize (Hampton and TeKrony, 1995). It subjects seeds placed on soil to a period of chilling temperature for several days (10°C for 7 days) followed by seedling emergence at warm (25°C) temperature. The cold test imposes both a temperature and pathogen stress on seeds. Because soil type and pathogen pressure can differ between laboratories, the cold test has given variable results, but it is useful for ranking seed lots within a laboratory. There has been limited use of the cold test to evaluate flower seeds. Fay *et al.* (1993b) included the cold test among the several tests used to evaluate seed vigour for black-eyed Susan seed lots. Although the cold test identified the lowest quality seed lot, cold test conditions were severe and germination

for all seed lots was 10% or less. This was comparable to outdoor field emergence and the authors concluded that soil and pathogen pressure limited the usefulness of the cold test.

GERMINATION ACROSS A THERMAL GRADIENT. One aspect of seed vigour is the ability of a seed lot to germinate across diverse environments. Most vigour tests do not directly test this aspect of seed vigour. It is possible to measure germination at variable temperatures using a thermal gradient table (Timbers and Hocking, 1971). A thermal gradient table (also called thermogradient table, temperature gradient bar or plate) allows for the simultaneous testing of seeds over a range of temperature extremes. Commercial units are available and there are several designs for the construction of thermal gradient tables (Hensley *et al.*, 1982). The table is an insulated plate of aluminium to which a constant temperature source is attached at opposite ends of the plate (Fig. 16.2). Commonly, the temperature source is recirculated water from controlled water baths located at either end of the aluminium plate. For example, setting one water source at 15°C and the other at 30°C provides a linear temperature gradient across the table to permit simultaneous testing of seed germination over a 15°C to 30°C temperature range. Commercial seed producers use thermal gradient tables to evaluate seed vigour (higher vigour seeds germinate faster or at higher percentages than lower vigour seeds at the temperature extremes) and to evaluate thermal dormancy in seed lots.

There are few published results using thermal gradient tables to test flower seeds. Fay *et al.* (1993a) demonstrated the utility of a thermal gradient table for establishing the optimal, minimal and maximal germination range for black-eyed Susan seed lots. In a more extensive study, thermal requirements for germination were determined in eight flower species (Magnani *et al.*, 1994). Neither study directly evaluated thermal gradient table germination as a vigour test. However, each study used mean time to germination as an indicator of vigour in determining optimal germination temperatures for each species. Jianhua and McDonald (1996) included thermal gradient germination to evaluate two impatiens seed lots differing in seed vigour. They showed that the higher vigour seed lot germinated at higher percentages (radicle protrusion) after 3 and 5 days at temperatures below 26°C.

ARTIFICIAL AGEING TESTS.   Controlled deterioration and accelerated ageing (AA) are established vigour tests for agronomic, horticultural and forestry species (TeKrony and Hampton, 1995). Both tests are based on the premise that vigour is a measure of seed deterioration. Hampton and Coolbear (1990) concluded that ageing tests were the most promising vigour tests for most agronomic species. Both methods are described in detail in the AOSA vigour testing methods (Hampton and TeKrony, 1995).

Controlled deterioration (Matthews, 1980) exposes seeds to high temperature (40°C or 45°C) for a short duration (24 or 48 h) after the moisture content has been raised to approximately 20%. Seed moisture is raised prior to exposure to high temperature and maintained by keeping seeds in sealed watertight packages. Germination is usually assessed as radicle emergence, but normal germination improves results in some cases (Wang *et al.*, 1994).

Accelerated ageing is similar to controlled deterioration, but differs in the way seed moisture is increased and therefore modifies the duration of the test (TeKrony, 1993). For this test, seeds are placed on nylon mesh screens suspended over water in specially designed accelerated ageing plastic boxes (Fig. 16.3). Seed moisture content is raised by exposure to near 100% relative humidity. Temperature is held between 40°C and 45°C, but the test usually runs for 72 h or more. Germination is usually assessed as normal

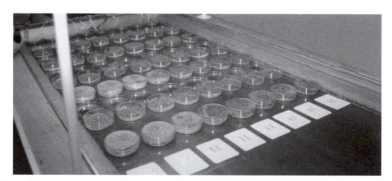

Fig. 16.2.   Petri dish germination studies using a thermal gradient table.

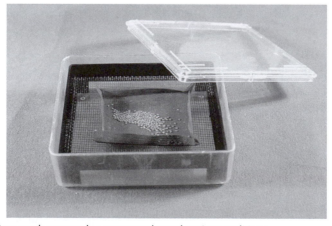

Fig. 16.3.   Impatiens seeds prepared to enter accelerated ageing conditions.

germination under standard germination conditions for a given species. In both controlled deterioration and accelerated ageing tests, temperature fluctuation affects test results. Specialized water-jacketed ageing chambers are preferred to maintain consistent temperature, although controlled deterioration was initially developed using a water bath to control temperature.

Published results for controlled deterioration as a vigour test for flower seeds are not currently available, but results from small-seeded vegetable crops suggest that controlled deterioration would be suitable for flower seed vigour testing. Powell and Matthews (1981) showed the utility of controlled deterioration on vigour testing in onion (*Allium cepa*), lettuce (*Lactuca sativa*), cabbage and Brussels sprouts (*Brassica oleracea*). They also demonstrated that it was an appropriate test for assessing initial vigour as a determinant of the ability of a seed lot to survive in storage (Powell and Mathews, 1984a,b). The major drawback to controlled deterioration is the difficulty in precisely equilibrating seed moisture prior to exposure to high temperature.

Accelerated ageing is a well-established test for large-seeded agronomic crops, but because seed moisture can be difficult to equilibrate during stress imposition, small-seeded crops often do not survive accelerated ageing conditions for extended periods regardless of their vigour (Powell, 1995). However, several small-seeded crops have been effectively evaluated for vigour using accelerated ageing. Examples include canola (*Brassica napus*) and onion. For canola, the test was run at 42°C for 48 h and was effective for determining vigour in seeds from different harvest dates (Elias and Copeland, 2001). For carrot (*Daucus carota*), accelerated ageing conditions were 45°C for 96 h (Elballa and Cantliffe, 1996). It effectively evaluated seed vigour in seeds produced at elevated temperature.

Accelerated ageing has only recently been applied to vigour testing in flower seeds. Fay *et al.* (1993b) used a modified accelerated ageing test to evaluate seed vigour in black-eyed Susan. Seeds were suspended over water at moderate ageing temperature (30°C) for

24–76 h. Radicle protrusion was used to measure germination. Seeds treated for 24 h showed moisture contents above 40%. However, even at these high moisture contents, they determined that seeds aged for 24 h correlated well with seedling emergence under greenhouse growing conditions for three seed lots.

The accelerated ageing test was further modified to accommodate small-seeded flower crops by reducing the relative humidity in accelerated ageing boxes by replacing water with various saturated salt solutions (Jianhua and McDonald, 1996; McDonald, 1997). Termed 'saturated salt accelerated ageing' (SSAA), the objective was to prevent excessive moisture gain by seeds during accelerated ageing. After 24 h at 41°C, water content in impatiens seeds suspended over water was approximately 65% (Jianhua and McDonald, 1996). This could be reduced to approximately 10% by using saturated solutions of KCl or NaCl and this moisture content could be maintained for 96 h. It was determined that for two seed lots with differing vigour, SSAA for 48 h at 41°C using saturated solutions of either KCl or NaCl was an appropriate stress for evaluating seed vigour in impatiens.

The SSAA test was also used in determining vigour in four pansy seed lots (McDonald, 1997). In this case, SSAA conditions of 72 h at 41°C maintained seed moisture at approximately 10–15%. In this study, eight laboratories participated in examining four seed lots of pansy. Germination counts varied between laboratories before and after SSAA. This was attributed to different conditions used between laboratories for the germination test (i.e. wetness of the medium) and differing criteria for evaluating normal seedlings.

NATURAL SEED AGEING.   There are numerous studies that report viability for flower seeds under different storage environments (see Carpenter *et al.*, 1995b, and references therein). Although, these are useful for recommendations on seed storage conditions, there are few studies with flower seeds that have attempted to model seed longevity in storage. One exception is the work by Kwong *et al.* (2001) for delphinium (*Delphinium elatum*) and salvia (*Salvia splendens*) seeds. From a seed vigour standpoint, there are no studies that have attempted to measure or model flower seed longevity as a function of initial seed lot vigour. Conversely, Ellis and Roberts (1981) have suggested that seed longevity models provide information on seed lot vigour. Seed survival can be predicted in a particular storage environment from a linear equation of probit-transformed viability percentages over time. The model contains an experimentally derived seed lot constant, K_i, which is a function of the genotype and the initial condition of the seed lot. They have suggested that K_i is an absolute measure of seed vigour. Although useful in concept, the disadvantage of using K_i as a seed vigour measurement is the time required to model seed survival in storage. It requires viability testing at several storage environments over many months or longer (see Kwong *et al.*, 2001, for conditions for determining K_i in flower seeds). Therefore, it cannot be recommended as a routine vigour measurement.

Vigour evaluation using seedling growth

GERMINATION RATE.   AOSA (1983) considers germination rate as an indicator of seed vigour. There are numerous ways to measure germination rate, but all use time to radicle protrusion as a parameter of germination rate. Caution must be used when calculating a single value to represent germination rate (see discussion by Brown and Mayer, 1986). There are numerous studies that use germination rate to evaluate seed lot performance between non-treated and primed flower seeds, but only a few that have used germination rate to compare seed lots that differ in vigour (Table 16.3). It is arguable whether priming improves all aspects related to seed vigour. Primed seeds show reduced time to seedling emergence, more uniform germination and better performance under less than optimal germination environments (Taylor *et al.*, 1998). These attributes certainly fit the accepted definitions of seed vigour. However, Hampton and Coolbear (1990) also include seed longevity in their discussion of vigour and emphasize that primed seeds have reduced seed survival in storage (Chapter 10,

this volume). They suggest that priming does not improve all of the 'seed properties' related to vigour.

For this discussion, primed seeds are considered as having high vigour. By far the most consistent measure of seed priming in flower seeds has been faster germination rate (reduced time to radicle protrusion). The most common measures are T_{50}, which determines the time to 50% germination in the population of germinating seeds (may be determined in log or probit-transformed data), and mean time to germination (MTG) based on calculations similar to Maguire's (1962) speed of germination calculation. It is calculated as $\Sigma T_i N_i / \Sigma N_i$, where N_i is the number of germinated seeds at time T_i.

Germination rate has not been used to directly estimate seed vigour in flower seeds that have not been primed. This is because of the lack of a definitive study comparing seed lots with high germination percentages but differing vigour confirmed by other standard vigour tests. However, there are a number of studies that use T_{50} as an indicator of seed vigour in seed lots stored at various temperatures and relative humidity (Table 16.3). However, in these studies, differences in T_{50} were seen only in seed lots with correspondingly reduced germination percentages, thereby confounding the significance of T_{50} as a true vigour index (Brown and Mayer, 1986).

In most of the studies reporting germination rate for primed or stored flower seeds, a measure of uniformity is also provided (Table 16.3). Intuitively, high-vigour seed lots should also have less spread in the time to radicle protrusion. However, it is not clear if uniformity provides necessary information about the vigour of a seed lot in addition to germination rate, because the two measures appear to be correlated (Finch-Savage *et al.*, 1991a). Orchard (1977) discusses using population parameters for germination rate and uniformity and reiterates that germination rate for a population approximates a normal distribution, but has a slightly skewed distribution frequency. He did not find it beneficial to use more complex equations to characterize a skewed population and suggested that a normal distribution could be assumed. Mean time to germination and its standard deviation

Table 16.3. Studies with flower seeds that have used germination rate, uniformity or seedling growth to measure vigour in primed or non-primed seeds.

Genus	Germination rate	Uniformity	Seedling growth	References
Priming				
Aquilegia	7-day germination %			Finnerty et al., 1992
Coreopsis	MTG[a]		Greenhouse emergence	Samfield et al., 1990, 1991
Echinacea	3-day germination %			Wartidiningsih and Geneve, 1994b
	MTG		Greenhouse emergence	Samfield et al., 1990, 1991
Impatiens	T_{50}[b]	T_{10-90}[c]		Frett and Pill, 1989; Simmonds, 1980
	MTG	Stand deviation[d]	Seedling length, greenhouse emergence	Finch-Savage, 1991; Finch-Savage et al. 1991b
Petunia	MTG	Stand deviation	Greenhouse emergence	Finch-Savage, 1991
Primula	MTG	Stand deviation	Greenhouse emergence	Finch-Savage, 1991
Salvia	T_{50}	T_{10-90}	Greenhouse emergence	Carpenter, 1989
	MTG	Stand deviation	Greenhouse emergence	Finch-Savage, 1991
Senecio	T_{50}			Carpenter, 1990
Verbena	MTG	Stand deviation	Greenhouse emergence	Finch-Savage, 1991
Viola	T_{50}	T_{10-90}		Carpenter and Boucher, 1991
Vigour				
Catharanthuus	T_{50}			Carpenter and Boucher, 1992b
Coreopsis	T_{50}	T_{10-90}		Carpenter and Ostmark, 1992
Delphinium	T_{50}	T_{10-90}		Carpenter and Boucher, 1992a
Gerbera	T_{50}	T_{10-90}		Carpenter et al., 1995a
Primula			Greenhouse seedling size	Khademi, et al. 1993
Rudbeckia			Greenhouse emergence	Fay et al., 1993b

[a]MTG is the mean time to germination calculated as $\Sigma T_i N_i / \Sigma N_i$, where N_i is the number of germinated seeds at time T_i.
[b]T_{50} is the time required for 50% germination in the germinating seed lot or inverse of germination rate.
[c]T_{10-90} is the span of germination as the days between 10% and 90% germination.
[d]Standard deviation of log-transformed germination percentages.

are easily calculated, allowing mathematical determinations to be developed that represent germination rate and uniformity. Hara (1999) arrived at the same conclusion after comparing Richard's function, which is used for population estimates in skewed populations, to calculate vigour. He suggested using dispersion (a similar calculation to standard deviation) as a measure of uniformity. Dispersion is calculated as one-half of the difference between the times for 25% and 75% of the seeds to germinate. This represents the time for 50% of the seeds closest to the mean to germinate. Given this discussion, only a few studies use standard deviation to represent uniformity during flower seed germination (Table 16.3). The most commonly reported measure is time span between 10% and 90% (T_{10-90}).

Similar values to germination rate can be calculated for seedling emergence in greenhouse studies. Few studies report seedling emergence for flower crops as an indication of vigour. In these studies, a single value for seedling emergence rates is rarely calculated (Finch-Savage, 1991); rather, data are recorded daily to determine emergence curves (Samfield *et al.*, 1991) or total emergence percentage after completion of the experiment (Fay *et al.*, 1993a) are reported.

SEEDLING GROWTH RATE.   In contrast to germination or seedling emergence rate, which are measures of time, seedling growth rate is a determination of growth parameters such as length, area or dry weight at periodic times after radicle protrusion. Seedling growth rate is a sensitive measure of seed vigour (Woodstock, 1969), but is difficult to incorporate in routine vigour testing because it is too labour-intensive to periodically evaluate seedling growth over time. Therefore, seedling growth rate has often been estimated by measuring seedling size after a set time. Radicle length measured at discrete intervals under controlled conditions has been used successfully to test for seed vigour in a number of small-seeded vegetable crops including carrot, lettuce (McCormac *et al.*, 1990; Smith *et al.*, 1973), cauliflower, onion and leek (Finch-Savage, 1986). Radicle length or growth rate using a slant-board test (Fig. 16.4)

was also correlated with field emergence in these crops. The method employs germination of seeds on a slanted board so that straight seedlings are obtained which are subsequently hand measured by an analyst as described for lettuce (Smith *et al.*, 1973). This test is used commercially for small-seeded horticultural crops, including flower crops.

Researchers have attempted to eliminate the time and potential errors inherent in hand analysis by using computer-aided measurements of seedling size. Initial studies included systems that used machine vision (Howarth and Stanwood, 1993), time sequence photography (Tomas *et al.*, 1992) and computerized automated seed analysis using a hand potentiometric caliper (Keys *et al.*, 1984). These tests have shown very good correlations with hand test evaluation for radicle length, but have failed to become routine tests used by commercial analysts for a variety of reasons related to standardized conditions for collecting images to be measured by the computer.

McCormac *et al.* (1990) captured images of small-seeded vegetable crops grown on a slant board using a CCD digital video camera prior to computer analysis of radicle length. Although this was an accurate measure of radicle length, problems with lighting and background colour made it difficult to evaluate the small roots of some horticultural crops. Paradigm Research Corporation (South Haven, Minnesota) developed an improved

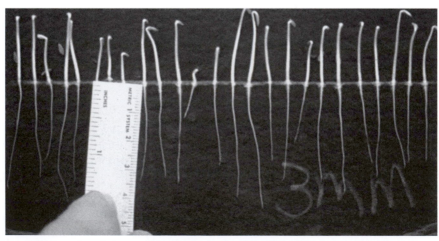

Fig. 16.4.   Slant board test in lettuce (provided by National Seed Storage Laboratory, Oregon).

digital video camera system with better control of lighting and contrast between the seedling and background, and this system uses standard Petri-dish germination test conditions rather than a slant board (McNertney, 1999).

An alternative to CCD digital video cameras is the use of flat bed scanners to capture digital images of seedlings in Petri dishes. Two systems have been reported. Sako *et al.* (2001) use an inverted flat bed scanner placed in a closed cabinet. It employs standard Petri-dish germination, but requires the lid to be removed before acquiring an image. Geneve and Kester (2001) also use a flat bed scanner, but rather than inverting the scanner, their system requires the use of a transparent (clear cellulose) germination medium for its Petri-dish germination and a top lighting attachment (transparency adapter) for the scanner. These conditions do not require removal of the Petri dish lid before imaging even with condensation on the lid, which reduces image resolution in other camera systems (Dell'-Aquila *et al.*, 2000). This may present a distinct advantage when robots are developed to automate delivery of Petri dishes to the scanner for imaging. The two systems also differ in the software used to evaluate seedling size. Sako *et al.* (2001) use proprietary software that can be customized to measure specific seedling components (radicle vs. hypocotyl), while Geneve and Kester (2001) use generic, commercially available software.

The flat bed scanner provides excellent resolution of small flower seeds that are currently difficult or impossible to achieve with digital video cameras. This is because the digital camera captures images as far as a metre from the seedlings. With the flat bed scanner, seedlings are very close to the image-capturing device, accounting for the high quality images. For example, lisianthus (*Eustoma*) is one of the smallest seeds grown by the bedding plant industry. Seedling detail is easily captured by a flat bed scanner (Geneve and Kester, 2001) and seedling length accurately measured (Fig. 16.5).

Although this technology is relatively new, Sako *et al.* (2001) used images acquired from a flat bed scanner to successfully evaluate lettuce seed lots with different vigour levels. This same research group has also evaluated images of impatiens seedlings with the same system (Hoffmaster *et al.*, 2003). Oakley *et al.* (2004) showed the utility of computer-aided analysis of digital images captured from a flat bed scanner for determining vigour. Using the system developed by Geneve and Kester (2001), they evaluated six impatiens seed lots that differed in vigour, but had standard germination of 95% or better. In general, seedling size measured by either length or area at a specific time or as cumulative growth rate was similar to using the Ball Vigor Index or SSAA (Table 16.4). These initial results suggest that computer-aided image analysis could satisfy the requirements for a quality vigour test, but additional studies are required for flower crops other than impatiens and there remains the demonstration that seedling growth correlates with seedling emergence over a range of environmental conditions for flower crops.

Commercial flower producers and brokers rely heavily on seedling emergence studies conducted under greenhouse or growth chamber conditions to evaluate seed vigour (Fig. 16.6). Usable seedlings are evaluated under conditions similar to those used by commercial seedling plug growers. Unfortunately, these tests are usually qualitative and rely on the experience of the analyst to rank seed lots for vigour. Ball Seed Company has developed a computer-aided analysis system that uses digital images of flower crop seedling emergence to provide a quantitative measure of usable seedlings. A CCD digital video camera images plug-grown seedlings under controlled environments. Commercially available software measures seedling leaf (cotyledon) area (Fig. 16.7) and calculates the Ball Vigor Index (Conrad, 1999). It has to be assumed that measuring seedling emergence between seed lots is an appropriate measure of seed vigour, but no supporting data have been published to compare seed lots using this system. However, Conrad (1999) states that correlative studies between the Ball Vigor Index and seedling emergence have been conducted with commercial greenhouse growers throughout the USA, and seeds with a high vigour rating have a longer storage life (an indicator of high vigour). The Ball Vigor

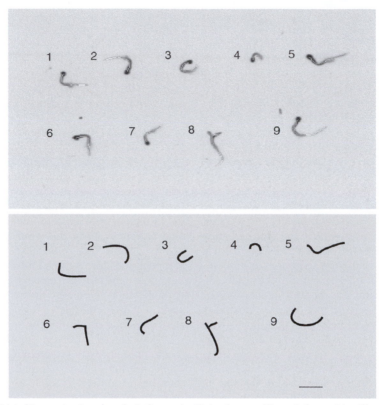

Fig. 16.5. Digital image (top) and analysis (bottom) of lisianthus seedlings. Software identifies the seedling and then draws a single pixel skeleton to measure seedling length. Reference bar = 5 mm.

Table 16.4. Standard germination and seed vigour determined by several methods for impatiens seeds varying in initial seed quality.

Seed lot	Standard germination (%)	Ball Vigor Index	Saturated salts accelerated ageing (%)	Seedling size after 9 days	
				Length (cm)	Rate (cm/day)
1	96 a[1]	651	81 a	1.05 b	0.25 b
2	97 a	642	84 a	1.44 a	0.36 a
3	97 a	561	69 bc	1.15 b	0.26 b
4	96 a	505	75 b	0.70 c	0.13 c
5	98 a	485	54 c	0.70 c	0.12 c
6	96 a	440	48 c	0.61 c	0.13 c

[1]Means with the same letter within a column were not different by Tukey's test ($P \le 0.05$).

Index is used commercially for several flower crops and the vigour rating is provided to growers on the seed package.

VIGOUR INDICES. Calculating a single vigour index value is an alternative for vigour that combines a growth measurement with aspects of uniformity. Several vigour indices have been proposed. The Ball Vigor Index (Conrad, 1999) uses total leaf area of seedlings in a plug flat divided by its standard deviation multiplied by the germination percentage. The germination percentage can be calculated based on simple emergence, but is more useful if calculated as those seedlings that have a leaf surface area that exceeds a predetermined

Fig. 16.6. Greenhouse grow-out test to compare seed lots for seedling emergence.

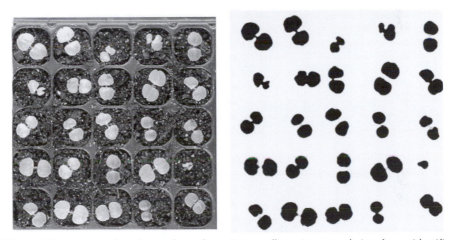

Fig. 16.7. Digital image and overlay analysis of impatiens seedlings. Image analysis software identifies the green colour of the cotyledon and creates a pixel overlay to measure its area.

threshold value which represents usable seedlings. Therefore, small but emerged seedlings are not counted in the germination percentage.

Sako *et al.* (2001) use a similar calculation for vigour index for seedlings germinated in Petri dishes. They calculate vigour as the sum of weighted values for growth and uniformity minus a penalty value for non-germinating seeds. The rationale for providing weighted values was to allow the analyst to emphasize growth or uniformity, depending on the seed being evaluated.

As discussed previously for germination rate, it is not clear whether uniformity adds significantly to the information provided by growth alone. Oakley *et al.* (2004) compared growth alone (seedling length or area) with vigour indices that included uniformity (standard deviation) to compare six seed lots that differed in vigour. Compared with other vigour tests, seed lots were ranked less efficiently for vigour by inclusion of uniformity in a vigour index compared with growth alone.

The Paradigm vigour rating (McNertney, 1999) uses both germination rate and seedling length growth rate. It is based on repeated measurements of individual seedlings over time. These two values are combined to generate a relative vigour value for each seed based on a rating scale of 0–10, 0 representing non-germinating seeds and 10 being the most vigorous seedlings. The percentage of seedlings that exceed a predetermined threshold value represent the vigour index or potential seedling emergence. It does not appear that this is a subjective rating scale, but rather is based on a calculation of germination rate and seedling growth rate. Therefore, it is not clear why a numerical vigour index is not computed as has been reported for other vigour tests. A non-replicated study using this rating scale was used to evaluate greenhouse emergence in 20 impatiens seed lots and was shown to adequately predict seedling emergence in 85% of the seed lots (McNertney, 1999).

COMPARISON OF GERMINATION RATE AND SEEDLING GROWTH RATE.   It still is not certain whether germination rate and seedling growth rate are related measures or if they measure different properties of seed vigour. The correlation between germination rate and subsequent

seedling growth was determined for several small-seeded vegetables (Finch-Savage, 1986). He concluded that faster germinating seeds within a seed lot produced a higher percentage of normal seedlings and showed increased seedling growth and uniformity. Although Heit (1957) came to the same conclusion for carnation seed lots, there have been few subsequent studies correlating flower seed germination rate and seedling growth. However, there are several studies that indicate, on a population basis, that flower seed lots with faster germination rates (attained by seed priming) have better seedling stand establishment (Simmonds, 1980; Carpenter, 1989; Samfield et al., 1991). This is due to earlier germinating seeds having larger sizes at a specified time. However, germination rate in individual impatiens seeds did not correlate with their subsequent seedling growth rate after germination (Fig. 16.8). In other words, faster germinating seeds do not necessarily have the fastest growth rates. This presents an interesting dichotomy between the behaviour of a seed lot as a population of seeds vs. the inherent germination potential of individual seeds when using germination rate as an indicator of seed vigour.

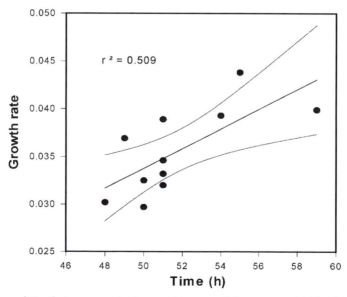

Fig. 16.8.   The correlation between germination rate (time to radicle emergence) with subsequent seedling growth rate in impatiens seedlings. Lines adjacent to regression line represent 95% confidence interval.

Future Challenges

From the previous discussions it is apparent that vigour testing for flower seeds presents numerous challenges to the seed analyst. No one single test is likely to be appropriate for the large number of species important to the greenhouse industry. It is also apparent that standards for flower seed testing lag behind those adopted for agronomic crops. There are a number of steps that scientists should pursue to advance vigour testing in flower seeds.

Since many of the traditional vigour tests use standard germination to evaluate seed lots, it is imperative to update the conditions for conducting standard germination for flower seeds. This is a daunting task because of the number of flower species routinely grown by the industry. It would be appropriate to emphasize the top 20 flower species by worldwide sales to initiate this effort. The current rules contain conditions for germination that may not fit current optimum germination conditions for modern flower hybrids. Commercial greenhouse recommendations for seed germination in flower crops should be the basis for evaluating standard germination temperatures in refereed studies. In addition, these crops need to be characterized for normal and abnormal seedling traits to aid analysts in the standardization of germination percentages.

Of the traditional vigour tests, ageing tests seem the most adaptable for small-seeded flower crops. Appropriate moisture content and temperature during artificial ageing needs to be determined for additional flower crops. Saturated salt accelerated ageing would seem the most adaptable test because many seed laboratories currently have the facilities to conduct accelerated ageing for agronomic seeds. Controlled deterioration would also be suitable, but care must be taken to obtain appropriate seed moisture content.

Germination rate and seedling growth are very promising tests for vigour in flower seeds. Computer-aided capture and analysis of seedling information appears to be an emerging technology that will facilitate developing industry standards for comparing seed lots for vigour. The fundamental relationship between germination rate, seedling growth and uniformity within a seed lot needs to be established in order to provide an industry standard for reporting a vigour index that best correlates with seedling emergence under a variety of germination conditions.

Commercial seed producers rely heavily on greenhouse or controlled environment grow-out tests to evaluate seedling emergence in plug flats as an indicator of seed vigour. In many cases, this is a qualitative evaluation rather than a quantitative evaluation. Standard environmental conditions and a quantitative measure of usable seedlings (i.e. a vigour index based on cotyledon area) would allow seed companies and growers to have a similar standard for evaluating seed lot performance.

References

[AOSA] Association of Official Seed Analysts. 1917. *Rules for Testing Seeds.* Association of Official Seed Analysts of North America.

[AOSA] Association of Official Seed Analysts. 1960. *Rules for Testing Seeds.* Association of Official Seed Analysts of North America.

[AOSA] Association of Official Seed Analysts. 1983. *Seed Vigor Testing Handbook.* Contribution No. 32. Association of Official Seed Analysts, Lincoln, Nebraska.

[AOSA] Association of Official Seed Analysts. 1992. *Seedling Evaluation Handbook.* No. 35. Association of Official Seed Analysts, Lincoln, Nebraska.

[AOSA] Association of Official Seed Analysts. 1998. *Rules for Testing Seeds.* Association of Official Seed Analysts, Lincoln, Nebraska.

Atwater, B.R. 1939. Notes on flower seed germination. *Proc. Association of Official Seed Analysts of North America* 31:111–113.

Aylsworth, J.D. 1996. Making seed cost effective. *Greenhouse Grower* Fall:28–29.

Bodger Seeds. 1935. *Valuable Information for Seedsmen,* 2nd edn. Bodger Seeds, El Monte, California.

Bratcher, C.B., J.M. Dole and J.C. Cole. 1993. Stratification improves seed germination of five native wildflower species. *HortSci.* 28: 899–901.

Brown, R.F. and D.G. Mayer. 1986. A critical analysis of Maguire's germination rate index. *Jour. Seed Technol.* 10:101–110.

Cantliffe, D.J. 1998. Seed germination for transplants. *HortTechnol.* 8:499–503.

Carpenter, W.J. 1989. *Salvia splendens* seed pregermination and priming for rapid and uniform plant emergence. *Jour. Amer. Soc. Hort. Sci.* 114:247–250.

Carpenter, W.J. 1990. Priming dusty miller seeds: role of aeration, temperature, and relative humidity. *HortSci.* 25:299–302.

Carpenter, W.J. and J.F. Boucher. 1991. Priming improves high-temperature germination of pansy seed. *HortSci.* 26:541–544.

Carpenter, W.J. and J.F. Boucher. 1992a. Temperature requirements for the storage and germination of *Delphinium × cultorum* seed. *HortSci.* 26:541–544.

Carpenter, W.J. and J.F. Boucher. 1992b. Geminating and storage of vinca seed is influenced by light, temperature, and relative humidity. *HortSci.* 27:993–996.

Carpenter, W.J. and E.R. Ostmark. 1992. Growth regulators and storage temperature govern germination of *Coreopsis* seed. *HortSci.* 27:1190–1193.

Carpenter, W.J., E.R. Ostmark and J.A. Cornell. 1995a. Temperature and seed moisture govern germination and storage of *Gerbera* seed. *HortSci.* 30:98–101.

Carpenter, W.J., E.R. Ostmark and J.A. Cornell. 1995b. Evaluation of temperature and moisture content during storage on the germination of flowering annual seed. *HortSci.* 30:1003–1006.

Clark, B.E. 1943. Comparative laboratory and field germination. *Proc Association of Official Seed Analysts* 34:90–99.

Conrad, R. 1999. Method and apparatus for assessing the quality of a seed lot. US Patent No. 5,901,237.

Dell'Aquila, A., J.W. van Eck and G.W.A.M. van der Heijden. 2000. The application of image analysis in monitoring the imbibition process of white cabbage (*Brassica oleracea* L.) seeds. *Seed Sci. Res.* 10:163–169.

Delouche, J.C. 1965. An accelerated aging technique for predicting relative storabiltiy of crimson clover and tall fescue seed lots. *Agronomy Abstracts* 40, American Society of Agronomy, Madison, Wisconsin.

Elballa, M.M.A. and D.J. Cantliffe. 1996. Alterations of seedstalk development, seed yield, and seed quality in carrot by varying temperature during seed growth and development. *Jour. Amer. Soc. Hort. Sci.* 121:1076–1081.

Elias, S.G. and L.O. Copeland. 2001. Physiological and harvest maturity of canola in relation to seed quality. *Ag. Jour.* 93:1054–1058.

Ellis, R.H. and E.H. Roberts. 1981. The quantification of ageing and survival in orthodox seeds. *Seed Sci. Technol.* 9:373–409.

Fay, A.M., S.M. Still and M.A. Bennett. 1993a. Optimum germination temperature of *Rudbeckia fulgida*. *HortTechnol.* 3:433–435.

Fay, A.M., M.B. McDonald and S.M. Still. 1993b. Vigor testing of *Rudbeckia fulgida* seeds. *Seed Sci. Technol.* 21:453–462.

Finch-Savage, W.E. 1986. A study of the relationship between seedling characters and rate of germination within a seed lot. *Ann. Appl. Biol.* 108:441–444.

Finch-Savage, W.E. 1991. Development of bulk priming/plant growth regulator seed treatments and their effect on the seedling establishment of four bedding plants. *Seed Sci. Technol.* 19:477–485.

Finch-Savage, W.E., D. Gray and G.M. Dickson. 1991a. Germination responses of seven bedding plant species to environmental conditions and gibberellic acid. *Seed Sci. Technol.* 19:487–494.

Finch-Savage, W.E., D. Gray and G.M. Dickson. 1991b. The combined effects of osmotic priming with plant growth regulator and fungicide soaks on the seed quality of five bedding plant species. *Seed Sci. Technol.* 19:495–503.

Finnerty, T.L., J.M. Zajicek and M.A. Hussey. 1992. Use of seed priming to bypass stratification requirements of three *Aquilegia* species. *HortSci.* 27:310–313.

Frett, J.J. and W.G. Pill. 1989. Germination characteristics of osmotically primed and stored *Impatiens* seeds. *Scientia Horticulturae* 40:171–179.

Geneve, R.L. 1998. Seed dormancy in commercial vegetable and flower species. *Seed Technol.* 20:236–250.

Geneve, R.L. and S.T. Kester. 2001. Evaluation of seedling size following germination using computer-aided analysis of digital images from a flat-bed scanner. *HortSci.* 36:1117–1120.

Hampton, J.G. and D.M. TeKrony. 1995. *Vigor Testing Methods*. 3rd edn. International Seed Testing Association. Zurich, Switzerland.

Hampton, J.G. and P. Coolbear. 1990. Potential versus actual seed performance – can vigour testing provide an answer? *Seed Sci. Technol.* 18:215–228.

Hara Y. 1999. Calculation of population parameters using Richard's function and application of indices of growth and seed vigor to rice plants. *Plant Production Sci.* 2:129–135.

Hartmann, H.T., D.E. Kester, F.T. Davies, Jr. and R.L. Geneve. 2002. *Plant Propagation: Principles and Practices*. Prentice-Hall, Inc., Englewood Cliffs, New Jersey. 7th edn.

Heit, C.E. 1946. Summarized laboratory germination data on over 100 flower seed species. *Association of Official Seed Analysts Newsletter* 20:13–16.

Heit, C.E. 1952. Laboratory germination of *Centaurea* and *Gypsophila* species, types and varieties. *Proc. Association of Official Seed Analysts* 42:101–104.

Heit, C.E. 1957. Laboratory germination and vigor as compared to soil tests and field performance of flower seed. *Proc. Association of Official Seed Analysts* 47:126–136.

Heit, C.E. 1972. Thirty years' testing flower seeds for germination. *Proc. International Seed Testing Association* 37:817–827.

Hensley, D.L., J.B. Masiunas and P.L. Carpenter. 1982. An inexpensive temperature gradient system. *HortSci.* 17:585–586.

Hoffmaster, A.L., K. Fujimura, M.B. McDonald and M.A. Bennett. 2003. An automated system for vigor testing three-day-old soybean seedlings. *Seed Sci. Technol.* 31:701–713.

Howarth, M.S. and P.C. Stanwood. 1993. Measurement of seedling growth rate by machine vision. *Transactions of the American Society of Agricultural Engineers* 36:959–963.

[ISTA] International Seed Testing Association. 1999. International rules for seed testing. *Seed Sci. Technol.* 27:1–299.

Jianhua, Z. and M.B. McDonald. 1996. The saturated salt accelerated aging test for small-seeded crops. *Seed Sci. Technol.* 25:123–131.

Karlovich, P.T. 1998. Flower seed testing and reporting needs of the professional grower. *Seed Technol.* 20:131–135.

Keys, R.D., R.G. Margapuram and G.A. Reusche. 1984. Automated seedling length measurement for germination/vigor estimation using a CASAS (computerized automated seed analysis system). *Jour. Seed Technol.* 9:40–53.

Khademi, M., D.S. Koranski and J. Peterson. 1993. Protein concentration and vigor of imbibed density-separated *Primula* seed. *HortSci.* 28:710–712.

Kwong, F., L. Stodolski, J. Mari, S.H. Gurusinghe and K.J. Bradford. 2001. Viability constants for *Delphinium* and *Salvia* seeds. *Seed Technol.* 23:113–125.

Leadley, P.R. and M.J. Hill. 1972. Use of the tetrazolium chloride method for determining viability of a range of flower seeds species. *Proc. International Seed Testing Association* 37:812–813.

Magnani, G., M. Macchia, M. Mazzarri and G. Serra. 1994. Thermal requirements during the germination stage in some new ornamental species. *Acta Horticulturae* 362:197–203.

Maguire, J.D. 1962. Speed of germination: aids in selection and evaluation for seedling emergence and vigor. *Crop Sci.* 2:176–177.

Mathews, S. 1980. Controlled deterioration: a new vigour test for crop seeds. *In*: P.D. Hebblethwaite (ed.). *Seed Production*. Butterworths, London. pp. 647–660.

McCormac, A.C., P.D. Keefe and S.R. Draper. 1990. Automated vigour testing of field vegetables using image analysis. *Seed Sci. Technol.* 18:103–112.

McDonald, M.B. 1980. Assessment of seed quality. *HortSci.* 15:784–788.

McDonald, M.B. 1993. The history of seed vigor testing. *Jour. Seed Technol.* 17:93–100.

McDonald, M.B. 1997. The saturated salt accelerated aging test of pansy and impatiens seeds. *Seed Technol.* 19:103–109.

McDonald, M.B. 1998. Seed quality assessment. *Seed Sci. Res.* 8:265–275.

McDonald, M.B. and L.O. Copeland. 1997. *Seed Production: Principles and Practices*. Chapman and Hall, New York.

McNertney, D.C. 1999. System and method for measuring seedlot vigor. US Patent No. 5,864,984.

Munn, M.T. 1926. Comparing laboratory and field viability tests of seed of garden peas. *Proc. Association of Official Seed Analysts* 18:55.

Munn, M.T. 1938. New and promising substrata for seed viability tests. *Proc. Association of Official Seed Analysts* 25:225.

Oakley, K., S.T. Kester and R.L. Geneve. 2004. Computer-aided digital image analysis of seedling size and growth rate for assessing seed vigour in impatiens. *Seed Sci. Technol.* 32:907–915.

Oluoch, M., R. Bino, H. Jalink and S. Groot. 1999. An overview of the technological advances in seed quality management. *Seed World* Jan:26–30.

Orchard, T.J. 1977. Estimating the parameters of plant seedling emergence. *Seed Sci. Technol.* 5:61–69.

Perry, D.A. 1981. *Handbook of Vigour Test Methods*. International Seed Testing Association, Zurich, Switzerland.

Powell, A.A. 1995. The controlled deterioration test. *In*: H.A. van de Venter (ed.). *Seed Vigour Testing Seminar*. International Seed Testing Association, Zurich, Switzerland. pp. 73–87.

Powell, A.A. and S. Matthews. 1981. Evaluation of controlled deterioration, a new vigour test for small seeded vegetables. *Seed Sci. Technol.* 9:633–640.

Powell, A.A. and S. Matthews. 1984a. Prediction of the storage potential of onion seed under commercial storage conditions. *Seed Sci. Technol.* 12:641–647.

Powell, A.A. and S. Matthews. 1984b. Application of the controlled deterioration vigour test to detect seed lots of Brussels sprouts with low potential for storage under commercial conditions. *Seed Sci. Technol.* 12: 649–657.

Sako, Y., M.B. McDonald, K. Fujimura, A.F. Evans and M.A. Bennett. 2001. A system for automated seed vigour assessment. *Seed Sci. Technol.* 29:625–636.

Samfield, D.M., J.M. Zajicek and B.G. Cobb. 1990. Germination of *Coreopsis lanceolata* and *Echinacea purpurea* seeds following priming and storage. *HortSci.* 25:1605–1606.

Samfield, D.M., J.M. Zajicek and B.G. Cobb. 1991. Rate and uniformity of herbaceous perennial seed germination and emergence as affected by priming. *Jour. Amer. Soc. Hort. Sci.* 116: 1–13.

Shadow W. and A. Carrasco. 2000. Practical single kernel NIR/visible analysis for small grains. *Cereal Foods World* 45:16–18.

Simmonds, J. 1980. Increased stand establishment of *Impatiens walleriana* in response to maximized germination rates. *Can. Jour. Plant Sci.* 60:559–564.

Smith, O.E., N.C. Welch and T.M. Little. 1973. Studies on lettuce seed quality. I. Effect of seed size and weight on vigor. *Jour. Amer. Soc. Hort. Sci.* 98:529–533.

Styer, R.C. and D.S. Koranski. 1997. *Plug and Transplant Production*. Ball Publishing, Batavia, Illinois.

Tang, S., D.M. TeKrony, D.B. Egli and P.L. Cornelius. 1999. Survival characteristics of corn seed during storage. II. Rate of seed deterioration. *Crop Sci.* 39:1400–1406.

Taylor, A.G., P.S. Allen, M.A. Bennett, K.J. Bradford, J.S. Burris and M.K. Misra. 1998. Seed enhancements. *Seed Sci. Res.* 8:245–256.

TeKrony, D.M. 1993. Accelerated aging test. *Jour. Seed Technol.* 17:111–120.

Timbers, G.E. and R.P. Hocking. 1971. A temperature gradient bar for seed germination and cold hardiness studies. *Can. Jour. Plant Sci.* 51: 434–437.

Tomas, T.N., A.G. Taylor and L.A. Ellerbrock. 1992. Time-sequence photography to record germination events. *HortSci.* 27:372.

[USDA] United States Department of Agriculture. 1897. Circular No. 34. *Rules and Apparatus for Seed Testing*. USDA Office of Experiment Stations, Washington, DC.

Wang, Y.R., J.G. Hampton and M.J. Hill. 1994. Red clover vigour testing: effects of three test variables. *Seed Sci. Technol.* 22:99–105.

Wartidiningsih, N. and R.L. Geneve. 1994a. Seed source and quality influence germination in purple coneflower [*Echinacea purpurea* (L.) Moench.]. *HortSci.* 29:1443–1444.

Wartidiningsih, N. and R.L. Geneve. 1994b. Osmotic priming or chilling stratification improves seed germination of purple coneflower. *HortSci.* 29:1445–1448.

Watkins, J.T. 1998. Seed quality problems commonly encountered during vegetable and flower seed production. *Seed Technol.* 20:125–130.

Woodstock, L.W. 1969. Seedling growth as a measure of seed vigor. *Proc. International Seed Testing Association* 34:273–280.

17   Conserving Herbaceous Ornamental Plant Germplasm

David Tay

Director, Ornamental Plant Germplasm Center, 670 Vernon Tharp Street, Columbus, OH 43210-1086, USA

Introduction

Plant genetic resources, also known as plant germplasm, provide the genetic variability that plant breeders use to breed new crop cultivars. These genetic materials include landraces, traditional or heirloom cultivars, wild and weedy forms, related wild species, genetic stocks, inbred lines and modern cultivars maintained as seed or other plant propagules for present and future use. These genetic materials contain individual genes, their alleles, linkage groups, epistatic gene combinations and combinations of different genomes. In this new era of gene identification, isolation and gene transformation, these genes are being identified and transferred from such germplasm found in nature into our crops. The conservation of plant germplasm therefore becomes even more crucial and important. This can be attributed to three main reasons: (i) gene transformation technologies have allowed us to introduce genes of one species into the genome of another; i.e. the tertiary gene-pool boundary (Harlan and de Wet, 1971) of a species has expanded to encompass other organisms. This calls for the collection and conservation of a wider range of species in order to capture greater genetic diversity; (ii) the rapid adoption of modern flower cultivars with improved adaptive and aesthetic characteristics globally is rapidly replacing many traditional heirloom cultivars, causing serious genetic erosion worldwide; and (iii) the accelerating rate of urban expansion and human disturbance in our natural environment is displacing and reducing natural genetic diversity. For example, wild species of African violet, *Saintpaulia*, in their natural habitats are facing environmental decline in East Africa because of forest clearing and urban and farm development.

Plant genetic resource conservation has in the last half century developed into a science and management discipline. Special facilities, commonly called genebanks, consisting of seed-storage freezer vaults maintained at $-18°C$ for preserving base collections and at $2°C$ for active collections of seed, and cryo-preservation tanks with liquid nitrogen at $-196°C$ for storing seed, tissue cultures and buds have been developed for plant germplasm conservation. However, the focus in the last half century has been on food, fibre and industrial crops. Globally, there are 6.1 million germplasm accessions of such crops (FAO, 1998) but very few of these are ornamental plants. The Consultative Group on International Agricultural Research – which is a joint effort of the United Nations, the World Bank and other agencies – has no mandate for ornamental species and most national programmes have no or limited focus on ornamentals. The conservation of this important aesthetic group of plants is often cared for by botanical gardens and arboreta, seed companies, plant nurseries and private individuals where the emphasis is on the conservation of species and genera rather than the entire genepool of a targeted species and its relatives. For example, the Kew Gardens Millennium Seed Bank Project in England sets a goal to conserve a representation of over 24,000 plant species rather than the whole range of variation in each species (Linington, 2000).

The genetic base of many of our modern flower cultivars is becoming increasingly narrow due to concentrated breeding efforts on aesthetic qualities such as flower and plant characteristics. Furthermore, the genetic backgrounds of many modern F_1 hybrid flower cultivars from different breeding programmes of a species are related. For example, most impatiens, pansy, petunia and marigold cultivars of the major international flower seed companies are quite similar, as evidenced in seed catalogues of these major flower seed companies. Coupled with global marketing and the adoption of these cultivars worldwide, many heirloom varieties of these and other ornamental crops are being replaced by modern cultivars with a relatively narrow genetic base, resulting in continuing loss of our traditional heirloom varieties and causing serious genetic erosion. Herbaceous ornamental plant genebanks are, therefore, crucial to collect and conserve the disappearing heirloom genetic materials, to explore, collect and conserve new genetic materials and species with more than just aesthetic characteristics, and to evaluate and distribute these germplasm collections.

The United States National Plant Germplasm System

The National Plant Germplasm System (NPGS) consists of a national network of seed and clonal genebanks/repositories. Its mission is to provide 'the genetic diversity necessary to improve crop productivity and to reduce genetic vulnerability in future food and agriculture development, not only in the United States but for the entire world' (formulated at the National Workshop on Plant Germplasm at Peoria, Illinois, 1981). These genebanks and repositories are thus mandated to acquire, maintain, evaluate and make readily accessible to plant scientists a wide range of genetic diversity in the form of seed and clonal materials of crops and potential new crops. The 450,000-accession inventory is the largest single plant germplasm collection in the world. These accessions are maintained in a decentralized network of 26 *ex situ* seed and clonal repositories as active collections in different parts of the country, representing a wide range of plant growing conditions from Mayaguez, Puerto Rico in the east and south to Hilo, Hawaii, in the west and Palmer, Alaska, in the north (Fig. 17.1, Table 17.1). The establishment of the four Regional

Fig. 17.1. The locations of the National Plant Germplasm System repositories and centres.

Table 17.1. Germplasm collections maintained by NPGS sites in 2002.

Site	Accessions	Countries	Genera	Species
Barley Genetic Stocks Center (GSHO)	3,044	3	1	1
Clover Collection (CLO)	246	30	1	118
Cotton Collection (COT)	9,308	124	1	41
Desert Legume Program (DLEG)	2,585	56	198	1279
Maize Genetic Stock Center (GSZE)	4,710	2	1	1
National Arboretum (NA)	1,909	51	257	861
National Arctic Plant Genetic Resources Unit (PALM)	493	31	50	145
National Arid Land Plant Genetic Resources Unit (PARL)	961	32	12	124
National Center for Genetic Resources Preservation (NSSL)	23,299	106	198	493
National Small Grains Collection (NSGC)	126,563	170	15	148
Natl. Germplasm Repository – Brownwood (BRW)	881	3	2	23
Natl. Germplasm Repository – Corvallis (COR)	11,687	92	56	757
Natl. Germplasm Repository – Davis (DAV)	5,105	79	19	202
Natl. Germplasm Repository – Geneva (GEN)	5,136	58	6	91
Natl. Germplasm Repository – Hilo (HILO)	675	41	22	76
Natl. Germplasm Repository – Mayaguez (MAY)	560	40	137	229
Natl. Germplasm Repository – Miami (MIA)	4,606	90	213	527
Natl. Germplasm Repository – Riverside (RIV)	1,167	30	38	152
North Central Regional PI Station (NC7)	47,032	177	319	1767
Northeast Regional PI Station (NE9)	11,730	126	32	196
Ornamental Plant Germplasm Center (OPGC)	967	58	62	287
Pea Genetic Stock Collection (GSPI)	501	3	1	2
Plant Germplasm Quarantine Office (PGQO)	4,827	59	19	76
Potato Germplasm Introduction Station (NR6)	5,503	38	1	168
Southern Regional PI Station (S9)	82,579	184	246	1433
Soybean Collection (SOY)	20,415	91	1	16
Tobacco Collection (TOB)	2,106	67	1	65
Tomato Genetic Stock Center (GSLY)	3,287	18	2	22
Western Regional PI Station (W6)	69,946	162	368	2423
Wheat Genetic Stocks Center (GSTR)	334	1	1	1
Total	452,162	NA		

Note: The 26 NPGS germplasm stations, centres and repositories officially listed on the NPGS webpage (http://www.ars-grin.gov/npgs) are as follows:
1. Barley Genetic Stock Center (GSHO), Aberdeen, Idaho – http://www.ars-grin.gov/ars/PacWest/Aberdeen/hang.html
2. C.M. Rick Tomato Genetics Resource Center, Davis, California – http://tgrc.ucdavis.edu/
3. Desert Legume Program, Tucson, Arizona – http://ag.arizona.edu/bta/bta20.html
4. Maize Genetics Cooperation – Stock Center (GSZE), Urbana, Illinois – http://w3.aces.uiuc.edu/maize-coop/
5. G.A. Marx Pea Genetic Stock Center (GSPI), Pullman, Washington – http://www.ars-grin.gov/ars/PacWest/Pullman/GenStock/pea/MyHome.html
6. National Clonal Germplasm Repository (COR), Corvallis, Oregon – http://www.ars-grin.gov/ars/PacWest/Corvallis/ncgr/ncgr.html
7. National Clonal Germplasm Repository for Citrus and Dates, Riverside, California – http://www.ars-grin.gov/ars/PacWest/Riverside/homepg1.htm
8. National Clonal Germplasm Repository for Tree Fruit/Nut Crops and Grapes (DAV), Davis California – http://www.ars-grin.gov/ars/PacWest/Davis/
9. National Germplasm Resources Laboratory (NGRL), Beltsville, Maryland – http://www.barc.usda.gov/psi/ngrl/ngrl.html
10. National Center for Genetic Resources Preservation (NSSL), Fort Collins, Colorado – http://www.ars-grin.gov/ars/NoPlains/FtCollins/nsslmain.html

Table 17.1. *Continued.*

11. National Small Grains Collection (NSGC), Aberdeen, Idaho – http://www.ars-grin.gov/ars/PacWest/Aberdeen/nsgc.html

12. National Temperate Forage Legume Genetic Resources Unit, Prosser, Washington – http://www.forage.prosser.wsu.edu/

13. North Central Regional Plant Introduction Station (NC7), Ames, Iowa – http://www.ars-grin.gov/ars/MidWest/Ames/

14. Ornamental Plant Germplasm Center (OPGC), Columbus, Ohio – http://opgc.osu.edu

15. Pecan Breeding & Genetics, Brownwood and Somerville, Texas – http://extension-horticulture.tamu.edu/carya/

16. Plant Genetic Resources Conservation Unit (S9), Griffin, Georgia – http://www.ars-grin.gov/ars/SoAtlantic/Griffin/pgrcu/

17. Plant Genetic Resources Unit (NE9), Geneva, New York – http://www.ars-grin.gov/ars/NoAtlantic/Geneva/

18. Plant Germplasm Quarantine Office (PGQO), Beltsville, Maryland – http://www.barc.usda.gov/psi/fl/pgqo.html

19. Soybean/Maize Germplasm, Pathology, and Genetics Research Unit, Urbana, Illinois – http://www.life.uiuc.edu/plantbio/ars/ppgru.html

20. Subtropical Horticulture Research Station (MIA), Miami, Florida – http://www.ars-grin.gov/ars/SoAtlantic/Miami/homeshrs.html

21. Tropical Agriculture Research Station, Mayagüez, Puerto Rico – http://www.ars-grin.gov/ars/SoAtlantic/Mayaguez/mayaguez.html

22. Tropical Plant Genetic Resource Management Unit (HILO), Hilo, Hawaii – http://pbarc.ars.usda.gov/pages/research/tpgrmu/germplasm.shtml

23. United States Potato Genebank – NRSP-6, Sturgeon Bay, Wisconsin – http://www.ars-grin.gov/ars/MidWest/NR6/

24. Western Regional Plant Introduction Station (W6), Pullman, Washington – http://www.ars-grin.gov/ars/PacWest/Pullman/

25. Wheat Genetic Stock Center (GSTR), Aberdeen, Idaho – http://www.ars-grin.gov/ars/PacWest/Aberdeen/hang.html

26. Woody Landscape Plant Germplasm Repository, Washington, DC – http://www.usna.usda.gov/Research/wlpgr.html

Plant Introduction Stations in Geneva, New York (Plant Genetic Resources Unit – NE9), Ames, Iowa (North Central Regional Plant Introduction Station – NC7), Griffin, Georgia (Plant Genetic Resources Conservation Unit – S9), and Pullman, Washington (Western Regional Plant Introduction Station – W6), in the 1946 Farm Bill marked the beginning of NPGS. In 1958, the National Seed Storage Laboratory (the present National Center for Genetic Resources Preservation – NCGRP) was dedicated to serve as the NPGS base collection site for long-term preservation of accessions and this facility was significantly expanded in 1992 to incorporate *in vitro* and cryopreservation facilities. In the 1980s, clonal repositories for conserving vegetatively propagated species were established. Since that time, the Ornamental Plant Germplasm Center (OPGC) has been the only repository added to NPGS.

The NPGS repositories are located across a range of climatic regions so the most suitable climatic environments are available for multiplying seed and maintaining plants in the field of all the targeted crops of both seed and clonally propagated species. The repositories are crop-specific and the allocation of mandated crop species is based on the comparative advantages of the site in terms of available expertise and facilities, state government and industry supports, proximity to where the crop is grown, location of key users, etc. Except for the NCGRP, all the repositories are active sites and are equipped to distribute either seed or vegetative propagules of clonal materials to researchers worldwide. Plant germplasm accessions are acquired by these active sites, and the collected germplasm is multiplied and stored, characterized and evaluated, documented and distributed to researchers. The NCGRP functions in the long-term security duplication of all plant materials conserved by NPGS repository nodes.

The NCGRP employs state-of-the-art, long-term storage facilities which include both −18°C freezer vaults (Fig. 17.2) for seed preservation and cryopreservation tanks (Fig. 17.3) with liquid nitrogen at −196°C or −160°C in the vapour phase for plant propagules and seed in base collections. The

Fig. 17.2.   The long-term storage seed vault for storing 'orthodox' seed at −18°C at the National Center for Genetic Resources Preservation, Fort Collins, Colorado, USA.

Fig. 17.3.   The cryopreservation tanks for storing vegetative plant tissues at −196°C in liquid nitrogen and for seed at −160°C in liquid nitrogen vapour phase at the National Center for Genetic Resources Preservation, Fort Collins, Colorado, USA.

plant propagules under preservation are plant tissue culture, dormant temperate fruit buds, recalcitrant seed and pollen. The mission of NCGRP Seed Viability and Storage Research Unit is 'to effectively document, preserve and maintain viable seed and propagules of diverse plant germplasm in long-term storage, to develop and evaluate procedures for determining seed quality of accessions, and to provide administrative support to allow for effective operation of the Unit. The mission also includes the distribution of seed, when not available from the active collections, for crop improvement throughout the world.' All the active repositories, with mandate to preserve seed, have medium-term seed storage cooler facilities operating at 2°C–5°C and below 40% relative humidity. The clonal repositories and other sites that have a mandate for clonal materials are equipped with living plant genebanks both in greenhouse and field genebank and often tissue culture facilities.

The NPGS is under the auspices of the United States Department of Agriculture (USDA) Agricultural Research Service in Beltsville, Maryland. It has a National Program Leader who administers the funding for the system and ensures that national policies and procedures are followed. The repositories are linked together by the Germplasm Resources Information Network (GRIN), a master database managed and maintained by the National Germplasm Resources Laboratory (NGRL) in Beltsville, Maryland. Germplasm information about NPGS collections is accessed via the GRIN Internet site at http://www.ars-grin.gov/npgs. The NGRL also manages the United States Plant Exchange Office, which provides coordination between NPGS curators and their counterparts and scientists in other countries to facilitate germplasm exchange and to ensure that both US and collaborating national policies, laws and regulations are followed. GRIN, designed and written by NGRL, is a relational database system used by all NPGS curators to input germplasm information such as passport, characterization, evaluation and seed storage data including inventory and seed quality. End-users can access GRIN from anywhere in the world using the Internet and submit germplasm requests. Documentation is therefore an important component of NPGS.

The NPGS invites crop-specific experts from universities, industries and USDA and other related organizations to form Crop Germplasm Committees (CGCs). Most major crops have their own CGC. There are currently 40 such committees. The role of a CGC is advisory, consultative and supportive, and it includes the establishment of priorities for germplasm acquisition, identification of weaknesses and gaps in a collection; recommendations on crop descriptors, germplasm evaluation methodologies, seed regeneration techniques, *in vitro* procedures and preservation; and advice on how to increase the value of a collection, such as the use of geographic information system (GIS) data, DNA technologies, etc. It also plays an important role in communication and as an interface between the crop scientific research community and its targeted industry. This system provides a means for capturing technical input from germplasm users, as well as a forum for feedback on the viability and quality of accessions received, reports on the value of the germplasm, its contributions to published research, etc. The relationship of a CGC to a repository will be further illustrated in the section on OPGC.

The OPGC is a germplasm repository of NPGS located at The Ohio State University (OSU), Columbus, Ohio, for the conservation and distribution of herbaceous ornamental plants. Of the more than 450,000 accessions of 4474 species conserved by NPGS, most are species that are food, fibre and industrial crops and only about 3,000 accessions are herbaceous ornamental plants, representing only 0.7% of the entire collection. Considering the estimated US$50 billion global floriculture industry in 1995 (de Groot, 1998) and US$4.88 billion alone in the USA in 2002 (USDA, 2003), there is a critical need for a specialized genebank to accept the important task of conserving herbaceous ornamental plants and this was strongly advocated by the American floricultural industry (GMPRO, 1999). The OPGC is a cooperative effort between USDA, the government of Ohio and the American floriculture industry through a joint agreement between USDA and OSU in August 1999 and its official inauguration in July 2001 marked the beginning of this effort to establish a specialized genebank for flowers (Tay *et al.*, 2004). It is the NPGS's newest germplasm repository.

The Mission, Goal and Objectives of the OPGC

The OPGC mission is 'to conserve and nurture the world's wealth of herbaceous ornamental plant diversity by systematically collecting, saving, evaluating and enhancing its use to bring happiness and health to humankind and to promote awareness in herbaceous ornamental crop germplasm conservation'. The goal is to build the world's leading herbaceous ornamental plant genebank and thus become a 'centre of excellence' in floricultural genetic resources conservation and research. The four main objectives to achieve this mission and goal are:

1. To establish an herbaceous ornamental plant genebank as an integrated component of NPGS.
2. To build a network of cooperators from the horticultural industry, universities, USDA, botanic gardens and arboreta, crop specific societies and individuals.
3. To establish a system that promotes industry-orientated, collaborative research to enhance germplasm utilization.
4. To institute academic programmes and industrial short courses in germplasm conservation and enhancement of flowers.

The OPGC has three main activities: (i) be a floricultural germplasm service provider, including germplasm acquisition, conservation, characterization, evaluation, enhancement of utilization, distribution and documentation; (ii) conduct floricultural germplasm and conservation research to develop more efficient and cost-effective germplasm conservation and utilization management systems, new techniques to preserve seed and clonally propagated germplasm, and germplasm enhancement; and (iii) provide floricultural germplasm education and training to fill an important gap in specific training in this field.

The OPGC benefits the floriculture industry and clientele by making available unique genetic materials essential for the improvement of present and future floral crops. These include pest and disease resistance genes, product-appeal genes to increase diversity in form, colour and fragrance, genes that contribute to the reduced use of production inputs such as more efficient water usage, and genes for pharmaceutical, nutriceutical, agrochemical and functional food properties.

OPGC Organizational Structure

The OPGC, as a NPGS centre, abides by NPGS policies and procedures and participates in all NPGS activities. All major operational issues are conducted in consultation with USDA through the project Authorized Departmental Officer's designated representatives. The everyday management of the Center is under the leadership of the OPGC Director with the support of crop curators. The Center also operates under the guidance and recommendation of the USDA Herbaceous Ornamental Crop Germplasm Committee (HOCGC) and its OPGC Subcommittee on technical issues relating to the OPGC mandate. The HOCGC consists of 37 prominent floriculturists and researchers from flower seed companies, plant nurseries, universities, USDA, botanic gardens and arboreta (see http://www.ars-grin.gov/cgi-bin/npgs/html/mbrship.pl?HERBACEOUS).

The OPGC crop-specific technical working groups are established to provide the following advice on the priority genera (see below):

- Germplasm diversity, natural distribution and existing collections, gaps in collections, taxonomic and evolutionary relationships, genetics, etc.
- Seed and plant material multiplication and storage.
- Descriptors, characterization and evaluation.
- Germplasm-related research areas.

Each group consists of five to seven members and its membership is composed of specialists in the crop or discipline, e.g. plant collector, taxonomist, geneticist, plant breeder, pathologist, entomologist, and business is conducted through e-mail, tele-conferencing, HOCGC meetings, industry conferences and professional meetings. The composition of the existing crop technical working groups is found at the OPGC website (http://opgc.osu.edu). The organizational structure of OPGC and USDA is summarized in Fig. 17.4.

The Priority Genera

Flowers are very diverse, consisting of many genera and species from all climatic regions of the world. While many of them have thousands of years of domestication history, many more are being collected, evaluated and introduced into gardens. In this situation, it is vital that the OPGC has a focused effort in order to create a positive research and economic impact. One of the important roles of the HOCGC is to develop a recommended priority genera list. Since 1994, three lists have been developed: in 1995, 1999 and 2001. The OPGC uses the 2001 list in its present work plan. A summary of the three HOCGC lists consisting of a total of 64 genera is given in Table 17.2. At the genus level, the 30 priority genera of the 2001 list consist of

about 6700 taxa based on Hortus Third (1976) as detailed in Table 17.2. The prioritization at the species level to decide which species to collect and preserve becomes even more critical. This effort poses a greater challenge due to the lack of stable taxonomic classifications in many ornamental plant genera. The function of the crop-specific technical working groups becomes extremely important. Surveys and consultations with scientists and the industry to rank the HOCGC 30 priority genera so that the OPGC can concentrate its efforts on the most important genera showed in descending order the first 14 genera as follows: *Begonia, Impatiens,* geranium (referring to *Pelargonium*), *Petunia, Salvia, Viola, Dianthus, Campanula, Hemerocallis, Alstroemeria, Lilium, Verbena, Rudbeckia* and *Phlox* (Tay, 2003). The criteria used in the survey were market potential, product life phase, needs for bottleneck genes, researchable, availability of germplasm and potential other uses.

The Strategic Plan and Networking Concept

The operation of a genebank is multidisciplinary. The germplasm conservation and utilization steps include plant

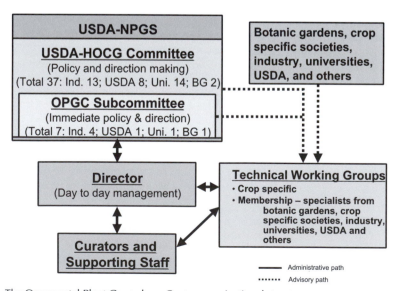

Fig. 17.4. The Ornamental Plant Germplasm Center organizational structure.

Table 17.2. Development of the USDA Herbaceous Ornamental Crop Germplasm Committee (HOCGC) priority genera list of herbaceous ornamentals.

Year	No. of genera	Genus
1995	24	*Ageratum, Alstroemeria, Aster, Begonia, Caladium, Catharanthus, Chrysanthemum, Dianthus, Dieffenbachia, Euphorbia, Eustoma, Hemerocallis, Hippeastrum, Impatiens, Kalanchoe, Liatris, Lilium, Lobelia, Pelargonium, Petunia, Spathiphyllum, Verbena, Zantedeschia* and *Zinnia*
1999	48	*Tagetes* (7), *Antirrhinum* (6), *Catharanthus* (6), *Chrysanthemum* (6), *Euphorbia* (6), *Impatiens* (6), *Iris* (6), *Pelargonium* (6), *Petunia* (6), *Verbena* (6), *Begonia* (5), *Dianthus* (5), *Salvia* (5), *Viola* (5), *Cyclamen* (4), *Geranium* (4), *Lilium* (4), *Penstemon* (4), *Aster* (3), *Crocus* (3), *Delphinium* (3), *Eustoma* (3), *Freesia* (3), *Gloxinia* (3), *Gypsophila* (3), *Hemerocallis* (3), *Lobelia* (3), *Orchidaceae* (3), *Saintpaulia* (3), *Alstroemeria* (2), *Campanula* (2), *Echinacea* (2), *Gazania* (2), *Gentiana* (2), *Heuchera* (2), *Hosta* (2), *Lantana* (2), *Liatris* (2), *Osteospermum* (2), *Paeonia* (2), *Papaver* (2), *Phlox* (2), *Primula* (2), *Rudbeckia* (2), *Senecio* (2), *Solidago* (2), *Veronica* (2) and *Vinca* (2) *Note*: The number in brackets indicates the number of committee members who selected the genus.
2001	30	*Algaonema* (50), *Alstroemeria* (60), *Anthurium* (600), *Aquilegia* (70), *Aster* (250–500), *Baptisia* (30–35), *Begonia* (1000), *Campanula* (300), *Chrysanthemum* (100–200), *Dianthus* (300), *Dieffenbachia* (30), *Euphorbia*/Poinsettia (10), *Geranium* (300), *Hemerocallis* (15), *Impatiens* (500), *Iris* (200), *Lilium* (80–90), *Narcissus* (26), *Pelargonium* (280), *Petunia* (30), *Phalaenopsis* (55), *Philodendron* (200), *Phlox* (60), *Rudbeckia* (25), *Salvia* (750), *Spathiphyllum* (35), *Tagetes* (30), *Verbena* (200), *Veronica* (250) and *Viola* (500) *Note*: The number in brackets indicates the number of species of the genus based on Hortus Third (1976) and the total number of taxa in the 30 genera is 6692.
Total		64 distinct genera from the three lists

exploration and collecting; seed production, drying, processing, packaging, storage and viability monitoring during storage; and germplasm characterization, evaluation, documentation and distribution. An efficient genebank is one, with fixed resources, that is able to put the maximum amount of genetic diversity of its targeted crops into safe storage without losing the individual accession genetic integrity but allowing for a tolerable change in the genetic composition of an accession. The aim of the OPGC is to operate a genebanking system that minimizes the loss of genetic variability of its germplasm accessions during seed regeneration and storage, i.e. maintaining both high genetic, physiological and physical seed quality standard. To meet these quality assurance requirements, the following technical issues were considered in developing the OPGC strategic plan:

- The many species and variants to be collected – wild relatives, heirloom and obsolete cultivars, landraces, genetic stocks, hybrid complexes, ploidy series, etc.
- The lack of taxonomic and evolutionary relationship treatments and knowledge.
- The wide geographical distribution, diverse centre of origin, and lack of information on genetic erosion.
- The lack of information on sexual and asexual propagation and conservation:
  - Seed germination, production, seed technology and storage methods.
  - Tissue culture and cryopreservation methods.
- The lack of plant quarantine risk and requirements in introduction and distribution.

- Duplicate identification, formulation of crop descriptors, prioritizing traits for evaluation, and DNA marker technologies.
- Establishment of core collections.

The conservation strategy adopted to maintain such a large number of diverse genetic materials is as follows:

- For seed producing heirloom cultivars – use seed of individual accessions.
- For wild related species – use seed of individual accessions.
- For seed producing modern cultivars – use seed of polycrosses within groups.
- For non-seeding species – use *in vitro* and live plants.
- For non-seeding heirloom cultivars – use *in vitro* and live plants.

Genetic erosion in a genebank, i.e. loss of genes while under conservation, is a serious problem. To minimize this, it is critical to prevent a shift in the genetic composition of an accession during seed regeneration and, similarly, during seed storage to monitor seed viability to prevent a decline in germination below a threshold where loss of genes might occur. However, every cycle of seed multiplication and storage will cause genetic erosion. When the number of accessions requiring seed regeneration increases and accumulates in a genebank, a decision must be made to move these accessions as quickly as possible into safe storage. A good plant genetic resources conservation programme is one where all the desired accessions are regenerated with acceptable genetic and seed quality standards, as compared with another where only a portion of the desired accessions is regenerated with very high quality seed, but where some accessions are lost completely.

Based on the needs of the floriculture industry, the technical considerations above and the project resources available, an OPGC development plan was formulated as shown in Fig. 17.5.

In this early stage of the Center's development, manpower development is being given top priority and team building through networking with scientists in different disciplines from universities, USDA, industry, botanic gardens and crop-specific societies. Similarly, these working relationships are also being developed and promoted in other countries. The OPGC networking model is shown in Fig. 17.6.

Establishment Phase (2001–2002)
- Building infrastructure
- Building teams – staff and external networking
- Setting priorities – consultation, survey and literature review
- Launching germplasm introduction and accessioning

Development Phase (2002–2004)
- Evaluating progress and planning session
- Perfecting genebanking techniques
- Expanding team building, staff development and graduate assistantships
- Seeking R&D funds
- Formulating a manageable OPGC programme

Operational Phase (2004 onward)
- Running a predictable yearly programme, including services and training activities with flexibility to introduce new activities
- Continuing to seek R&D funds and maintaining cooperator and clientele base

Fig. 17.5.   The Ornamental Plant Germplasm Center development plan for the period from 2001 to 2004.

The Genebank Infrastructure

The OPGC physical structure was constructed in 1993 as a bio-containment facility to quarantine ornamental plant introduction and was remodelled for use as a genebank in 2001 (Fig. 17.7). The 550 m² building currently consists of a 25,000-accession capacity medium-term seed-storage cooler, a floral seed research laboratory and a seed-processing plant (Fig. 17.8). Adjoining the building is a 1060 m² computerized, environment controlled greenhouse complex for year-round seed regeneration and research. The specialized herbaceous ornamental seed research laboratory is equipped with a thermo-gradient germinator, two accelerated ageing water-jacket incubators, four germinators, a custom-built single-seed weight sorting machine, a STS-MACS four-channel aspirator, a KA-K gravity separator, a LA-H laboratory brush machine, single deck vibratory seed separator, a Clipper office tester and cleaner, a belt thresher, a

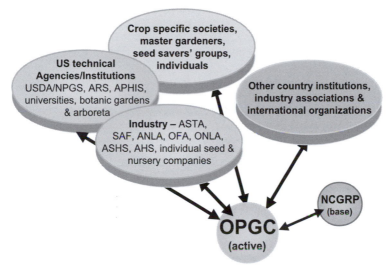

Fig. 17.6. The Ornamental Plant Germplasm Center Networking Model.

Fig. 17.7. The Ornamental Plant Germplasm Center building showing the front view.

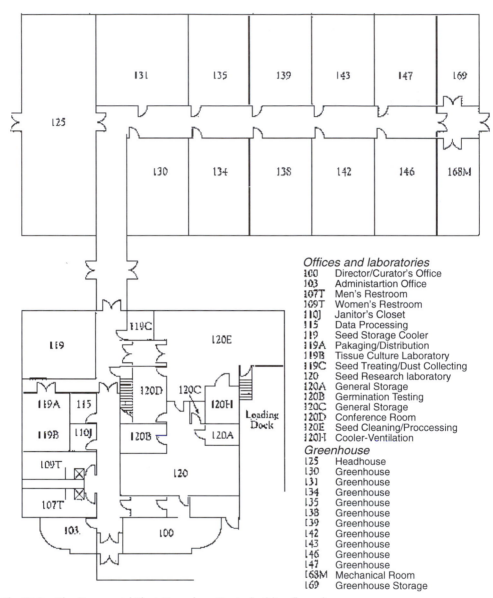

Fig. 17.8.   The Ornamental Plant Germplasm Center building floor plan.

seed scarifier, a STS-1B-30°C cabinet system dryer and a STS temperature and time controlled precision dryer (Fig. 17.9).

The core of the building is the seed-storage cooler, maintained at 2–5°C and 30% relative humidity (Fig. 17.10). This combination of temperature and humidity regime allows seed to be stored and dried to a low seed moisture content of 5–7% (wet weight basis), and under these conditions most 'orthodox' seed are expected to have a storage life of 10–15 years. The low humidity of the cooler allows seed dried in a room environment to be put into storage directly in non-airtight containers because the cooler in this case functions as a slow dryer in further drying the seed during storage. The seed conditioning facilities provide a free-flowing production line system with the belt thresher at one end and the electronically

Fig. 17.9.   The Ornamental Plant Germplasm Center (OPGC) STS temperature- and time-controlled precision dryer.

Fig. 17.10.   The inside of OPGC medium-term storage cooler maintained at 2°C–5°C and 30% relative humidity.

controlled four-channel seed aspirator at the other end.

The OPGC tissue culture laboratory is housed in another building for the maintenance of clonally propagated germplasm as meristem cultures under slow-growth conditions in precision incubators. In addition to *in vitro* storage, a field genebank has been established for clonally propagated germplasm at the OPGC greenhouse and experimental field. Back-up collections will be established in NPGS sites. In addition, the OPGC has a 2.74 × 3.05 m bulb and tuber storage room operating at 8°C and this room is also used for seed and plant vernalization.

The OPGC Operational Procedures and Seed Science and Technology

The activities in a genebank include germplasm acquisition, regeneration, conservation, characterization, evaluation, documentation and distribution. An efficient genebank has three essential components – a high-quality facility, a functional operational system and a knowledgeable dedicated team with skills in seed science and technology to accomplish these sequential activities. This is described below to demonstrate the close connection between plant germplasm conservation activities and seed science and technology.

Starting with germplasm collection in the field, field inspection procedures and seed sampling techniques as used in official seed certification protocols are used so that representative seed subsamples are collected to form an accession. Seed of an accession collected in a standing field is therefore a composite sample made up of small subsamples collected from many individual plants across transects as used by a seed inspector in seed certification. An accession collected from seed

bins is made up of representative sub-samples from different parts of a seed bin and from different bins. The ability of a collector to identify seed of different species and varieties, the knowledge to differentiate between mature, good quality and healthy seed compared with poor-quality seed, and the practical skill to estimate seed moisture content, will increase the success of a germplasm exploration mission. Knowledge about national seed permits, international treaty and convention documentations, and phytosanitary certification is extremely important in germplasm collecting overseas.

During seed regeneration, i.e. the multiplication of new seed lots for long-term storage, a genebank operational manual takes into consideration the growing season, seed germination, field selection, crop rotation, seedling growth, cultural techniques, pollination, harvesting, threshing, cleaning, drying, packaging and storage. The aims are to maintain the original genetic composition of the accession and to produce high-quality seed. The targeted number of plants to be grown per accession in order to capture the original

Fig. 17.11.   The laboratory scarifier for seed with hard seed coat or thick dried fruit at the Ornamental Plant Germplasm Center.

genepool of the collection is derived based on best estimations. Seed dormancy-breaking treatments, e.g. using GA and KNO$_3$, stratification at 8°C and hard seed scarification with a scarifier (Fig. 17.11) are applied to ensure uniform germination to prevent genetic drift. The seedlings are randomly selected during transplanting and the best growing conditions are provided to allow all the plants to grow and produce seed. All detectable and suspected off-type plants are rogued using accepted field inspection methods of seed certification. When in doubt whether a plant is true-to-type, the philosophy is to cull it. This process is straightforward in inbred and pure-line situations. In the case of wild and cross-pollinated heirloom populations, the variation of an accession is often a continuum with small differences between one plant and another. The strategy is to retain the entire range of genetic variability in order to retain the complete genepool. Insect or hand pollination in greenhouses or pollination cages to provide isolation is used for cross-pollinating species. Vernalization, photoperiod treatments and cultural manipulations are applied to synchronize and concentrate flowering to promote inter-pollination between plants of

an accession. This also allows for concentrated seed maturation and harvesting. Accessions with self-incompatibility and male-sterility systems must be maintained with specific pollination techniques such as bud-pollination and carbon dioxide treatments as in brassicas. When an accession has to be maintained as a clone to remain true-to-type, meristem culture is used to back up the living plant collection as an *in vitro* collection.

In addition to genetic purity, equal attention has to be placed on seed physiological and physical quality. The good storage conditions in a genebank will not improve a seed lot's storage quality since it only slows down its deterioration rate. Seed regeneration should be implemented so that harvesting occurs in a cool, dry environment to minimize field weathering. At the OPGC, greenhouse seed regeneration is emphasized to take advantage of the computer-controlled greenhouse environments (Fig. 17.12). At harvesting, the aim is to harvest the same amount of seed from every plant to form a representative composite sample for storage. Multiple harvestings are done for species with different stages of seed maturation and immediately after physiological maturity to obtain

Fig. 17.12. Seed production of *Antirrhinum* genetic stock in the Ornamental Plant Germplasm Center computerized-controlled greenhouse using bumblebees as pollinator. (In the next compartment, honeybees were used and note the blue honeybee nuclear hive in that compartment.)

high-quality seed. Harvesting a mixture of mature and immature seeds together should be avoided because it is difficult to clean unfilled seed and excessive trash. A system of labelling and crosschecking is also used to prevent mislabelling, which would result in mechanical contamination. Pest-infested and diseased seed are not harvested. Any free

Fig. 17.13. The custom-built threshing board measuring 39 cm × 39 cm × 10 cm (H) with rubber base and the rubber rubbing block used at the Ornamental Plant Germplasm Center.

water due to morning dew or rain on the harvested seeds and pods must be immediately dried with a high volume airflow dryer.

During seed threshing and cleaning, mechanical injuries and contamination are main considerations. Common injuries are the broken seed, visible cracks and micro-cracks on seed. In germplasm conservation, the small seed lots are preferably threshed by hand or by using a hand-threshing board (Fig. 17.13) which is gentle and prevents injuries. Hand cleaning is efficient because seed lots are small and are relatively clean as they are hand-harvested. Large and small pieces of extraneous material are removed by sieves and light material of the same size as the seed is removed using a seed blower/aspirator. Seeds that accidentally tumble from a seed-cleaning tray are discarded to prevent inadvertently adding seed of another accession. The OPGC uses a spring-loaded belt thresher to clean bigger seed lots (Fig. 17.14). In this case, the harvested pods and flower heads must be dried until they can be easily broken for efficient threshing.

Seed drying is an important operational step in maintaining seed quality of seed lots in a genebank where seed are placed in long-term storage. Generally, a two-phase drying

Fig. 17.14. The spring-loaded belt thresher for cleaning bigger seed lots at the Ornamental Plant Germplasm Center.

regime is applied to gently remove seed moisture down to a safe level for cold storage at −18°C. Phase I drying includes the period from harvesting to threshing and processing, and phase II is the final drying before packaging for storage. Phase I drying is probably the most critical stage, especially when harvesting has be done in wet weather. The wet seed and plant parts, often with free surface water, have to be dried as fast as possible to prevent seeds from germinating in warm, moist conditions. The OPGC has an efficient high volume airflow ventilator/ dryer in its seed processing room with a temperature of around 25°C and 40% RH (Fig. 17.15). In this condition, the equilibrium seed moisture content of most seeds is approximately 8% wet basis (w.b.). Phase II drying is usually achieved by equilibrating seeds at low temperature and relative humidity; the International Plant Genetic Resources Institute recommendation is 15°C and 15% RH (FAO and IPGRI, 1994). The OPGC medium-term seed storage cooler at 5°C and 30% RH is used efficiently as a slow dryer to decrease seed moisture content to below 7% w.b. by simple passive equilibration.

Seeds in the medium-term seed storage cooler at the OPGC are kept in non-airtight 226.8 g (8 oz) screw-top glass jars for active collections so that seed can be easily withdrawn for distribution. During seed withdrawal, the seed container is allowed to equilibrate in the packaging room at 20°C and 40% RH to prevent moisture condensation on seed when the container is opened (Tay, 1988). In medium-term storage good quality seed can generally be stored for 10–15 years based on experience with other agronomic and horticultural seeds. Seeds with high seed moisture content suffer from freezing injury in sub-zero temperatures. All the seed for long-term storage at the NCGRP are further dried to low seed moisture contents of 4–6% (w.b.) before being hermetically sealed for storage at −18°C. Agronomic crop seeds in long-term storage freezers at −18°C generally remain of acceptable quality with 80–85% germination after 15–30 years, storage if the initial quality is high (approximately 100% germination). Seeds under storage therefore must be monitored at set intervals so that seed lots that deteriorate below the viability threshold limit can be detected in time for regeneration to prevent genetic erosion in storage. At the OPGC, standard germination procedures (ISTA, 1999) are adopted but with 200 seeds because

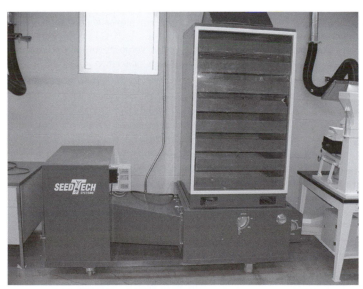

Fig. 17.15. The high volume airflow ventilator/dryer makes use of the dry air-conditioned environments of 25°C and 40%–50% relative humidity for drying wet seed.

of the small seed lots. The monitoring interval is every 3–5 years during the first 15 years for lots with high initial seed quality, followed by more frequent intervals thereafter. Table 17.3 summarizes seed science and technology activities in relation to OPGC germplasm conservation tasks as discussed above.

The OPGC Genebank Management System

As part of the NPGS, OPGC uses the GRIN database (http://www.ars-grin.gov/npgs/) for managing its germplasm passport, characterization, evaluation, taxonomy, seed inventory and distribution information. The OPGC management system is therefore built on to the GRIN database, which is curator-based, i.e. crop-based as in other NPGS repositories. A curator is responsible for all the genebanking activities of a genus or a group of genera and these include acquisition, post-entry quarantine, multiplication, characterization, identification of duplicates, accessioning, seed conditioning, viability testing, packaging, storage, further characterization and evaluation and documentation and distribution. A natural split of responsibilities between curators is either seed-based genera or clonal-based/recalcitrant seed genera. A curator is responsible for the proper acquisition of an accession according to the Convention on Biological Diversity and other international treaty requirements, the intellectual property rights on the germplasm and the national quarantine requirements.

Table 17.3. Seed science and technology in plant germplasm conservation and distribution.

Genebanking activity	Seed science and technology
Acquisition	
1. Collecting – local	1. Sampling strategy, sample size and seed health
2. Collecting – overseas	2. As in (1) + plant collecting and transfer agreements, import and export permits and phytosanitary certification
Regeneration	
3. Seed germination and seedling raising	3. Germination, dormancy breaking, stratification and vernalization treatment and random sampling
4. Direct sowing	4. Production site selection and crop rotation
5. Seedling transplanting	5. As in (4)
6. Synchronizing flowering	6. Cultural, vernalization and photoperiod treatments
7. Pollination	7. Isolation, pollination (insect and hand) and pest and disease control
8. Harvesting	8. Harvesting and seed extraction techniques
9. Seed threshing and cleaning	9. Hand vs. machinery techniques
10. Final drying	10. Slow drying equipment and m.c. testing
Conservation	
11. Packaging	11. Hermetic vs. non-hermetic, container types, sealer type and packaging environment
12. Seed storage	12. Genebank design, machinery, control and safety
13. Viability monitoring	13. Sampling, number of seed, monitoring interval
14. Seed for regeneration	14. Threshold viability, seed sampling
Characterization and evaluation	
15. Seed germination	15. As in (3)
16. Sowing	16. As in (4)
17. Seedling transplanting	17. As in (5)
Distribution	
18. Withdrawing sample	18. Pre-packed vs. re-closable container
19. Packaging for shipment	19. Protecting against temperature and moisture
20. Documentations	20. Barcode vs. manual

Similarly, germplasm distribution to fill researcher requests is implemented according to USDA policies and procedures and the quarantine requirements of the recipient country. To date, USDA maintains the policy of free exchange and distribution of plant germplasm except for embargo countries. The detailed activities of the different genebanking steps are summarized in Fig. 17.16.

The Germplasm Collection Status

The OPGC currently has 966 accessions of 62 genera on the GRIN database (Table 17.4). The number of accessions is increasing rapidly and there are more than 800 newly acquired accessions with 'OPGC' numbers, the internal OPGC working numbering system (abbreviated as 'OPGC#') that will be added to the GRIN database. Most of these accessions have low seed numbers and are being regenerated for both base and active collection. Crop-specific descriptors have to be formulated for each of the crops so that germplasm characterization and evaluation data are standardized and studied.

The OPGC Research Programme

The OPGC has a service-orientated mandate as described and its research programme is focusing on areas that will improve its operational efficiency. In the seed science and technology discipline, the ability to produce high-quality seed, both physiological and physical, and to maintain the genetic integrity or composition of the accessions, is a matter of urgency because of the lack of existing specific information in OPGC priority genera. For example, the challenge to get all viable seed to germinate in the wild and semi-domesticated accessions, and all randomly selected seedlings to grow, flower and set an equal amount of seed is difficult to achieve because of uneven seed dormancy in an accession and the need to synchronize flowering to promote concentrated flowering and inter-crossing. There is no objective information on how many plants per accession to grow in order to capture an acceptable amount of the original genepool. The seed production of cross-pollinated and out-crossing species by controlled pollination with insect pollinators in small pollination cages vs. hand pollination has to be compared and a decision-making production

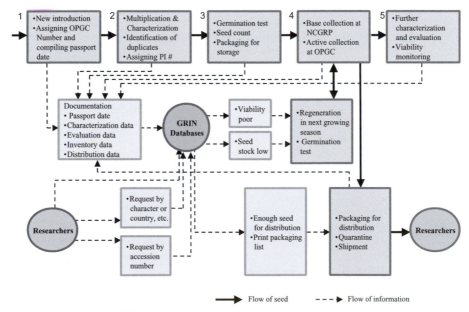

Fig. 17.16. The Ornamental Plant Germplasm Center genebank management system.

Table 17.4. Number of OPGC herbaceous ornamental plant accessions in the USDA GRIN database as of December, 2003 (Acc. = number of accessions). (*Note*: There are, currently, about 1000 accessions in OPGC working number under assessment and seed production before entering into the GRIN database.)

Genus	Acc.	Genus	Acc.	Genus	Acc.
Adiantum	1	Consolida	6	Penstemon	56
Alstroemeria	1	Convallaria	1	Petrorhagia	1
Antirrhinum	88	Coreopsis	1	Phlox	2
Aquilegia	4	Delphinium	31	Portulaca	4
Argyranthemum	3	Dianthus	102	Portulacaria	2
Arum	1	Echinops	1	Primula	1
Aruncus	3	Engelmannia	1	Pseudolysimachion	4
Asclepias	20	Filipendula	1	Rudbeckia	2
Astilbe	6	Geranium	4	Sanguisorba	1
Astilboides	1	Glandularia	2	Stokesia	38
Begonia	3	Gypsophila	29	Tagetes	114
Bredia	1	Impatiens	1	Trollius	1
Campanula	34	Iris	36	Turnera	1
Canna	1	Junellia	1	Veratrum	1
Cardiocrinum	6	Leonotis	1	Verbena	17
Caulophyllum	1	Leucanthemum	90	Veronica	1
Chelidonium	1	Liatris	7	Veronicastrum	1
Chrysanthemum	9	Ligularia	6	Viola	15
Clematis	4	Lilium	35	Xylanthemum	2
Cleome	38	Osteospermum	43	Zinnia	74
Clintonia	1	Pavonia	2	Total	966

system established for future use. The use of a high-density population, smaller plants, fewer flowers per plant and concentrated flowering, and thus harvesting, such as multi-plants per hill in the field and close planting in seedling flats or small pots in greenhouse year round, is being investigated to study the yield and seed quality. Studies in its 2002 seed production season indicated the efficacy of bumblebees as pollinators in small pollination cages and greenhouse compartments compared with honeybees. Another challenge is to determine when to harvest the seed, and how to dry, thresh and clean many small seed lots of many species.

Genetic erosion in a genebank was reported to be common and it is a focus issue (FAO, 1998). The main causes are attributed to the lack of genebank operational resources to maintain the collection and seed storage facility, and technical skills of genebank personnel. Research on conservation methodology and genebank management to streamline the operation to improve efficiency is given

priority at OPGC. For example, the use of a rapid, low-cost, accurate seed vigour test such as a computer-assisted seed imaging method (Sako *et al.*, 2001) to estimate a seed lot's initial quality and its projected storage life would be a valuable development. Another concern is that seeds are being used during seed viability monitoring tests in the small seed lots that are stored. Research to find a rapid, low-cost, non-destructive viability test will be a central research theme. The use of ultra-low-temperature cryopreservation techniques and ultra-dried seed storage methods are other research areas that may advance seed germplasm storage.

Global Status and Future Development in Herbaceous Germplasm Conservation

So far, the conservation of herbaceous ornamental plants is maintained mainly as breeder collections by public and private flower breeders on crops that they are breeding, as private collections by passionate plant collectors, members of crop-specific societies and specialized plant nurseries, and as collections in botanical gardens. The more than $US50 billion per annum global floriculture industry does not have a network of specialized comprehensive genebanks in the world as there is for food crops. To date, the FAO and IPGRI have no mandate in ornamental crop conservation.

In 1987, the founding of Botanic Gardens Conservation International (BGCI) with the mission 'to build and maintain a world network of Botanic Gardens for plant conservation' and 'to educate and promote conservation awareness and sustainability by providing technical guidance, data and support for Botanic Gardens worldwide' marked the beginning of organized conservation endeavour in ornamental plants (BGCI, 2004). Its 500+ member institutions in 112 countries have adopted a worldwide Botanic Gardens Conservation Strategy for plant conservation to promote exchange and sharing of information and experience between members and to create public awareness in plant conservation. Its International Agenda for Botanic Gardens in Conservation (Wyse Jackson and Sutherland, 2000) is the first global policy framework for the conservation of ornamental plants. BCGI, through its 2003 Memorandum of Understanding with the Convention on Biological Diversity (CBD), further strengthens its leadership in policy matters and implementation procedures for the fulfilment of CBD requirements (see CBD, 2004) in international germplasm collecting, transfer, conservation and utilization. The world is in a transitional stage to adopt CBD from where plant germplasm was a common heritage of humankind before the Convention in 1993 to that of national natural resources where permissions are required in germplasm collection, transfer and use (D. Tay, approved for publication). BGCI continues to play the leading role in this global undertaking. Regional and national networks of botanic gardens have been organized under the BGCI network.

Regional and national plant collections have been established with responsibility to conserve world plant diversity: the North American Plant Collections Consortium (NAPCC) of the American Association of Botanic Gardens and Arboreta (AABGA), the National Council for the Conservation of Plants and Gardens (NCCPG) in the UK, the Ornamental Plant Conservation Association in Australia (OPCAA, 2004) etc. This development is facilitating the conservation of herbaceous ornamental plants. For example, the current 644 NCCPG collections consist of two aquatic crops including *Iris*, 34 bulbs crops, 22 cacti and succulents, eight climbers, 21 exotics, 13 ferns, 110 herbaceous and 16 orchids, and the remainder are shrubs and trees (NCCPG, 2004). Out of the total 30 NAPCC collections, three are herbaceous plants, namely, native wild ginger, *Trillium* and *Hosta* (with provisional status) (AABGA, 2004). The *Hosta* collection is a joint initiative between Toledo Botanic Garden and the OPGC to put forth a combined NAPCC and NPGS collection so that the same collection will be served by two conservation systems. The roughly 50 collections of OPCAA include about six herbaceous taxa (OPCAA, 2004).

These regional and national collections are essentially maintained as living collections. There is thus no permanent base collection in long-term safe storage in a genebank. However, there are a few conventional genebanks that conserve herbaceous ornamental plants. Table 17.5 shows the number of accessions of the USDA HOCGC 30 priority genera at Kew Royal Botanic Gardens' genebank (Kew), the Institute of Plant Genetics and Crop Plant Research (IPK) genebank (Gatersleben), the Ministry of Agriculture, Forestry and Fisheries genebank (Japan) and OPGC. The establishment of OPGC serves to fill this gap in the conservation of herbaceous ornamental species. In 2½ years, OPGC has developed into a functional unit and has been able to acquire 1049 accessions of its priority genera. These 30 genera consist of some 6700 species and most of them are not in cultivation. Germplasm exploration missions in the centres of origin of the targeted species are therefore vital to collect this germplasm. Global networking and close collaboration with botanic gardens, public research institutions, the floriculture industry, universities

Table 17.5. Number of OPGC priority genera accessions at Kew Royal Botanic Gardens genebank (Kew), the Institute of Plant Genetics and Crop Plant Research (IPK) genebank (Gatersleben), the Ministry of Agriculture, Forestry and Fisheries genebank (Japan) and OPGC in 2003.

Genus	No. of accessions			
	Kew	Gatersleben	Japan	OPGC
Aglaonema	1	0	0	0
Alstroemeria	10	0	0	1
Anthurium	2	0	0	0
Aquilegia	19	16	0	8
Aster	40	8	0	8
Baptisia	9	2	0	3
Begonia	12	102	0	136
Campanula	38	17	0	35
Chrysanthemum (Dendranthema)	16	99	43	12
Dianthus	39	58	1	110
Dieffenbachia	1	0	0	0
Euphorbia (Poinsettia)	76	8	0	2
Geranium	25	0	0	9
Hemerocallis	0	2	0	20
Impatiens	12	17	0	50
Iris	22	11	0	44
Lilium	0	0	0	54
Narcissus	0	0	0	12
Pelargonium	7	10	0	362
Petunia	5	82	0	4
Phalaenopsis	3	0	0	0
Philodendron	2	0	0	0
Phlox	12	3	0	2
Rudbeckia	12	6	0	13
Salvia	76	43	0	0
Spathiphyllum	1	0	0	0
Tagetes	7	65	0	123
Verbena	22	27	0	19
Veronica	36	6	0	4
Viola (pansy)	39	58	0	18
Total	544	640	44	1049

and researchers in other disciplinary areas such as pharmacy and the health industry have to be fostered to take on this enormous task.

This chapter deals only with 'orthodox' seed species. Many tropical ornamental species, especially foliage plants, have recalcitrant seed and the problems presented and conservation techniques applied are different from those discussed here. *In vitro* slow growth and cryopreservation techniques in liquid nitrogen have to be used.

References

AABGA. 2004. North American Plant Collections Consortium – Current NAPCC Participants. http://www.aabga.org/napcc/napcc2.htm

BGCI. 2004. Botanic Gardens Conservation International. http://www.bgci.org.uk

CBD. 2004. Convention on Biological Diversity. http://www.biodiv.org/default.aspx

de Groot, N.S.P. 1998. Floriculture worldwide trade and consumption patterns. *World Conference on Horticultural Research*, 17–20 June 1998, Rome, Italy.

[FAO] Food and Agriculture Organization of the United Nations. 1998. *The State of the World's Plant Genetic Resources for Food and Agriculture*. FAO, Rome, Italy. 98 pp.

FAO and IPGRI. 1994. *Genebank Standards*. Food and Agriculture Organization of the United Nations, Rome, Italy and International Plant Genetic Resources Institute, Rome, Italy. 13 pp.

[GMPRO] Greenhouse Management & Production. 1999. Terril Nell: on the Floriculture and Nursery Research Initiative. *Greenhouse Management and Production* (GMPRO) 19(2):48–50.

Harlan, J.R. and J.M.J. de Wet. 1971. Towards a rational classification of cultivated plants. *Taxon* 20:509–517.

[Hortus Third] The staff of the Liberty Hyde Bailey Hortorium. 1976. *Hortus Third – a Concise Dictionary of Plants Cultivated in the United States and Canada*. Macmillan General Reference, New York.

[ISTA] International Seed Testing Association. 1999. *International Rules for Seed Testing 1999*. Seed Science and Technology. Vol. 27, Supplement, 1999. ISTA, Zurich, Switzerland.

Linington, S. 2000. The Millennium Seed Bank Project. *In*: B.S. Rushton, P. Hackney and C.R. Tyrie (eds). *Biological Collections and Biodiversity*. Linnean Society of London Special Publication No 3. 326 pp.

NCCPG. 2004. The National Council for the Conservation of Plants and Gardens. http://www.nccpg.com/

OPCAA. 2004. The Ornamental Plant Conservation Association of Australia. http://opcaa.rbg.vic.gov.au/index.html

Sako, Y., M.B. McDonald, K. Fujimura, A.F. Evans and M.A. Bennett. 2001. A system for automated seed vigour assessment. *Seed Sci. Technol.* 29:625–636.

Tay, C.S. 1988. Present status, management and utilization of tropical vegetable genetic resources at AVRDC. *In*: S. Suzuki (ed.). *Crop Genetic Resources of East Asia*. International Board for Plant Genetic Resources, Rome, Italy. pp. 41–51.

Tay, D. 2003. The Ornamental Plant Germplasm Center – ranking priority genera for conservation. *HortSci.* 38:678.

Tay, D., M.P. Widrlechner and J.L. Corfield. 2004. Establishment of a new gene bank for herbaceous ornamental plants. *Plant Genetic Resources Newsletter* 137:26–33.

[USDA] United States Department of Agriculture. 2003. *Floriculture Crops 2002 Summary*. National Agricultural Statistics Service, Agricultural Statistics Board, USDA.

Wyse Jackson, P.S. and L.A. Sutherland. 2000. *International Agenda for Botanic Gardens in Conservation*. Botanic Gardens Conservation International, London.

18   Conclusions

Francis Y. Kwong[1] and Miller B. McDonald[2]

[1]PanAmerican Seed Company, 622 Town Road, West Chicago, IL 60185-2698, USA;
[2]Seed Biology Program, Department of Horticulture and Crop Science,
Ohio State University, 2021 Coffey Road, Columbus, OH 43210-1086, USA

Flowers are often described as food for the soul. They contribute to increasing the quality of life rather than sustaining our mere existence. Aesthetic values vary between societies and change with time. Consequently, there is great diversity in the use of flowers from both a geographical and a historical perspective. The broad use of bedding plants in our gardens can be traced back to the early 19th century, when colourful annuals and tender ornamentals were imported to Western Europe from the Americas, South Africa and Asia. From the onset, domestication of exotic plant species has been a mainstay in the development of garden designs (Hobhouse, 1992). The seed is a perfect natural package that facilitates the migration of plant species across land and sea. Early developments in the flower and seed industry generally involved importing exotic plant materials and improving them through selection in accordance to the local environments and prevalent tastes. Mass migration of peoples throughout the world, starting in the 18th century, contributed to the broader distribution of such novel ornamental varieties. We have a glimpse of how the seed industry and primary uses of flower seeds have changed in recent times in the first three chapters of this book. New trends in the desirable flower types and how flower seeds are used will continue to develop, and the flower seed trade will evolve with them.

The vast diversity of products available is a unique feature of the flower seed industry and presents more technical challenges to it than to other agricultural industries. Globally the flower seed industry is expanding and diversifying. For example, developments in flower seed production in New Zealand were recently reviewed by Hill and Hampton (2003).

Scientific breeding has been the primary means of new variety development since the 19th century. From the advent of Mendelian inheritance through the elucidation of the genetic code, plant breeders now have more tools with which to perform their work and predict the outcomes of their labour. Advances in molecular biology further provide the common language in genetic and physiological studies. Most of our current knowledge in flower seed biology (Chapters 5–9) is obtained from research done at the whole plant level. While we have a good general understanding of the physiology of flowering, seed development, seed dormancy and seed germination, specific data on individual flower species, even the most commercially significant ones, are still fragmentary. This general lack of specific details impedes the speed of adaptation of the newer, laboratory-based techniques in product development. There remains a strong need for descriptive work on plant morphology and anatomy, as well as observations on

©CAB International 2005. *Flower Seeds: Biology and Technology*
(eds M.B. McDonald and F.Y. Kwong)   357

physiological behaviours of the major flower classes on the whole plant, organ and subcellular levels.

In addition to the development of new flower varieties that add colour and texture to our landscape, there has been a general focus on seed quality in recent years. The industrialization of the horticultural industry has increased the demand for reliable germination performance. The seed trade responded by developing better germination and vigour testing procedures. There is more scientific input into the traditional arts of seed production and seed cleaning. New treatments are developed to make the seeds easier to use in a highly mechanized crop production system. Advances in seed technology rely not only on knowledge of seed biology; innovations in instrumentation also play a pivotal role. Much of the equipment used in the seed trade today is adopted from the food, mining and pharmaceutical industries. The seed industry can continue to benefit by looking for solutions developed in other industries. While the general seed quality levels of flower seeds have risen through the years, we are still far from realizing the goal of turning each seed into a marketable plant. There are no standardized viability and vigour tests in the trade. And there is much debate on the relevance of controlled laboratory test results for the practical growers. Given the diversity of crop species and their production sources, this will remain an issue in the foreseeable future.

The Future of the Flower Seed Industry

Man will continue to associate himself with the beauty of flowers in future generations. However, no-one can foretell the face of the flower industry, and the role of flower seeds in it, beyond the next decade. Changes in industry structure generally follow the overall societal trends and new product development often relies on available resources and technological advances. We see a number of significant cultural and technological developments that will be likely to shape the future of the flower world, as discussed below.

Structure of the industry and product offerings

Historically the ornamental industry, both at the grower and seed company levels, was populated by a large number of small, mainly family-held enterprises. Considerable industry consolidation has occurred in the last two decades, resulting in fewer, but larger, corporations. Company consolidation may result in improvements in business efficiency, but often at the expense of the breadth of product offerings. This is odd in an industry that at its core thrives on speciality products. As the more experienced companies consolidate, niches soon open in the market-place and a new wave of small, specialist companies is established to occupy them. These small companies will grow, then thrive or perish. At the industry level, despite periodic variations in percentage composition, small specialists, medium-sized independents, and large conglomerates will continue to coexist.

Since the establishment of a significant plug production segment in the ornamental industry 20 years ago, the functional separation between seed companies and growers has shifted. Many leading seed companies are also plug producers now and they focus on delivering starting materials, as seeds or seedlings, to the finishers. The traditional concept of seed has been extended to include unique genetic materials delivered in the form of young plants. Philosophically, enhanced seed products like primed and pregerminated seeds are products situated along the same path as biological seed and young plants. It is conceivable that seed companies will also offer vegetatively propagated products, e.g. synthetic seed based on tissue culture multiplication, either as a 'seed' product or in the form of young plants. Indeed, a number of seed breeders have established breeding programmes to develop flower products that cannot be seed propagated. These products are propagated vegetatively and sold as rooted or unrooted cuttings. This market segment has been occupied by other specialists in the nursery trade. The large companies are evolving from the traditional 'seed' companies to 'horticultural' companies. While this may create confusion and often heated discussions within the industry, the consumers who buy the finished products may ultimately benefit from having multiple sources of garden plant supply. The overall economy of developing the technical capabilities to meet the identified market demands will determine how extensively this type of horizontal integration will proceed within the ornamental trade.

Gardening trends

Penelope Hobhouse (1992) observed that throughout the history of gardening there is a constant conflict between 'a delight in logical rhythm of planting and in more natural free effect – between considering the art of gardening as 'nature perfected' and using gardening to re-interpret the roles of plants in order to imitate the wild'. It is the result of this conflict that we have both formal gardens and natural landscapes to enjoy. Today, the major use of flower seeds is in the bedding plant segment. Even within bedding plants, there is continuous ebb and flow in the preference for more tidy, colourful patterns or scruffier, earthy looks. Population growth and increasing urbanization over the last century has severely diminished the amount of natural resources available to the average household. The concept of a garden has now been extended from a plot of land around the house to include container gardens and interior-scapes catering to city and suburban dwellers. Space limitations, low quality water, and poor soil conditions are providing key challenges to breeders and gardeners alike to bring the sense of nature closer to man's immediate environment. Mixed plantings of trees, shrubs, foliage and flowering plants in confined spaces are likely to become more important. In this case, ornamental plants will be used primarily as temporary decorative items, fitting into the overall designs of residential and business real estate developments. As a result, flower breeders have to devote more attention to selecting plants that will thrive in unnatural conditions such as compact soil, salty water and artificial lights.

Speed of new product development

Flowers appeal to the aesthetic senses of the consumers. They are analogous to works of art, created by the skilful hands of the breeders. In reality, breeders only create blueprints of the pictures. It is growers who colour the canvas and produce the final products. It is also the grower who sells the flowers to retailers. A beautiful product that is difficult to produce is unlikely to gain widespread acceptance. Introduction of innovative products that require major changes in production methods, either at the grower or consumer level, will demand additional time and marketing effort. Time is probably the most limiting resource in the 21st century. High-speed communication technologies and fiercely competitive marketing are nurturing a culture of instantaneous gratification. This places significant pressure on the speed of new product development. It is no longer acceptable for the seed company to boast the 10 years that it takes to introduce a new variety, which depends solely on the effort of its own internal resources. As suggested by a recent Dutch study on horticultural industry trends (Krinkels, 2002), strategic alliances between breeders, growers and retailers will be needed to bring forth a continuous stream of new products to satisfy the interests of the public.

Biotechnology

Biotechnology is seen as a major tool to reduce product development time in breeding. While commercial successes are reported for the use of genetically modified varieties in row crops in the last decade, no parallel examples can be found in flowers. There are two main reasons for this observation. The first reason is that desirable traits in ornamentals are difficult to modify. Characters such as flower colour, fragrance and plant habit are generally phenotypic expressions of multiple gene interactions, not readily amenable to simple genetic manipulations. This difficulty is further compounded by numerous knowledge gaps in the genetics and physiology of any specific flower crop. The second reason is that even when the genes of interest are identified, the cost-to-benefit ratio for biotechnology is too high. Today, the economic value of all flower seeds combined is still much smaller than that of maize alone. No single flower item has the market size to generate a reasonable financial return for the research work and licensing costs required to develop a new variety. The cost of genetic engineering, like other emerging technologies (e.g. computers a few decades ago), will eventually be reduced in the future. But it will have to decrease substantially before broad-scale adaptation of biotechnology in flowers is realistic. In the meantime, breeders will turn to induced mutation, embryo rescue and other more traditional tissue culture techniques for the creation of unique germplasms to shorten product development time. The horticultural skills applied to crop improvement will be increasingly a collaborative effort of laboratory manipulation and field selection.

Ecological movements

There is increasing awareness of the destruction of natural habitats worldwide due to man's activities in the late 20th century. This awareness has brought about a trend in wildflower gardens (Chapter 4) and ecological landscaping designs focusing on the use of native plant materials. While large commercial flower companies are concentrating their efforts on introducing domesticated improved varieties, there is increasing concern that we are losing our valuable natural resources. Germplasm preservation is now a major function of botanical gardens and germplasm centres (see Chapter 17). This is a daunting task, given the wide diversity of species with potential ornamental value in the wild. It is unrealistic to expect that the limited resources available can save the majority of the wild species from extinction. The relatively small number of species that are conserved will depend on individual judgements of the ornamental value of the materials, which are again subject to cultural and economic biases.

Another ecological movement is ongoing within the agricultural industries in the form of organic farming. The idealistic vision of organic production is the creation of a self-sustaining local ecology within the production field. By necessity, this involves a multi-cropping scheme, which may include using ornamental plants as companion plants in the production of food crops (Jeavons, 2002). Large commercial crop productions are generally mono-crop cultures. In the popular commercial version, organic production focuses on the elimination of synthetic chemicals and other harmful input materials in cultural routines. Concerns about the negative environmental and human health impacts from the overuse of chemicals have increased the market demand for organic products. Organic production in both Western Europe and the Americas has seen double-digit annual increases for the last decade. Most of these increases are accounted for in food crops. There is very little commercial organic flower production today. But production of organically grown flowers will increase because of the favourable social and economic environment. Consumers are willing to pay up to a 20% premium for organically grown products (French and Emerich, 2002). On the grower level, the rate of conversion from conventional methods to organic production can be accelerated by the availability of technical advice and natural products for crop management locally. Many universities and governmental extension services are now offering classes and conducting active research in sustainable agriculture. Seed companies can respond to this organic trend by developing varieties that are well suited for organic production and initiate programmes to produce seed organically. Many organic certification agencies are already in the process of making organically grown starting materials, i.e. seeds or cuttings, a requirement for consumer product certification.

Non-ornamental uses of flowers

Other than the vibrant colours that appeal to our senses, flowers also provide unique chemicals that are useful to man. Recent medical studies have highlighted the anti-oxidant values of various carotenoids and flavonoids. These colourants are highly concentrated in the petals. Lutein is commercially extracted from marigolds, and commonly used as an additive in poultry feed. The same compound can potentially be used in food colourants and dietary supplements to counter macular degeneration. Other natural pharmaceuticals or nutraceuticals can be extracted from various plant parts of some ordinary flower plants. For example, the vinca (*Catharanthus rosea*) plant is a rich source of vinblastine and vincristine, alkaloids employed in cancer chemotherapy (Mouton, 1977). Root extracts of coneflower (*Echinacea angustofolia*) are used as dietary supplements against the common cold. Bio-pesticides can be extracted from ornamental plants. An example is *Pyrethrum* extracts that have insecticidal activities. Marigold is also known to exude nematicidal compounds from its roots and is used in companion plantings for this reason. With the emergence of a market segment that focuses on health and sustainability in the new millennium (French and Emerich, 2002), the search for natural products that have human health benefits, as well as natural products useful for organic food production, is likely to intensify. An increasing amount of traditional flower crops will probably be used as raw materials for these nutraceutical, pharmaceutical and agricultural products. Special varieties of these important crops can be developed through classical breeding or genetic manipulation to increase the yield of desirable biochemicals. The success of such new product development efforts will depend on collaboration of specialists across different industries.

In the next decade, the flower seed industry may remain a special segment of the overall ornamental trade, or it may become a key part in an emerging environmental health industry, which encompasses environmental remediation, natural food and natural health care products. No matter which commercial direction it takes, it will remain a fertile ground for the interaction of scientific specialists and practical horticulturists.

References

French, S. and M. Emerich. 2002. The LOHAS consumer identified! *Natural Business LOHAS Journal*, Spring 2002.

Hobhouse, P. 1992. *Plants in Garden History*. Pavilion Books, Ltd, London.

Jeavons, J. 2002. *How to Grow More Vegetables*. Ten Speed Press, Berkeley, California.

Krinkels, M. 2002. Cooperation will dominate the horticulture scene. *Prophyta Annual* 2002: 26–27.

Morton, J.F. 1977. *Major Medicinal Plants. Botany, Culture and Uses*. Charles C. Thomas Publishers, Springfield, Illinois.

Hill, M. and J.G. Hampton. 2003. The potential for flower seed production in New Zealand. *AgScience* 12 May:4–5.

Index